J & M Merefield
53 Rollestone Crescent
Sylvania Park
Exeter
EX4 5EB

Sediment and Water Quality in River Catchments

Sediment and Water Quality in River Catchments

Edited by
IAN FOSTER
Coventry University, UK

ANGELA GURNELL
University of Birmingham, UK

BRUCE WEBB
University of Exeter, UK

JOHN WILEY & SONS
Chichester · New York · Brisbane · Toronto · Singapore

Copyright © 1995 by John Wiley & Sons Ltd,
Baffins Lane, Chichester,
West Sussex PO19 1UD, England
Telephone National (01243) 779777
International +44 1243 779777

All rights reserved.

No part of this book may be reproduced by any means,
or transmitted, or translated into a machine language
without the written permission of the publisher.

Other Wiley Editorial Offices

John Wiley & Sons, Inc, 605 Third Avenue,
New York, NY 10158-0012, USA

Jacaranda Wiley Ltd, 33 Park Road, Milton,
Queensland 4064, Australia

John Wiley & Sons (Canada) Ltd, 22 Worcester Road,
Rexdale, Ontario M9W 1L1, Canada

John Wiley & Sons (SEA) Pte Ltd, 37 Jalan Pemimpin #05-04.
Block B, Union Industrial Building, Singapore 2057

Library of Congress Cataloging-in-Publication Data

Sediment and water quality in river catchments / edited by Ian Foster,
  Angela Gurnell, Bruce Webb.
     p.   cm.
  Includes bibliographical references and index.
  ISBN 0-471-95728-3
  1. River sediments. 2. Water quality. I. Foster, Ian (Ian D.
L.) II. Gurnell, A. M. (Angela M.) III. Webb, Bruce.
GB1399.6.S43 1995
551.48′3—dc20   95-3054
                                                            CIP

British Library Cataloguing in Publication Data

A catalogue record for this book is available from the British Library

ISBN 0-471-95728-3

Typeset in 10/12pt Times by Mathematical Composition Setters Ltd, Salisbury, Wiltshire
Printed and bound in Great Britain by Bookcraft (Bath) Ltd

This book is printed on acid-free paper responsibly manufactured from sustainable forestation,
for which at least two trees are planted for each one used for paper production.

Contents

List of Contributors		ix
Preface		xiii
Foreword *K.J. Gregory*		xv
1	Hydrology, Water Quality and Sediment Behaviour *B.W. Webb, I.D.L. Foster and A.M. Gurnell*	1
SECTION I	**QUANTITY AND QUALITY DIMENSIONS**	**31**
2	The Role of Geographical Information Systems in Hydrology *T.J. Browne*	33
3	Modelling Nitrate Leaching at the Catchment Scale *H.J.E. Rodda*	49
4	Regulation and Thermal Regime in a Devon River System *B.W. Webb*	65
SECTION II	**SEDIMENT DYNAMICS AND YIELDS**	**95**
5	A Conceptual Model of the Instantaneous Unit Sedimentgraph *K. Banasik*	97
6	Magnitude and Frequency of Fluvial Sediment Transport Determined From Recent Lake Sediment Cores *R.H.F Curr*	107
7	Sediment Sources and Their Environmental Controls *C. Clark*	121
SECTION III	**SEDIMENT QUALITY**	**143**
8	Sediment Mineralogy and the Environmental Impact of Mining *J.R. Merefield*	145
9	Long-term Dispersal of Metals in Mineralised Catchments by Fluvial Processes *S.B. Bradley*	161
10	Fingerprinting Sediment Sources: An Example From Hong Kong *M.R. Peart*	179

SECTION IV	SEDIMENT SOURCES AND SINKS	187

11 The Identification of Catchment Sediment Sources 189
    R.J. Loughran and B.L. Campbell

12 Determination of Suspended Sediment Provenance Using Caesium-137,
    Unsupported Lead-210 and Radium-226: A Numerical Mixing Model
    Approach 207
    Q. He and P. Owens

13 Processes of River Bank Erosion and Their Contribution to the Suspended
    Sediment Load of the River Culm, Devon 229
    D. Ashbridge

14 The Rates and Patterns of Overbank Deposition on a Lowland Floodplain 247
    D.J. Simm

15 Lake and Reservoir Bottom Sediments as a Source of Soil Erosion and
    Sediment Transport Data in the UK 265
    I.D.L. Foster

SECTION V	RADIONUCLIDE STUDIES	285

16 The Development and Application of Caesium-137 Measurements in
    Erosion Investigations 287
    D.L. Higgitt

17 Estimation of Erosion Rates from Caesium-137 Data: The Calibration
    Question 307
    T.A. Quine

18 The Erosional Transport of Radiocaesium in Catchment Systems: A Case
    Study of the Exe Basin, Devon 331
    J.S. Rowan

19 Use of Caesium-137 to Investigate Sediment Sources in the
    Hekouzhen–Longmen Basin of the Middle Yellow River, China 353
    X. Zhang and Y. Zhang

SECTION VI	NATIONAL AND GLOBAL PERSPECTIVES	363

20 Patterns of Erosion and Suspended Sediment Yield in Mediterranean
    River Basins 365
    J.C. Woodward

21 Soil Erosion and Sediment Yield in the Philippines 391
    S. White

22 Sediment Yield from Alpine Glacier Basins 407
    A.M. Gurnell

23 Sediment Transport and Deposition in Mountain Rivers 437
    J. Bogen

24	Sediment Dynamics in the Polish Flysch Carpathians *W. Froehlich*	453
Index		463

Contributors

D. Ashbridge
Lob House, Lob Lane, Stamford Bridge, York YO4 1BN, UK; formerly: Geography Department, University of Exeter, UK

K. Banasik
Department of Hydraulic Structures, SGGW—Warsaw Agricultural University, ul. Nowoursynowska 166, 02-776 Warsaw, Poland

J. Bogen
Norwegian Water Resources and Energy Administration, PO Box 5091, Maj. 0301 Oslo, Norway

S.B. Bradley
Environmental Science, Westlakes Research, The Princess Royal Building, Westlakes Science and Technology Park, Moor Row, Cumbria CA24 3LN, UK

T.J. Browne
The Computing Centre, University of Sussex, Falmer, Brighton, East Sussex BN1 9QJ, UK

B.L. Campbell
Department of Geography, University of Newcastle, Callaghan, New South Wales 2308, Australia; formerly: Australian Nuclear Science and Technology Organisation, Lucas Heights Research Laboratories, Menai, NSW, Australia

C. Clark
Charldon Hill Research Station, Shute Lane, Bruton, Somerset, UK

R.H.F. Curr
Faculty of Applied Science, Bath College of Higher Education, Newton Park, Bath BA2 9BN, UK

I.D.L. Foster
Centre for Environmental Research and Consultancy, Geography Division, Coventry University, Priory Street, Coventry, CV1 5FB, UK

W. Froehlich
Institute of Geography and Spatial Organisation, Polish Academy of Sciences, Department of Geomorphology and Hydrology Research Station, Frycowa 113, PL 33-335 Nawojowa, Poland

K.J. Gregory
(Warden), Goldsmiths' College, University of London, New Cross, London SE14 6NW, UK

A.M. Gurnell
School of Geography, University of Birmingham, Birmingham B15 2TT, UK

Q. He
Geography Department, University of Exeter, Rennes Drive, Exeter EX4 4RJ, UK

D.L. Higgitt
Department of Geography, University of Durham, Science Laboratories, South Road, Durham DH1 3LE, UK

R.J. Loughran
Department of Geography, University of Newcastle, Callaghan, New South Wales 2308, Australia

J.R. Merefield
Earth Resources Centre, University of Exeter, North Park Road, Exeter EX4 4QE, UK

P.N. Owens
Geography Department, University of Exeter, Rennes Drive, Exeter EX4 4RJ, UK

M.R. Peart
Department of Geography and Geology, University of Hong Kong, Pokfulam Road, Hong Kong

T.A. Quine
Geography Department, University of Exeter, Rennes Drive, Exeter EX4 4RJ, UK

H.J.E. Rodda
New Zealand National Institute of Water and Atmospheric Research, Ecosystems Division, 100 Aurora Terrace, PO Box 11-115, Hamilton, New Zealand

J.S. Rowan
Environmental Science Division, Institute of Environmental and Biological Sciences, Lancaster University, Lancaster LA1 4YQ, UK

D.J. Simm
Geography Section, Department of Science, St Mary's University College, Waldegrave Road, Strawberry Hill, Twickenham, Middlesex TW1 4SX, UK

B.W. Webb
Geography Department, University of Exeter, Rennes Drive, Exeter EX4 4RJ, UK

S. White
Instituto Pirenaico de Ecología, CSIC, Campus de Aula Dei, Avda. Montañana 177, 50080 Zaragoza, Spain

J.C. Woodward
School of Geography, University of Leeds, Leeds, LS2 9JT, UK

X. Zhang
Chengdu Institute of Mountain Hazards and Environment, Academia Sinica, PO Box 417, Chengdu, Sichuan Province, China

Y. Zhang
Department of Physics, Sichuan University, Chengdu, Sichuan Province, China

Preface

The idea for these books arose at an IAHS meeting in Oslo, co-convened by Des Walling and Jim Bogen, when two of us (IDLF and AMG) were sharing a small bottle of very expensive Norwegian beer and discussing the remarkable contributions made by Ken Gregory and Des Walling to catchment research over more than 30 years. Indeed, 1994 marked the 21st anniversary of the first publication of their book, *Drainage Basin: Form and Process* (Gregory, K.J. and Walling, D.E. (1973), Edward Arnold, London). Their contributions have not only been to the understanding of catchment processes but also to the inspiration of future generations of researchers, who are now occupying a number of academic positions in UK and overseas universities, colleges and research organisations.

The very strong presence of their former postgraduate students at conferences throughout the world testifies to the important impact that both Ken and Des have had not only on the global academic community but also on the research training from which all four of us, and many others, have benefited in the Geography Departments at the Universities of Exeter and Southampton. Their contribution, however, goes well beyond formal training of ourselves and other postgraduate students. At Exeter and Southampton, postgraduates benefited from the presence of many international visitors, who were passing through or spending time working with Des and Ken. It is a testament to their academic stimulus and friendship that many of these visitors have been keen to contribute to these two volumes.

It seems appropriate that four former PhD students whom Ken Gregory and Des Walling successfully supervised to completion should have co-edited these books. Though we have all developed our own areas of research since the 1970s, it is undoubtedly the case that our approach to catchment research has been influenced to a great extent by three years of research training under their guidance.

The long-kept secret of the publication of these volumes has benefited from the availability of electronic mail (which neither of them have yet learned to break into!), to an anonymous postman delivering parcels to Bruce at home in Exeter rather than to the Geography Department, and to Edna Pellett's undercover operation at Goldsmiths' College. As a result of everybody's efforts to keep this project a secret, we are confident that, while both Ken and Des were enthusiastically aware of the books for the *other*, neither knew anything of their *own* book until they received copies at a joint reception on publication of the two volumes. This secrecy, up to the presentation, has also been due to the tremendous support given by John Wiley & Sons, who agreed not to advertise these volumes before the official launch.

We are indebted to many individuals at the Universities of Birmingham, Coventry, Exeter and Southampton, particularly the cartographers who worked so hard to produce the final versions from less-than-perfect original diagrams. Special thanks go to the cartographers at Coventry, Shirley Addleton, Paul Bird, Ruth Gaskell, Kate Muir and Steve Turnbull.

Ian Foster	Angela Gurnell
Geoffrey Petts	Bruce Webb

Foreword: The Potential, Progression, Participation, Positive Impact and Prospect of Professor D.E. Walling

K.J. GREGORY
Goldsmiths' College, University of London

It is a great pleasure to write the 'Foreword' to this volume in which the contribution that Des Walling has made is so clearly reflected in the range and quality of the following pages, provided by the research students developed and inspired by Des, and by experts from across the world. All testify to the breadth of the creative, national and international influence that Des has had upon research. His influence can be considered in terms of his personal potential, his contribution to the progression of research, his participation in national and international research, and the positive impact that he has had, leading to a few thoughts about the prospect.

PERSONAL POTENTIAL

In a chapter contributed to a volume of which I was an editor in 1987 (Walling, 1987), Des referred to a satirical jingle presented at the annual dinner of the University of Exeter Geographical Society in 1964 when Des was a first-year undergraduate. I had first met him over a year previously when, as a sixth-former, he came to Exeter for interview and I was one of two staff who interviewed him that day. He had applied as his first choice to the University of London, but during the course of the interview we inevitably discussed the merits of studying at the University of Exeter. He subsequently changed his mind and became an undergraduate at Exeter in 1966, reading for a BA in geography with an additional study of archaeology. He proved to be an excellent undergraduate student, exhibiting a lively sense of humour and some of the broad range of interests that have continued in the subsequent 30 years. In his third year, one of the optional courses he selected was in geomorphology; it became clear in the latter part of this year that he was keen to consider research at the University of Exeter, and so became my first research student. In turn, his first research student was Dr Angela Gurnell, one of the editors of this volume.

It is important to remember that in the 1960s movement of individuals between universities from first degree to higher degree was not as clearly developed in geography as it has been in subsequent decades. It was important for Exeter to begin to develop its research school, so as far as I was concerned, and from the point of view of the Department, Des Walling was an excellent acquisition because he had many of the

---

Sediment and Water Quality in River Catchments. Edited by I.D.L. Foster, A.M. Gurnell and B.W. Webb.
© 1995 John Wiley & Sons Ltd.

necessary qualities. He was very conscientious, hardworking, showed considerable imagination and, perhaps most important, was willing to combine technical knowledge with computing, being happy to take on any new techniques to achieve the necessary goals. He obtained his PhD in 1971 but previously, in 1968, against tough competition, he had been appointed to the nationally advertised post of Assistant Lecturer in Geography in Exeter. He became Lecturer in 1971, Reader in Physical Geography in 1977 and a Professor of Physical Geography in the University in 1986. After I left the University of Exeter, in 1976, I was delighted to see Des appointed to a Readership the following year and to become, quite deservedly, probably the youngest Reader in any Geography Department in Britain at that time. The University of Exeter is fortunate that he has decided to stay, being appointed to a Personal Chair in 1986 and subsequently resisting invitations to move to other universities. Professor Des Walling has become an ambassador for the University of Exeter and for British geography and hydrology so that the contribution he has made and the distinctive achievements realised at the University of Exeter are now recognized and acclaimed throughout the world.

PROGRESSION AND RESEARCH DEVELOPMENT

His PhD submitted in 1971 was entitled 'Instrumented catchments in south east Devon' and was based upon results from five instrumented catchments, of different sizes, which were headwaters of the Otter and Sid drainage basins. It is important to recall that at the time there was comparatively little research on hydrological and fluvial processes in the UK, whereas in North America there were at least two strands of developing research in fluvial geomorphology. The first, concentrating on particular, small catchments, was more hydrological and in sympathy with the objectives of the Vigil Network. There was, secondly, a move in geomorphology towards research much more concerned with processes, epitomised by the publication in 1964 of *Fluvial Processes in Geomorphology* (Leopold *et al.*, 1964). This was also reflected in the significant number of research students emerging from Schools in the USA such as those from the Columbia University School under Professor Arthur Strahler. In Britain, however, there had been comparatively little research into processes and this required the development of studies of hydrology, which were assimilated within Geography Departments.

Studies of the five catchments in south-east Devon were significant in developing a way of analysing results from small catchments, relating the results from one to another and encompassing both hydrological (Gregory and Walling, 1968; Walling, 1971a) and sedimentological (Walling, 1971b) investigations. At that time, although hydrological research techniques were being developed in the USA and elsewhere, there were two distinctive characteristics evidenced by the research on catchments in south-east Devon: first, they were established with continuous records, and secondly, they included measurements, not only of water flow, but also of sediment and of solutes. This meant that a major contribution in the PhD by Des Walling was a systematic attempt to compare the five catchments in terms of runoff, to evaluate the differences in sediment and solute production, and to suggest reasons for variations between the five catchments. These results were the product of considerable investment in field instrumentation and I remember the way that we sometimes worked together to install V-notch weirs and gauging stations. The ingenuity shown by Des Walling at that time and his determination to overcome obstacles, whether environmental, financial or departmental, are characteristics that have been continued in the subsequent 25 years.

From that foundation of doctoral research, the network of instrumentation in southeast Devon was extended so that ultimately the drainage basin of the Exe must have become one of the most intensively studied, continuously monitored and well-published basins anywhere in the world. In 1974 we produced a special publication of the Institute of British Geographers (Gregory and Walling, 1974), following a British Geomorphological Research Group meeting organised in Exeter in May 1973. This focused on the development and adoption of new techniques necessary for the assessment of fluvial processes. Appropriate methods had already been established in Exeter, and subsequently Des proceeded to develop his research to make a number of extremely significant contributions. It is not appropriate to go into the detail of these because many of them are cited in subsequent chapters, but a major landmark was surely the inclusion of solutes as well as of suspended sediment in analyses of catchment budgets. This is well illustrated by data from a number of catchments and exemplified in very specific investigations of the consequences of the 1976 drought (Walling, 1979; Foster and Walling, 1978). Central to the analysis of sediment and solute yields from catchments was the previous dependence that had been placed upon rating curves; Des clearly demonstrated in a number of papers (e.g. Walling, 1977) the inadequacies of such reliance on rating curves, and proceeded to propose new methods of sediment and solute load calculation. Subsequently, there have been other equally important contributions, which include the use of caesium-137 as a tracer to enable erosion rates and sediment budgets to be tied down very precisely (e.g. Walling *et al.*, 1979) together with detailed analyses of spatial variations in solute production (Walling and Webb, 1975, 1980) and in sediment production. This has been extended into research involving the collation of results of studies from gauging stations throughout several continents to produce revised maps of sediment yield (e.g. Walling and Webb, 1983). More recently, the links between spatial patterns of sediment production and remote sensing have been explored.

As these successive research developments have evolved, they have required the development of appropriate laboratory instrumentation. Des has pioneered the development of appropriate techniques, often imported from other disciplines and applied to environmental science. He has been as equally at home in the laboratory with field instrumentation and in the field collecting data as he is in international meetings delivering the results of research, always based upon substantial data sets. In the days in the early 1970s, when we both lived near the Rosebarn catchment, we met one night by accident at a gauging station in the dark, both with torches and sample bottles!

PARTICIPATION IN RESEARCH

In the late 1960s, knowledge of processes in the fluvial system was limited, but considerable progress has now been made in the two subsequent decades, with Des Walling contributing significantly to that progress. One of the major reasons for a greater investigation of the dynamics of the fluvial system was that the results of studies of process could contribute to our understanding of landscape change. That prospect is now becoming evident from some of the products of the most recent research from the Walling group because, as the several strands of research have proceeded, they have indeed been integrated to better inform our understanding of landscape behaviour. A particularly good example of this is the way in which the use of tracers, fingerprinting of sediment, and investigations of rates of erosion have led to greater understanding of floodplain sedimentation and development (Walling *et al.*, 1992). The volume of research inspired by Des Walling has contributed to the realisation of the original aim that was so much a pipe-dream in the 1960s (e.g. Gregory, 1985, ch. 5, especially pp. 97–100). In the course of making progress towards the application on studies of landscape dynamics to landscape evolution, Des Walling has made significant contributions to particular aspects of investigation, all of which have been acknowledged in their own right. These include improved understanding of solute and sediment behaviour, the methods of analysis and interpretation of rating curves, the impact of urbanisation, the analysis of spatial patterns of process at the drainage basin scale and at the world scale, and the use of Chernobyl fallout radionuclides to document sediment transport and redistribution very precisely (Walling *et al.* 1989). Specific research contributions have led to very important position papers such as those on the analysis of river loads (Walling, 1978), on sediment yield research (Walling, 1980) and on the sediment delivery problem (Walling, 1983), and these have been supplemented by well-known book chapters and books. Since *Drainage Basin Form and Process* was first published in 1973 (Gregory and Walling, 1973), books have included the themes of human activity (Gregory and Walling, 1980, 1987) and of erosion and sediment yield (Walling, 1982; Hadley and Walling, 1984; Hadley *et al.*, 1985; Bordas and Walling, 1988) — all providing a context for the theme of this volume on sediment and water quality in river catchments.

A POSITIVE IMPACT

With such a substantial contribution marked not only by research investigations achieved in great depth but also by contributions made in several related areas, which have then been fruitfully brought together, it is inevitable that Des, achieving this by the age of 50, has made a very positive impact on the course and direction of research. The six major sections of this volume indicate six broad areas in which Des has made significant contributions. Recognition for these contributions has been shown not only through citation, where he was, from 1984 to 1988, in the top 13 Physical Geographers according to Citation Indices (Bodman, 1991), but also by a number of awards, including the Back award of the Royal Geographical Society in 1985 for contributions to international hydrological research. Internationally, it is perhaps with the International Association of Hydrological Sciences, International Commission on Continental Erosion that he has made the greatest impact, being Secretary from 1975 to 1983 and President from 1983 to 1991. Many other indicators of outstanding performance could be mentioned, including membership of national and international committees, contributions as a member of research council committees, and work as a member of the editorial board of the *Hydrological Sciences Journal* since 1980 and of *Beitrage zur Hydrologie* since 1984. In addition, he was one of the founding editors of

Hydrological Processes, a Wiley journal produced first in 1985. These are all explicit indications of the positive impact that he has made, but perhaps the greatest contribution is less obvious but no less important, and that is the impact made by a geographer. In the late 1960s and early 1970s, geographers were not taken as seriously by other disciplines as they are today, and it is a testimony to the calibre of Des' research that this situation has changed. In the past it was not realised that geographers could use and develop the hardware and software for the advances needed in investigations of environmental processes. Professor Des Walling has demonstrated by example how things have changed so that no longer are particular researchers categorised, or sometimes stigmatised, according to discipline, in the way that was prevalent some two or three decades ago.

PROSPECT

The contribution that Des Walling has made since 1971 when his first paper was published (Walling, 1971a) could have been exemplified very systematically by publishing a detailed CV with well over 150 publications. What such a CV would not show, however, is the humanity, friendship, passion, humour, enthusiasm and insight that Des has exhibited as a leading academic who has already been responsible for significant progress at the frontiers of knowledge. When I met with John Davey in 1970 and outlined the structure of a possible *Drainage Basin Form and Process*, I agreed to write it only if an excellent young member of staff could join me in the venture. Des' contribution to that book set high standards, which he has maintained in all that he has done subsequently. Throughout his research career he has fostered relationships with other researchers, both research students and research scientists from other organisations, and he has facilitated research collaboration of a kind that is fundamental to the improvement in understanding of sediment and water quality in river catchments.

It has been a pleasure to write this 'Foreword' and to reflect upon some of the achievements over the last two decades, the more so because I can look forward to the continuing contribution that Professor Des Walling will make and to the impact that it will certainly continue to have.

REFERENCES

Bodman, A.R. (1991) Weavers of influence: the structure of contemporary geographic research. *Trans. Inst. Br. Geogr.* **16**, 21–37.

Bordas, M. and Walling, D.E. (1988) *Sediment Budgets*, IAHS Publication no. 174, 591pp.

Foster, I.D.L. and Walling, D.E. (1978) The effects of the 1976 drought and autumn rainfall on stream solute levels. *Earth Surf. Processes* **3**, 393–406.

Gregory, K.J. (1985) *The Nature of Physical Geography*, Edward Arnold, London, 262pp.

Gregory, K.J. and Walling, D.E. (1968) Instrumented catchments in south-east Devon. *Trans Devon Assoc.* **100**, 247–262.

Gregory, K.J. and Walling, D.E. (1973) *Drainage Basin Form and Process*, Edward Arnold, London, 458pp.

Gregory, K.J. and Walling, D.E. (eds) (1974) *Fluvial Processes in Instrumented Watersheds*, Institute of British Geographers Special Publication no. 6, 196pp.

Gregory, K.J. and Walling, D.E. (1980) *Man and Environmental Processes*, Butterworth, London, 276pp.

Gregory, K.J. and Walling, D.E. (eds) (1987) *Human Activity and Environmental Processes*, John Wiley & Sons, Chichester, 466pp.

Hadley, R.F. and Walling, D.E. (eds) (1984) *Erosion and Sediment Yield: Some Methods of Measurement and Modelling*, Geo Books, Norwich, 218pp.

Hadley, R.F., Lal, R., Onstad, C.A., Walling, D.E. and Yair, A. (1985) *Recent Developments in Erosion and Sediment Yield Studies*, UNESCO, Technical Document in Hydrology, 127pp.

Leopold, L.B., Wolman, M.G. and Miller, J.P. (1964) *Fluvial Processes in Geomorphology*, Freeman, San Francisco, 522pp.

Walling, D.E. (1971a) Streamflow from instrumented catchments in south east Devon. In: Gregory, K.J. and Ravenhill, W.L.D. (eds), *Exeter Essays in Geography*, University of Exeter, pp. 55–81.

Walling, D.E. (1971b) Sediment dynamics of small instrumented catchments in south east Devon. *Trans. Devon Assoc. Advnt. Sci.* **103**, 147–165

Walling, D.E. (1977) Assessing the accuracy of suspended sediment rating curves for a small basin. *Water Resour. Res.* **13**, 531–538.

Walling, D.E. (1978) Reliability considerations in the evaluation and analysis of river loads. *Z. Geomorphol.* **29**, 29–42.

Walling, D.E. (1979) The 1976 drought and solute levels in two Devon catchments. In: Doornkamp, J.C., Gregory, K.J. and Burn, A.S. (eds), *Atlas of Drought in Britain*, Institute of British Geographers, London, p. 49.

Walling, D.E. (1980) Sediment yield research: an overview. In: *Proceedings of the International Symposium on River Sedimentation (Beijing, China)*, Chinese Society of Hydraulic Engineering, Guanghua Press, Beijing, pp. 1137–1147.

Walling, D.E. (1982) Modification of hydrological processes by building activity. In: *Application of the Results from Representative and Experimental Basins*, UNESCO, Studies and Reports in Hydrology, no. 32, pp. 383–401.

Walling, D.E. (1983) The sediment delivery problem. *J. Hydrol.* **65**, 209–237.

Walling, D.E. (1987) Hydrological and fluvial processes: revolution and evolution, In: Clark, M.J., Gregory, K.J. and Gurnell, A.M. (eds), *Horizons in Physical Geography*, Macmillan, Basingstoke, pp. 106–120.

Walling, D.E. and Webb, B.W. (1975) Spatial variation of river water quality: a survey of the river Exe. *Trans. Inst. Br. Geogr.* **65**, 155–172.

Walling, D.E. and Webb, B.W. (1980) The spatial dimension in the interpretation of stream solute behaviour. *J. Hydrol.* **47**, 129–149.

Walling, D.E. and Webb, B.W. (1983) Patterns of sediment yield. In: Gregory, K.J. (ed.), *Background to Palaeohydrology*, John Wiley & Sons, Chichester, pp. 69–100.

Walling, D.E., Peart, M.R., Oldfield, F. and Thompson, R.. (1979) Suspended sediment sources identified by magnetic measurements. *Nature* **281**, 110–113.

Walling, D.E., Rowan, J.S. and Bradley, S.B. (1989) Sediment-associated transport of Chernobyl fallout radionuclides. In: Hadley, R.F. and Ongley, E.D. (eds), *Sediment and the Environment (Proceedings of Baltimore Symposium)*, IAHS Publication no. 184, pp. 37–45.

Walling, D.E., Quine, T.A. and He, Q. (1992) Investigating contemporary sites of floodplain sedimentation. In: Carling, P.A. and Petts, G.E. (eds), *Lowland Floodplain Rivers, Geomorphological Perspectives*, John Wiley & Sons, Chichester, pp. 165–184.

1 Hydrology, Water Quality and Sediment Behaviour

BRUCE W. WEBB,[1] IAN D.L. FOSTER[2] AND ANGELA M. GURNELL[3]

[1]*Geography Department, University of Exeter, UK;* [2]*Centre for Environmental Research and Consultancy, Coventry University, UK; and* [3]*School of Geography, University of Birmingham, UK*

INTRODUCTION

In April 1995, Professor Des Walling celebrated his 50th birthday. This book has been put together not only to mark this milestone, but also as a tribute to the contribution that he has made to river catchment research over a 30-year period. The purpose of this chapter is to explain the structure of the volume, and to provide a wider context for the chapters that follow.

All of the contributors to the book have been associated with Des Walling: either as former research students, or as academic collaborators in Britain and overseas. It is inevitable, however, that not everyone linked to Des, during the course of a career now entering its fourth decade, has been able to make a contribution. For example, it was decided at the outset of the project (1993) that research students then being supervised by Des would not be involved. It should be noted that some of these students have neared or reached the end of their research during the gestation of this volume, and are publishing the work done under Des's supervision (e.g. Nicholas and Walling, 1994; Phillips and Walling, 1995).

This book very much reflects the hydrological and geomorphological interests pursued by Des Walling and inspired by him in others. The individual sections have been organised not only to represent, as far as possible, the breadth of his interest, but also to mirror its development. The latter began in the late 1960s when process geomorphology and catchment hydrology were first attracting UK geographers into drainage basin research. It has continued up until the present day when geographers working in British universities, such as Des Walling, command respect as leading world experts on many different aspects of river systems.

The Foreword to this volume, which provides an overview of Des Walling's career and academic contribution, is fittingly written by Ken Gregory, who supervised Des in his PhD research and with him wrote the book *Drainage Basin Form and Process: A Geomorphological Approach*. This book was first published in 1973, and is known worldwide as a benchmark text on river catchments.

Subsequent chapters of the present volume are arranged into six sections. The first of these echoes the interest that Des Walling has had in the quality, as well as the quantity, of river runoff since his doctoral research in south-east Devon. Although he has worked on different chemical and physical properties of water quality in rural and urban catchments, including solute behaviour and thermal regime, his primary focus has been

---

Sediment and Water Quality in River Catchments. Edited by I.D.L. Foster, A.M. Gurnell and B.W. Webb.
© 1995 John Wiley & Sons Ltd.

on fluvial sediments and especially on suspended sediment carried by rivers. In the early stages of his career, Des concentrated on quantifying and explaining suspended sediment dynamics and yields at the outlet of river catchments, and chapters dealing with this theme are arranged in Section II of this book. He soon recognised that it was also important to investigate the properties, as well as the amount, of suspended sediment transported by rivers, and this concern is addressed in Section III of the present volume. Before long, it became evident to Des that, in order fully to understand suspended sediment dynamics and transport, the drainage basin should not be viewed as a 'black box'. Rather, the complex linkages between the various sources and sinks of sediment, involving temporary and more permanent stores of material within the drainage basin that control delivery of sediment to the catchment outlet, required investigation. The fourth section of the book includes studies that address sediment sources and sinks, and some links between them. One of the techniques pioneered and strongly developed in the UK by Des Walling has been the use of ^{137}Cs measurements to elucidate patterns of erosion and deposition, and to quantify sediment budgets for drainage basins. This theme is illustrated by the chapters collected together in Section V. Although the catchment research undertaken by Des Walling has had a firm foundation in the long-term water quality and sediment monitoring programme that he set up in the Exe basin of Devon, his interests have been far from parochial. Indeed, they have not only extended to other UK rivers and to studies overseas, but also have frequently embraced a global perspective. The last section of the book, therefore, includes several sediment studies in river environments beyond the United Kingdom.

The remainder of this chapter provides further context for these themes, and also aims to cover those areas of Des Walling's research that are not directly addressed by the contributed chapters. Since the framework of this book is based on the development of Des Walling's research, it is inevitable that this chapter makes frequent reference to his work. A bibliography of Des Walling's main publications is therefore included at the end of this chapter.

SECTION I: QUANTITY AND QUALITY DIMENSIONS

Hydrological research in the 1950s and 1960s concentrated mainly on the quantity dimension and generated a better understanding of not only the volumes and rates of movement of water through the hydrological cycle, but also the controlling processes. Early studies on the quantity dimension undertaken by geographers are exemplified well by Des Walling's analysis of the streamflow characteristics of five small catchments in east Devon (Walling, 1971a). This research highlighted the role of different catchment characteristics in controlling different aspects of the discharge regime. The impact of catchment characteristics on hydrology at the regional scale of Devon and south-west England was further explored through subsequent research at Exeter on the nature of the annual water balance (Foyster, 1973, 1975), water resources (Buckett, 1976) and low flows (Browne, 1978). In more recent years, Des Walling has cooperated with the Institute of Grassland and Environmental Research and its predecessors in studies of the effects of drainage on the hydrology of grassland areas (Hallard, 1988).

The 1970s marked an upsurge of interest in the quality aspects of the hydrological cycle. At Exeter, research led by Des Walling focused on 'background' water quality of streams and rivers in rural areas free from major sources of industrial and domestic pollution. A catchment-wide view (Figure 1.1) of the mechanisms influencing the quality of water moving through the major phases of the hydrological cycle was developed, and water quality was studied very much from the perspective of the hydrologist and geomorphologist (Walling, 1980a; Walling and Webb, 1986a). Water quality research at Exeter has been underpinned by work in the local rivers. In particular, installation of a network of monitoring sites in the River Exe during the mid-1970s, with the aim of providing detailed and long-term data, has yielded much information on spatial and temporal variations in water quality at the catchment scale.

In common with other UK and overseas studies of water chemistry during the 1970s and 1980s, results from the Exe basin and from other catchments in south and east Devon studied by Des Walling have shown contrasts between total dissolved solids and individual solute species with respect to spatial patterns of concentration, annual cycles of variation, and rating relationships with discharge (e.g. Walling and Webb, 1975; Troake and Walling, 1975; Troake et al., 1976; Foster, 1978a; Webb and Walling, 1983). In addition, combination of information on solute concentrations and discharge allowed dissolved transport to be calculated, from which estimates of catchment chemical denudation rates could be derived by making corrections for the non-denudational components of solute loads (e.g. Walling and Webb, 1978; Foster, 1980). Detailed data collected from Devon rivers have particularly highlighted the complexity of stream solute behaviour, especially during storm periods. It soon became clear, for example, that individual ions differed in their response to flood events because different chemical species had contrasting origins, were stored in different locations within the drainage basin and were accessed by runoff from various sources (e.g. Walling, 1975a; Foster, 1978b). Hysteretic behaviour, whereby solute concentrations differ markedly between the rising and falling limbs of the flood hydrograph, was observed as a common feature in Devon rivers and attributed to both flushing and exhaustion effects (Walling and Foster, 1975; Webb and Walling, 1983). Furthermore, strong seasonal contrasts in storm-period responses were noted in some catchments for some solute species. For example, NO_3-N concentrations in the River Dart tributary of the Exe basin were strongly diluted in winter storms but were raised during summer events (Webb and Walling, 1985a).

The Exe basin monitoring network has also shown that storm-period solute responses in sizable drainage basins should not be interpreted only in terms of runoff originating from vertically differentiated sources in the catchment, such as surface, soil and groundwater flows. A spatial dimension, which considers the aggregation of contrasting responses from different tributaries and the routeing down the main channel of these responses, may provide a better explanation of storm-period solute behaviour at downstream locations (Walling and Webb, 1980). The Exe basin network has also provided information on solute behaviour during rare hydrological events, such as those at the end of the 1975–76 drought when conditions of extreme dryness were rapidly terminated by exceptional rainfall. In the case of NO_3^- ions, concentrations in a small agricultural catchment evidenced a 50-fold increase in the immediate post-drought period, which was attributed not only to flushing out of nitrogen accumulated in the preceding protracted dry spell but also to changes in the interrelationship of organic and inorganic soil nitrogen. The latter were caused by high temperatures and dry conditions of the drought followed by rapid soil wetting (Foster and Walling, 1978; Walling and

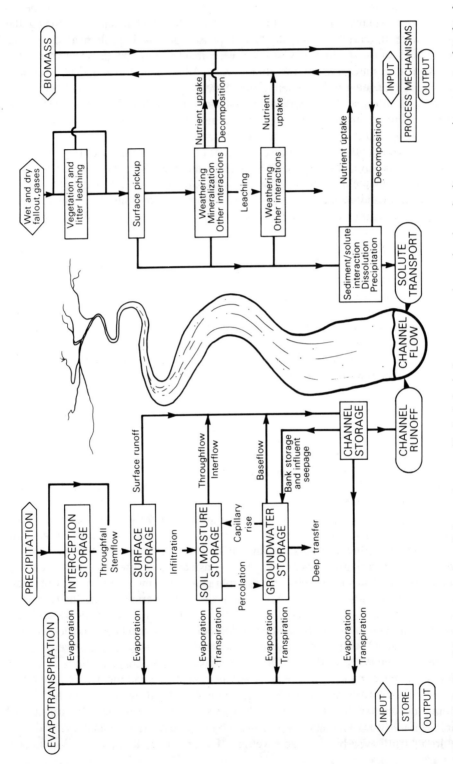

Figure 1.1 Simplified representation of the catchment-wide hydrological processes and associated mechanisms governing stream solute levels (after Walling and Webb, 1986a)

Foster, 1978). An understanding of the processes of nitrogen cycling in catchments is vital to a proper explanation of stream nitrate dynamics and transport. Collaboration with the Institute of Grassland and Environmental Research referred to earlier also has involved studies of how processes in the nitrogen cycle, such as mineralisation and denitrification, affect NO_3^- variations in runoff from drained and undrained grassland plots (Blick, 1989; Blantern, 1991). In Chapter 3 of this volume, Harvey Rodda shows how a model of nitrogen cycling developed at the scale of individual grassland plots can be extended to predict nitrate behaviour in catchments with dominantly pasture land use. The value of geographical information systems (GIS) as a tool for more general modelling of solutes in drainage basins is explored by Tom Browne in Chapter 2.

The research led by Des Walling has investigated physical as well as chemical properties of water quality. For example, investigations of different aspects of the thermal regime, including extreme values and annual and diurnal water temperature cycles, have revealed significant differences between tributary and mainstream sites in the Exe basin (Walling and Carter, 1980; Webb and Walling, 1985b, 1986a). Detailed records of water temperature fluctuations in this catchment are among the longest collected for any British river system, and have allowed the nature of trends over the last 15 and more years to be investigated (Webb and Walling, 1992a, 1993a). In Chapter 4, Bruce Webb uses data from the Exe basin monitoring network to examine the impact over a 17-year period of the Wimbleball Scheme of reservoir construction and river regulation on downstream thermal regime and, in turn, aspects of river ecology.

Although investigations at Exeter have addressed many issues relating to water chemistry and thermal regime, Des Walling's main focus in catchment research has been on suspended sediment. A research interest in this physical property of water quality began during his doctoral study of five small catchments in south-east Devon, which considered the dynamics of suspended sediment behaviour (Walling, 1971b).

SECTION II: SEDIMENT DYNAMICS AND YIELDS

Early investigations of the behaviour of fine-grained particulate material in rivers focused attention on the form of the suspended sediment concentration–discharge relationship (e.g. Colby, 1956; Guy 1964, Hall, 1967). Building on these studies, and extending them by implementing very detailed measurement programmes, Des Walling was able to highlight the complexity of concentration–discharge relationships. Early research in the Rosebarn catchment on the margins of Exeter, for example, demonstrated that such relationships typically evidence seasonal contrasts, differences depending on stage conditions and shifts between individual flood events (Walling, 1974a). This and subsequent work, based on the results of long-term and intensive monitoring using continuous recording photoelectric turbidity meters set up in Devon rivers (Walling and Webb, 1987a), have also emphasised the complicated nature of storm-period suspended sediment dynamics, on both an intra- and inter-flood basis. For example, it is rare during a storm event for fluctuations in sediment concentrations to be in phase with discharge variations, and the precise nature of the hysteresis exhibited will reflect the relative timing and form of the sediment and water responses (e.g. Gurnell, 1987; Williams, 1989). Equally, suspended sediment behaviour can vary markedly between isolated storms occurring at different times of the year or as a sequence of flood events (e.g. Walling, 1978a). Such complications may be related to exhaustion of easily erodible sediment and the effects of baseflow dilution in small catchments (Walling and Webb, 1982a; Carling, 1983), and to mixing and routeing of contrasting water and sediment responses from individual tributaries in larger drainage basins (e.g. Heidel, 1956; Walling and Woodward, 1995).

The nature of suspended sediment behaviour during storms, documented in Devon and other rivers, clearly highlights the fact that the wash-load of silt and clay carried by a river cannot be treated as a capacity load, which is simply related to hydraulic conditions and the energy available for transport. Suspended sediment loads of rivers are commonly very much less than the transport capacity of the flow, and are influenced by a large number of factors, other than water discharge, which control the supply of fine material to a river. The latter include rainfall intensity, antecedent temperature, moisture and discharge conditions, hydrograph shape and temporal variations in surface condition and vegetation cover (Walling, 1995a). In Chapter 7 of this volume, which provides a wide-ranging review of sediment production and dynamics based on studies of small catchments in east Devon, the New Forest and east Somerset, Colin Clark shows how storm-period sediment concentrations and loads are related to a range of factors in a multivariate, rather than a simple, manner.

The importance of supply considerations to storm-period behaviour of fine particulate material increases the difficulty of successfully modelling suspended sediment dynamics of rivers. Different approaches have been tried, including tank cascade models (Okunishi *et al.*, 1990) and those which take account of time-variant sediment supply (VanSickle and Beschta, 1983; Moore, 1984). In Chapter 5, Kazimierz Banasik discusses a conceptual model of the instantaneous unit sedimentgraph, which has been developed by combining the instantaneous unit hydrograph with a dimensionless sediment concentration distribution. A sediment routeing coefficient in the model was estimated empirically from field data available for rainfall, runoff and suspended sediment concentration in small Carpathian catchments.

Because of the influence that sediment supply exerts on the dynamics and yield of fine particulate material carried in many rivers, it is the case that suspended sediment transport is often highly biased to storm events, which are capable of mobilising sediment and transferring it to the drainage network. For example, detailed records collected for Devon rivers show that 90% of the total suspended sediment transport takes place in 10% or less of the time, and reveal that flows around the bankfull stage are most effective in removing fine particulate material (Webb and Walling, 1982a, 1984a). Conclusions drawn from river records regarding the magnitude and frequency aspects of sediment transport are inevitably limited by the relatively short periods for which detailed information on sediment concentrations is usually available. In Chapter 6, Rick Curr provides a longer-term perspective for a small catchment near Bath, UK, by correlating the pattern of particle-size variations in sediments trapped in a lake with stream responses simulated from a 103-year daily rainfall record at a nearby station. This analysis revealed that discharges just in excess of bankfull stage, and associated with the development of overland flow on cultivated fields of the catchment, were the most effective for transporting suspended particulate material over a period of more than 100 years.

Bias in suspended sediment transport towards high flows and the lack of a clearly defined relationship between fine sediment concentration and discharge in most

Table 1.1 Estimates of total suspended sediment load for three Devon rivers using rating relationships fitted to entire sets of hourly concentration and discharge values (after Walling and Webb, 1988)

River	Period	Actual load (t)	Rating estimate (t)	Bias-corrected estimate[a] (t)
Dart	1975–85	24 499	862	2 145
Creedy	1972–80	82 863	15 443	39 579
Exe	1978–80	41 402	2 754	9 212

[a] Rating estimate corrected for bias according to the method proposed by Ferguson (1986, 1987)

catchments creates problems for estimating the yield of suspended sediment carried in rivers. The reliability of traditional techniques to estimate suspended sediment yields is one area of Des Walling's work that is not directly represented in this volume. Several papers based on detailed suspended sediment records available for Devon rivers (Walling, 1977a,b, 1978b; Walling and Webb, 1981a, 1985) have revealed that yield estimates derived from interpolation or extrapolation procedures for combining concentration and flow information are likely to suffer from serious problems of inaccuracy, imprecision or both. The traditional approach of constructing a sediment rating curve from weekly or less frequent sampling, and applying it to a detailed time series of river flows, has been found to give a serious underestimate of yields in many rivers (e.g. Olive et al., 1980; Farr and Clarke, 1984). Although it has been suggested that this underestimation can be largely removed by applying simple correction factors to account for the bias inherent in the use of log-transformed regression to derive sediment rating relationships (Ferguson, 1986, 1987), tests on data from the Exe basin (Table 1.1) do not support the efficacy of such procedures (Walling and Webb, 1988). There remains considerable scope for investigating how load calculation procedures can be improved to take into account the complexity of suspended sediment behaviour when deriving sediment yields from concentration data collected at infrequent intervals (e.g. Gurnell and Fenn, 1984; Thomas, 1986; Clarke, 1990; Littlewood, 1990)

SECTION III: SEDIMENT QUALITY

Although palaeolimnological research in the UK has long been concerned with the chemistry of lake sediments (e.g. Mackereth, 1966), and studies of particulate-associated contaminant transport were conducted for European and North American rivers during the 1960s and 1970s (e.g. Kopp and Kroner, 1968; Förstner and Müller, 1974; Shear and Watson, 1977), most research in British catchments before the early 1980s focused on the amount and behaviour of sediment being carried in the river rather than on the chemical and physical properties of the material itself. Des Walling was among the first to pioneer studies of the properties of suspended sediments in British rivers (e.g. Oldfield et al., 1979). Investigation of sediment quality was seen as a logical extension to studies of sediment dynamics and yields, and had the potential to further the understanding of suspended sediment behaviour in rivers in three main ways. First, because the characteristics of suspended sediment will reflect the properties of the source soil materials from which it is derived, investigation of sediment quality may provide a powerful technique for 'fingerprinting' the sources of suspended sediment in river systems (e.g. Walling et al., 1979). Secondly, comparison of the properties of source materials and suspended sediments often reveals significant differences, such as the enrichment of river sediments with respect to fines and secondary minerals (e.g. Peart and Walling, 1982). Such changes shed light on the selectivity of the erosion, transport and deposition processes that are involved in the transfer of sediment between soil source and river channel. Thirdly, studies of sediment quality illuminate the important role that sediment-associated transport has for the flux of nutrients and contaminants through drainage basins, and which may significantly influence river quality through interactions between chemical species in particulate-associated and dissolved forms (e.g. Walling and Kane, 1982).

The studies of sediment quality established at Exeter by Des Walling have investigated a number of properties of fluvial sediments and source materials, including sediment mineralogy, particle size, major-element compositions, organic matter and nutrient content (C, N and P) and mineral magnetic characteristics in addition to radionuclide signatures, which are considered in more detail below. This research has identified the dominant sources of suspended sediments in catchments characterised by contrasting land uses (e.g. Walling and Kane, 1984; Peart and Walling, 1986) and has begun to assess the relative merits for sediment fingerprinting of different diagnostic tracers, used individually or in combination (Walling et al., 1993). Seasonal and storm-period variability in the particle-size characteristics of suspended sediment has been studied in detail for the Exe basin, where the occurrence of complex and contrasting responses between tributaries in a relatively small area (Figure 1.2) has highlighted the inadequacy of hydraulic conditions alone to provide explanations of the behaviour of this property (Walling and Moorehead, 1987, 1989). Furthermore, the importance of particle aggregation has been underlined for erosion, transfer and in-channel transport of suspended sediment (Walling, 1988a; Droppo and Ongley, 1992). Work in Devon rivers using a field-based water elutriation system and a field-portable laser backscatter particle sizer has provided reliable information on the difference for riverine suspended sediments between the *in situ* or effective grain-size distribution and the absolute or ultimate particle-size characteristics determined for chemically dispersed samples in the laboratory (Walling and Woodward, 1993; Phillips and Walling, 1995). Such differences indicate that much of the clay-sized material is transported as composite particles, and confirms the importance of particle aggregation in the freshwater environment. Some of the first research in British rivers to investigate problems of estimating sediment-associated transport and its importance to nutrient fluxes was also conducted in the Exe basin (Thornton, 1985; Walling and Webb, 1985; Garrett, 1992; Walling et al., 1992d).

The approaches to studies of sediment quality developed at Exeter are well illustrated by three contributions to the present volume. Two of these are concerned with tracing the impact of mining activity on fluvial sediments, but they use different fingerprinting techniques. In Chapter 8, John Merefield employs information on sediment mineralogy to trace the movement of sediment derived from mining activity in the River Teign of south Devon, while in Chapter 9, Steve Bradley focuses on sediment-associated metal content to investigate processes of sediment redistribution in a number of British

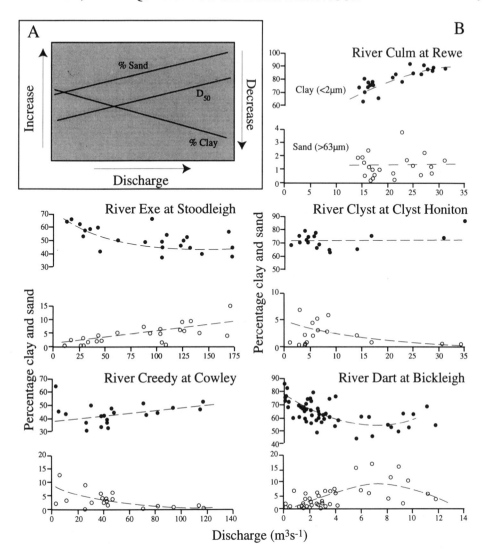

Figure 1.2 The traditional hydraulic interpretation of the relationship between the grain-size composition of suspended sediment and river flow reflecting an increasing median diameter and percentage sand at higher discharges and flow velocities (A), and examples of the more diverse and complex relationships found in Devon rivers (B) (after Walling, 1995a)

catchments. By way of contrast, Mervyn Peart discusses in Chapter 10 how the sediment properties of C and N contents, C:N ratios and loss on ignition may be used to fingerprint sediment sources in a Hong Kong catchment.

SECTION IV: SEDIMENT SOURCES AND SINKS

As noted in the context of sediment quality, the use of fingerprinting techniques based on sediment properties provides a potentially powerful tool for quantifying the contribution of different catchment sources to the suspended sediment carried in river systems. This approach to studying sediment sources in Devon rivers has been developed under the leadership of Des Walling, and an example of its application is provided by Qingping He and Phil Owens in Chapter 12. By devising a numerical mixing model, which was based on the behaviour of the radionuclides ^{137}Cs, ^{210}Pb and ^{226}Ra, and took into account enrichment in fines and other processes affecting the transfer of material between source and river channel, He and Owens suggested that 53, 35 and 12% of the suspended sediment transported by the River Culm originated from cultivated topsoil, uncultivated topsoil and channel bank erosion, respectively. Fingerprinting techniques, however, are not the only methods available to the geomorphologist for identifying sediment sources and quantifying their contribution in drainage basins. In Chapter 11 of this volume, Bob Loughran and Bryan Campbell review a variety of different approaches, while in Chapter 13, Dave Ashbridge draws conclusions regarding the processes, controls and contribution of bank erosion to suspended sediment transport in the River Culm by using information from erosion pins and from detailed monitoring of bank moisture conditions, water discharge and suspended sediment concentration for the river channel.

One of the major goals of sediment research to emerge since the late 1970s has been the need to understand better the nature of catchment–river linkages (Wolman, 1977; Walling, 1995a). Research by Des Walling and others has highlighted the importance of delivery and conveyance processes, which are interposed between, on the one hand, the mechanisms of sediment erosion and the character of soil source materials on the slopes of drainage basins, and, on the other, the yield and properties of suspended sediment transported in the river channel (Walling, 1983a, 1990a). It is clear that there are many opportunities for sediment storage within the drainage basin, although the amount of eroded sediment that is delivered beyond the local environment of the field will vary between and within different soil type categories (Table 1.2). For many catchments only a small proportion of eroded sediment will be readily transported to the catchment outlet. However, it is also apparent that simple concepts, such as the sediment delivery ratio, are inadequate to account for spatial and temporal variability and complexity in delivery and conveyance processes, which include attenuation and discontinuity in transport associated with sediment remobilisation over a range of time-scales (e.g. Verhoff *et al.*, 1979; Duijsings, 1986; Trimble, 1995). It is therefore necessary to view sediment in the drainage basin in terms of a budget that can quantify the complex interrelationships between the different sources and sinks of material involved (Dietrich and Dunne, 1978; Swanson, *et al.*, 1982). Studies led by Des Walling in Devon and other rivers have attempted to define budgets for catchments of different scale (Walling, *et al.*, 1986b; Walling and Quine, 1993), and have investigated temporary and more permanent storage of sediment on slopes (Quine and Walling, 1991), in the river channel (Lambert and Walling, 1986, 1988) and in floodplains (Lambert and Walling, 1987; Walling and Bradley, 1989a). In Chapter 14 of this book, David Simm reports the results of a detailed investigation of short- and longer-term deposition of sediment on the floodplain of the River Culm, an east-bank tributary of the River Exe which is prone to frequent inundation.

Natural and artificial lakes also represent long-term sinks for sediments in drainage basins. In Chapter 15, Ian Foster discusses the information that bottom sediments of

Table 1.2 Delivery ratio of sediment eroded from arable fields on various soil types in southern Britain (after Quine and Walling, 1991)

Soil type and site	Delivery ratio (%)
Brown sand	
Hole, Norfolk	48
Dalicott, Shropshire	64
Rufford Forest, Nottinghamshire	86
Brown rendzina	
Manor House, Norfolk	38
Lewes, Sussex	33
Brown calcareous earth	
West Street, Kent	56
Brown earth	
Mountfield, Somerset	48
Yendacott, Devon	36
Argillic brown earth	
Wootton, Herefordshire	44
Fishpool, Gwent	37
Calcareous pelosol	
Higher, Dorset	60
Brook End, Bedfordshire	33

lakes and reservoirs can provide on soil erosion and sediment transport in the United Kingdom.

SECTION V: RADIONUCLIDE STUDIES

Despite the early use of ^{137}Cs for dating lake and reservoir sediments in Britain (Pennington et al., 1973), few UK researchers during the 1970s and early 1980s had recognised the potential of using radionuclides in studies of catchment sediment systems, although rapid developments were taking place at this time in the United States (e.g. Ritchie et al., 1975) and in Australia (e.g. McCallan et al., 1980; Loughran et al., 1982). Des Walling was among the first researchers in Britain to appreciate that studies of ^{137}Cs provide a means of determining sediment behaviour in the drainage basin over the last 40 years, and catchment research at Exeter under his leadership in the last decade has focused on the development and refinement of this approach. Measurement of ^{137}Cs in soils, in river suspended sediments, and in slope, channel and floodplain deposits has provided new and detailed insights into sediment sources, patterns of slope erosion and deposition, delivery processes, catchment sediment budgets, fluvial conveyance and redistribution of sediments, and rates of floodplain accretion (e.g. Walling *et al.*, 1986b; Loughran *et al.*, 1987; Walling and Bradley, 1988a,b, 1989b, 1990; Walling and Quine, 1990a,b, 1991a; Walling and He, 1992, 1993a; Walling, *et al.*, 1992c; Walling and& Woodward, 1992; Quine and Walling, 1993a; Foster *et al.*, 1993; Foster and Walling, 1994). Studies of ^{137}Cs profiles in floodplain sediments have proved to be particularly useful in revealing the patterns of deposition in relation to microtopographic features and local hydraulic conditions, and in deconvoluting the contribution of radiocaesium from atmospheric fallout to the surface and in sediment-associated deposition during storm-period inundation (Walling *et al.*, 1992b).

The basis of using ^{137}Cs in catchment sediment studies is essentially simple. This radioisotope does not occur naturally but was produced as a by-product of the atmospheric testing of thermonuclear weapons in the 1950s, 1960s and 1970s, and also through occasional releases from nuclear reactors, such as occurred in the 1986 Chernobyl accident. Deposition of ^{137}Cs from the stratosphere is effected by rainout and washout, and the annual variation in fallout in the Northern Hemisphere (Figure 1.3) is related to the pattern of annual rainfall and to a decline following treaties between the major powers in the 1960s to ban atmospheric testing of nuclear weapons. On reaching the ground surface, ^{137}Cs becomes attached to the finer fraction of the soil, and its subsequent redistribution occurs through the erosion and deposition of sediment.

Use of ^{137}Cs in sediment studies has made assumptions concerning the spatial distribution of fallout, the initial and subsequent sorption to clay minerals, the identification of stable sites to establish reference inventories, and the relationship between ^{137}Cs loss or gain and physical processes of soil erosion, fluvial transport, and

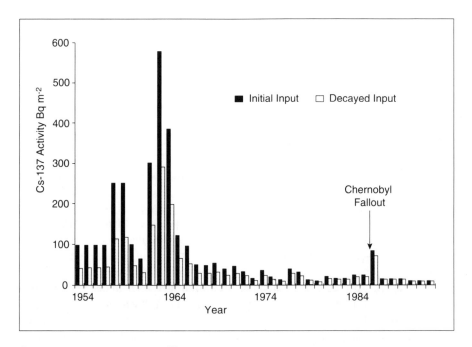

Figure 1.3 The initial input of ^{137}Cs recorded at Chilton, Oxfordshire, UK, over the period 1954–93, and the input corrected for decay up to 1992 (data by courtesy of AERE, Harwell)

alluvial and colluvial deposition. In Chapter 16 of this volume Dave Higgitt examines some of these assumptions in more detail, while in Chapter 17, Tim Quine discusses the validity of mass-balance models, which are used to derive rates of soil erosion from ^{137}Cs measurements. Examples of the erosional transport and redistribution of radiocaesium in fluvial systems are provided by John Rowan in Chapter 18.

Although the first studies of ^{137}Cs at Exeter were made in local catchments, Des Walling has applied this approach to studies of soil erosion, ground retreat and river sediment budgets; in other parts of the UK (e.g. Quine and Walling, 1991; Walling and Quine, 1993; Higgitt et al., 1994). The Exeter research group has also evaluated the use of ^{137}Cs investigations in environments outside of the UK, including the High Veldt of Zimbabwe (Quine et al., 1993a), glacierised basins of Greenland (Hasholt and Walling, 1992) and semi-arid and more humid mountain regions in Spain and Poland, respectively (Higgitt et al., 1992; Navas and Walling, 1992). Considerable attention has been given to the use of radiocaesium in studying areas suffering very high soil erosion in the Loess Plateau region of China (Zhang et al., 1990, 1994), and Xinbao Zhang and Yiyun Zhang describe some of this research in Chapter 19 of this volume.

Although studies at Exeter have made most use of ^{137}Cs measurements, other radionuclides, including ^{134}Cs, ^{210}Pb and ^{226}Ra, have also been employed, and their application to catchment sediment investigations is being extended (e.g. Foster et al., 1993; Walling et al., 1993).

SECTION VI: NATIONAL AND GLOBAL PERSPECTIVES

A commitment to long-term and detailed investigations of a single river system will produce insights regarding water quality and sediment behaviour that are not possible from shorter-term studies conducted in a range of catchments. At the same time, however, focusing alone on a particular river, however varied its catchment characteristics, may provide a limited perspective and make generalisation of research findings difficult. Des Walling has avoided the latter problems by complementing his long-term interest in the Exe basin with studies of other UK catchments, such as the River Severn (e.g. Higgitt et al., 1993; Walling and Quine, 1993), with fieldwork based in other countries, including Iraq, Spain, Poland, Greenland, Zimbabwe and China (e.g. Al-Ansari et al., 1988; Navas et al., 1992; Froehlich and Walling, 1992a; Hasholt and Walling, 1992; Quine et al., 1992; Froehlich et al., 1993; Quine et al., 1993a), with investigations using remote-sensing techniques (Aranuvachapun and Walling, 1987, 1988), and with data compilations at countrywide (e.g. Walling and Webb, 1986b), continental (e.g. Walling, 1984a) and global scales (e.g. Walling and Kleo, 1979).

In the context of British rivers, increasing standardisation of analytical procedures and sampling schedules between the regional sections of statutory authorities responsible for river surveillance and monitoring has encouraged a countrywide perspective concerning water quality. For example, the advent of the Harmonized Monitoring Scheme during the 1970s (Simpson, 1978, 1980) has yielded data for about 250 sites, which allow the variability of water quality conditions across mainland Britain to be mapped (e.g. Brown et al., 1982; Roberts and Marsh, 1987). Although valuable, maps based on Harmonized Monitoring data essentially show the variation in the quality of large watercourses close to the tidal limit. In order not only to investigate the finer detail of national patterns of water quality, but also to explain meaningfully the variation in terms of catchment characteristics, there is a need to collect data on the quality of runoff close to source. The potential of routine surveys undertaken by the statutory authorities for providing such information was seen by Des Walling. Although sampling is typically less frequent than in the Harmonized Monitoring Scheme, routine surveys cover very many more sites across mainland Britain. For example, careful collation of data for more than 1500 sites revealed not only contrasting patterns and controls of total solute levels, pH and chloride and nitrate-nitrogen concentrations in British rivers, but also allowed combination with national patters of river discharge in order to assess countrywide patterns of solute yields and, through further analysis, chemical denudation (Walling and Webb, 1981b, 1986a). Statistical relationships between water quality and upstream catchment characteristics may help to explain national patterns. For example, Betton *et al.* (1991) found that multiple regression relationships, based on farm type, land capability, soil type, hydrogeology, relief, rainfall and runoff, accounted for more than 80% of the variation in NO_3-N concentrations and loads across mainland Britain.

In the case of suspended sediment in British rivers, it is more difficult to establish countrywide patterns of behaviour because, unlike some other countries, routine measurements are not included in the national hydrometric monitoring network. Some information is available through careful collation of the results of localised research projects (Walling and Webb, 1981b, 1987a). Notwithstanding problems of erratic spatial coverage of sampling sites, generally short-term records and variability in data reliability, the results of such investigations suggest first that suspended sediment concentrations in British rivers rarely exceed 5000 mg l^{-1}, and often do not rise above 500 mg l^{-1}, secondly that annual suspended sediment yields range from <1.0 to nearly 500 t km^{-2} yr^{-1}, with 50 t km^{-2} yr^{-1} being a typical value, thirdly that the sand fraction of the suspended sediment is typically 5% or less, and fourthly that the organic matter fraction is often in the range 10–30%.

Seen in a world context, sediment concentrations and yields of British rivers are low. This fact is highlighted by several of the chapters in the last section of this volume. In Chapter 20, Jamie Woodward emphasises that soil erosion and sediment transfer in rivers draining the Mediterranean basin are encouraged by high erosivity of precipitation, an active tectonic setting, steep topography, fissile sedimentary geology and easily erodible soil mantles. The natural vulnerability of the Mediterranean region to erosion has been exacerbated by a long history of human disturbance to the landscape, and it has been calculated that the anthropogenic component of river sediment yields is higher in this morphogenetic zone than in any other (Dedkov and Mozzherin, 1992). In consequence, some rivers in the Mediterranean area have suspended sediment yields of >4000 t km^{-2} yr^{-1}. Sediment yields of almost an order of magnitude greater (39 700 t km^{-2} yr^{-1}) are reported by Sue White in Chapter 21 of this volume for a small basin in the Magat catchment of the Philippines. In this case, the very high transport rates can be attributed to the effects of rainfall generated in tropical cyclones and thunderstorms on soils exposed through the disturbance of natural vegetation cover. Sediment loads well above the global average may also be found in proglacial rivers, where the availability of sediment for fluvial transport is generally high. In Chapter 22, Angela Gurnell shows from long-term studies of four Alpine glacier basins located in the Val d'Hérens, Switzerland, that annual yields of suspended

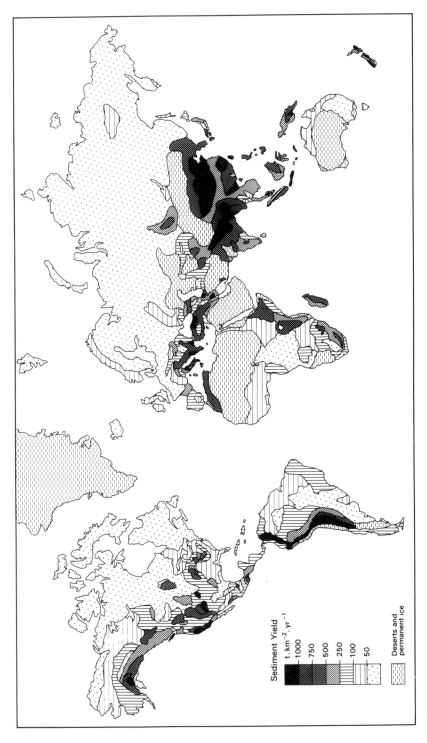

Figure 1.4 A map of global sediment yields based on values of specific suspended sediment yields for drainage basins with areas of 1000–10 000 km² (after Walling and Webb, 1983a)

sediment may reach 3000 t km^{-2} yr^{-1}, although loads as low as 280 t km^{-2} yr^{-1} were also recorded in this study. The complexity of sediment behaviour in the Alpine environment, demonstrated by this work, reflects seasonal and diurnal variability in river discharge, the interaction of subglacial drainage systems and subglacial sediment source area, and the wide range of sediment sources in the proglacial zone. The complexities of erosion, delivery and transport of river sediments in mountain environments is further highlighted by Jim Bogen's studies of Norwegian catchments and Wojciech Froehlich's long-term investigation of the Homerka catchment in the Polish Carpathians, which are reported in Chapters 23 and 24, respectively.

Increasing availability of information on sediment and solute yields of rivers in recent years has stimulated a growing interest in the patterns and controls of fluvial transport and denudation at the global scale (e.g. Meybeck, 1977, 1982, 1983; Jansson, 1982, 1988; Milliman and Meade, 1983; Dedkov and Mozzherin, 1984; Milliman and Syvitski, 1992). Des Walling has been very much involved in conflating and interpreting global databases of sediment and solute concentration and yield data, which currently reveal, for example, that suspended sediment yields range globally from a minimum of 0.9 to a maximum of >53 500 t km^{-2} yr^{-1} (Walling, 1995a). He has played a leading role not only in mapping worldwide variations (Figure 1.4), but also in interpreting global patterns in terms of climate, topography, geology, tectonic setting, soil type, land use and land-use history, and in understanding changes in world water quality and sediment behaviour over a range of timescales (e.g. Walling and Webb, 1983a,b, 1986a, 1987b; Walling, 1995b).

REFERENCES

References to publications by Des Walling in this chapter may be found in the following bibliography. The first part of the reference list contains all the other text citations.

Blantern, P.J. (1991) Factors affecting nitrogen transformations in grazed grassland soils with specific reference to the effects of artificial land drainage and N-fertilization. Unpublished PhD thesis, University of Exeter.
Blick, C. (1989) The impact of drainage and reseeding on nitrogen losses from grassland soils. Unpublished MSc thesis, University of Exeter.
Brown, C., Rodda, J.C. and Williams, M. (1982) A countrywide study of river water quality for Great Britain. In: *Effects of Waste Disposal on Groundwater and Surface Water (Proceedings of Exeter Symposium, 1982)*, IAHS Publication no. 139, pp. 195–205.
Browne, T.J. (1978) An analysis of streamflow recession curves in Devon. *Rep. Trans. Devon. Assoc. Advmt. Sci., Lit. Art* **100**, 81–94.
Buckett, J. (1976) The water resources of Dartmoor and the surrounding area: a study using limited hydrometric data. Unpublished PhD thesis, University of Exeter.
Carling, P. (1983) Particulate dynamics, dissolved and total load, in two small basins, northern Pennines, UK. *Hydrol. Sci. J.* **28**, 355–375.
Clarke, R.T. (1990) Bias and variance of some estimators of suspended sediment load. *Hydrol. Sci. J.* **35**, 253–261.
Colby, B.R. (1956) *Relationship of Sediment Discharge to Streamflow*, US Geological Survey Open File Report.
Dedkov, A.P. and Mozzherin, V.I. (1984) *Eroziya I Stok Nanasov na Zemle*, Izdatelstvo Kazanskogo Universiteta.

Dedkov, A.P. and Mozzherin, V. (1992) Erosion and sediment yield in mountain regions of the world. In: *Erosion, Debris Flows and Environment in Mountain Regions (Proceedings of Chengdu Symposium, July 1992)*, IAHS Publication no. 209, pp. 29–36.

Dietrich, W.E. and Dunne, T. (1978) Sediment budget for a small catchment in mountainous terrain. *Z. Geomorphol.* **29**, 191–206.

Droppo, I.G. and Ongley, E.D. (1992) The state of suspended sediment in the freshwater fluvial environment: a method of analysis. *Water Res.* **26**, 65–72.

Duijsings, J.J.H.M. (1986) Seasonal variations in the sediment delivery ratio of a forested drainage basin in Luxembourg. In: *Drainage Basin Sediment Delivery (Proceedings of Albuquerque Symposium, August 1986)*, IAHS Publication no. 159, pp. 153–164.

Farr, I.S. and Clarke, R.T. (1984) Reliability of suspended load estimates in chalk streams. *Arch. Hydrobiol.* **102**, 1–19.

Ferguson, R.I. (1986) River loads underestimated by rating curves. *Water Resour. Res.* **22**, 74–76.

Ferguson, R.I. (1987) Accuracy and precision of methods for estimating river loads. *Earth Surf. Processes Landforms* **12**, 95–104.

Förstner, U. and Müller, G. (1974) *Schwermetalle in Flüssen und Seen als Ausdruck der Umweltverschmutzung*, Springer-Verlag, Berlin.

Foster, I.D.L. (1978a) Seasonal solute behaviour of stormflow in a small agricultural catchment. *Catena* **5**, 151–163.

Foster, I.D.L. (1978b) A multivariate model of storm-period solute behaviour. *J. Hydrol.* **39**, 339–353.

Foster, I.D.L. (1980) Chemical yields in runoff, and denudation in a small arable catchment, East Devon, England. *J. Hydrol.* **47**, 349–368.

Foyster, A.M. (1973) Application of the grid square technique to mapping of evapotranspiration. *J. Hydrol.* **19**, 205–226.

Foyster, A.M. (1975) Mapping runoff by the grid square technique. *Nordic Hydrol.* **6**, 207–221.

Garrett, K.J. (1992) The role of sediment-associated transport in the nutrient budget of a small Devon catchment. Unpublished PhD thesis, University of Exeter.

Gurnell, A.M. (1987) Suspended sediment. In: *Glacio-fluvial Sediment Transfer: An Alpine Perspective*, Gurnell, A.M. and Clark, M.J. (eds), John Wiley & Sons, Chichester, pp. 305–354.

Gurnell, A.M. and Fenn, C.R. (1984) Box–Jenkins transfer function models applied to suspended sediment concentration–discharge relationships in a proglacial stream. *Arctic Alpine Res.* **16**, 93–106.

Guy, H.P. (1964) *An Analysis of Some Storm-Period Variables Affecting Stream Sediment Transport*, US Geological Survey Professional Paper no. 462E.

Hall, D.G. (1967) *The pattern of sediment movement in the River Tyne*, IAHS Publication no. 75, pp. 117–140.

Hallard, M. (1988) The effects of agricultural drainage on the hydrology of a grassland site in South-West England. Unpublished PhD thesis, University of Exeter.

Heidel, S.G. (1956) The progressive lag of sediment concentration with flood waves. *Trans. Am. Geophys. Union* **37**, 56–66.

Jansson, M.B. (1982) *Land Erosion by Water in Different Climates*, UNGI Rapport no. 57, Department of Physical Geography, Uppsala University.

Jansson, M.B. (1988) A global survey of sediment yield. *Geogr. Ann. A* **70**, 81–98.

Kopp, J.F. and Kroner, R.C. (1968) *Trace Metals of the United States*, Division of Pollution Surveillance, US Federal Water Pollution Control Administration.

Littlewood, I.G. (1990) *Estimating Loads Transported in Rivers*, Final Report DoE Contract PECD 7/7/317, Natural Environment Research Council.

Loughran, R.J., Campbell, B.L. and Elliott, G.L. (1982) The identification and quantification of sediment sources using ^{137}Cs. In: *Recent Developments in the Explanation and Prediction of Erosion and Sediment Yield (Proceedings of Exeter Symposium, July 1982)*, IAHS Publication no. 137, pp. 361–369.

Mackereth, F.J.H. (1966) Some chemical observations on post-glacial lake sediments. *Phil. Trans. R. Soc. Lond. B* **250**, 165–213.

McCallan, M.E, O'Leary, B.M. and Rose, C.W. (1980) Redistribution of caesium-137 by erosion and deposition on an Australian soil. *Aust. J. Soil Res.* **18**, 119–128.

Meybeck, M. (1977) Dissolved and suspended matter carried by rivers: composition, time and space variations and world balance. In: *Interactions Between Sediments and Fresh Water*, Golterman, H.L. (ed.), Junk, The Hague, pp. 25–32.

Meybeck, M. (1982) Carbon, nitrogen, and phosphorus transport by world rivers. *Am. J. Sci.* **282**, 401–450.

Meybeck, M. (1983) Atmospheric inputs and river transport of dissolved substances. In: *Dissolved Loads of Rivers and Surface Water Quantity/Quality Relationships (Proceedings of Hamburg Symposium, August 1983)*, IAHS Publication no. 141, pp. 173–192.

Milliman, J.D. and Meade, R.H. (1983) World-wide delivery of river sediment to the oceans. *J. Geol.* **91**, 1–21.

Milliman, J.D. and Syvitski, J.P.M. (1992) Geomorphic/tectonic control of sediment discharge to the ocean: the importance of small mountainous rivers. *J. Geol.* **100**, 325–344.

Moore, R.J. (1984) A dynamic model of basin sediment yield. *Water Resour. Res.* **20**, 89–103.

Olive, L., Rieger, W.A. and Burgess, J.S. (1980) Estimation of sediment yields in small catchments: a geomorphic guessing game? *Proceedings of Conference of Institute of Australian Geographers (Newcastle, NSW)*.

Pennington, W., Cambray, R.S. and Fisher, E.M. (1973) Observations on lake sediments using fallout caesium-137 as a tracer. *Nature* **242**, 324–326.

Ritchie, J.C., Hawkes, P.H. and McHenry, J.R. (1975) Deposition rates in valleys determined using fallout caesium-137. *Geol. Soc. Am. Bull.* **86**, 1128–1130.

Roberts, G. and Marsh, T. (1987) The effects of agricultural practices on the nitrate concentrations in the surface water domestic supply sources of western Europe. In: *Water for the Future: Hydrology in Perspective (Proceedings of Rome Symposium, April 1987)*, IAHS Publication no. 164, pp. 365–380.

Shear, H. and Watson, A.E.P. (eds) (1977) *The Fluvial Transport of Sediment-Associated Nutrients and Contaminants*, International Joint Commission on the Great Lakes, Windsor, Ontario.

Simpson, E.A. (1978) The harmonization of the monitoring of the quality of inland fresh water. *J. Inst. Water Eng. Scient.* **32**, 45–56.

Simpson, E.A. (1980) The harmonization of the monitoring of the quality of rivers in the United Kingdom. *Hydrol. Sci. Bull.* **25**, 13–23.

Swanson, F.J., Janda, R.J., Dunne, T. and Swanson, D.N. (eds) (1982) *Sediment Budgets and Routing in Forested Drainage Basins*, US Forest Service General Technical Report PNW-141.

Thomas, R.B. (1986) Calibrating SALT: a sampling scheme to improve estimates of suspended sediment yield. In: *Monitoring to Detect Changes in Water Quality Series (Proceedings of Budapest Symposium, July 1986)*, IAHS Publication no. 157, pp. 79–88.

Thornton, R.C. (1985) Sediment-associated nutrients and their contribution to the nutrient loads of Devon catchments. Unpublished PhD thesis, University of Exeter.

Trimble, S.W. (1995) Catchment sediment budgets and change. In: *Changing River Channels*, Gurnell, A.M. and Petts, G.E. (eds), John Wiley & Sons, Chichester (in press).

VanSickle, J. and Beschta, R.L. (1983) Supply-based models of suspended sediment transport in streams. *Water Resour. Res.* **19**, 768–778.

Verhoff, F.H., Melfi, D.A. and Yaksich, S.M. (1979) Storm travel distance calculations for total phosphorus and suspended material in rivers. *Water Resour. Res.* **15**, 1345–1360.

Williams, G.P. (1989) Sediment concentration versus water discharge during single hydrologic events in rivers. *J. Hydrol.* **111**, 89–106.

Wolman, M.G. (1977) Changing needs and opportunities in the sediment field. *Water Resour. Res.* **13**, 50–54.

Bibliography of Des Walling's main publications

This list contains many works cited in the text of this chapter. The entries here are arranged by year, in the order: Walling and ...; Walling *et al.*; other works.

1971

Walling, D.E. (1971a) Streamflow from instrumented catchments in South East Devon. In: *Exeter Essays in Geography*, Gregory, K.J. and Ravenhill, W.L.D. (eds), University of Exeter, Exeter, pp. 55–81.

Walling, D.E. (1971b) Sediment dynamics of small instrumented catchments in South East Devon. *Trans. Devon Assoc. Advmt. Sci.* **103**, 147–165.

Walling, D.E. and Teed, A. (1971) A simple pumping sampler for research into suspended sediment transport from small catchments. *J. Hydrol.* **13**, 325–337.

Gregory, K.J. and Walling, D.E. (1971) Field measurements in the drainage basin. *Geography* **56**, 227–292.

1973

Gregory, K.J. and Walling, D.E. (1973a) *Drainage Basin Form and Process: A Geomorphological Approach*, Edward Arnold, London, 456pp.

Gregory, K.J. and Walling, D.E. (1973b) *Field Measurements in the Drainage Basin*, Warner Modular Publication, 16pp.

Gregory, K.J. and Walling, D.E. (1973c) Fluvial processes in small instrumented watersheds in the British Isles. *Area*, **5**, 297–302.

Troake, R.P. and Walling, D.E. (1973) The hydrology of the Slapton Wood stream. *Field Stud.* **3**, 719–740.

1974

Walling, D.E. (1974a) Suspended sediment and solute yields from a small catchment prior to urbanisation. In: *Fluvial Processes in Instrumented Watersheds*, Institute of British Geographers Special Publication no. 6, pp. 169–192.

Walling, D.E. (1974b) Suspended sediment production and building activity in a small British basin. In: *Proceedings of the Paris Symposium on the Effects of Man on the Interface of the Hydrological Cycle with the Physical Environment*, IAHS Publication no. 113, pp. 137–144.

Gregory, K.J. and Walling, D.E. (eds) (1974a) *Fluvial Processes in Instrumented Watersheds* Institute of British Geographers Special Publication no. 6, 196pp.

Gregory, K.J. and Walling, D.E. (1974b) The geomorphologists's approach to instrumented watersheds in the British Isles. In: *Fluvial Processes in Instrumented Watersheds*, Institute of British Geographers Special Publication no. 6, pp. 1–6.

Webb, B.W. and Walling, D.E. (1974) Local variation in background water quality. *Sci. Total Environ.* **3**, 141–153.

1975

Walling, D.E. (1975a) Solute variations in small catchment streams: some comments. *Trans. Inst. Br. Geogr.* **64**, 141–147.

Walling, D.E. (1975b) Three hundred years of scientific hydrology. *Area* **7**, 36–37.

Walling, D.E. and Foster, I.D.L. (1975) Variations in the natural chemical concentration of river water during flood flows and the lag effect: some further comments. *J. Hydrol.* **26**, 237–244.

Walling, D.E. and Webb, B.W. (1975) Spatial variation of river water quality: a survey of the River Exe. *Trans. Inst. Br. Geogr.* **65**, 155–171.

Troake, R.P. and Walling, D.E. (1975) Some observations on stream nitrate levels and fertilizer application at Slapton, South Devon. *Trans. Devon Assoc. Advmt. Sci.* **107**, 77–90.

1976

Walling, D.E. (1976a) Modification of hydrological processes consequent upon suburbanisation. In: *Man-Made Transformation of Water Balance*, Keller, R. (ed), pp. 126–130.

Walling, D.E. (1976b) The effects of building activity on drainage basin dynamics. In: *Proceedings of International Geoqraphical Union Symposium on the IHP (Leningrad)*, pp. 109–115.

Troake, R.P., Troake, L.E. and Walling, D.E. (1976) Nitrate loads of South Devon streams. In: *Agriculture and Water Quality*, Technical Bulletin no. 32, Ministry of Agriculture, Fisheries and Food, pp. 340–351.

1977

Walling, D.E. (1977a) Assessing the accuracy of suspended sediment rating curves for a small basin. *Water Resour. Res.* **13**, 531–538.

Walling, D.E. (1977b) Limitations of the rating curve technique for estimating suspended sediment loads, with particular reference to British Rivers. In: *Erosion and Solid Matter Transport in Inland Waters (Proceedings of Paris Symposium)*, IAHS Publication no. 122, pp. 34–48.

Walling, D.E.(1977c) Natural sheet and channel erosion of unconsolidated source material: geomorphic control, magnitude and frequency of transfer mechanisms. In: *Proceedings of a Workshop on the Fluvial Transport of Sediment-Associated Nutrients and Contaminants*, Shear, H. and Watson, A.E.P. (eds), International Joint Commission on the Great Lakes, Windsor, Ontario, pp. 11–33.

Walling, D.E. (1977d) Derivation of erosion and sediment yield parameters in areas with deficient data-reconnaissance measurements. *Hydrol. Sci. Bull.* **22**, 517–520.

Walling, D.E. (1977e) Physical hydrology. *Prog. Phys. Geogr.* **1**, 143–151.

Gardiner, V., Gregory, K.J. and Walling, D.E. (1977) Further notes on the drainage density/area relationship. *Area* **9**, 117–121.

1978

Walling, D.E. (1978a) Suspended sediment and solute response characteristics of the River Exe, Devon, England. In: *Research in Fluvial Geomorphology (Proceedings of 5th Guelph Symposium on Geomorphology)*, Davidson-Arnott, R. and Nickling, W. (eds), Geo Books, Norwich, pp. 169–197.

Walling, D.E. (1978b) Reliability considerations in the evaluation and analysis of river loads. *Z. Geomorphol.* **29**, 29–42.

Walling, D.E. (1978c) Physical hydrology. *Prog. Phy. Geogr.* **2**, 143–149.

Walling, D.E. and Foster, I.D.L. (1978) The 1976 drought and nitrate levels in the River Exe Basin. *J. Inst. Water Eng. Scient.* **32**, 341–352.

Walling, D.E. and Webb, B.W. (1978) Mapping solute loadings in an area of Devon, England. *Earth Surf. Processes* **3**, 85–99.

Foster, I.D.L. and Walling, D.E. (1978) The effects of the 1976 drought and autumn rainfall on stream solute levels. *Earth Surf. Processes* **3**, 393–406.

1979

Walling, D.E. (1979a) The hydrological impact of building activity. In: *Man's Impact on the Hydrological Cycle in the United Kingdom*, Hollis, G.E. (ed), Geo Books, Norwich, pp. 135–152.

Walling, D.E. (1979b) Hydrological processes. In: *Man and Environmental Processes*, Gregory, K.J. and Walling, D.E. (eds), Dawson, Folkestone, pp. 57–81.

Walling, D.E. (1979c) Physical hydrology. *Prog. Phys. Geogr.* **3**, 141–150.

Walling, D.E. (1979d) Spatial variations in streamflow: or why no two rivers are alike. *Teach. Geogr.* **4**, 162–167.

Walling, D.E. and Foster, I.D.L. (1979) The impact of the 1976 drought on nutrient concentrations and loadings in the River Exe Basin, UK. In: *Proceedings of the International Symposium on Hydrological Aspects of Droughts (New Delhi, India)*, pp. 151–161.

Walling, D.E. and Kleo, A.H.A. (1979) Sediment yields of rivers in areas of low precipitation: a global view. In: *The Hydrology of Areas of Low Precipitation (Proceedings of Canberra Symposium)*, Publication no. 128, pp. 479–483.

Walling, D.E., Peart, M.R., Oldfield, R. and Thompson, R. (1979) Suspended sediment sources identified by magnetic measurements. *Nature* **281**, 110–113.

Gregory, K.J. and Walling, D.E. (eds) (1979) *Man and Environmental Processes*, Dawson, Folkestone, 276pp.

Oldfield, F., Rummery, T.A., Thompson, R. and Walling, D.E. (1979) Identification of suspended sediment sources by means of magnetic measurements. *Water Resour. Res.* **15**, 211–218.

1980

Walling, D.E. (1980a) Water in the catchment ecosystem. In: *Water Quality in Catchment Ecosystems*, Gower, A.M. (ed), John Wiley & Sons, Chichester, pp. 1–47.

Walling, D.E. (1980b) The effects of agriculture. In: *General Reports, International Symposium on the Influence of Man on the Hydrological Regime (Helsinki)*.

Walling, D.E. (1980c) The 1976 drought and solute levels in two Devon catchments. In: *Atlas of Drought in Britain*, Doornkamp, J. C., Gregory, K.J. and Burn, A.S. (eds), Institute of British Geographers, London, p. 50.

Walling, D.E. (1980d) Sediment yield research: an overview. In: *Proceedings of the International Symposium on River Sedimentation (Beijing, China)*, Chinese Society of Hydraulic Engineering, Guanghua Press, Beijing, pp. 1137–1147.

Walling, D.E.(1980e) Physical hydrology. *Prog. Phys. Geogr.* **4**, 107–117.

Walling, D.E. and Carter, R. (1980) River water temperatures. In: *Atlas of Drought in Britain*, Doornkamp, J.C., Gregory, K.J. and Burn, A.S. (eds), Institute of British Geographers, London, p. 49.

Walling, D.E. and Peart, M.R. (1980) Some quality considerations in the study of human influence on sediment yields. In: *The Influence of Man on the Hydrological Regime with Special Reference to Representative and Experimental Basins (Proceedings of Helsinki Symposium)*, IAHS Publication no, 130, pp. 293–302.

Walling, D.E, and Webb, B.W. (1980) The spatial dimension in the interpretation of stream solute behaviour. *J. Hydrol.* **47**, 129–149.

Webb, B.W. and Walling, D.E. (1980) Stream solute studies and geomorphological research: some examples from the Exe Basin, Devon, UK. *Z. Geomorphol.* **36**, 245–263.

1981

Walling, D.E. (1981a) Yellow River which never runs clear. *Geogr. Mag.* **53** (9), 568–575.

Walling, R.E.(1981b) Physical hydrology. *Prog. Phys. Geogr.* **5**, 123–131.

Walling, D.E. (1981c) Some research trends in the study of erosion and sediment yield. In: *A Compilation of the Lectures on River Sedimentation*, Sedimentation Committee, Chinese Society of Hydraulic Engineering, Beijing, pp. 231–251.

Walling, D.E. (1981d) Some instrumentation and experimental problems in the study of erosion and sediment yield. In: *A Compilation of the Lectures on River Sedimentation*, Sedimentation Committee, Chinese Society of Hydraulic Engineering, Beijing, pp. 253–276.

Walling, D.E. (1981e) The quality dimension in sediment studies. In: *A Compilation of the Lectures on River Sedimentation*, Sedimentation Committee, Chinese Society of Hydraulic Engineering, Beijing, pp. 277–301.

Walling, D.E. (1981f) The dissolved loads of rivers, their interpretation and prediction. In: *A Compilation of the Lectures on River Sedimentation*, Sedimentation Committee, Chinese Society of Hydraulic Engineering, Beijing, pp. 303–351.

Walling, D.E. (1981g) Prediction and modelling of suspended sediment yields. In: *A Compilation of the Lectures on River Sedimentation*, Sedimentation Committee, Chinese Society of Hydraulic Engineering, Beijing, pp. 353–383.

Walling, D.E. and Webb, B.W. (1981a) The reliability of suspended load data. In: *Erosion and*

Sediment Transport Measurement (Proceedings of Florence Symposium), IAHS Publication no. 133, pp. 177–194.
Walling, D.E. and Webb, B.W. (1981b) Water quality. In: *British Rivers*, Lewin, J. (ed.), Allen and Unwin, Hemel Hempstead, pp. 126–69.

1982

Walling, D.E. (ed.) (1982a) *Recent Developments in the Explanation and Prediction of Erosion and Sediment Yield (Proceedings of Exeter Symposium, July 1982)*, IAHS Publication no. 137, 430pp.
Walling, D.E. (1982b) Modification of hydrological processes by building activity. In: *Application of the Results from Representative and Experimental Basins*, UNESCO Studies and Reports in Hydrology no. 32, pp. 383–401.
Walling, D.E. and Kane, P. (1982) Temporal variation of suspended sediment properties. In: *Recent Developments in the Explanation and Prediction of Erosion and Sediment Yield (Proceedings of Exeter Symposium, July 1982)*, IAHS Publication no. 137, pp. 409–419.
Walling, D.E. and Webb, B.W. (1982a) Sediment availability and the prediction of storm-period sediment yields. In: *Recent Developments in the Explanation and Prediction of Erosion and Sediment Yield (Proceedings of Exeter Symposium, July 1982)*, IAHS Publication no. 137, pp. 327–337.
Walling, D.E. and Webb, B.W. (1982b) The design of sampling programmes for studying catchment nutrient dynamics. In: *Hydrological Research Basins and their Use in Water Resources Planning (Proceedings of Berne Symposium)*, pp. 747–758.
Peart, M.R. and Walling, D.E. (1982) Particle size characteristics of fluvial suspended sediment. In: *Recent Developments in the Explanation and Prediction of Erosion and Sediment Yield (Proceedings of Exeter Symposium, July 1982)*, IAHS Publication no. 137, pp. 397–407.
Webb, B.W. and Walling, D.E. (1982a) Magnitude and frequency characteristics of fluvial transport in a Devon drainage basin and some geomorphological implications. *Catena* **9**, 9–23.
Webb, B.W. and Walling, D.E. (1982b) Catchment scale and the interpretation of water quality behaviour. In: *Hydrological Research Basins and their Use in Water Resources Planning (Proceedings of Berne Symposium)*, pp. 759–770.

1983

Walling, D.E. (1983a) The sediment delivery problem. *J. Hydrol.* **65**, 209–237.
Walling, D.E. (1983b) Physical hydrology. *Prog. Phys. Geogr.* **7**, 97–112.
Walling, D.E. and Webb, B.W. (1983a) Patterns of sediment yield. In: *Background to Paleohydrology*, Gregory, K.J. (ed.), John Wiley & Sons, Chichester, pp. 69–100.
Walling, D.E. and Webb, B.W. (1983b) The dissolved loads of rivers: a global overview. In: *Dissolved Loads of Rivers and Surface Water Quantity/ Quality Relationships (Proceedings of Hamburg Symposium, August 1983)*, IAHS Publication no. 141, pp. 3–20.
Webb, B.W. and Walling, D.E. (1983) Stream solute behaviour in the River Exe Basin, Devon, UK. In: *Dissolved Loads of Rivers and Surface Water Quantity/Quality Relationships (Proceedings of Hamburg Symposium, August 1983)*, IAHS Publication no. 141, pp. 153–169.

1984

Walling, D.E. (1984a) The sediment yields of African Rivers. In: *Challenges in African Hydrology and Water Resources*, Walling, D.E., Foster, S.S.D. and Wurzel, P. (eds), IAHS Publication no. 144, pp. 265–283.
Walling, D.E. (1984b) Dissolved loads and their measurement. In: *Erosion and Sediment Yield: Some Methods of Measurement and Modelling*, Hadley, R.F. and Walling, D.E. (eds), Geo Books, Norwich, pp. 111–117.
Walling, D.E. (1984c) Sediment delivery from drainage basins. In *Drainage Basin Erosion and*

Sedimentation (Proceedings of a Conference on Erosion, Transportation and Sedimentation in Australian Drainage Basins), Loughran, R.J. (ed.), University of Newcastle/Soil Conservation Service of NSW, pp. 71–80.

Walling, D.E. (1984d) The quality dimension in the study of sediment yields. In: *Drainage Basin Erosion and Sedimentation (Proceedings of a Conference on Erosion, Transportation and Sedimentation in Australian Drainage Basins)*, Loughran, R.J. (ed.), University of Newcastle/Soil Conservation Service of NSW, pp. 127–138.

Walling, D.E. (1984e) Physical hydrology. *Prog. Phys. Geogr.* **8**, 129–138.

Walling, D.E. (1984f) Muddy waters move mountains. *Geogr. Mag.* **41**, (May), 262–267.

Walling, D.E. and Kane, P. (1984) Suspended sediment properties and their geomorphological significance. In: *Catchment Experiments in Fluvial Geomorphology*, Burt, T.P. and Walling, D.E. (eds), Geo Books, Norwich, pp. 311–334.

Walling D.E. and Thornton, R. (1984) The role of suspended sediment in catchment nutrient budgets. *J. Sci. Food Agri.* **35**, 855–856.

Walling, D.E. and Webb, B.W. (1984) Local variation of nitrate levels in the Exe Basin, Devon, UK. *Beitr. Hydrol.* **10** (1), 71–100.

Walling, D.E., Foster, S.S.D. and Wurzel, P. (eds) (1984) *Challenges in African Hydrology and Water Resources (Proceedings of Harare Symposium, July 1984)*, IAHS Publication no. 144, 587pp.

Burt, T.P., Trudgill, S.T., Walling, D.E., Arkell, B. and Thomas, A.D. (1984) Hillslope hydrology and its relationships to nitrate sources in a catchment of mixed land use. *J. Sci. Food Agri.* **35**, 857–859.

Burt, T.P. and Walling, D.E. (eds) (1984a) *Catchment Experiments in Fluvial Geomorphology*, Geo Books, Norwich, 593pp.

Burt, T.P. and Walling, D.E. (eds)(1984b) Catchment experiments in fluvial geomorphology: a review of objectives and methodology. In: *Catchment Experiments in Fluvial Geomorphology*, Burt, T.P. and Walling, D.E. (eds), Geo Books, Norwich, pp. 3–18.

Hadley, R.F. and Walling, D.E. (eds)(1984) *Erosion and Sediment Yield: Some Methods of Measurement and Modelling*, Geo Books, Norwich, 218pp.

Webb, B.W. and Walling, D.E. (1984a) Magnitude and frequency characteristics of suspended sediment transport in Devon rivers. In: *Catchment Experiments in Fluvial Geomorphology*, Burt, T.P. and Walling, D.E. (eds), Geo Books, Norwich, pp. 399–415.

Webb, B.W. and Walling, D.E. (1984b) Nitrate levels in Devon catchments. *J. Sci. Food Agri.* **35**, 851–852.

1985

Walling, D.E. (1985) Physical hydrology. *Prog. Phys. Geogr.* **9**, 97–103.

Walling, D.E. and Webb, B.W. (1985) Estimating the discharge of contaminants to coastal waters by rivers: some cautionary comments. *Marine Pollut. Bull.* **16**, 488–492.

Hadley, R.F., Lal, R., Onstad, C.A., Walling, D.E. and Yair, A. (1985) *Recent Developments in Erosion and Sediment Yield Studies*, UNESCO Technical Document in Hydrology, 127pp.

Webb, B.W. and Walling, D.E. (1985a) Nitrate behaviour in streamflow from a grassland catchment in Devon, UK. *Water Res.* **19**, 1005–1016.

Webb, B.W. and Walling, D.E. (1985b) Temporal variation of river water temperatures in a Devon river system. *Hydrol. Sci. J.* **30**, 449–464.

Webb, B.W. and Walling, D.E. (1985c) Temperature characteristics of Devon rivers. *Proc. Ussher Soc.* **6**, 237–245.

1986

Walling, D.E. (1986a) Land degradation and sediment yields in rivers: a background to monitoring strategies. In: *Proceedings of the SADCC Symposium on Environmental Monitoring (Gaborone, Botswana, November 1986)*, SADCC, Lesotho, pp. 64–105.

Walling, D.E. (1986b) Environmental monitoring: What is desired, what is feasible? In:

Proceedings of the SADCC Symposium on Environmental Monitoring (Gaborone, Botswana, November 1986), SADCC, Lesotho, pp. 7–20.
Walling, D.E. (1986c) Sediment yields and sediment delivery dynamics in Arab countries: some problems and research needs. *J. Water Resour.* **5**, 775–799.
Walling, D.E. (1986d) Physical hydrology. *Prog. Phys. Geogr.* **10**, 69–80.
Walling, D.E. and Webb, B.W. (1986a) Solutes in river systems. In Solute Processes, Trudgill, S.T. (ed.), John Wiley & Sons, Chichester, pp. 251–327.
Walling, D.E. and Webb, B.W. (1986b) Countrywide patterns of background water quality in the UK and their control. *Beitr. Hydrol.* **5**, 299–315.
Walling, D.E. and Webb, B.W. (1986c) Solute transport by rivers in arid environments: an overview. *J. Water Resour.* **5**, 800–822.
Walling, D.E., Bradley, S.B. and Lambert, C. (1986a) Conveyance losses of suspended sediment within a floodplain system. In: Drainage Basin Sediment Delivery (Proceedings of Albuquerque Symposium, August 1986), IAHS Publication no. 159, pp. 119–131.
Walling, D.E., Bradley, S.B. and Wilkinson, C.J. (1986b) A caesium-137 budget approach to the investigation of sediment delivery from a small agricultural drainage basin in Devon, UK. In: *Drainage Basin Sediment Delivery (Proceedings of Albuquerque Symposium, August 1986)*, IAHS Publication no. 159, pp. 423–435.
Lambert, C.P. and Walling, D.E. (1986) Suspended sediment storage in river channels: a case study of the River Exe, Devon, UK. In: *Drainage Basin Sediment Delivery (Proceedings of Albuquerque Symposium, August 1986)*, IAHS Publication no. 159, pp. 263–276.
Peart, M.R. and Walling, D.E (1986) Fingerprinting sediment source: the example of a drainage basin in Devon, UK. In: *Drainage Basin Sediment Delivery (Proceedings of Albuquerque Symposium, August 1986)*, IAHS Publication no. 159, pp. 41–55.
Webb, B.W. and Walling, D.E. (1986a) Spatial variation of river water temperature characteristics and behaviour in a Devon river system. *Freshwater Biol.* **16**, 585–608.
Webb, B.W. and Walling, D.E. (1986b) Regional variation of temperature in Devon streams and rivers. *Beitr. Hydrol.* **5**, 277–297.

1987

Walling, D.E. (1987a) Hydrological and fluvial processes: revolution and evolution. In *Horizons in Physical Geoqraphy*, Clark, M.J., Gregory, K.J. and Gurnell, A.M. (eds), Macmillan, Basingstoke, pp. 106–20.
Walling, D.E. (1987b) Rainfall, runoff and erosion of the land: a global view. In: *Energetics of Physical Environment*, Gregory, K.J. (ed.), John Wiley & Sons, Chichester, pp. 98–117.
Walling, D.E. (1987c) Basin sediment systems. In: *International Geomorphology 1986*, Gardiner, V. (ed.), John Wiley & Sons, Chichester, pp. 747–750.
Walling, D.E. (1987d) Hydrological processes. In: *Human Activity and Environmental Processes*, Gregory, K.J. and Walling, D.E. (eds), John Wiley & Sons, Chichester, pp. 53–85.
Walling, D.E. (1987e) Physical hydrology. *Prog. Phys. Geogr.* **11**, 112–120.
Walling, D.E. (1987f) Physical hydrology. *Prog. Phys. Geogr.* **11**, 590–597.
Walling, D.E. and Gregory, K.J. (1987) A perspective. In: *Human Activity and Environmental Processes*, Gregory, K.J. and Walling, D.E. (eds), John Wiley & Sons, Chichester, pp. 445–460.
Walling, D.E. and Moorehead, P.W. (1987) Spatial and temporal variation of the particle-size characteristics of fluvial suspended sediment. *Geogra.. Ann. A* **69**, 47–59.
Walling, D.E. and Webb, B.W. (1987a) Suspended load in gravel-bed rivers: UK experience. In: *Sediment Transport in Gravel-Bed Rivers*, Thorne, C.R., Bathurst, J.C. and Hey, R.D. (eds), John Wiley & Sons, Chichester, pp. 691–732.
Walling, D.E. and Webb, B.W. (1987b) Material transport by the world's rivers: evolving perspectives. In: *Water for the Future: Hydrology in Perspective (Proceedings of Rome Symposium, April 1987)*, Rodda, J.C. and Matalas, N.C. (eds), IAHS Publication no. 164, pp. 313–329.
Aranuvachapun, S. and Walling, D.E. (1987) The use of a microcomputer for image analysis. *Int. J. Remote Sensing* **8**, 1385–1397.

Gilbey, J., Bradley, S.B. and Walling, D.E. (1987) The deposition of caesium-137 on grassland at a site in SW England following the Chernobyl accident. *Grass Forage Sci.* **42**, 439–442.

Gregory, K.J. and Walling, D.E. (eds) (1987) *Human Activity and Environmental Processes*, John Wiley & Sons, Chichester, 466pp.

Lambert, C.P. and Walling, D.E. (1987) Floodplain sedimentation: a preliminary investigation of contemporary deposition within the lower reaches of the River Culm, Devon, UK. *Geogr.. Ann.* **69**, 393–404.

Loughran, R.J. Campbell, B.L. and Walling, D.E. (1987) Soil erosion and sedimentation indicated by caesium-137: Jackmoor Brook catchment, Devon, England. *Catena* **14**, 201–212.

1988

Walling, D.E. (1988a) Erosion and sediment yield research — some recent perspectives. *J. Hydrol.* **100**, 113–141.

Walling, D.E. (1988b) Measuring sediment yield from river basins. In: *Soil Erosion Research Methods*, Lal, R. (ed.), Soil and Water Conservation Society, Ankeny, Iowa, pp. 39–73.

Walling, D.E. and Bradley, S.B. (1988a) Transport and redistribution of Chernobyl fallout radionuclides by fluvial processes: some preliminary evidence. *Environ. Geochem. Health* **10**, 35–39.

Walling, D.E. and Bradley, S.B. (1988b) The use of caesium-137 measurements to investigate sediment delivery from cultivated areas in Devon, UK. In: *Sediment Budgets (Proceedings of Porto Alegre Symposium, December 1988)*, Bordaso, M.P. and Walling, D.E. (eds), IAHS Publication no. 174, pp. 325–335.

Walling, D.E. and Webb, B.W. (1988) The reliability of rating curve estimates of suspended sediment yield: some further comments. In: *Sediment Budgets (Proceedings of Porto Alegre Symposium, December 1988)*, Bordas, M.P. and Walling, D.E. (eds), IAHS Publication no. 174, pp.337–350.

Al-Ansari, N.A., Asaad, N.M., Walling, D.E. and Hussan, S.A. (1988) The suspended sediment discharge of the River Euphrates at Haditha, Iraq. *Geogr.. Ann. A* **70**, 203–213.

Aranuvachapun, S. and Walling, D.E. (1988) Landsat-MSS radiance as a measure of suspended sediment in the Lower Yellow River (Hwang Ho). *Remote Sensing Environ.* **25**, 145–165.

Bordas, M. and Walling, D.E. (eds) (1988) *Sediment Budgets (Proceedings of Porto Alegre Symposium, December 1988)*, IAHS Publication no. 174, 591pp.

Burt, T.P., Arkell, B.P., Trudgill, S.T. and Walling, D.E. (1988) Stream nitrate levels in a small catchment in South West England over a period of 15 years (1970–1985). *Hydrol. Processes* **2**, 267–284.

Lambert, C.P. and Walling, D.E. (1988) Measurement of channel storage of suspended sediment in a gravel-bed river. *Catena* **15**, 65–80.

Peart, M.R. and Walling, D.E. (1988) Techniques for establishing suspended sediment sources in two drainage basins in Devon, UK: a comparative assessment. In *Sediment Budgets (Proceedings of Porto Alegre Symposium, December 1988)*, Bordas, M.P. and Walling, D.E. (eds), IAHS Publication no. 174, pp. 269–279.

Webb, B.W. and Walling, D.E. (1988) Modification of temperature behaviour through regulation of a British river system. *Regul. Rivers: Res. Manage.* **2**, 103–116.

1989

Walling, D.E. (1989a) The erosion problem. *Int. J. Sediment Res.* **4**(1), 1–11.

Walling, D.E. (1989b) Linking erosion and sediment yield: some problems of interpretation. *Int. J. Sediment Res.* **4**(1), 13–26.

Walling, D.E. (1989c) Physical and chemical properties of sediment: the quality dimension. *Int. J.Sediment Res.* **4**(1), 27–39.

Walling, D.E. and Bradley, S.B. (1989a) Rates and patterns of contemporary floodplain sedimentation: a case study of the River Culm, Devon, UK. *Geojournal* **19**, 53–62.

Walling, D.E. and Bradley, S.B. (1989b) Use of caesium-137 measurements to investigate rates and patterns of recent floodplain sediment. *Proceedings of Fourth International Symposium on River Sedimentation (Beijing, China, November 1989)*, pp. 1451–1458.

Walling, D.E. and Moorehead, P.M. (1989) The particle size characteristics of fluvial suspended sediment. *Hydrobiologia*, **176/177**, 125–149.

Walling, D.E., Rowan, J.S. and Bradley, S.B. (1989) Sediment-associated transport of Chernobyl fallout radionuclides. In: *Sediment and the Environment (Proceedings of Baltimore Symposium, May 1989)*, Hadley, R.F. and Ongley, E.D. (eds), IAHS Publication no. 184, pp. 37–45.

1990

Walling, D.E. (1990a) Linking the field to the river: sediment delivery from agricultural land. In: *Soil Erosion on Agricultural Land*, Boardman, J., Foster, I.D.L. and Dearing, J.A. (eds), John Wiley & Sons, Chichester, pp. 129–152.

Walling, D.E. (1990b) Water: provision and control. In: Forty Years of the People's Republic of China, *Geography* **74**, 356–358.

Walling, D.E. (1990c) Monitoring contemporary geomorphological processes. In: *Global Change Regional Research Centres: Scientific Problems and Concept Developments*, Breymeyer, A. (ed.), Institute of Geography and Spatial Organisation, Polish Academy of Sciences, Warsaw, pp. 7–32.

Walling, D.E. (1990d) The struggle against erosion and a perspective on recent research. In: *Water Erosion*, Ivanov, K. and Pechinov, D. (eds), UNESCO Technical Document in Hydrology, UNESCO, Paris, pp. 39–60.

Walling, D.E. and Bradley, S.B. (1990) Some application of caesium–137 measurements in the study of erosion, transport and deposition. In: *Erosion, Transport and Deposition Processes (Proceedings of Jerusalem Workshop, April 1987)*, Walling, D.E., Yair, A. and Berkowicz, S. (eds), IAHS Publication no. 189, 179–203pp.

Walling, D.E. and Quine, T.A. (1990a) Use of caesium–137 to investigate patterns and rates of soil erosion on arable fields. In: *Soil Erosion on Agricultural Land*, Boardman, J., Foster, I.D.L. and Dearing, J.A. (eds), John Wiley & Sons, Chichester, pp. 129–152.

Walling, D.E. and Quine, T.A. (1990b) Calibration of caesium–137 measurements to provide quantitative erosion rate data. *Land Degrad. Rehab.* **2**, 161–175.

Walling, D.E., Yair, A. and Berkowicz, S. (eds) (1990) *Erosion, Transport and Deposition Processes (Proceedings of Jerusalem Workshop, March–April 1987)*, IAHS Publication no. 189, 204pp.

Okunishi, K., Walling, D.E. and Saito, T. (1990) Discharge of suspended sediment and solutes from a hilly drainage basin in Devon, UK, as analysed by a cascade tank model. *Bull. Disast. Prev. Res. Inst. Kyoto Univ.* **40**, 143–160.

Zhang, X., Higgitt, D.L. and Walling, D.E. (1990) A preliminary assessment of the potential for using caesium–137 to estimate rates of soil erosion in the Loess Plateau of China. *Hydrol. Sci. J.* **35**, 243–252.

1991

Walling, D.E. (1991a) Drainage basin studies. In: *Field Experiments and Measurement Programmes in Geomorphology*, Slaymaker, O. (ed.), Balkema, Rotterdam, pp. 17–59.

Walling, D.E. (1991b) Sediment yield investigations: a perspective on recent developments and future needs. In: *Proceedings of the Fourth International Symposium on River Sedimentation, Post Symposium Volume*, Beijing, China, pp. 37–64.

Walling, D.E. and Quine, T.A. (1991a) Use of ^{137}Cs measurements to investigate soil erosion on arable fields in the UK: potential applications and limitations. *J. Soil Sci.* **42**, 147–165.

Walling, D.E. and Quine, T.A. (1991b) Recent rates of soil loss from areas of arable cultivation in the UK. In: *Sediment and Water Quality in a Changing Environment: Trends and Explanation*, IAHS Publication no. 203, pp. 123–131.

Walling, D.E. and Quine, T.A. (1991c) Fluvial redistribution of Chernobyl radionuclides. *NERC News* July pp. 22–25.

Betton, C., Webb, B.W. and Walling, D.E. (1991) Recent trends in NO_3-N concentrations and loads in British Rivers. In: *Sediment and Water Quality in a Changing a Environment: Trends and Explanation*, IAHS Publication no. 203, pp. 169-180.

Froehlich, W. and Walling, D.E. (1991) Zastosowanie Ceza 137 do badan procesow erozji gleb transportu i sedymentacji fluwialnej w zlewniach karpackich. In: *I Zjazd Geomorfoloqow Polskich*, Poznan, pp. 34-35.

Peters, N.E. and Walling, D.E. (eds) (1991) Sediment and Water Quality in a Changing Environment: Trends and Explanation (Proceedings of Vienna Symposium, August 1991), IAHS Publication no. 203, 374pp.

Quine, T.A. and Walling, D.E. (1991) Rates of soil erosion on arable fields in Britain: quantitative data from caesium-137 measurements. *Soil Use Manage.* **7**, 169-176.

1992

Walling, D.E. and He, Q. (1992) Interpretation of caesium-137 profiles in lacustrine and other sediments: the role of catchment-derived inputs. *Hydrobiologia* **235/236**, 219-230.

Walling, D.E. and Quine, T.A. (1992) The use of caesium-137 measurements in soil erosion surveys. In: *Erosion and Sediment Transport Monitoring Programmes in River Basins*, IAHS Publication no. 210, pp. 143-152.

Walling, D.E. and Webb, B.W. (1992) Water quality I: physical characteristics. In: *The Rivers Handbook: Hydrological and Ecological Principles*, Blackwell, Oxford, pp. 48-72.

Walling, D.E. and Woodward, J.C. (1992) Use of radiometric fingerprints to derive information on suspended sediment sources. In: *Erosion and Sediment Transport Monitoring Programmes in River Basins*, IAHS Publication no. 210, pp. 153-164.

Walling, D.E., Davies, T.R. and Hasholt, B. (eds) (1992a) *Erosion, Debris Flows and Environment in Mountain Regions (Proceedings of Chengdu Symposium, July 1992)*, IAHS Publication no. 209, 485pp.

Walling, D.E., Quine, T.A. and He, Q. (1992b) Investigating contemporary rates of floodplain sedimentation. In: *Lowland Floodplain Rivers*, Petts, G. and Carling, P.A. (eds), John Wiley & Sons, Chichester, pp. 165-184.

Walling, D.E., Quine, T.A. and Rowan, J.S. (1992c) Fluvial transport and redistribution of Chernobyl fallout radionuclides. *Hydrobiologia* **235/236**, 231-246.

Walling, D.E., Webb, B.W. and Woodward, J.C. (1992d) Some sampling considerations in the design of effective strategies for monitoring sediment-associated transport. In: *Erosion and Sediment Transport Monitoring Programmes in River Basins*, IAHS Publication no. 210, pp. 279-288.

Bates, P.D., Anderson, M.G., Baird, L., Walling, D.E. and Simm, D.(1992) Modelling floodplain flows using a two-dimensional finite element model. *Earth Surf. Processes Landforms* **17**, 575-588.

Bogen, J., Walling, D.E. and Day, T. (eds) (1992) Erosion and Sediment Transport Monitoring Programmes in River Basins (Proceedings of Oslo Symposium, August 1992), IAHS Publication no. 210, 538pp.

Froehlich, W. and Walling, D.E. (1992a) The use of fallout radionuclides in investigations of erosion and sediment delivery in the Polish Flysch Carpathians. In: *Erosion, Debris Flows and Environment in Mountain Regions*, IAHS Publication no. 209, pp. 61-76.

Froehlich, W. and Walling, D.E. (1992b) Badania proceszow erozji i sedymentacji przy uzyciu izotopu cezu 137. In: *Sesfa Naukowa*, IGi PZ, PAN 1993.

Hasholt, B. and Walling, D.E. (1992) Use of caesium-137 to investigate sediment sources and sediment delivery in a small glacierized mountain drainage basin in eastern Greenland. In: *Erosion, Debris Flows and Environment in Mountain Regions*, IAHS Publication no. 209, pp. 87-100.

Higgitt, D.L., Froehlich, W. and Walling, D.E. (1992) Applications and limitations of Chernobyl radiocaesium measurements in a Carpathian erosion investigation. *Land Degrad. Rehab.* **3**, 15-26.

Navas, A. and Walling, D.E. (1992) Using caesium-137 to assess sediment movement on slopes in a semi-arid upland environment in Spain. In: *Erosion, Debris Flows and Environment in Mountain Regions*, IAHS Publication no. 209, pp. 129–138.

Navas, A., Walling, D.E., Quine, T.A. and Machim, J. (1992) Estimacion de la erosion mediante medidas de cesio–137 en una vertiente de las Bardenas. *Proceedings of Pamplona Symposium*, pp. 552–557.

Quine, T.A. and Walling, D.E. (1992a) Patterns and rates of contemporary soil erosion using caesium–137: measurement, analysis and archaeological significance. In: *Past and Present Soil Erosion*, Bell, M. and Boardman, J. (eds), Oxbow Books, Oxford, pp. 185–196.

Quine, T.A. and Walling, D.E. (1992b) Caesium work at Dalicott. In: *First International ESSC Congress, Post Congress Tour Guide*, European Society for Soil Conservation, pp. 56–62.

Quine, T.A., Walling, D.E., Zhang, X. and Wang, Y. (1992) Investigation of soil erosion on terraced fields near Yanting, Sichuan Province, China, using caesium–137. In: Erosion, Debris Flows and Environment in Mountain Regions, IAHS Publication no. 209, pp. 155–168.

Rowan, J.S. and Walling, D.E. (1992a) The transport and fluvial redistribution of Chernobyl-derived radiocaesium within the River Wye basin, UK. *Sci. Total Environ.* **121**, 109–131.

Rowan, J.S. and Walling, D.E. (1992b) Monitoring the sediment-associated transmission of weapons-test caesium–137 in the Exe basin, Devon, UK. In: *Erosion and Sediment Transport Monitoring Programmes in River Basins*, NVE, Oslo, pp. 126–131.

Rowan, J.S., Bradley, S.B. and Walling, D.E. (1992) Fluvial redistribution of Chernobyl fallout: reservoir evidence in the Severn basin. *J. Inst. Water Environ. Manage.* **6**, 659–666.

Webb, B.W. and Walling, D.E. (1992a) Long term water temperature behaviour and trends in a Devon, UK, river system. *Hydrol. Sci. J.* **37** (6), 567–580.

Webb, B.W. and Walling, D.E. (1992b) Water quality II: chemical characteristics. In: *The Rivers Handbook: Hydrological and Ecological Principles*, Calow, P. and Petts, G.E. (eds), Volume 1, Blackwell, Oxford, pp. 73–100.

Woodward, J.C. and Walling, D.E. (1992) A field sampling method to obtain representative samples of composite fluvial suspended sediment particles for SEM analysis. *J. Sediment. Petrol.* **64**, 742–744.

1993

Walling, D.E. and He, Q. (1993a) Use of caesium–137 as a tracer in the study of rates and patterns of floodplain sedimentation. In: *Tracers in Hydrology*, IAHS Publication no. 215, pp. 319–328.

Walling, D.E. and He, Q. (1993b) Towards an improved interpretation of ^{137}Cs profiles in lake sediments. In: *Geomorphology and Sedimentology of Lakes and Reservoirs*, McManus, J. and Duck, R.W. (eds), John Wiley & Sons, Chichester, pp. 31–53.

Walling, D.E. and Quine, T.A. (1993) Using Chernobyl-derived fallout radionuclides to investigate the role of downstream conveyance losses in the suspended sediment budget of the River Severn, UK. *Phys. Geogr.* **14** (3), 239–253.

Walling, D.E. and Woodward, J.C. (1993) Use of a field-based water elutriation system for monitoring the *in-situ* particle size characteristics of fluvial suspended sediment. *Water Res.* **27** (9), 1413–21.

Walling, D.E., Woodward, J.C. and Nicholas, A.P. (1993) A multi-parameter approach to fingerprinting suspended sediment sources. In: Tracers in Hydrology, IAHS Publication no. 215, pp. 329–338.

Foster, I.D.L., Walling, D.E. and Owens, P.N. (1993) Sediment yields and budgets in the Start valley. In: *The Geomorphology of the Slapton Region*, Burt, T.P. (ed.), Field Studies Council, Preston Montford, Shropshire, UK, pp. 25–30.

Froehlich, W., Higgitt, D.L. and Walling, D.E. (1993) The use of caesium–137 to investigate soil erosion and sediment delivery from cultivated slopes in the Polish Carpathians. In: *Farm Land Erosion in Temperate Plains and Hills*, Wicherek, S. (ed.), Elsevier, Amsterdam, pp. 271–283.

Govers, G., Quine, T.A. and Walling, D. E. (1993) The effect of water erosion and tillage movement on hillslope profile development: a comparison of field observation and model results. In: *Farm Land Erosion in Temperate Plains and Hills*, Wicherek, S. (ed.), Elsevier, Amsterdam, pp. 285–300.

Higgitt, D.L. and Walling, D.E. (1993) The value of caesium–137 measurements for estimating soil erosion and sediment delivery in an agricultural catchment, Avon, UK. In: *Farm Land Erosion in Temperate Plains and Hills*, Wicherek, S. (ed.), Elsevier, Amsterdam, pp. 301–315.

Higgitt, D.L., Rowan, J.S. and Walling, D.E. (1993) Catchment-scale deposition and redistribution of Chernobyl radiocaesium in upland Britain. *Environ. Int.* **19**, 155–166.

Quine, T.A. and Walling, D.E. (1993a) Assessing recent rates of soil loss from areas of arable cultivation in the UK. In: *Farm Land Erosion in Temperate Plains and Hills*, Wicherek, S. (ed.), Elsevier, Amsterdam, pp.357–371.

Quine, T.A. and Walling, D.E. (1993b) Use of caesium–137 measurements to investigate relationships between erosion rates and topography. In: *Landscape Sensitivity*, Thomas, D.S.G. and Allison, R.J. (eds), John Wiley & Sons, Chichester, pp. 31–48.

Quine, T.A., Walling, D.E. and Mandiringana, O.T. (1993a) An investigation of the influence of edaphic, topographic, and land-use controls on soil erosion on agricultural land in the Borrowdale and Chinamara areas, Zimbabwe, based on caesium–137 measurements. In: *Sediment Problems: Strategies for Monitoring, Prediction and Control*, IAHS Publication no. 217, pp. 185–196.

Quine, T.A., Walling, D.E. and Zhang, X. (1993b) The role of tillage in soil redistribution within terraced fields on the Loess Plateau, China: an investigation using caesium–137. In: *Runoff and Sediment Yield Modelling*, Banasik, K. and Zbikowski, A. (eds), Warsaw Agricultural University Press, pp. 149–155.

Peters, N.E., Hoehn, E., Leibundgut, Ch., Tase, N. and Walling, D.E. (eds) (1993) *Tracers in Hydrology (Proceedings of Yokahama Symposium, July 1993)*, IAHS Publication no. 215, 350pp.

Rowan, J.S., Higgitt, D.L. and Walling, D.E. (1993) Incorporation of Chernobyl-derived radiocaesium into reservoir sedimentary sequences. In: *Geomorphology and Sedimentology of Lakes and Reservoirs*, McManus, J. and Duck, R.W. (eds), John Wiley & Sons, Chichester, pp. 31–54.

Webb, B.W. and Walling, D.E. (1993a) Longer-term water temperature behaviour in an upland stream. *Hydrol. Processes* **7**, 19–32.

Webb, B.W. and Walling, D.E. (1993b)Temporal variability in the impact of river regulation on thermal regime and some biological implications. *Freshwater Biol.* **29**, 167–182.

1994

Walling, D.E. (1994) Measuring sediment yield from river basins. In: *Soil Erosion Research Methods* Lal, R. (ed.), Soil and Water Conservation Society, Ankeny, Iowa, pp. 39–80.

Walling, D.E. and He, Q. (1994) Rates of overbank sedimentation on the floodplains of several British rivers during the past 100 years. In: *Variability in Stream Erosion and Sediment Transport (Proceedings of Canberra Symposium, December 1994)*, IAHS Publication no. 224, pp. 203–210.

Foster, I.D.L. and Walling, D.E. (1994) Using reservoir deposits to reconstruct changing sediment yields and sources in the catchment of the Old Mill reservoir, South Devon, UK, over the past 50 years. *Hydrol. Sci. J.* **39**, 347–368.

Froehlich, W. and Walling, D.E. (1994) Use of Chernobyl-derived radiocaesium to investigate contemporary overbank sedimentation on the floodplains of Carpathian Rivers. In: *Variability in Stream Erosion and Sediment Transport (Proceedings of Canberra Symposium, December 1994)*, IAHS Publication no. 224, pp. 161–170.

Higgitt, D.L., Walling, D.E. and Haigh, M.J. (1994) Estimating rates of ground retreat on mining spoils using caesium–137. *Appl. Geogr.* **14**(4), 294–307.

Nicholas, A.P. and Walling, D.E. (1994) Modelling contemporary overbank sedimentation on floodplains: some preliminary results. In: *Proceedings of the Third International Geomorphology Conference (Hamilton, Ontario, August 1993)* (in press).

Quine, T.A., Desmet, P.J.J., Govers, G., Vandaele, K. and Walling, D.E. (1994a) A comparison of the roles of tillage and water erosion in landform development and sediment export on agricultural land near Leuven, Belgium. In: *Variability in Stream Erosion and Sediment Transport (Proceedings of Canberra Symposium, December 1994)*, IAHS Publication no. 224, pp. 77–86.

Quine, T.A., Navas, A. and Walling, D.E. (1994b) Soil erosion on cultivated and uncultivated land near Las Bardenas in the Central Ebro River Basin, Spain. *Land Degrad. Rehab.* **5**, 41–55.

Zhang, X., Quine, T.A. and Walling, D.E. (1994) Application of the caesium–137 technique in a study of soil erosion on gulley slopes in a Yuan area on the Loess Plateau near Xifeng, Gansu Province, China. *Geogr. Ann. A* **76**, 103–120.

1995

Walling, D.E. (1995a) Suspended sediment transport by rivers: a geomorphological and hydrological perspective. In: *Proceedings of the International Symposium on Particulate Matter in Rivers and Estuaries (Reinbek/Hamburg, March 1994)*, (in press).

Walling, D.E. (1995b) Suspended sediment yields in a changing environment. In: *Changing River Channels*, Gurnell, A.M. and Petts, G.E. (eds), John Wiley & Sons, Chichester (in press).

Walling, D.E. and Quine, T.A. (1995) The use of fallout radionuclides in soil erosion investigations. In: *Nuclear and Elated Techniques in Soil/Plant Studies and Sustainable Agriculture and Environmental Preservation*, International Atomic Energy Agency, Vienna (in press).

Walling, D.E. and Woodward, J.C. (1995) Tracing suspended sediment sources in river basins: a case study of the River Culm, Devon, UK. *J. Marine Freshwater Res.* **46** (in press).

Phillips, J.M. and Walling, D.E. (1995) *In-situ* measurement of the effective particle size characteristics of fluvial sediment using a field-portable laser backscatter probe: some preliminary results. *J. Marine Freshwater Res.* **46** (in press).

Section I

QUANTITY AND QUALITY DIMENSIONS

2 The Role of Geographical Information Systems in Hydrology

TOM J. BROWNE

The Computing Centre, University of Sussex, UK

INTRODUCTION

A catchment is regarded by hydrologists as a fundamental unit geomorphologically, morphometrically and hydraulically, and it is also significant environmentally, culturally and ecologically. It provides a logical unit within which to calculate the water balance, to determine the transport characteristics of a river network (sediments and solutes in erosion and deposition), to estimate the probabilities of extreme events and to manage the water resources more effectively. Unlike political boundaries, which are rarely useful environmentally, catchment boundaries identify units that can most readily be apportioned responsibility, accountability and management (CERL, 1993).

Assisted by the dramatic increase in the computational power of modern computers, hydrologists have become very effective at constructing a variety of models of hillslope and catchment behaviour. There has been a move away from *empirical* models, where the initial conditions are noted and changes observed but the underlying causes are not understood, to *deterministic* models, where the processes taking place in a catchment can be described in terms of known physical or chemical laws. Many models define their parameters in terms of land-use, topography, soils, geology, etc, and they simulate catchment behaviour for different combinations of these parameters. The lack of suitable data, however, to determine these parameters for physically-based hydrologic process models often requires that they are spatially averaged for the whole catchment, thus treating processes in a 'lumped' form.

Increasingly, the aim has been to break a heterogeneous catchment down into a number of more homogeneous elements, treating each of these as a series of connecting, lumped models. Briefly, such a distributed model calculates the water budget in each element, with excess runoff presumed to be routed into a neighbouring downhill one, often using the kinematic wave assumption of the cascade model. If catchments upstream of a gauging station can be described by numerical indices that influence their flow regimes, then these models can be used with greater reliability upstream of ungauged locations, which is regarded as one of the ultimate tests of physically-based hydrology (Beven, 1983). Early models focused primarily on estimating runoff volumes, but with the growth of environmental concerns, particularly from the mid-1970s, sediment and nutrient modelling has been incorporated. Todini (1988) provides a useful historical review of some of these developments.

Sediment and Water Quality in River Catchments. Edited by I.D.L. Foster, A.M. Gurnell and B.W. Webb.
© 1995 John Wiley & Sons Ltd.

HOW CAN A GIS HELP?

A major difficulty with the distributed approach is the enormous data demands it makes, both in the requirement to identify values throughout the catchment for each element and also in the enormous computer-storage and processing demands. As spatial variability increases, so the size of the computational element must decrease in order to capture the variability of the input parameters.

A GIS is often described, somewhat prosaically, as an information system that uses geographically referenced data, and it can be viewed as a chain of operations involving data collection, input, storage, analysis and display, within the framework of defining an objective and ultimate use of that analysis for some decision-making purpose. But there are many permutations on this definition, including those that emphasise it as a set of tools or as a database management system for both spatial and non-spatial (attribute) data, or primarily as a means of analysis in order to help solve real-world problems, or as an infrastructure that incorporates human organisational features, and each betrays the culture and background of the proponent. Indeed, to be effective, it must be holistic and interdisciplinary in approach, as reflected in the similarly interdisciplinary nature of hydrology. One of its defining characteristics is its ability to integrate diverse information representing a variety of geophysical or socio-economic variables, with spatial location as the common denominator linking the data sets together. The accruing benefits, outlined by Shepherd (1991), include the ability to perform a broader range of operations on these combined data than is possible on separate, fragmented, disparate data sets. Each spatial data set, e.g. land-use, soils, can be considered as representing a separate layer within the database and they will each have associated attribute information. Parameters can be calculated from within a layer and by operations between different layers, and the results can be directed to physically-based hydrological models. The results from the model can then be used by the GIS for overlay, display and interpretation. In this way, a GIS can help to develop hypotheses and stimulate the search for causal factors, and different management or land-use scenarios can be simulated and their impact on the hydrology of an area evaluated. Zelt (1991) cites a typical example of using 25 layers in a water quality study!

The diffusion of information technology in general and of GIS in particular, within hydrology, with its technical and cultural challenges to previously well-established working practices, mirrors its development within other disciplines, although this association is still relatively young, only really gaining momentum from the mid-1980s. The development of physically-based models, however, which began to stress the data storage and computational challenges, began a decade earlier (see Gupta and Solomon, 1977). In the United Kingdom, the 'Chorley Report' (see Chorley and Buxton, 1991) also proved to be a tremendous catalyst, particularly in promoting the availability of geographically referenced data and their integration.

MODELLING REALITY IN A GIS

GIS have been described, in a hydrological context, as a 'modelling opportunity' (Abrahart et al., 1994), but geographical variation in the world is infinitely complex and it would take a similarly infinitely large database to capture the real world precisely, so data must somehow be reduced to a finite and manageable quantity. The *vector* and *raster* data models (and their associated structures as implemented in a computer) both attempt to do this. With the vector model, reality, in terms of points, lines and areas, must be known before data collection can take place. Implicit in this process is intelligent filtering of reality. While this can also be true of the raster model, increasingly such gridded data are becoming available from satellite imagery, where no prior knowledge of the phenomenon is required. The raster model is also becoming increasingly popular through modern display technologies and because of the relative computational advantages of using raster rather than vector data, at least on PCs. Some disadvantages are the considerable storage implications of raster data, which has led to research in data compression and filtering techniques, the apparent loss of accuracy (although transitional or 'fuzzy' boundaries may reflect reality, e.g. soil boundaries) and the requirement to interpret the remotely sensed image. Another criticism is that such an arbitrary raster geometry cannot reflect morphologically homogeneous units, often called hydrological response units. A methodological distinction is sometimes made on the basis that the raster model emphasises location and its surroundings while the vector model emphasises the attributes of locations, but in any case, conversion tools between the two approaches are now widely available.

SOME CONTRIBUTORY DISCIPLINES

Many disciplines have contributed to the union between physically-based, distributed hydrological models and GIS. Both have greatly benefited from advances in computing technology, but computer-assisted cartography, relational databases and remote sensing, all of which have drawn upon each other, deserve particular mention.

Computer-assisted cartography

Primarily driven by the vector model, the initial stimulus was to replicate the manually produced product while gaining the flexibility obtained from manipulating a spatial digital database. Beran (1982) and Gustard *et al.* (1989) discuss this specifically within a hydrological context. But if such data can be digitally obtained in a suitably topologically structured organisation, whereby it is possible to define relationships between spatial features, rather than as 'spaghetti' or unstructured data, then GIS become powerful visualisation tools, with maps becoming a means to an end rather than an end in themselves, i.e. they are no longer merely an effective means of visually communicating the relative location of features but they now also highlight and assist in analysing spatial relationships. Indeed, the spatial analysis techniques developed for analytical digital cartography still dominate the functionality in much GIS software. Visualisation focuses on exploring ideas and 'what-if' scenarios through temporary maps, three- and four-dimensional representation and, in association with multimedia, such visualisation can provide a pathway through the ever-increasing firehose syndrome of data availability (see Buttenfield and Mackaness, 1991; Hearnshaw and Unwin, 1994). Powerful workstations with high-quality displays permit initial rapid qualitative visual evaluation of the outcome of different distributed model calibrations. In large water organisations, only one digital database need be stored, with multiple visual access points, allowing just a single central spatial database to be updated, so avoiding repetition and redundancy.

Satellite remote sensing

The raster basis of satellite imagery has been a contributory factor in this model and data structure being widely used in hydrological applications. Although aerial photography was an initial stimulus, remotely sensed images from satellites, recording the non-visible parts of the electromagnetic spectrum, have become of primary interest, especially as their spectral coverage and resolution have improved. There has been considerable research into calibrating and classifying these satellite surrogate measurements, to provide indices for physically-based models. Satellite technology is providing an enormous global source of data, particularly for vegetation and soil indices and in improving their definition, especially where ground survey may be very difficult to perform (e.g. see Broers et al., 1990; Roberts et al., 1993). In the USA, the Soil Conservation Service (SCS) curve number is widely used to estimate the peak or flood discharge for a drainage basin. The model requires a parameter called the runoff curve number (CN), and this has been calculated from remotely sensed data, for each class of vegetation. Drayton et al. (1993), Rango (1985) and Stuebe and Johnson (1990) provide reviews of such applications. The frequency of repeat measurements has greatly assisted temporal analyses and, with radar, measurements of precipitation are available for hydrological modelling almost in real time. New parameters can be generated using remote sensing, but to accommodate them, existing models have to be modified. For example, Ottle et al. (1989) illustrate how surface temperature can be derived from thermal infrared measurements, in order to provide surrogate estimates of soil moisture, but hydrological models need to be restructured to accommodate this new index. Engam and Gurney (1991) and Shultz (1993) provide useful reviews on a range of different satellites and their spectral bands, resolutions and orbital or geostationary character.

Databases

It has already been noted that the storage and management of spatial information is pivotal to analytical cartography. What distinguishes GIS from other software that handles spatial data, however, is the communication between the database of spatial entities and any associated non-spatial attributes. Commonly, a hybrid approach is used whereby the spatial entities are stored in their own database, with their own data structure, but linked to an attribute database that is part of a Database Management System (DBMS), although in simple implementations of the raster model, the attributes are often merely directly the values of the cells. The most common DBMS is the Relational DBMS (RDBMS) (see Healey, 1991). In a large organisation, various databases may also be distributed at different nodes on a network. This is particularly useful in a water management context where it is possible to maintain the spatial entities on one processor, but linked to non-spatial or attribute entities located on other processors, which in other contexts are dedicated to other applications (Annand, 1989). The use of a query language such as the Structured Query Language (SQL), though it has limitations in the querying of spatial data, is widely used.

DIGITAL TERRAIN AND ELEVATION MODELS

The spatial distribution of topographic attributes often reflects and is a major control upon the spatial variability of hydrologic processes (Moore et al., 1991). A Digital Elevation Model (DEM) represents the spatial distribution of height, while a Digital Terrain Model (DTM) represents the spatial distribution of terrain attributes. These direct and derived topographic attributes such as catchment area, slope, aspect and profile curvature are used for numerous model indices (see Morris and Heerdegen, 1988). The ability to derive these indices throughout a catchment has been a major catalyst to the development of distributed models as well as giving credence to their physically-based claims, both in surface and groundwater flow routeing and in water quality and transport modelling, and so DEMs have become an essential component of any hydrological GIS database. The knowledge of the spatial variability of topographic variables has even been used to derive surrogate measures for precipitation, evapotranspiration and soil characteristics, where more direct data are unavailable or too difficult to obtain. There is a basic assumption that the DEM is a reflection of current dominant hydrological processes, which is unlikely to be true geomorphologically, but for practical purposes, can be considered true morphologically (Beven et al., 1988).

Constructing a DEM can be very demanding computationally, and so it has been the subject of considerable research to develop efficient computer-based algorithms. Conventionally, these have been based upon a grid of regularly spaced data, from a Triangulated Irregular Network (TIN) or, more rarely, from a contour-based network (Moore and Grayson, 1991). TIN models are sometimes considered to be less arbitrary than a grid (see Gandoy-Bernasconi and Palacios-Velez, 1990; Vieux, 1991) because triangles can be selected that more closely reflect the hydrological variability of the catchment. A grid structure can incorporate considerable redundancy if the terrain contains areas of extreme ruggedness and smoothness, as the size of the grid element must reflect the former. Also, it is sometimes argued that, while they are well-suited to terrain modelling, they are less applicable to geomorphological and hydrological applications. For the reasons stated earlier, however, grid-based DEMs now predominate. In associated attribute files, for each cell, in addition to information determined from the DEM, climatic, geophysical and socio-economic data can be registered to the same grid, so building up many layers of information of immense value to a distributed hydrologic model. Terrestrial survey data from spot heights and contours will gradually be superseded by remotely sensed data from aerial stereophotography and, currently, primarily SPOT satellite imagery, in order to generate DEMs. Many national agencies now provide DEMs, for example, the Ordnance Survey provide them for the United Kingdom at a resolution of 50 m, which compares favourably with the data provided by many other agencies (see Gittings, 1994).

Many models use the area of a catchment and indices reflecting the drainage network and flow direction. Manually deriving or even digitising them is very time-consuming, and appropriate source maps may not be available on a regional or world scale. Historically, the 'blue line' was determined from topographic maps but this is a highly subjective procedure and is influenced by the scale and accuracy of the map. Also its length cannot portray seasonal variability. Nevertheless, in the absence of adequate alternatives at the time, Moore (1983) describes how this 'blue line' network was digitised from Ordnance Survey 1:50 000 scale maps for the United Kingdom, in part to assist in the more accurate presentation and updating of river pollution maps. This digitised network was subsequently used by Sekulin *et al.* (1992), who, by allocating river stretches to cells in a raster, identified hydrologically contributory cells and thereby catchment boundaries. However, DEMs are increasingly being used for this purpose. The drainage network is usually derived using neighbourhood operations, identifying, for every cell, the eight adjacent cells and determining which ones are uphill and thereby contribute flow to their downhill neighbour. Derivation of catchment boundaries is an obvious corollary. CERL (1993) is ambitiously attempting to use satellite imagery to derive catchment networks and boundaries globally, as part of their climatic change action project. Band (1993), Jenson and Dominique (1988), Moore *et al.* (1991) and Tribe (1992) provide useful reviews of various techniques to derive a DEM, and discuss the problems of determining the ends of the network and the assignment of drainage directions across flat areas and depressions, which need to be resolved in order to produce a continuous network, which can then be used in hydrological modelling.

SAMPLING, DATA MEASUREMENT, SCALE AND ACCURACY

In hydrological modelling, catchment characteristics are assigned numerical indices that describe factors that affect flow regimes. To interpret adequately the results of any modelling, it is important to collect the right sort of data, both current and antecedent, to the most appropriate resolution, taking account of spatial and temporal variability of the phenomenon in question, and to identifiable standards of accuracy. The ever-increasing volumes of environmental and socio-economic data are creating challenging problems in data storage, and are obliging practitioners to filter their data. Conversely, as hydrologic models become ever more realistic, the pressure to obtain more data, to a finer resolution and of higher quality, also grows, leading to what Burrough (1989) calls a 'parameter crisis'. Also, the problem of inter-correlation between variables becomes increasingly acute, with different combinations of parameters giving equivalent results, and it is not always easy to propose a physical reason why one combination should be chosen in preference to another.

With the possible exception of remote sensing, it is usually necessary, when measuring geographic phenomena, to adopt some sampling strategy. In many instances, data sets in hydrology are only available as sample point measurements, e.g. from rainfall gauges or spot heights, and from these an interpolated areal measurement has to be derived. Gupta and Solomon (1977), Gustard *et al.* (1989) and Wiltshire *et al.* (1986) provide examples where a variety of indices thought to be significant in hydrological analysis were interpolated from irregularly spaced data to obtain derived data distributed over a uniform grid. This process involves many approximations and assumptions, which are not always considered in the modelling and deduction process. Also, rarely is there an assessment of the accuracy of the sample data, or of the boundaries, often derived from thematic and topographic maps at varying scales (and where the analogue cartographers' objectives are not always sympathetic to the requirements expected of digital accuracy), or any assessment of the affect of the classification process of attributes, e.g. soil types, on the parametrisation of variables. GIS are often assessed on the quality of their graphics, but attractive maps can convey a spurious indication of accuracy and errors in individual layers can be propagated as a result of multiple overlays (garbage in, pretty map out!). Water managers are under growing pressure to comply with increasingly stringent environmental standards and this implies high-quality data for a range of hydrological, biological, chemical and socio-economic variables. But data are now highly marketable products and the impediment of high cost may result in less satisfactory, surrogate data being used or, for spatial data, performing time-consuming in-house digitising.

Many distributed models have been developed for small research catchments, but invariably these are much smaller than the size of catchments requiring management decisions. In a small catchment, a model may correctly identify Hortonian, saturation excess surface flow or subsurface stormflow as the dominant mechanism within a model (see Beven and Binley, 1992), but these need not be mutually exclusive or, conversely, apply in catchments other than from where the model was tested. Also, scale thresholds, which are particularly pertinent at a global perspective, may require simplifying assumptions that undermine the original model, because as catchments get larger, response times become increasingly linear, so that a catchment can be regarded as a single linear store (Kirkby, 1988). This assumption gives validity to the initial formulation of the lumped 'geomorphological instant unit hydrograph' (Rodriguez-Iturbe and Valdes, 1979), although Maidment (1993a) has used a GIS to derive a spatial unit hydrograph, so relaxing the condition for uniform excess rainfall over the whole catchment.

Anderson and Rogers (1987), Beven (1989) and Grayson *et al.* (1992) question the validity of distributed models, arguing that, even at the research scale, the physical basis for the parametrisation is suspect and, at the management scale, is completely inappropriate because of the coarse resolutions that have to be employed. Jenson (1991) has investigated the suitability of different resolutions as part of a geochemical sediment survey, and Hoover *et al.* (1991) caution against attempting to use DEMs derived from satellite imagery of a resolution too coarse to reflect the runoff and transport processes that are operating within the catchment. Lumped parameters and multivariate regionalisation techniques that work, even though they may have little meaning in terms of the physical processes, may provide results that are of comparable validity and with much lower requirements for data and computation. A distinction should therefore be made between the explanatory and predictive usefulness of a model.

Clark (1993) outlines the compromises that often have to made when data are sparse, too expensive, or otherwise unavailable. This is also highlighted in the CORRINE project (Co-ordination of Information on the Environment) (see Rhind, 1991). Beginning in 1985, the aim was to collect, coordinate and ensure the consistency of information on the state of the environment and natural resources in the old European Community, and to bring together existing data holdings, to develop methods for holding, analysing and presenting data, and also to encourage the exchange of data. In addition to data unavailability owing to administrative difficulties and copyright restrictions, the lack of harmonisation was considerable. For example, there were eight different definitions of evapotranspiration and four different definitions of maximum temperature. Indeed, D.W. Rhind (1994, personal communication) notes that major temperature gradients seem to occur at national boundaries! Data were also available at a variety of different scales and resolutions, hindering the compilation of a coherent database. Such impediments are often of greater concern than technological bottlenecks. It is possible, nevertheless, to become overly exercised with the problems of data accuracy, as any definition of quality must always be related to its relevance to the problem in hand.

SOFTWARE

The current generation of readily available GIS are sometimes criticised for not having true modelling capabilities. It is argued that they are primarily concerned with the functions of data storage, retrieval and presentation of spatial data, and the spatial analysis tools are limited to those commonly found in cartographic models (Burrough, 1989). They are also very poor at handling the time dimension, which to hydrologists is often of primary concern, especially when dealing with very small time increments, because in the current generation of GIS, for each time step, the relevant equations have to be recalculated using rather inefficient macro language code. Also, for repetitive simulation and visualisation, the minimum of delay in the cycle of modelling, display and interpretation, which implies a close coupling between the model and the GIS at the level of the data structure, is preferred. The modelling of hydrological processes invariably exploits the finite-differences (grid mesh) model or the finite-elements (irregular mesh) model, which could map onto the two most common data structures used by GIS (see Maidment, 1993b), but the absence of a data structure that adequately incorporates time has resulted in many researchers producing their own customised GIS software, based on the toolbox approach of integrating highly tailored modules to address a specific task. Applying current terminology, this is similar to using 'application builders', which combine powerful computational and visualisation facilities in a networked, modular environment. Such software is gaining increasing currency in many fields of scientific visualisation. The alternative approach, usually using mainstream GIS, is to provide better external links between GIS and hydrological models, allowing each to concentrate on what it does best.

It would be an invidious task to identify and list all relevant GIS software, not least because the claims of some vendors regarding the functionality of their products is of doubtful validity. Weatherbee (1994) has provided a recent list and brief review of just the major commercial and public-domain software; but in broad terms, the hydrological literature indicates that GRASS (a raster GIS) and Arc/Info (until recently, exclusively a vector GIS) are arguably the two most widely used. GRASS is public domain and, although this may initially have influenced some researchers to select it, it has been successfully coupled to many hydrological models and its functionality has been improved to respond to the needs of hydrologists. Arc/Info has been widely used for TIN modelling and cartographic display, but is now developing raster modules and also has some highly tailored hydrological tools. Within UK higher education, Arc/Info and, more recently, Imagine (a raster GIS widely used in remote sensing) are available through financially favourable nationally negotiated deals (Kitmitto, 1994). The idea of a 'desktop' PC is fashionable but the data storage, computational and display considerations often favour a networked UNIX workstation configuration, although this can present many challenges to the way GIS is promoted and exploited, as is illustrated by Browne (1992). Arc/Info and Imagine also have PC DOS-based implementations and GRASS can run under PC UNIX implementations.

COUPLING GIS WITH HYDROLOGICAL MODELS — SOME EXAMPLES

One model that is gaining widespread attention is the Système Hydrologique Européen (SHE) model (Abbot et al., 1986). It claims to be fully distributed but is so demanding in its range of parameter specification and requirements for fine spatial resolution of data that currently its use with GIS is highly constrained. Models that are widely used are more accurately described as semi-distributed, because they make pragmatic choices regarding the selection of variables and are less exacting in resolution, though they still claim to be physically based.

Lengthy reviews could be written on all the diverse themes within hydrology, and there is clearly considerable interconnectivity between them all. A recent broad compilation can be found in Kovar and Nachtnebel (1993) and a number of them are specifically cited in this review. In keeping with the spirit of this book, however, the following primarily provides a few brief examples on the use of some models used in sediment and nutrient modelling but which are also exploiting GIS. Some earlier models that were designed to simulate surface and subsurface flow have been modified to incorporate water quality issues. The need for this for solutes as well as contaminants was recognised by Walling and Webb (1980), and now modified or specific non-point pollution distributed models are increasingly being used to determine the pollution potential of different land-use management activities (e.g. see Kaden, 1993; Sokel et al., 1993), as it is now widely recognised that agricultural practices are a major contributor to sediment and nutrient water pollution (see Gilliland and Baxter-Potter, 1987; Hession and Shanholtz, 1988; Vieux, 1991). Hydrologic processes are fundamental in water quality analyses because pollutants are transported by runoff, drainflow and throughflow to surface and ground water by percolation. But the spatial variability of this pollution will be influenced by the spatial variability of soils, topography, climate, land management practices, etc, which affect hydrological pathways and therefore erosion, sedimentation and leaching. While the following outline is, of necessity, very brief, many of the references cited in other contexts should be perused for additional examples.

AgNPS (Agricultural Non Point Source Pollution) (Young et al., 1987) is used as a grid-based tool for assessing the impacts of agriculture on surface water quality. To calculate impacts of non-point source pollution, it predicts runoff volume, peak rates, soil erosion, and nitrogen, phosphorus and chemical oxygen demand concentration in the runoff. About one-third of all parameters can be obtained from a DEM. Klaghofer et al. (1993) used AgNPS with the IDRISI GIS to estimate the sediment and nutrient export from a small lower alpine drainage basin. Data were needed on topography, soil types, climate, cropping pattern and tillage, and were overlaid in IDRISI to identify areas of different soil erosion conditions. (Complementary soil erosion analysis was performed with EPIC, the Erosion Productivity Impact Calculator.) Needham and Vieux (1989) obtained 21 different parameters covering soil conditions, land cover and topography, using a TIN DEM and additional parameters derived from these, as input to AgNPS. Using interpolation techniques, data from a range of irregular boundaries were mapped onto the grid structure required by AgNPS. From the model, 23 output parameters were produced and fed back into Arc/Info for display and interpretation. Engel et al. (1993) provide a comprehensive study of the use of AgNPS with GRASS.

MINDER (Model Input Nutrients Determine Eutrophication Risk) (Woodrow, 1993) is a GIS system being developed at the Water Research Centre, UK. It examines the effects of nutrients such as phosphorus and nitrogen, which are extensively used agriculturally, particularly in their encouragement of excessive plant growth in the surface waters. One version of the model uses a distributed approach and a grid-based model, using the raster GRID module in Arc/Info. Utilising flow direction, as calculated in the grid, it attempts to calculate the runoff and quantity of phosphorus that accumulates through each cell on a monthly time step and predicts such concentrations for anywhere in the river network. A similar model has been used by the Institute of Hydrology, UK, and applied to the Ouse catchment in Bedfordshire (Finch, 1992).

REGIS (the REgional Geohydrological Information System) (see Hoogendoorn and Boswinkel, 1991) is both a data management system and an environment for a variety of groundwater models, and it is still undergoing rapid development. It has been developed by staff at the Institute of Applied Geoscience in The Netherlands, who are researching the impact of nitrate pollution caused by agricultural practices and heavy-metal pollution from former and current industrial activity. An essential preliminary step was to digitise a variety of data from existing archives of paper groundwater maps. In this context, REGIS is being coupled with Arc/Info for data management and visualisation (De Lange and Van der Meij, 1993; Hoogendoorn et al., 1993). It models and monitors geohydrological properties of the subsurface and groundwater, both their level and their chemical content, as part of groundwater management.

TOPMODEL, proposed initially by Beven and Kirkby (1979), for flood forecasting, but continually being modified, is a physically-based, semi-distributed model that is widely cited in the literature, though as yet its usefulness in the context of non-point pollution and sediment transport is mostly only alluded to. It requires two categories of parameters, namely topographic, obtained from a grid-based DEM (area, channel network and lengths, slope, elevation and aspect), and hydrological (infiltration, interception storage, rainfall, evapotranspiration, and hydrological properties of soil and land cover). It conceptualises soil water storage as a sequence of storages with different properties. Fundamental parameters required by TOPMODEL are a topographic and a topographic/soils index, because the model assumes a close relationship between topographic form and subsurface flow and the production of surface runoff. TOPMODEL has been implemented as one of a number of application modules within the Water Information System (WIS) developed at the UK Institute of Hydrology (Romanowicz et al., 1993), and it has been integrated with GRASS (Chairat and Delleur, 1993; Abrahart et al., 1994) and the grid-based (using a quadtree data structure) SPANS (Stuart and Stocks, 1993). European legislation such as the 1991 Nitrate Directive requires all members of the Economic Union to identify zones vulnerable to nitrate leaching by the end of 1993, and this is providing considerable momentum to the further development of linking GIS with hydrological models. In this context, Wade et al. (1994) are linking TOPMODEL to SPANS to develop a catchment-scale nitrate vulnerability model, relating water quality to land use, soils and topographic data, in order to predict nitrate concentrations in both ground and surface waters.

ANSWERS (Areal Nonpoint Source Watershed Environment Response Simulation) (Beasley *et al.*, 1980) has been referred to by a wide number of researchers but has been explicitly used with GIS by De Roo *et al.* (1989) with the Map Analysis Package and later with GENMAP by De Roo (1993). A grid-based model with many similarities to TOPMODEL, ANSWERS is designed to simulate hydrological behaviour of catchments having agriculture as the primary land use, before and after a rainfall event. The primary application is in evaluating various strategies for controlling surface runoff and sediment transport from intensively cropped areas. Later versions of the model permit, for each grid element, the slope, aspect and channel variables to be obtained from a DEM, together with a range of soil and crop variables.

MANAGEMENT OF WATER RESOURCES

> Natural resource management problems have reached a level of complexity where no one category can be realistically considered in a vacuum. It is no longer possible to isolate the management of any one natural resource from that of others. (Eipper, 1973)

> We have reached the conclusion that the major difficulty is not in measuring particular aspects of natural or human systems but in integrating such measurements into a comprehensive scheme. (Richerson and McEvoy, 1973)

The above quotes were made in the context of a systems analysis approach and before GIS, as we now generally recognise that term, was regularly propounded as part of the armoury of resource management, and fuel the debate as to whether GIS offers a new methodology or is an old methodology in new clothes. Nevertheless, they cogently express the urgent need for an integrated approach to problem-solving. Growing environmental awareness is increasingly forcing water companies and their regulators to evaluate the potential of GIS in management and pollution modelling. In a research environment, complexity may be welcomed as added functionality, but in a support and maintenance environment, the challenge is to make the system as easy to operate as possible. Decisions will no doubt always be influenced by individual personalities and varying degrees of rational argument, but Decision Support Systems (DSS), which are a set of tools that support quick and comprehensive generation, simulation and analysis of a series of scenarios, are increasingly being used to assist in evaluating the consequences of water management decisions. There are many interactions and conflicts between water users and this has created more complex problems for water managers, who now have to estimate the economic, social and ecological impacts as well as any hydrological consequences (Stanbury *et al.*, 1991). These systems are now incorporating expert systems and have highly customised Graphical User Interfaces (GUIs) in order to assist in improving the efficiency and quality of hydrological information processing. In fact, to be most effective, they must be customised to address the institutional, management and planning objectives of the particular organisation (Fedra, 1993) and allow operators to involve their own experience and judgement. Indeed, caution must be exercised in assuming that any unskilled operator can use such a system. The uncertainty in much of the data and the modelling processes will ensure that the intervention and control of the skilled human agency will, for the foreseeable future, be of paramount importance (see Rhind, 1991). In fact, developments in multimedia will enable hydrological information to be used that cannot readily be incorporated into current models, and so, as part of the visualisation experience, the human skills of interrogation and analysis will be further exploited (Shepherd, 1991).

In corporate environments, it has been shown as imperative that personnel are integrated into the new culture at a very early stage in order to draw on their full potential and gain their full support (see Corbin, 1991). Previously, much valuable geo-referenced information was held on paper, microfiche, etc, and in no small part in employees' heads. Following privatisation in the UK in 1989, the motivation in the water industry has been driven increasingly by the business environment, and the integrating of hydrological data together with financial services databases in a geo-referenced manner is seen as providing new marketing opportunities (see Elkins, 1990), and the hydrological imperatives may become secondary to other more short-term commercial information technology opportunities (Turner, 1993). At one level, much activity has been principally at the level of asset management, initially digitising, as the first stage in identifying what resource holdings an authority has. But for GIS to become an integral part of an employee's day, it has to be accessible wherever that employee is currently located. To this end, IT Southern, responsible for water management in part of the south-east of England, is now providing up-to-date data available to vehicles, and from information provided in the field, the central database can be updated within 24 h.

THE FUTURE

Developments in information technology have enabled progress in hydrological distributed modelling and in spatial database management. These two parallel developments are increasingly being combined and have enabled the hydrologist to study spatial patterns as well as processes. But there are still many research agendas that need to be actively pursued, arguably the most pressing being to incorporate more adequately the temporal component of hydrological processes. Others include strategies for better integrating the raster and vector models and associated structures, often called Integrated GIS (IGIS), taking account of the lumping effect at the element level, be it a grid or a triangular facet, accounting for data error and model uncertainty, full three-dimensional modelling, especially for groundwater studies (where they are sometimes called 3D Geoscientific Information Systems, or 3D GSIS) and better spatial analysis tools. Object-oriented models are gaining increasing attention, because, although their definition raises more controversy than illumination, in general terms, they attempt to define an abstraction that is similar to that used by humans to interpret and understand reality, in that they provide less of a distinction between spatial and attribute components. Developments in parallel processing will permit more realistic modelling, as, in reality, events do not occur sequentially but simultaneously and in parallel (Burrough, 1989). Finally, as high-bandwidth networking is permitting a greater *technologically* distributed environment, there is an urgent need for better data transfer standards and pricing, and copyright considerations must encourage rather than inhibit data exchange, as without the ability to integrate data from many different sources, the full promise of the role of GIS within hydrology will not be realised.

REFERENCES

Abbot, M.B., Bathurst, J.C., Cunge, J.A., O'Connell, P.E. and Rasmussen, J. (1986) An introduction to the European Hydrological System — Système Hydrologique Européen, 'SHE', 2: Structure of a physically-based, distributed modelling system. *J. Hydrol.* **87**, 61–77.

Abrahart, R.J., Kirkby, M.J. and McMahon, M.L. (1994) MEDRUSH — a combined geographical information system and large scale distributed process model. In: Fisher, P. (ed.), *Proceedings of the GIS Research UK (University of Leicester, April)*, pp. 67–75.

Anderson, M.G. and Rogers, C.C.M. (1987) Catchment scale distributed hydrological models: a discussion of research directions. *Prog. Phys. Geogr.* **11**(1), 28–52.

Annand, K. (1989) TAMESIS — a geographic database for river management. In: *Managing Geographical Information Systems and Databases (University of Lancaster, Sept)*, 5 pages.

Band, L.E. (1993) Extraction of channel networks and topographic parameters from digital elevation data. In: Beven, K. and Kirkby, M.J. (eds), *Channel Network Hydrology*, ch. 2, pp. 13–42, Wiley, Chichester.

Beasley, D.B., Huggins, L.F. and Monke, E.J. (1980) ANSWERS: a model for watershed planning. *Trans. ASAE* **23**(4), 938–944.

Beran, M.A. (1982) Hydrology and automated cartography. *Cartographica* **19**(2), 56–61.

Beven, K. (1983) Surface water hydrology — runoff generation and basin structure. *Rev. Geophys. Space Phys*. US National Report to International Union of Geodesy and Geophysics, 1979–1982. **1**(3), 721–730.

Beven, K.J. (1989) Changing ideas in hydrology — the case of physically-based models. *J.Hydrol.* **105**, 157–172.

Beven, K. and Binley, A. (1992) The future of distributed models: model calibration and uncertainty prediction. *Hydrol. Processes* **6**, 279–298.

Beven, K.J. and Kirkby, M.J. (1979) A physically based variable contributing area model of basin hydrology. *Hydrol. Sci. Bull.* **24**(1), 43–69.

Beven, K.J., Wood, E.F. and Sivapalan, M. (1988) On hydrological heterogeneity and catchment response. *J. Hydrol.* **100**, 353–375.

Broers, H.P., Peters, S.W.M. and Biesheuvel, A. (1990) Design of a groundwater quality monitoring network with GIS and remote sensing. *EGIS '90 Proceedings (Amsterdam)*, **1**, pp. 95–105.

Browne, T.J. (1992) The great divide: central versus departmental support for teaching GIS. *J. Geogr. Higher Educ.* **16**(20), 225–230.

Burrough, P.A. (1989) GIS — more than a spatial database. In: *Managing Geographical Information Systems and Databases (University of Lancaster, Sept)*, pp. 1–16.

Buttenfield, B.P. and Mackaness, W.A. (1991) Visualisation. In: Maguire, D.J., Goodchild, M.F. and Rhind, D.W. (eds), *Geographical Information Systems, Principles and Applications*, Vol. 1, ch. 28, pp. 445–56, Longmans, Harlow.

CERL (1993) CERL's global GRASS CD-ROMs. *CERL Newslett.*, April.

Chairat, S. and Delleur, J.W. (1993) Integrating a physically based hydrological model with GRASS. In: Kovar, K. and Nachtnebel, H.P. (eds), *Application of Geographic Information Systems in Hydrology and Water Resources (Proceedings of Vienna Conference, April 1993)*, IAHS Publication no. 211, pp. 143–150.

Chorley, R. and Buxton, R. (1991) The Government setting of GIS in the United Kingdom. In: Maguire, D.J., Goodchild, M.F. and Rhind, D.W. (eds), *Geographical Information Systems, Principles and Applications*, vol. 1, ch. 5, pp. 67–79, Longmans, Harlow.

Clark, M.J. (1993) Data constraints on GIS application development for water resources management. In Kovar, K. and Nachtnebel, H.P. (eds), *Application of Geographic Information Systems in Hydrology and Water Resources (Proceedings of Vienna Conference, April 1993)*, IAHS Publication no. 211, pp. 451–463.

Corbin, C.E.H. (1991) A geographical information system for managing the assets of a water company. *ICL Tech. J.* 515–536.

De Lange, W.J. and Van der Meij, J.L. (1993) A national groundwater model combined with a GIS for water management in the Netherlands. In: Kovar, K. and Nachtnebel, H.P. (eds), *Application of Geographic Information Systems in Hydrology and Water Resources (Proceedings of Vienna Conference, April 1993)*, IAHS Publication no. 211, pp. 333–343.

De Roo, A.P.J. (1993) Validation of the ANSWERS catchment model for runoff and soil erosion simulation in catchments in the Netherlands and the United Kingdom. In: Kovar, K. and Nachtnebel, H.P. (eds), *Application of Geographic Information Systems in Hydrology and Water Resources (Proceedings of Vienna Conference, April 1993)*, IAHS Publication no. 211, pp. 465–474.

De Roo, A.P.J., Hazelhoff, L. and Burrough, P.A. (1989) Soil erosion modelling using ANSWERS and geographical information systems. *Earth Surf. Processes Landforms* **14**, 517–532.

Drayton, R.S., Wilde, B.M. and Harris, J.H.K. (1993) Geographical information system approach to distributed modelling. In: Beven, K.J. and Moore, I.D. (eds), *Terrain Analysis and Distributed Modelling in Hydrology*, ch. 11, pp. 193–200, Wiley, Chichester.

Eipper, A.W. (1973) The role of the technical expert in decision-making. In: Goldman, C.R., McEvoy, III, J. and Richerson, P.J. (eds), *Environmental Quality and Water Development*, ch. 21, pp. 393–397, Freeman, San Francisco.

Elkins, P. (1990) Water under the bridge — the background to Southern Water's experience with GIS. *Mapping Awareness* **4**(8), 38–40.

Engam, E.T. and Gurney, R.J. (1991) *Remote Sensing in Hydrology*, Chapman and Hall, London.

Engel, B.A., Srinivasan, R. and Rewerts, C. (1993) A spatial decision support system for modeling and managing non-point-source pollution. In: Goodchild, M.F., Parks, B.O. and Steyaert, L.T. (eds), *Environmental Modeling with GIS*, ch. 20, pp. 231–237, Oxford University Press, New York.

Fedra, K. (1993) Models, GIS and expert systems: integrated water resources models. In: Kovar, K. and Nachtnebel, H.P. (eds), *Application of Geographic Information Systems in Hydrology and Water Resources (Proceedings of the Vienna Conference, April 1993)*, IAHS Publication no. 211, pp. 296–308.

Finch, J. (1992) Spatial data and GIS at the Institute of Hydrology. *Mapping Awareness* **6**(3), 17–20.

Gandoy-Bernasconi, W. and Palacios-Velez, O. (1990) Automated cascade numbering of unit elements in distributed hydrological models. *J. Hydrol.* **112**, 375–393.

Gilliland, M.W. and Baxter-Potter, W. (1987) A geographic information system to predict non-point source pollution potential. *Water Resour. Bull.* **23**(2), 281–291.

Gittings, B. (1994) Digital elevation models. List sent to comp.infosystem gis, on *Internet Network News*, April.

Grayson, R.B., Moore, I.D. and McMahon, T.A. (1992) Physically based hydrologic modeling. 2. Is the concept realistic? *Water Resour. Res.* **26**(10), 2659–2666.

Gupta, S.K. and Solomon, S.I. (1977) Distributed numerical model for estimating runoff and sediment discharge of ungauged rivers 1. The information system. *Water Resour. Res.* **13**(3), 613–618.

Gustard, A., Roald, S., Lumadjeng, H.S. and Gross, R. (1989) *Flow Regimes from Experimental and Network Data (FREND)*. Hydrological Studies, vol. I, Institute of Hydrology, Wallingford, UK.

Healey, R.G. (1991) Database management systems. In: Maguire, D.J., Goodchild, M.F. and Rhind, D.W. (eds), *Geographical Information Systems, Principles and Applications*, Vol. 1, ch. 18, pp. 251–267, Longmans, Harlow.

Hearnshaw, H. and Unwin, D. (eds) (1994) *Visualisation in Geographical Information Systems*, John Wiley & Sons, Chichester.

Hession, W.C. and Shanholtz, V.O. (1988) A geographic information system for targeting nonpoint-source agricultural pollution. *J. Soil Water Conserv.* May-June, 264–266.

Hoogendoorn, J.H. and Boswinkel, J.A. (1991) GIS as a tool in data processing geohydrology. *EGIS '90 proceedings (Amsterdam)*, vol 5, pp. 463–472.

Hoogendoorn, J.H., Van der Linden, W. and Te Stroet, C.B.M. (1993) The importance of GIS in regional geohydrological studies. In: Kovar, K. and Nachtnebel, H.P. (eds), *Application of Geographic Information Systems in Hydrology and Water Resources (Proceedings of Vienna Conference, April 1993)*, IAHS Publication no. 211, pp. 375–383.

Hoover, K.A., Foley, M.G. and Heasler, P.G. (1991) Sub-grid-scale characterisation of channel lengths for use in catchment modeling. *Water Resour. Res.* **27**(11), 2865–2873.

Jenson, S.K. (1991) Applications of hydrologic information automatically extracted from digital elevation models. *Hydrol. Processes* **5**, 31–44.

Jenson, S.K and Dominique, J.O. (1988) Extracting topographic structure from digital elevation data for geographic information system analysis. *Photogram. Eng. Remote Sensing* **54**(11), 1593–1600.

Kaden, S.O. (1993) GIS in water-related environmental planning and management: problems and solutions. In: Kovar, K. and Nachtnebel, H.P. (eds), *Application of Geographic Information Systems in Hydrology and Water Resources. (Proceedings of Vienna Conference, April 1993)*, IAHS Publication no. 211, pp. 385–397.

Kirkby, M. (1988) Hillslope runoff processes and models. *J. of Hydrol.* **100**, 315–339.

Kitmitto, K. (1994) CHEST the ticket for academic software: provision of spatial sciences software and datasets to higher education in Britain. *Mapping Awareness* **8**(2), 14–15.

Klaghofer, E., Birnhaum, W. and Summer, W. (1993) Linking sediment and nutrient export models with a geographic information system. In: Kovar, K. and Nachtnebel, H.P. (eds), *Application of Geographic Information Systems in Hydrology and Water Resources (Proceedings of Vienna Conference, April 1993)*, IAHS Publication no. 211, pp. 501–506.

Kovar, K. and Nachtnebel, H.P. (eds) (1993) *Application of Geographic Information Systems in Hydrology and Water Resources (Proceedings of Vienna Conference, April 1993)*, IAHS Publication no. 211.

Maidment, D.R. (1993a) Developing a spatially distributed unit hydrograph by using GIS. In: Kovar, K. and Nachtnebel, H.P. (eds), *Application of Geographic Information Systems in Hydrology and Water Resources (Proceedings of Vienna Conference, April 1993)*, IAHS Publication no. 211, pp. 181–192.

Maidment, D.R. (1993b) GIS and hydrologic modeling. In: Goodchild, M.F., Parks, B.O. and Steyaert, L.T. (eds), *Environmental Modeling with GIS*, ch. 14, pp. 148–167, Oxford University Press, New York.

Moore, I.D. and Grayson, R.B. (1991) Terrain-based catchment partitioning and runoff prediction using vector elevation data. *Water Resour. Res.* **27**(6), 1177–1191.

Moore, I.D., Grayson, R.B. and Ladson, A.R. (1991) Digital terrain modelling: a review of hydrological, geomorphological and biological applications. *Hydrol. Processes* **5**, 3–30.

Moore, R.V. (1983) Digitising the United Kingdom river network. In: Weller, B.S. (ed), *Auto Carto (Proceedings of 6th International Symposium on Automated Cartography, Ottowa, Ontario)*, 331–338.

Morris, D.G. and Heerdegen, R.G. (1988) Automatically derived catchment boundaries and channel networks and their hydrological applications. *Geomorphology* **1**, 131–141.

Needham, S. and Vieux, B.E. (1989) A GIS for AgNPS parameter input and mapping output. *ASAE Winter Meeting (Chicago, Illinois)*, Paper no. 89-2673.

Ottle, C., Videl-Madjar, D. and Girad, G. (1989) Remote sensing applications to hydrological modelling. *J. Hydrol.* **105**, 369–384.

Rango, A. (1985) Assessment of remote sensing input to hydrologic models. *Water Resour. Bull.* **21**(5) 423–432.

Rhind, D. (1991) Environmental modelling and prediction. In: Masser, I. and Blakemore, M. (eds), *Handling Geographical Information: Methodology and Potential Applications*, Longman, Harlow, ch. 9.

Richerson, P. and McEvoy, III, J. (1973) The measurement of environmental quality and its incorporation into the planning process. In: Goldman, C.R., McEvoy, III, J. and Richerson, P.J. (eds), *Environmental Quality and Water Development*, ch. 6, pp. 111–135, Freeman, San Francisco.

Roberts, G., France, M., Johnson, R.C. and Law, J.T. (1993) The analysis of remotely sensed images of the Balquhidder catchment for the estimation of land cover types. *J. Hydrol.* **145**, 259–265.

Rodriguez-Iturbe, I. and Valdes, J.B. (1979) The geomorphological structure of hydrologic response. *Water Resour. Res.* **15**, 1409–1420.

Romanowicz, R., Beven, K., Freer, J. and Moore, R. (1993) TOPMODEL as an application model within WIS. In: Kovar, K. and Nachtnebel, H.P. (eds), *Application of Geographic Information Systems in Hydrology and Water Resources (Proceedings of Vienna Conference, April 1993)*, IAHS Publication no. 211, pp. 211–223.

Sekulin, A., Bullock, A. and Gustard, A. (1992) Rapid calculation of catchment boundaries using automated river network overlay techniques. *Water Resour. Res.* **28**(8), 2101–2109.

Shepherd, I.D.H. (1991) Information integration and GIS. In: Maguire, D.J. Goodchild, M.F. and Rhind, D.W. (eds), *Geographical Information Systems, Principles and Applications*, vol. 1, ch. 22, pp. 337–360, Longman, Harlow.

Shultz, G.A. (1993) Applications of GIS and remote sensing in hydrology. In Kovar, K. and Nachtnebel, H.P. (eds), *Application of Geographic Information Systems in Hydrology and Water Resources (Proceedings of Vienna Conference, April 1993)*, IAHS Publication no. 211, pp. 127–140.

Sokel, G., Leibundgut, Ch., Shulz, K.P. and Weinzierl, W. (1993) Mapping procedures for assessing groundwater vunerability to nitrate and pesticides. In: Kovar, K. and Nachtnebel, H.P. (eds), *Application of Geographic Information Systems in Hydrology and Water Resources (Proceedings of Vienna Conference, April 1993)*, IAHS Publication no. 211, pp. 631–639.

Stanbury, J., Woldt, W., Bogardi, I. and Bleed, A. (1991) Decision support system for water transfer evaluation. *Water Resour. Res.* **27**(4), 443–451.

Stuart, N. and Stocks, C. (1993) Hydrological modelling within GIS: an integrated approach. In: Kovar, K. and Nachtnebel, H.P. (eds), *Application of Geographic Information Systems in Hydrology and Water Resources (Proceedings of Vienna Conference, April 1993)*, IAHS Publication no. 211, pp. 319–329.

Stuebe, M.M. and Johnson, D.M. (1990) Runoff volume estimation using GIS techniques. *Water Resour. Bull.* **26**(4), 611–620.

Tribe, A. (1992) Automated recognition of valley lines and drainage networks from grid digital elevation models: a review and a new method. *J. Hydrol.* **139**, 263–293.

Todini, E. (1988) Rainfall–runoff modeling — Past, present and future. *J. of Hydrol.* **100**, 341–352.

Turner, P. (1993) Northumbian Water: managing for change. *Mapping Awareness GIS Europe* **7**(6), 49–50.

Vieux, B.E. (1991) Geographic information systems and non-point source water quality and quantity modelling. *Hydrol. Processes* **5**, 101–113.

Wade, S., Baban, S. and Foster, I. (1994) The development of a GIS for mapping nitrate vunerability: alternative modelling approaches for supporting the implementation of nitrate action programmes. In: Fisher, P. (ed). *Proceedings of the GIS Research UK (University of Leicester, April)* pp. 308–314.

Walling, D.E. and Webb, B.W. (1980) The spatial dimension in the interpretation of stream solute behaviour. *J. Hydrol.* **47**, 129–49.

Weatherbee, O. (1994) Listing of GIS packages. List sent to comp.infosystem.gis, on *Internet Network News*, March.

Wiltshire, S.E., Morris, D.G. and Beran, M.A. (1986) Digital data capture and automated overlay analysis for basin characteristic calculation. Cartogr. J. **23**, 60–65.

Woodrow, D. (1993) Environmental applications of GIS in the water industry. *Mapping Awareness GIS Europe* **7**(6) 22–23.

Young, R.A., Onstad, C.A., Bosh, D.B. and Anderson, W.P. (1987) *AGNPS, Agricultural Non-Point-Source Pollution Model*, USDA Agricultural Research Conservation Research Report no. 35.

Zelt, R.B. (1991) GIS technology used to manage and analyse hydrologic information. *GIS World* **4**(5), 70–73.

3 Modelling Nitrate Leaching at the Catchment Scale

HARVEY J.E. RODDA
New Zealand National Institute of Water and Atmospheric Research, Hamilton, New Zealand

INTRODUCTION

During recent years a considerable amount of attention has been devoted to studying the leaching of nitrate from agricultural land. Such studies were initially undertaken in response to concern in the farming community that nitrogen is not being used efficiently in agricultural systems. More recently, concern over health and environmental effects of high nitrate concentrations in drinking water has been the driving force behind these experiments, particularly in response to the EC Directive on Drinking Water (80/788) and the EC Nitrate Directive (91/676). The first directive set the maximum admissible concentration for nitrate-nitrogen (NO_3-N) in drinking water as 11.3 $mg\,l^{-1}$, while the second directive required member states to designate areas where water sources were vulnerable to contamination by nitrate.

Two conditions need to be satisfied in order for nitrate to be leached from the soil. First, a pool of nitrate must be available to be leached and secondly, there must be sufficient flow of water to remove this nitrate in solution. Nitrate leaching is largely a result of modern farming practices, in that large quantities of nitrogen (N) are applied to agricultural land as fertiliser. Nitrogen, in the form of nitrate, is highly soluble in water and not readily adsorbed onto soil particles, and is therefore easily transported by the movement of water through the soil. Most high-grade agricultural land is artificially drained, which increases the rate at which nitrate is removed from the soil into streams and rivers. There is, however, a common misconception that nitrate is directly leached from unused fertiliser. This is rarely the case in the climatic conditions experienced in the UK, where a soil moisture deficit over summer ensures that no nitrate is actually lost from fertiliser. Instead, the increased organic matter in the soil/crop system generated by the extra N is mineralised (the transformation of organic N into inorganic N) into nitrate during the autumn and winter and subsequently leached. Under extremely wet conditions during the time of the first application of fertiliser in the spring, some nitrate can be directly leached from the fertiliser, but this is less than 10% of the total N that is leached during a year (Addiscott and Powlson, 1989).

MODELLING NITRATE LEACHING

The growing importance and concern of nitrate pollution has led to the widespread use of models to simulate and predict nitrate leaching from agricultural land. The complexity of nitrate leaching has meant that modelling studies have been undertaken within a variety of scientific disciplines, with the most prominent being hydrology, soil science and agricultural science. Early hydrological models were largely based on empirical catchment studies relating the changes in river nitrate concentration to changes in land use and fertiliser application rates (e.g. Owens, 1970; Edwards and Thornes, 1973). Models developed from an agricultural or biological perspective considered leaching as one of the processes linked with the biochemical transformations of N within an agricultural system (e.g. Reuss and Innis, 1977). The initial models that were developed within the field of soil science generally just considered the movement of nitrate, as a solute, through the soil profile using physical transport equations (e.g. Bresler, 1967; Burns, 1975).

The use of microcomputers has accelerated the development of nitrate leaching models in recent years, but studies have been criticised for using models that did not include some important aspects of the leaching process. For example, a report on river and groundwater nitrate concentrations (Department of the Environment, 1986) was widely criticised from the field of agricultural research (e.g. Addiscott and Powlson, 1989) because only simple empirical relationships were used between the fertiliser applied and the amount of nitrate leached. On the other hand, agriculturally based models did not consider the transport of nitrate through the groundwater, so there was a general requirement for the integration of models.

Complex simulation models were subsequently developed, which included a combination of the key elements from the different disciplines outlined above. Equations describing the flow of water and nitrate through the soil were combined with equations that accounted for the biochemical transformations of N during the growing season. Examples of these models include LEACHN (Wagenet and Hutson, 1990), SOILN (Bergstrom and Jarvis, 1991), DAISY (Hansen *et al.*, 1991) and SACFARM (Addiscott *et al.*, 1991). These models were also dynamic, so that transformations were described as rate processes, and the model simulation could take the form of results predicted for discrete time steps (e.g. days) over the course of the growing season. The model development and testing required the continuous monitoring of the processes at the experimental site and detailed records of meteorology and agricultural management.

Catchment-scale modelling

All of the complex models described above were restricted to plot-scale simulations. The detailed nature of the input parameters and internal state variables, and the degree of site monitoring that was required to test the model predictions, made them difficult to apply over a whole catchment. Also, the use of a plot-scale model for an entire catchment area would not represent many catchment-scale processes. One of the most important is the hydrological component, in that all of the plot-scale models mentioned earlier only predict the concentration or load of nitrate that is leached below the root zone, and do not incorporate the different pathways through which the nitrate is transported to the river. These include rapid flow from field drains, macropores and surface runoff, and a much slower response from groundwater seepage. Differences in land use are also important at the catchment scale and these were not included in plot-scale models. Studies have highlighted the need to distinguish between nitrate leaching from grazing and arable land (Ryden *et al.*, 1984), and that the age and history of agricultural land is particularly important in determining the organic N content of the soil and N mineralisation rates (Scholefield *et al.*, 1991). Another important distinction between plot-scale and catchment-scale models is that many of the input variables such as the amount of fertiliser and meteorological data, are not uniform or even available over the whole catchment.

Some very good catchment-scale models have been developed that incorporate most of the characteristics of the dynamic plot-scale models and the necessary components that enable a catchment-scale simulation, such as the simulation of streamflow and flow through the saturated zone. One such model was developed by Cooper *et al.* (1992), which simulated nitrate leaching and N transformations for the Bourne Brook catchment in the English Midlands. The model operated on a grid square basis, with groundwater, surface water and N components included for each grid square. Nitrogen transformations were simulated using the routines from the SACFARM model (Addiscott and Whitmore, 1987), and allowed simulation for different land-use types. A similar study was undertaken in southern Sweden (Brandt, 1990), where models describing the generation of runoff, solute transport processes, biochemical N transformations and water quality were combined to simulate runoff and nitrate leaching from a mixed land-use catchment.

The problem with catchment-scale models such as these is that a large number of parameters are required in order to run the model for a given catchment, which involves an extensive program of site monitoring, land-use evaluation and mapping of soil and geology. Such a model is therefore difficult to apply for a range of catchments and over a number of years, attributes which are particularly relevant in respect of the proposed changes in agricultural management required to reduce leaching and comply with the nitrate concentration limits set by the EC. The requirement for catchment-scale models to predict nitrate leaching, for a number of catchments under different land management scenarios, led to the development of management models that were less complex, but able to predict nitrate leaching for different catchments under a range of climatic and agricultural conditions to a reasonable degree of accuracy.

A CASE STUDY OF THE N-CATCH MODEL

One such management-based model was developed from a field-scale model entitled N-CYCLE (Scholefield *et al.*, 1991). The N-CYCLE model was devised by researchers at the Institute of Grassland and Environmental Research, North Wyke, Devon, from the results of long-term field drainage experiments under grassland. N-CYCLE uses only a small number of input parameters: soil type, drainage conditions, land-use history, climatic zone and the amount of N deposited from the atmosphere. The model predicts the annual fluxes of N within a beef grazing system that would occur under average weather conditions (Figure 3.1).

Model development

The development of N-CYCLE into N-CATCH was based on utilising the leaching output from N-CYCLE, to predict the total N load and nitrate concentration for small predominantly grassland catchments. This development was undertaken in four stages:

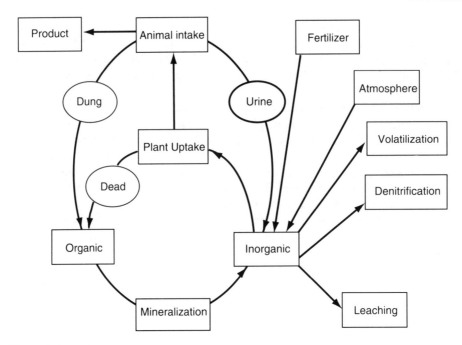

Figure 3.1 A flow diagram illustrating the processes incorporated in the N-CYCLE model

the development of the model from plot scale to catchment scale; the incorporation of subroutines to account for year-to-year variations in weather conditions; the calculation of leaching output in terms of concentration; and the development of a subroutine to incorporate differences in catchment hydrogeology.

First, the model was modified to work on a catchment scale. This consisted of lumping together the model simulations for each individual field unit within a given catchment so that the total catchment N load could be calculated.

The effect of annual climatic conditions

The second stage of the model development was to incorporate subroutines that were sensitive to annual weather patterns. The original N-CYCLE model employed a climatic classification in which Great Britain was divided into three zones, according to average temperature and rainfall, with the leaching output not affected by any variations in the annual conditions. River water quality studies have demonstrated the effect of the annual climatic conditions on the amount of nitrate leached. Walling and Foster (1979) recorded N loads for the Yendacott catchment in east Devon, which increased from 10.28 kg ha^{-1} in 1975–76 to 73.31 kg ha^{-1} in 1976–77, owing to summer drought conditions in 1976 followed by a very wet autumn.

The weather conditions can affect the amount of nitrate leached in two different ways. Firstly, both the temperature and the moisture conditions affect the biochemical transformations of N within the soil. The plant uptake of N is increased with a higher temperature, particularly during the early part of the growing season (April–June), whereas the N mineralisation is affected by both the temperature and soil moisture

(MacDuff and White, 1985). Higher temperatures and a moist soil are more favourable for mineralisation, which can occur during the growing season and throughout the autumn and winter. In addition, the gaseous loss of N through denitrification and ammonia volatilisation is increased under wet and water-logged conditions during the growing season (Ryden, 1986); therefore dry conditions would allow more N to be lost through leaching in the following autumn.

In order to incorporate the effect of the temperature and moisture conditions on these processes, the specific algorithms within the model that calculated the plant uptake of N, the N mineralisation and the denitrification were modified. Such modifications were possible since measured data from the North Wyke drainage experiment were available over a nine-year period, covering a range of climatic conditions (Scholefield *et al.*, 1993). The maximum soil moisture deficit (SMD) was taken as the index with which to classify the characteristics of a particular growing season, since it gave a quantitative description of both how hot and how dry the conditions were, and had been shown to influence the amount of N leached from the experimental sites (Figure 3.2).

The second climatic factor that influences nitrate leaching is the amount of winter drainage. The combination of the effect of three biochemical transformations, mentioned above, will generate a pool of potentially leachable N in the soil. The degree of drainage through the soil in the autumn and winter will determine how much of this pool will actually be lost from the soil into the rivers or groundwater. This relationship was incorporated into the model using regressions between the drainage volume and

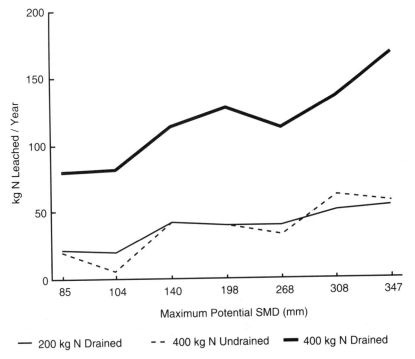

Figure 3.2 The relationship between maximum potential soil moisture deficit and the amount of N leached per year for three of the plots at North Wyke

proportion of nitrate leached from the soil for different soil types and drainage conditions (Figure 3.3). The drainage volume for a particular year was taken as the effective rainfall, obtained from Meteorological Office MORECS square data. Any nitrate that was not leached was assumed to remain in the soil and be available as inorganic N for the start of the following growing season.

Concentration calculation

The third stage in the model development was to modify the output so that the maximum and average NO_3-N concentrations could be predicted. The original model output gave N leached in terms of an annual load, but a concentration output was desirable to enable a direct comparison between the model predictions and the limit of NO_3-N concentration as set by the EC. Work by Tyson (1989) showed that the maximum NO_3-N concentration could be predicted from a simple empirical relationship between the total N load and the maximum concentration for the experimental sites at North Wyke. An examination of the concentration of NO_3-N in the drainage during the autumn and winter in a well-structured clay loam (Figure 3.4), and the knowledge of the different solute flow mechanisms incorporated into other models, underlined the effect of soil structure on the peak concentration. Well-structured clay soils generate preferential and bypass flow (Addiscott and Whitmore, 1991), whereas simple displacement flow occurs in sandy soils, and these give rise to differences in the

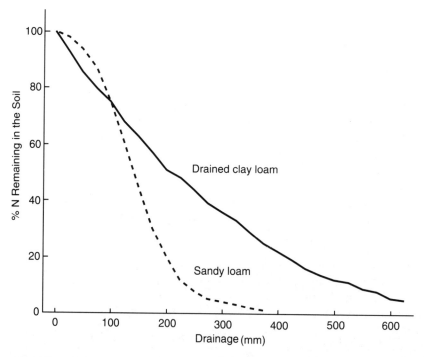

Figure 3.3 The relationship between the proportion of N remaining in the soil profile and winter drainage for different soil types

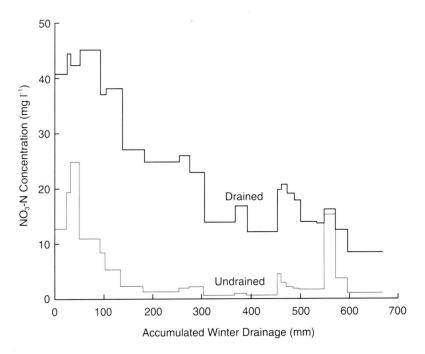

Figure 3.4 The variation of NO_3-N concentration with drainage from the 400 kg N plots at North Wyke

magnitude and timing of the peak concentration. Therefore, if empirical relationships were to be used, it was clear that individual relationships should be derived for soils with different structures. Well-fitted linear regressions of peak NO_3-N concentration on total N load were obtained for an artificially drained clay loam, an undrained clay loam, and a highly permeable sandy loam, each under grassland. Interpolation between data for these soil textural and drainage conditions gave analogous regressions for a loam and for a clay loam with moderately efficient field drainage. All of these regression equations were incorporated into the model, along with similar linear regressions relating average concentration and the total load for different drainage conditions. Average concentrations were not dependent on soil structure, and therefore not calculated for the different soil types.

The effect of catchment hydrogeology

The final phase in model development was to incorporate a subroutine that would account for differences in catchment hydrogeology. Research has shown the rate of movement of nitrate through groundwater to be slow. Oakes (1988) reported average travel times of 1 m yr^{-1} in chalk and 2 m yr^{-1} in sandstone. Therefore, nitrate that is leached from soil overlying permeable strata is going to remain in the groundwater for a number of years before being returned to the river through baseflow. Some catchment models have overcome this problem by using a time delay factor (e.g. Onstad and Blake, 1980), but N-CATCH employed a groundwater-balance submodel (Figure 3.5).

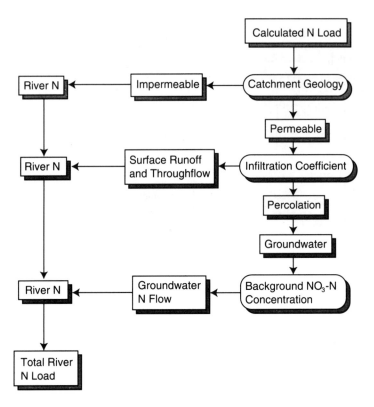

Figure 3.5 A flow diagram of the hydrogeological procedure incorporated into N-CATCH. The elliptical boxes represent parameters stipulated by the user

The submodel assumed that a proportion of the nitrate leached over permeable rock was infiltrated into the groundwater, depending on the rock type. Infiltration coefficients were estimated for the three major permeable rock types in the UK, based on the transmissivity values (Downing, 1991). A value of 0.8 was adopted for chalk, 0.65 for sandstone (including Triassic sandstones and greensand) and 0.5 for limestone (including Jurassic, Carboniferous and Magnesian limestone).

In order to maintain a groundwater balance, it was assumed that the same volume of water that percolated into the aquifer was returned to the river through baseflow with a background NO_3-N concentration. The background concentration values were estimated from borehole data and land-use history for a given catchment. Comprehensive borehole concentration data were not available, so the background concentrations were limited to three classes comprising low, moderate and high NO_3-N concentrations of 1.0, 3.5 and 7.0 mg l^{-1} respectively.

Model testing

The completed model was tested for 11 catchments that were included in a large database of river nitrate concentrations and discharge measurements for the period 1974–87 held at Exeter University (Betton, 1990). The catchments, listed in Table 3.1,

Table 3.1 The characteristics of the study catchments. Standard average annual rainfall (SAAR) is interpolated from Meteorological Office isohyetal maps, 1950–81

Catchment	Location	Area (km^2)	SAAR (mm)	Altitude (m)	Geology
Biggar	Borders	7.8	1200	224–363	Andesite and basalt
Crimple	North Yorkshire	16.9	800	108–203	Carboniferous shales and sandstones
Hooke	Dorset	9.6	1000	137–241	Chalk over Upper Greensand
Sydling	Dorset	12.7	1000	103–265	Chalk over Upper Greensand
Drimpton	Somerset	4.8	1000	107–195	Jurassic limestones and sandstones
Otter	East Devon	15.4	1200	162–289	Upper Greensand over Keuper Marl
Tale	East Devon	11.3	1200	114–283	Upper Greensand over Keuper Marl
Gara	South Devon	0.6	1400	159–195	Devonian shales and slates
Cole	North Devon	8.5	1200	108–193	Carboniferous shales and sandstones
Camel	Cornwall	13.5	1500	185–308	Devonian shales
Helford	Cornwall	13.7	1200	3–136	Granite and Devonian slates

were selected to give a range of climatic and hydrological conditions, but were all small and predominantly under grassland. The input data required to run the model were obtained from either fieldwork and questionnaire surveys, or from published information such as the Soil Survey maps and memoirs (Findlay et al., 1984). The model was run to generate output on total annual N load and maximum and average NO$_3$-N concentrations for the time period covered by the database of flow and nitrate measurements.

Results from the simulations reveal that a reasonable degree of success was obtained for the prediction of the N load. For about one-third of the catchments studied, there was generally a good agreement between the measured and predicted values (Figure 3.6a); for another one-third there were discrepancies only in some years (Figure 3.6b); and for the remaining third there were larger systematic discrepancies over most years (Figure 3.6c). Reasons for the discrepancies between the predicted and measured load

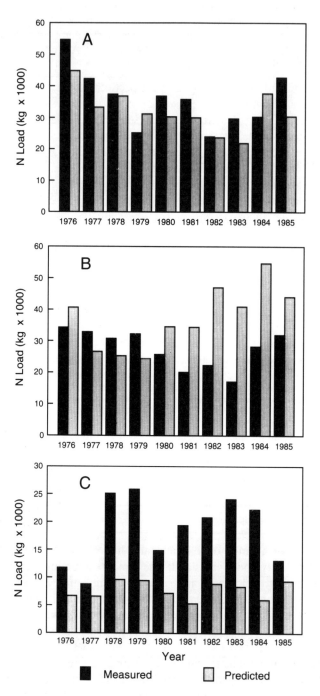

Figure 3.6 Comparison of measured and predicted N load for the Tale (A), Sydling (B) and Biggar (C) catchments

values could be attributed to errors in the measured data, in that a certain degree of interpolation was required to obtain complete spatial and temporal coverage for the input data in some of the catchments. In addition, some of the limitations of the original N-CYCLE model were still present in N-CATCH. In particular, leaching could only be calculated for a beef grazing system, and in some catchments, especially the Sydling and the Helford, about half of the agricultural land was under arable land uses. Limitations associated with the model developments incorporated into N-CATCH were also a possible reason for poor model performance in some catchments. For example, the hydrogeology sub-model was applied to the catchment as a whole, not to individual fields, owing to the lack of geological data at a more finely resolved scale.

Simulation of nitrate concentration was not as successful as the nitrate load results. The predicted peak concentrations were invariably higher than the measured values for all of the study catchments, and the predicted average concentrations were greater for all but two of the catchments (Table 3.2). The systematic nature of these discrepancies suggests that sources of dilution may have been omitted from the model. The model calculates the concentration of the water leached below the root zone and not the actual river concentration. The peak concentrations of nitrate from each field would not necessarily have been coincident in reaching the river, and furthermore significant reduction in concentration may occur during the transport process as a result of denitrification, which can occur in the lower soil horizons and groundwater (Fustec *et al.*, 1991) and in riparian buffer zones (Haycock and Burt, 1992). Surface runoff low in N from paved surfaces and non-agricultural land is also a significant source of dilution, particularly in the small upland study catchments, where the headwaters are often under moorland or rough grazing and generate most runoff in the catchment.

Despite the lack of accuracy in some of the results, N-CATCH is generally user-friendly, and can be used to investigate the effect of various grassland management scenarios on river water quality. For example, one of the policy implications of the EC Nitrate Directive was to designate 'Nitrate Sensitive Areas' where land previously under intensive agriculture was to be set-aside as rough grazing. In a hypothetical study, N-CATCH was used to investigate the effects of land-use change for three catchments,

Table 3.2 A comparison of measured and predicted maximum and average NO_3-N concentrations (mg l^{-1})

Catchment	Measured average	Predicted average	Measured maximum	Predicted maximum
Biggar	4.3	3.9	7.9	10.7
Crimple	2.2	5.6	9.6	15.4
Hooke	4.1	5.4	6.4	16.6
Sydling	3.8	8.2	13.3	34.6
Drimpton	2.5	7.4	6.9	33.9
Otter	2.7	5.1	5.1	18.2
Tale	4.5	6.4	8.6	22.9
Gara	4.4	6.0	11.0	18.6
Cole	1.9	7.5	3.6	28.7
Camel	3.1	6.5	4.2	23.7
Helford	6.3	5.9	9.0	19.9

the Wyre in north-west England, the Avon in central southern England and the Cleddau in Dyfed. The results showed that, in order to comply with the EC concentration limit for NO_3-N of 11.3 mg l^{-1}, the maximum amount of fertiliser N applied, as a catchment average, was between 122 and 165 kg ha^{-1}. It was also found that maintaining a uniform limit of fertiliser application was far more effective at controlling the NO_3-N concentration, while maintaining a sufficient level of production, than the use of set-aside areas and zones receiving particularly high levels of N under a polarisation of land use. For example, when fertiliser N application was limited to 125 kg ha^{-1} in the Wyre catchment and 90 kg ha^{-1} in the Avon and Cleddau catchments, maximum NO_3-N concentration estimated for 1990 was just over 10 mg l^{-1} for each catchment, and N production for the Avon, Cleddau and Wyre was approximately 400, 450 and 500 t respectively. By contrast, when half the land area in each catchment was put under set-aside (i.e. received no fertiliser), but the total catchment fertiliser input remained the same as the previous scenario, maximum NO_3-N concentrations increased to 23, 18 and 27 mg l^{-1} for the Avon, Cleddau and Wyre, with no significant increase in the amount of product N.

FURTHER DEVELOPMENTS IN CATCHMENT-SCALE NITRATE MODELLING

The results from the N-CATCH simulations have highlighted the potential for further model development. One notable observation was the need for models that allow the effects of different catchment land uses to be integrated, since the poor performance of N-CATCH in some of the study catchments was due to the mixed land use. N-CATCH was unable to simulate nitrate leaching from arable land, so the combination of N-CATCH with the equivalent arable model would allow simulation for a wider range of catchments and undoubtedly improve the accuracy of the predictions. As yet, no arable models comparable to N-CATCH exist, although a number of management-based field-scale arable models exist, including Sundial (Bradbury *et al.*, 1993). The combination of different land-use models and perhaps a more distributed hydrogeology sub-model could lead to the development of an expert system for predicting nitrate leaching at the catchment scale.

The use of concentration–load relationships

Another aspect of the N-CATCH simulations, which resulted in further studies, was the use of the empirical relationship to predict the maximum nitrate concentration from the total annual N load (Rodda, 1993). Models that can predict nitrate concentrations at the catchment and regional scales will be required to assess the effects of land-use changes necessary to reduce leaching and to comply with the EC Directive limit of nitrate in water. N-CATCH has already demonstrated its ability to predict the N load fairly accurately, and a number of other models can produce a similar level of accuracy for different agricultural systems (e.g. cereals, vegetables). Therefore, the use of an empirical relationship to predict maximum NO_3-N concentrations from these load values would be most useful. The empirical relationships that were used in N-CATCH were also derived for other leaching experiments, conducted on a range of soil types and agriculture (Scholefield et al., 1993). The results of these linear regressions are presented in Table 3.3. As such relationships have been derived for field-scale plots, it is feasible that the approach could be extended to estimate the peak NO_3-N concentrations in river catchments under predominant soil–crop combinations for different regions of the UK.

The use of GIS in catchment-scale modelling

A significant limitation of the N-CATCH simulation was that the catchments had to be divided into individual areal units based on the full range of input parameters, which include climate, atmospheric deposition of N, soil type, drainage conditions, sward history, age of grassland, amount of fertiliser N applied, volume of drainage, maximum SMD, underlying geology and groundwater nitrate concentration. The employment of a GIS interface could greatly simplify the creation of such datasets and simulation at the catchment scale (Browne, this volume). Once a catchment has been mapped and the information stored within a GIS, then, for any given grid square or areal unit within the catchment, the parameters could easily be ascertained and a model simulation could be undertaken. The output, in the form of a colour-coded map, would also be very desirable for management purposes. Cook (1991) used a system of overlays to highlight areas at risk from nitrate leaching in the North Downs of Kent. In this study, a number of key parameters (e.g. geology, soil type, land use, distance from borehole) were mapped using GIS. A scale of risk factors were attributed to each of the individual parameters, and a final map was produced combining the risk factor rating.

Other useful information could easily be interpolated from topographical maps to aid with the identification of the flow pathways of nitrate, such as the proximity of a given land unit to the stream channel and the degree of slope.

Table 3.3 Linear regressions of peak NO_3-N concentration on the total N leached, where peak (mg l^{-1}) = $a \times$ N load (kg ha^{-1}) + b

Soil type	Depth of sampler	Type of sampler	Agriculture	a	b	r^2
Loamy sand	90	Porous cup	Grassland	0.99	1.4	0.88
Loamy sand	90	Porous cup	Arable	0.92	2.4	0.93
Loam	90	Porous cup	Grassland	0.62	3.5	0.99
Clay loam, undrained	10	Surface drain	Grassland	0.58	3.9	0.96
Clay (cracking), mole-drained	90	Tile drain	Arable	0.37	2.8	0.33
Clay loam, mole-drained	85	Tile drain	Grassland	0.28	5.6	0.97

CONCLUDING REMARKS

This chapter has attempted to trace the development of modelling nitrate leaching at the catchment scale, and in particular examined the development of a management-based model, N-CATCH. Although a number of models have been mentioned, no comparison of the performance of these different models has been made. Comparisons of the actual model results must take the nature and the purpose of the model into consideration. Detailed plot-scale models such as SACFARM have produced very accurate and comprehensive simulations of nitrate leaching for given plot-scale studies (Addiscott et al., 1988), and complex catchment-scale models (e.g. Cooper et al., 1992) have produced simulations that are limited by some inaccuracies. The performance of N-CATCH was mixed, with systematic errors relating to the prediction of nitrate concentrations. The requirements of these models were all different, in that the purpose of the plot-scale model was to investigate in detail the mechanisms of nitrate transport within the soil–crop system. The purpose of the catchment-scale model, however, was to evaluate the performance of N transformation processes used at the plot scale combined with hydrological models over the catchment scale. By contrast, the main aim of N-CATCH was to apply a management-based approach over the catchment scale. Each of these models has its own advantages and disadvantages, and perhaps the best way to use them is in different levels of the same study. On a regional basis, N-CATCH could predict which catchments were likely to generate high N losses through leaching; a more detailed catchment model could then be employed to analyse specific catchments and highlight the areas within a catchment where nitrate leaching was greatest. In turn, a plot-scale model could be used to simulate leaching and N transformations for the defined area within a given catchment.

REFERENCES

Addiscott, T.M. and Powlson, D.S. (1989) Laying the ground rules for nitrate. *New Scient.* April, 28–29.

Addiscott, T.M. and Whitmore, A.P. (1987) Computer simulation of changes in soil mineral nitrogen and crop nitrogen during autumn, winter and spring. *J. Agric. Sci. Cambs.* **109**, 141–157.

Addiscott, T.M. and Whitmore, A.P. (1991) Simulation of solute leaching in soils of different permeabilities. *Soil Use Manage.* **7**, 94–102.

Addiscott, T.M., Whitmore, A.P. and Bland, G.J. (1988) *SACC-FARM: Soil and Crop Nitrogen Computer Simulation Model*, AFRC Institute of Arable and Crops Research Publication, Rothamsted Experimental Station, Harpenden, Herts.

Addiscott, T.M., Bailey, N.J., Bland, G.J. and Whitmore, A.P. (1991) Simulation of nitrogen in soil and winter wheat crops: a management model that makes the best use of limited information. *Fert. Res.* **27**, 305–312.

Bergstrom, L. and Jarvis, N.J. (1991) Prediction of nitrate leaching losses from arable land under different fertilization intensities using the SOIL-SOILN models. *Soil Use Manage.* **7**, 79–85.

Betton, C. (1990) Nitrate-N levels in British streams and rivers — a countrywide perspective. Unpublished PhD thesis, University of Exeter.

Bradbury, N.J., Whitmore, A.P., Hart, P.B.S. and Jenkinson, D.S. (1993) Modelling the fate of nitrogen in crop and soil in the years following the application of 15-N labelled fertilizer to winter wheat. *J. Agric. Sci. Cambs.* **121**, 363–379.

Brandt, M. (1990) Simulation of runoff and nitrate transport from mixed basins in Sweden. *Nordic Hydrol.* **21**, 13–34.

Bresler, E. (1967) A model for tracing salt distribution in the soil profile and estimating the efficient combination of water quality and quantity under varying field conditions. *Soil Sci.* **104**, 227–233.

Burns, I.G. (1975) A model for predicting the redistribution of salts applied to shallow soils after excess rainfall or evaporation. *J. Soil Sci.* **25**, 165–178.

Cook, H.F. (1991) Nitrate protection zones targeting and land use over an aquifer. *Land Use Policy* **12**, 16–28.

Cooper, D.M., Ragab, R., Lewis, D.R. and Whitehead, P.G. (1992) Modelling Nitrate Leaching to Surface Waters, MAFF/NERC Final Report, Institute of Hydrology, Wallingford, Oxfordshire.

Department of the Environment (1986) *Report on Nitrate in Water*, HMSO, London.

Downing, R.A. (1991) Groundwater supply in Britain. In: Downing, R.A. and Wilkinson, W.B. (eds), *Applied Groundwater Hydrology—A British Perspective*, Clarendon Press, Oxford, pp. 149–163.

Edwards, A.M.C. and Thornes, J.B. (1973) Annual cycle in river water quality: a time series approach. *Water Resour. Res.* **9**(5), 1286–1295.

Findlay, D.C., Colbourne, G.J.N., Cope, D.W., Harrod, T.R., Hogan, D.V. and Staines, S.J. (1984) *Soils and Their Use in South-West England*, Soil Survey of England and Wales, Bulletin no. 14.

Fustec, E., Mariotti, A., Grillo, X. and Sajus, J. (1991) Nitrate removal by denitrification in alluvial groundwater: role of a former channel, *J. Hydrol.* **123**, 337–354.

Hansen, S., Jensen, H.E., Nielsen, N.E. and Svendsen, H. (1991) Simulation of nitrogen dynamics and biomass production in winter wheat using the Danish simulation model DAISY. *Fert. Res.*, **27**, 245–259.

Haycock, N. and Burt, T.P. (1992) Floodplains as Nitrate buffer zones. *NERC News* **21**, 28–29.

MacDuff, J.M. and White, R.C. (1985) Net mineralization rates in permanent grassland. *Plant Soil* **86**, 151–172.

Oakes, D.B. (1988) *Modelling Nitrate in Groundwater Systems*, Report of the Special Topic Review on Nitrate Modelling, MAFF Publication, HMSO, London.

Onstad, C.A. and Blake, J. (1980) Thames basin nitrate and agricultural relations. *Proceedings of the International Symposium on Watershed Management*, ASCE, Boise, Idaho.

Owens, M. (1970) Nutrient balances in rivers. *Water Treat. Exam.* **19**, 239–247.

Reuss, J.O. and Innis, G.S. (1977) A grassland nitrogen flow simulation model. *Ecology* **58**, 379–388.

Rodda, H.J.E. (1993) The development and application of a nitrogen cycle model to predict nitrate leaching from grassland catchments within the UK. Unpublished PhD thesis, University of Exeter.

Ryden, J.C. (1986) Gaseous losses of nitrogen from grassland. In: Van der Meer, H.G., Ryden, J.C. and Ennick, G.C. (eds), *Nitrogen Fluxes in Intensive Grassland Systems*, Martinus Nijhoff, Dordrecht, pp. 99–113.

Ryden, J.C., Ball, P.R. and Garwood, E.A. (1984) Nitrate leaching from grassland. *Nature* **331**,(5981), 50–53.

Scholefield, D., Lockyer, D.R., Whitehead, D.C. and Tyson, K.C. (1991) A model to predict transformations and losses of nitrogen in UK pastures grazed by beef cattle. *Plant Soil* **132**, 165–177.

Scholefield, D., Lord, E.I. and Rodda, H.J.E. (1993) Empirical relationships describing the effects of preferential flow on nitrate leaching. *Proceedings of the 15th World Congress on Soil Science (Acapulco, Mexico)*.

Tyson, K.C. (1989) *The Rowden Moor Drainage Experiment Annual Report*, ADAS Publication, North Wyke, Devon.

Wagenet, R.J. and Hutson, J.L. (1990) *LeachM: A Process Based Model of Water and Solute Movement, Transformations Plant Uptake, and Chemical Reactions in the Unsaturated Zone*, Version II, Department of Agronomy, Cornell University, Ithaca, NY.

Walling, D.E. and Foster, I.D.L. (1979) The impact of the 1976 drought on nutrient concentration and loadings in the River Exe Basin, UK, *Proceedings of the International Symposiums on Hydrological Aspects of Droughts* (New Delhi), vol. 1, 151–162.

4 Regulation and Thermal Regime in a Devon River System

B.W. WEBB
Geography Department, University of Exeter, UK

INTRODUCTION

One of the most important ways by which human activity can modify the thermal regime of river systems is through reservoir construction and associated regulation of the downstream watercourse. Interest in the effects of impoundment on water temperature behaviour has been long-standing, especially in countries, such as Japan and the United States, where the practical consequences of thermal modification associated with large-scale regulation schemes have been keenly appreciated (e.g. Jaske and Goebel, 1967; Williams, 1968; Nishizawa and Yambe, 1970; Bolke and Waddell, 1975).

In recent years, the impacts of reservoirs on thermal regime have been more widely recognised and studies have been conducted in many different parts of the world (e.g. Petts, 1984; Crisp, 1987; Casado *et al.*, 1989; Tuch and Gasith, 1989; Liu and Yu, 1992; Gippel and Finlayson, 1993; Tvede, 1994). The influence of reservoirs constructed at different scales (e.g. Neel, 1963; Penaz *et al.*, 1968), used for different purposes (e.g. Raney, 1963; Collings, 1973) and located at different positions and in different numbers within the river system (e.g. Ward and Stanford, 1983; O'Keeffe *et al.*, 1990) has been investigated. Consideration has also been given to how different aspects of thermal regime, including general characteristics (e.g. Crisp, 1977; Webb and Walling, 1988a), seasonal and diurnal cycles (Pitchford and Visser, 1975; Ward and Stanford, 1980) and short-term responses (e.g. Crisp, 1976; Webb and Walling, 1988b), are modified through impoundment and river regulation. Furthermore, the extent to which the impact persists beyond the immediate environment of the reservoir has been studied (e.g. Walker *et al.*, 1978; Edwards, 1984; Palmer and O'Keeffe, 1989), together with the role of various natural and human factors which promote or retard downstream recovery of the thermal regime (e.g. Stanford and Ward, 1983; Byren and Davies, 1989).

A multiplicity of controlling factors has been invoked to account for the impact of regulation on water temperature behaviour (e.g. Sylvester, 1963; Petts, 1986), although the influence of thermal stratification within a reservoir, especially in relation to drawoff level, has been highlighted when explaining changes below impoundments (e.g. Ward, 1982; Mackie *et al.*, 1983; Cowx *et al.*, 1987). Several models have also been developed to simulate the nature of thermal regime in impounded rivers (e.g. Delay and Seaders, 1963; Troxler and Thackston, 1977; Jobson and Keefer, 1979;

Sediment and Water Quality in River Catchments. Edited by I.D.L. Foster, A.M. Gurnell and B.W. Webb.
© 1995 John Wiley & Sons Ltd.

Zimmerman and Dortch, 1989). Much of the interest in understanding and predicting the thermal impacts of reservoirs reflects the importance of water temperature as a fundamental control in freshwater biology. Many studies have documented how aquatic ecosystems, and especially invertebrate and fish biology, have responded to changes in thermal regime resulting from regulation (e.g. Ward and Stanford, 1979; Brooker, 1981; Edwards and Crisp, 1982; Boon, 1987; Rader and Ward, 1988; Brittain and Saltveit, 1989; Marchant, 1989; Voeltz and Ward, 1989; Saltveit, 1990).

Despite the wide interest in the impact of impoundment and regulation on river thermal regime, many of the conclusions concerning the nature of the modification, its downstream persistence and its biological implications have been drawn from studies limited in one or more of the following ways. First, there has been a dearth of investigations based on monitoring before, during and after construction, and the majority of studies have compared temperature records below reservoirs with those on unregulated tributaries above or neighbouring the impounded water body. In consequence, there may be problems of determining what changes in temperature would have occurred under natural conditions between inflow and outflow sites, especially in studies of large reservoirs (Moore, 1967), or difficulties in finding a control catchment with physiographic characteristics similar enough to allow meaningful comparison with the regulated river. A short-term perspective is a second limitation of many studies of the thermal impacts of impoundment. In the UK, for example, the major published investigations of water temperature behaviour in regulated rivers of northern England and Wales (Lavis and Smith, 1972; Cowx et al., 1987; Crisp, 1987) are based on between one and five years of data. In the absence of long-term records, it may be difficult to differentiate between the effects of impoundment and those of climate changes, especially for studies based on monitoring before and after regulation (Moore, 1967). Furthermore, a few years of data make it difficult to assess the stability of the impact of impoundment in response to factors such as the maturation of water quality conditions in reservoirs, significant inter-annual variability in hydrometeorological conditions and changes in the operation schedules of regulation schemes. A third limitation to some studies of temperature in regulated rivers is the absence of detailed monitoring. Although spot measurements or the use of maximum–minimum thermometers may provide valuable information on the general impact of impoundment, it is only through detailed records that some aspects of thermal modification below reservoirs, such as changes in duration characteristics and the short-lived but often dramatic effects of releases, can be properly recognised and quantified.

The present chapter aims to synthesise the results from a reservoir study in southwest England, which has been designed, as far as possible, to be free of the limitations just described. The study is based on continuous monitoring at sites located immediately below and at increasing distances downstream from the impoundment, as well as on a control catchment that neighbours the regulated river. Records have been kept for the periods before, during and after reservoir construction, and the study, which has now been in progress for more than 17 years, is probably the longest detailed investigation of the thermal consequences of impoundment to be published. The study was designed and initiated by Des Walling in 1976 as part of his efforts to develop a network of stations in the Exe basin, Devon, UK dedicated to long-term and detailed monitoring of water quality conditions in rural areas. It is a tribute to his skill and foresight that not only was water temperature selected as one of the parameters for continuous monitoring, but also the opportunity was grasped to implement a sampling scheme that allowed the effects of the then-proposed Wimbleball Scheme to be thoroughly investigated.

THE WIMBLEBALL STUDY

Wimbleball Lake, which is located in the River Haddeo subcatchment of the Exe basin on the eastern margins of Exmoor (Figure 4.1A), was constructed as a dual-purpose reservoir to regulate the mainstream of the Exe and also to provide a direct supply for part of Somerset (Battersby et al., 1979). Impounding began in mid-December 1977, and the first overflow was recorded in November, 1980 (Figure 4.1B). The catchment upstream of the reservoir has an area of 29 km^2, and mean annual rainfall and runoff of 1330 and 910 mm, respectively. Details of the reservoir, dam, spillway and drawoffs are given in Table 4.1. The scheme is operated so that, when the River Exe falls to a prescribed flow level of 3.157 m^{3}s^{-1} at the Thorverton gauging station about 40 km downstream, a volume of water equivalent to abstractions in excess of licences of right is released from the reservoir. A compensation release of 9.1 Ml d^{-1} (0.105 m^{3}s^{-1}) is also required at all times and is made up by seepage from springs lying just downstream from the dam, as well as water taken directly from the reservoir. Most of the released water is abstracted from the top 25 m of the water body via the upper, middle and extra drawoffs (Table 4.1). There is also provision for releasing artificial freshets and drawing off bottom water at rates of up to 45.5 and 910 Ml d^{-1} (0.527 and 10.5 m^{3}s^{-1}), respectively.

In the network of temperature monitoring stations established for the Exe basin, several sites were incorporated specifically to investigate the impact of the Wimbleball Scheme (Figure 4.1A). Water temperatures are recorded in the regulated River Haddeo at Upper Haddeo, which lies 0.4 km downstream from the dam at the end of an improved channel section, and at Lower Haddeo, which is situated a further 4.9 km downstream and close to the confluence with the main channel of the River Exe. Measurements are also made on the River Pulham, a neighbouring tributary that is similar to the River Haddeo in topographic and other characteristics, but is unregulated. Commercially available 'Cambridge' mercury-in-steel thermographs were installed at these sites during the spring of 1976, some 21 months before impounding of the River Haddeo commenced. The measuring bulbs are anchored to the stream bed to prevent exposure to the air, and the thermographs produce a continuous record of water temperature, which is checked against weekly readings taken with a standard laboratory-calibrated thermometer. No systematic drift in the performance of the thermographs has been detected, and spot readings made during particularly cold weather revealed that subfreezing temperatures recorded by the thermographs were genuine. The accuracy of this equipment is considered to be typically ±0.5°C (Stevens et al., 1975), but routine checking of the instruments used in the present study rarely reveals discrepancies between recorder and calibrated thermometer readings of greater than +0.2°C. In processing thermograph charts, hourly values of temperature were abstracted. Corrections were made for differences between thermometer and recorder readings, while occasional small gaps in the temperature records were filled by

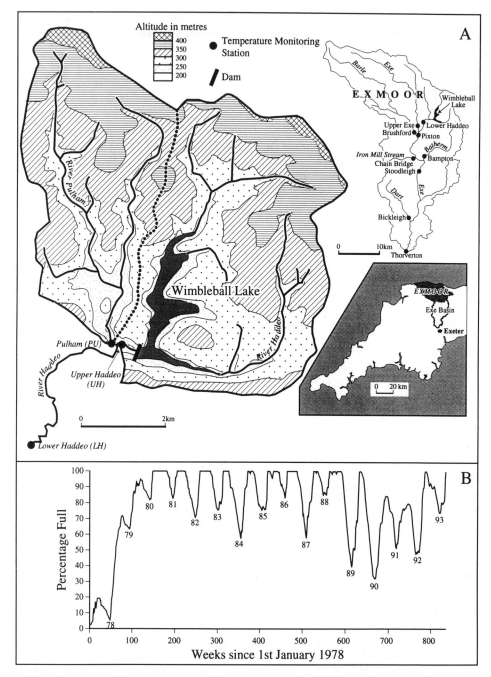

Figure 4.1 The study reservoir and location of the study sites (A) and weekly variation in drawdown of Wimbleball Lake throughout the study period (B). Numbers refer to individual summers

Table 4.1 Characteristics of Wimbleball Lake

Reservoir	
Top water level	235.61 m OD[a]
Area of surface at top water level	162 ha
Nett storage capacity	21 562 Ml
Dam	
Length of crest	300 m
Level of crest	238 m OD
River bed level	188 m OD
Spillway	
Length of crest	51.3 m
Design flood (inflow)	300 m^{3}s^{-1}
Design flood (outflow)	210 m^{3}s^{-1}
Drawoffs	
Upper	229.3 m OD, controlling 5637 Ml
Middle	220.9 m OD, controlling 14 411 Ml
Lower	208.5 m OD, controlling 20 116 Ml
Extra	Discharges into top of spillway

[a] Ordnance Datum

correlation between these and other stations in the Exe basin monitoring network. The latter, which includes one site on the mainstream above (Exe) and three below (Pixton, Stoodleigh and Thorverton) the confluence of the River Haddeo, provide data on the extent to which the effects of impoundment and regulation persist downstream. Information is also available for the major unregulated tributaries in the middle and upper Exe basin (Figure 4.1A), and water temperatures at these and the mainstream sites are recorded mainly by using purpose-built sensors linked to strip-chart recorders and, latterly, data-loggers.

Information on river flow immediately below Wimbleball Dam and further downstream, on water temperature within the lake and on the extent of reservoir drawdown was available to this study from South West Water Plc and the National Rivers Authority (NRA) South-West Region.

LOCAL EFFECTS

The impact of the Wimbleball Scheme on the thermal environment immediately downstream of the reservoir can be evaluated from a number of different perspectives.

Overall impact

It is clear that, since construction of Wimbleball Lake, the thermal regime of the River Haddeo has been altered. The main effect of reservoir construction and flow regulation

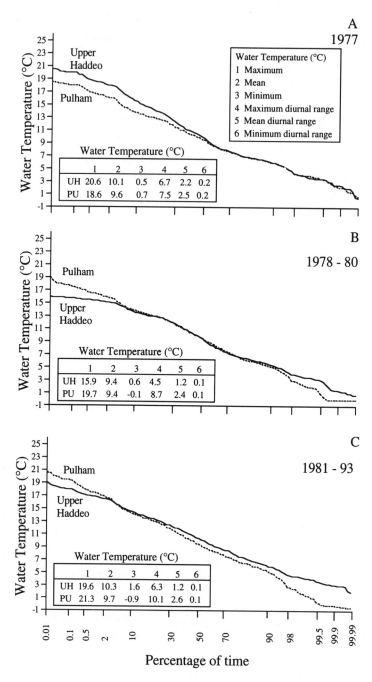

Figure 4.2 Temperature statistics and duration curves for the regulated (Upper Haddeo) and unregulated (Pulham) river stations in the pre-impoundment (A), filling (B) and post-first overflow (C) phases of Wimbleball Lake

on water temperatures has been to increase the mean value, eliminate freezing conditions, depress summer maxima, delay the annual cycle and reduce diurnal fluctuation. The evolution of this impact can be seen when hourly temperature records are used to construct duration curves and compute summary statistics for the monitoring stations at Upper Haddeo and Pulham for a 17-year period, divided into pre-impoundment (1997), filling (1978–80) and post-first overflow (1981–93) phases (Figure 4.2). Results reveal that higher summer water temperatures occurred in the River Haddeo compared with the control catchment during the year before the dam was closed. Duration curves show temperatures exceeded 60% of the time in 1977 were higher at Upper Haddeo than Pulham, and the maximum value recorded was 2°C greater at the former site (Figure 4.2A). This difference may reflect the effects of preliminary clearance of channel vegetation from the River Haddeo. In other respects, pre-impoundment thermal conditions in the Rivers Haddeo and Pulham were very similar. Data for the period 1978–80 show that water temperatures in the River Haddeo were affected while filling of Wimbleball Lake took place (Figure 4.2B). Winter and summer extreme temperatures were ameliorated and the duration curve flattened for the regulated compared with the control stream. There was no effect apparent on mean water temperature in this phase, but the maximum and mean diurnal range were significantly lower at Upper Haddeo than at Pulham. Results based on a 13-year period following the attainment of top water level in Wimbleball Lake (Figure 4.2C) suggest that impoundment and regulation have caused a long-term increase in mean water temperature of more than 0.5°C. The duration curve reveals that the reservoir has had a greater effect in increasing temperatures in the low to intermediate range than in reducing high values.

The delay in the annual cycle of water temperature change at Upper Haddeo compared with Pulham is evident from mean values for individual months, based on the post-first overflow phase (Figure 4.3A). The peak in the summer cycle occurs on average in July for the control catchment but one month later in the regulated stream. Lowest mean water temperature is recorded in February for both catchments, but the spring rise and autumn fall are delayed at Upper Haddeo, with particularly strong differences in temperature during the period from September to December. Similar contrasts between Upper Haddeo and Pulham are evident for the annual cycle of mean maximum water temperatures, although values are more markedly lower during the spring months in the regulated river (Figure 4.3B). Mean minimum values of water temperature are higher for Upper Haddeo than Pulham in all months, but the contrast is small in the period between May and July (Figure 4.3C).

Plotting of mean daily range against the mean value of water temperature for each month, based on records for the period following the first overflow (Figure 4.3D), highlights contrasts in the average seasonal variation of diurnal water temperature behaviour between the control and regulated rivers. The sites at Upper Haddeo and Pulham both exhibit a clockwise hysteresis between these variables. This trend has been recorded for several British catchments (Crisp, 1988a; Webb and Walling, 1993a) and indicates stronger diurnal fluctuations in spring than in autumn months, despite similar mean water temperatures in these seasons. However, the form of the loop for Upper Haddeo, in comparison to that for Pulham, reveals that the regulated river is characterised by only modest contrasts in diurnal water temperature behaviour throughout the year.

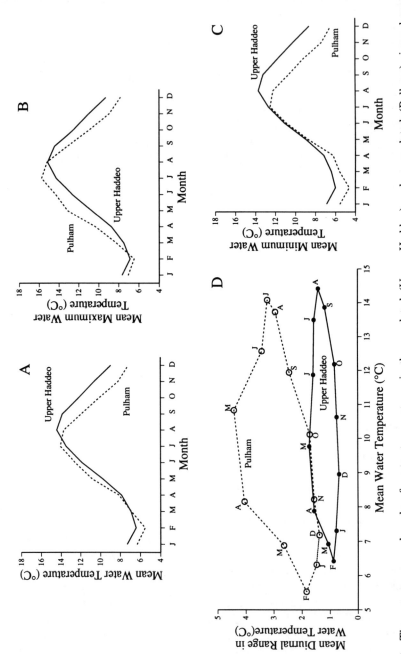

Figure 4.3 The average annual cycle of water temperatures in the regulated (Upper Haddeo) and unregulated (Pulham) rivers based on the period 1 January 1981 to 31 December 1993. Mean (A), mean maximum (B) and mean minimum (C) values plotted against time, and mean diurnal range plotted against mean temperature (D) for individual months

Inter-annual variability

Long-term records not only reveal the overall modification of the thermal regime in the River Haddeo, but also highlight considerable year-to-year variations in the impact of the Wimbleball Scheme. For example, although mean water temperatures in the River Haddeo have generally increased following regulation, annual statistics (Table 4.2) show this effect was absent for 1980 and 1982 but was greatest (+1.2°C) in 1989, 1991 and 1992. Similarly, no depression of the maximum water temperature recorded at Upper Haddeo compared with Pulham was evident in 1991 and 1992, but large reductions (>3.5°C) were observed in 1980, 1982 and 1984. Water temperature minima were higher in the regulated river than in the control catchment for every year following impoundment (Table 4.2), but the magnitude of the difference varied from +0.5°C in 1980 to +5.9°C in 1992. Annual statistics also show that mean and maximum daily ranges in water temperature recorded in the River Haddeo after closure of the dam did not attain levels observed in 1977 (Table 4.3). However, differences between the regulated and unregulated rivers also varied considerably from year to year, and ranged from values of −0.6 to −6.6°C and −0.9 to −1.9°C in the cases of annual maximum and mean daily fluctuation, respectively. The impact of regulation on the minimum daily water temperature range was less marked and less variable from year to year. In regulated and unregulated catchments alike, it is common for water temperature fluctuations to be restricted, at least for some winter days, because of overcast or freezing conditions.

Table 4.2 Annual water temperature statistics for Upper Haddeo (regulated) and Pulham (unregulated)[a]

Year[b]	Upper Haddeo			Pulham			Difference (UH − PU)		
	Mx	Mn	Mi	Mx	Mn	Mi	Mx	Mn	Mi
1977	20.6	10.1	0.5	18.6	9.6	0.7	+2.0	+0.5	−0.2
1978	15.9	9.7	0.6	17.9	9.5	−0.1	−2.0	+0.2	+0.7
1979	15.3	9.5	3.0	17.8	9.3	0.9	−2.5	+0.2	+2.1
1980	15.2	9.1	1.9	19.7	9.5	1.4	−4.5	−0.4	+0.5
1981	17.0	9.9	3.6	17.6	9.5	0.6	−0.6	+0.4	+3.0
1982	16.5	9.8	1.6	20.2	10.1	0.5	−3.7	−0.3	+1.1
1983	18.0	10.2	3.4	21.3	9.8	0.2	−3.3	+0.4	+3.2
1984	16.0	10.1	4.1	19.7	9.9	2.1	−3.7	+0.2	+2.0
1985	16.0	10.2	3.0	18.5	9.1	−0.3	−2.5	+1.1	+3.3
1986	15.9	9.2	2.5	19.4	8.8	−0.5	−3.5	+0.4	+3.0
1987	17.4	9.8	2.8	19.1	9.3	−0.9	−1.7	+0.5	+3.7
1988	17.5	10.1	4.9	17.6	9.5	1.0	−0.1	+0.6	+3.9
1989	19.6	11.5	5.7	19.9	10.3	1.8	−0.3	+1.2	+3.9
1990	18.3	11.4	6.2	19.9	10.4	2.0	−1.6	+1.0	+4.2
1991	17.7	10.8	4.1	17.6	9.6	−0.2	+0.1	+1.2	+4.3
1992	17.7	11.0	6.8	17.7	9.8	0.9	0.0	+1.2	+5.9
1993	16.7	10.3	5.0	17.4	9.4	1.1	−0.7	+0.9	+3.9

[a] Temperature values given in °C; Mx, Mn, Mi = maximum, mean and minimum values, respectively
[b] Data for the pre-impoundment (1977), filling (1978–80) and post-first overflow (1981–93) phases are grouped together

Table 4.3 Annual water temperature statistics at Upper Haddeo (regulated) and Pulham (unregulated)[a]

Year[b]	Upper Haddeo			Pulham			Difference (UH − PU)		
	Mx	Mn	Mi	Mx	Mn	Mi	Mx	Mn	Mi
1977	6.7	2.2	0.2	7.5	2.5	0.2	−0.8	−0.3	0.0
1978	3.3	0.9	0.2	7.3	2.4	0.1	−4.0	−1.5	+0.1
1979	4.1	1.2	0.2	6.9	2.3	0.3	−2.8	−1.1	−0.1
1980	4.5	1.6	0.1	8.7	2.6	0.3	−4.2	−1.0	−0.2
1981	3.9	1.4	0.3	8.0	2.3	0.3	−4.1	−0.9	0.0
1982	3.9	1.5	0.3	8.6	2.7	0.4	−4.7	−1.2	−0.1
1983	6.3	1.4	0.2	6.9	2.5	0.3	−0.6	−1.1	−0.1
1984	2.6	1.1	0.1	8.6	3.0	0.4	−6.0	−1.9	−0.3
1985	2.8	0.9	0.1	9.4	2.6	0.4	−6.6	−1.7	−0.3
1986	4.4	1.1	0.1	7.4	2.4	0.2	−3.0	−1.3	−0.1
1987	5.5	1.2	0.1	9.9	2.7	0.2	−4.4	−1.5	−0.1
1988	3.8	1.0	0.1	7.7	2.3	0.2	−3.9	−1.3	−0.1
1989	3.4	1.2	0.1	8.3	2.9	0.2	−4.9	−1.7	−0.1
1990	4.0	1.2	0.2	10.1	3.0	0.3	−6.1	−1.8	−0.1
1991	3.5	1.1	0.1	7.5	2.4	0.1	−4.0	−1.3	0.0
1992	4.4	1.1	0.2	7.6	2.5	0.5	−3.2	−1.4	−0.3
1993	3.8	1.0	0.2	7.0	2.4	0.3	−3.2	−1.4	−0.1

[a] Temperature values given in °C; Mx, Mn, Mi = maximum, mean and minimum diurnal fluctuation, respectively
[b] Data for the pre-impoundment (1977), filling (1978–80) and post-first overflow (1981–93) phases are grouped together

Plotting of individual monthly mean values of water temperature at Upper Haddeo and Pulham throughout the 17-year study period (Figure 4.4A) reveals that the effect of impoundment on the shape and timing of the seasonal cycle also varied in character from year to year. For example, the time lag between the peak of the annual cycle at Pulham and Upper Haddeo ranged from three months in 1988 and 1993 to zero months in 1981, 1984, 1987, 1989 and 1992. The seasonal cycles in monthly mean values of the daily water temperature range were broadly in phase for the regulated and control catchments in many of the years following impoundment (Figure 4.4B). However, time lags of between one and four months between the peak of this cycle at Pulham and at Upper Haddeo were observed in 1979, 1981, 1986, 1988 and 1992.

Flow-related responses

The detailed monitoring carried out in the Wimbleball study has allowed the effects of changes in the volume of flow leaving the reservoir on water temperatures in the River Haddeo to be investigated. Small adjustments to flow released from Wimbleball Lake have contrasting effects in summer and winter months. Typical responses are presented in Figure 4.5A, which shows that an increase in discharge from the reservoir causes temperature to rise in summer but fall in the winter and early spring period. Reduction in the volume of release engenders the opposite reaction of a drop in temperature

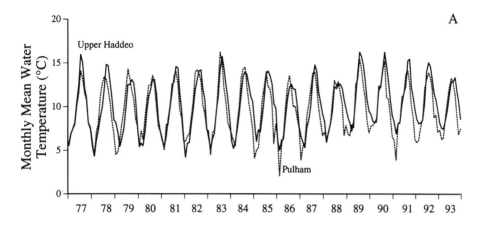

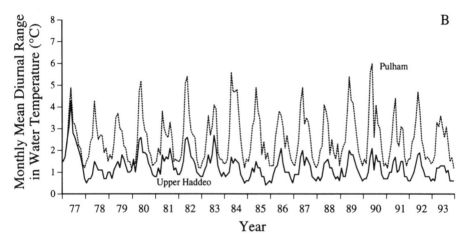

Figure 4.4 Inter-annual variability in the seasonal cycle of mean water temperature (A) and mean diurnal range in water temperature (B) for the regulated (Upper Haddeo) and unregulated (Pulham) rivers

during summer but an increase in winter months. These flow-related responses in temperature are sharp, often involving changes of 2 or 3°C, are synchronous with discharge alteration and are independent of diurnal changes recorded in the control catchment. Records for a two-day period in June 1986 (Figure 4.5B) demonstrate how closely temperature changes at Upper Haddeo may follow flow variations recorded near the tailbay of Wimbleball Lake.

Larger but more short-lived releases from Wimbleball Lake also have a varying impact on water temperature in the River Haddeo. Immediate and sharp rises in water temperature of more than 2°C occur when releases, which peak at 0.5 m^{3}s^{-1} or more, are made in summer months (Figure 4.6A). However, the increase is not sustained and water temperature declines rapidly as flow from the reservoir wanes. Larger and more complex releases made in the spring months may cause substantial temperature reductions of about 3°C at Upper Haddeo (Figure 4.6B). Responses to these events are longer-lived and vary in their detail with the pattern of flow change (Webb and

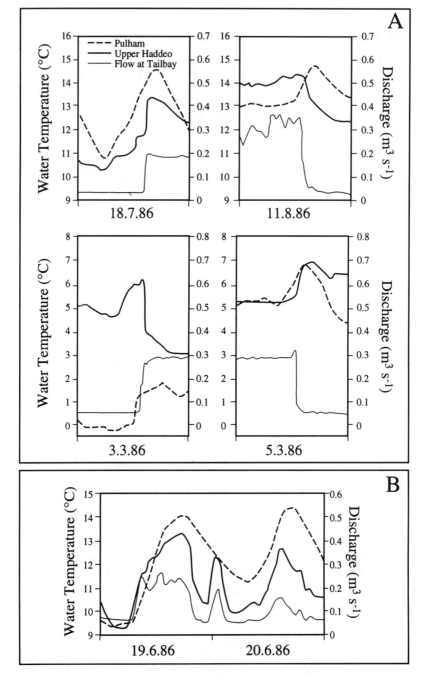

Figure 4.5 Water temperature responses in the regulated river (Upper Haddeo) to alteration of flow from Wimbleball Lake. Water temperature behaviour in the unregulated river (Pulham) is shown for comparison

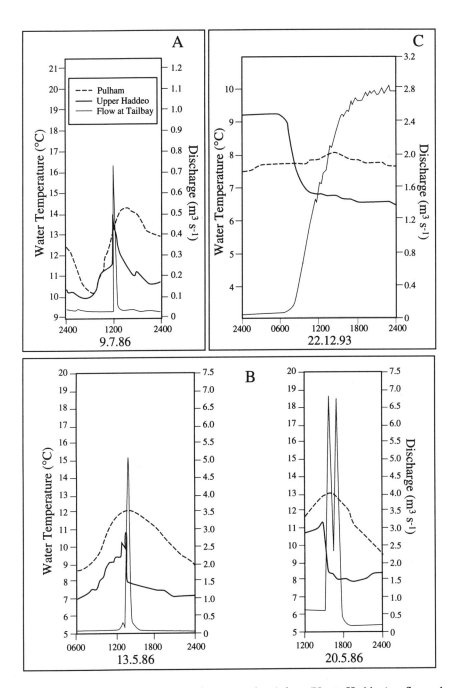

Figure 4.6 Water temperature responses in the regulated river (Upper Haddeo) to flow releases from Wimbleball Lake in summer (A) and spring (B) periods, and to overflow from the dam in winter months (C). Water temperature behaviour in the unregulated river (Pulham) is shown for comparison

Walling, 1988a). The effect of individual releases at any time of the year is to modify the normal diurnal water temperature cycle.

Flow changes associated not with artificial adjustments or releases, but with natural events, may also have a significant impact on water temperature behaviour at Upper Haddeo. An example of this effect occurred during 22 December 1993 when Wimbleball Lake began to overflow following a period of heavy rainfall (Figure 4.6C). Top water level was reached around 0600 hours, and overflow from the reservoir increased discharge from less than 0.1 to more than 2.0 $m^3 s^{-1}$ in approximately eight hours. Water temperature at Upper Haddeo in the early hours of 22 December was stable at 9.2°C, but arrival of overflow water from the reservoir caused a reduction to less than 7°C within four hours and a further decline of 0.5°C as flow built up in the Haddeo throughout the day. Temperatures in the control catchment exhibited only a small diurnal fluctuation typical of the winter period.

CAUSES OF THERMAL MODIFICATION

Past studies of regulated rivers have attributed thermal modification to a number of causes. These include greater thermal inertia and exposure of an impounded water body, release of water from particular layers (and especially the hypolimnion) in thermally well-stratified reservoirs, and alteration of the flow regime below the impoundment, which in turn causes changes in the thermal capacity of the regulated watercourse (e.g. Crisp, 1977; Ward, 1982; Petts, 1986; Palmer and O'Keeffe, 1989).

In the case of the River Haddeo, it is unlikely that thermal modification has arisen through stratification of Wimbleball Lake and release of hypolimnial water. The reservoir is not particularly deep (<50 m) and more significantly it is also equipped with a destratification system. Periodic measurements made by South West Water Plc of reservoir water temperature at different depths have shown that strong thermal gradients are absent (Webb and Walling, 1988b). Furthermore, most of the water released from the reservoir is taken from the near-surface layers. A programme of weekly spot measurements, made in the vicinity of Wimbleball Dam during 1991, 1992 and 1993, has revealed that the temperature of water discharging from the tailbay of the reservoir may be considerably in excess of temperatures recorded in the unregulated River Pulham during the summer months, but lower on occasions during the winter period (Figure 4.7A). These difference reflect the larger surface area of the lake compared with the control stream, which encourages greater solar heating in summer but higher radiative losses in winter. Water issuing from the reservoir passes through a concrete-lined channel of about 100 m length before mixing with flow routed from springs below the dam. Spot measurements show that the temperature of water released from the reservoir is generally very similar at the tailbay and at the end of the concrete-lined channel, although significantly higher values may be observed at the latter site under conditions of low flow and strong solar heating (Figure 4.7A). Such changes are consistent with the effects of channelisation reported in other studies (e.g. Parrish et al., 1978).

It is clear from measurements made within Wimbleball Lake and at the tailbay of the dam that water issuing directly from the reservoir cannot be responsible for summer lowering and winter raising of water temperatures at Upper Haddeo relative to the

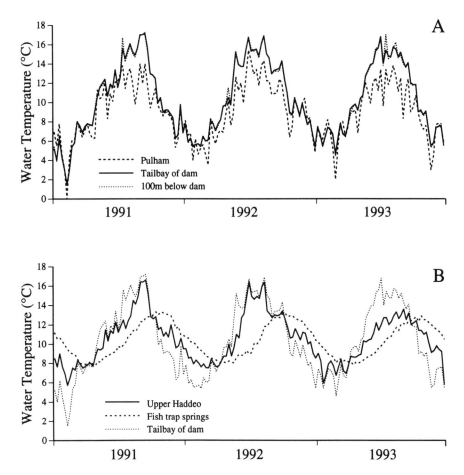

Figure 4.7 Variation in water temperature measured at the tailbay of Wimbleball Dam, downstream at the end of a concrete-lined channel and in the River Pulham (A), and in springflow, at Upper Haddeo and at the dam tailbay (B) based on weekly spot measurements in 1991–93

unregulated River Pulham. It appears that springflow, which comes from the southern side of the River Haddeo just below the dam and is counted as a substantial portion of the compensation water in the regulated stream, plays a major role in controlling water temperature behaviour recorded at Upper Haddeo. The latter site lies about 300 m below the confluence of the channels carrying reservoir runoff and springflow. Construction of the Wimbleball Scheme has greatly increased flow from the springs, which are thought to be fed by water from the base of Wimbleball Lake (C. Tubb, personal communication). The volume of springflow is typically 100 $1s^{-1}$ in winter and 70–80 $1s^{-1}$ in summer. Spot measurements show that the temperature fluctuation of the springflow throughout the year is limited (*ca.* 8–13°C) and lags well behind that of the unregulated river and reservoir runoff (Figure 4.7B). The increased groundwater flow has provided the River Haddeo since impoundment with a significant source of 'cool' runoff in summer months and 'warm' discharge in the winter period, which helps to explain not only the general modification of thermal regime, but also flow-related temperature responses in the regulated watercourse.

Essentially, temperature behaviour at Upper Haddeo is controlled by the mixing of two thermally contrasting sources of water, namely springflow and reservoir runoff. The annual march of water temperature at Upper Haddeo tends to be intermediate between that of the reservoir and the groundwater (Figure 4.7B). The latter influence acts to reduce water temperature in summer but to have the opposite effect in winter. Details of the seasonal cycle at Upper Haddeo will reflect how the volume and temperature of the two sources change within a particular year. Short-term temperature responses at Upper Haddeo can also be explained in terms of changes in the proportion of flow originating from the reservoir and the springs. An increase in the volume of relatively warm water released from the surface layers of the reservoir in the summer will increasingly cancel out the influence of a more constant inflow of relatively cool springflow, thus promoting a rise in water temperature. In winter and early spring, flow from the reservoir is cooler than the groundwater inflow and any increase in the volume released from the lake will reduce water temperature at Upper Haddeo. A reduction in runoff from the reservoir will clearly cause opposite effects to those just described. The response to larger artificial releases and natural overflows from the lake may not only reflect temperature differentials between reservoir runoff and springflow. Large releases in spring may tap somewhat cooler layers at greater depth within the reservoir, while marked increases in discharge may promote temperature changes through rapid alteration of thermal capacity in the regulated river.

Inter-annual variability in the effects of impoundment recorded in the Wimbleball Study may also be related to changes in how the reservoir has been operated with respect to release volumes. Clear relationships exist when the difference between mean temperature at Upper Haddeo and Pulham is plotted for a given month against the

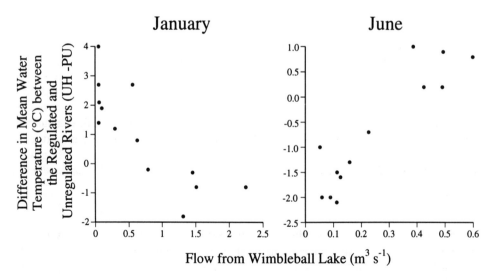

Figure 4.8 Difference in mean water temperature between the regulated (Upper Haddeo) and unregulated (Pulham) rivers in relation to mean flow from Wimbleball Lake for the months of January and June in the period 1 January 1981 to 31 December 1993

corresponding monthly mean flow from the reservoir for the 13 years of record following first overflow from the dam (Figure 4.8). In the winter months, exemplified by the results for January, the higher temperatures experienced at Upper Haddeo decline and disappear in years with greater runoff from the reservoir, which increasingly cancels out the amelioration provided by the springflow. In contrast, the relationship for June, which is typical of the summer period, shows that the cooling effect of groundwater is increasingly overriden in years when greater volumes of water are released from the reservoir.

DOWNSTREAM RECOVERY

The persistence downstream of thermal modification below an impoundment will depend on the volume and temperature of water released from the reservoir, the temperature and discharge of tributary inflows and the magnitude of heat exchange between the river, ground and atmosphere (e.g. Edwards and Crisp, 1982; Palmer and O'Keeffe, 1989). Results from the present study suggest that recovery distance depends upon which facet of the thermal regime is being considered, varies with the volume of flow from the reservoir and is more strongly influenced by heat exchange than by tributary inflow.

Tributary and mainstream changes

Water temperature statistics based on hourly records for the decade 1981–90 (Table 4.4) show that the Wimbleball Scheme clearly influences water temperature behaviour in the regulated River Haddeo downstream to the station at Lower Haddeo, which lies *ca.* 5 km distant from the dam and is situated close to the confluence with the mainstream of the Exe (Figure 4.1A). Although mean temperatures have been raised, maximum temperatures reduced, minimum temperatures ameliorated and diurnal ranges suppressed at Lower, as well as Upper Haddeo, relative to the unregulated River Pulham, there is also evidence of some recovery in thermal behaviour between the upstream and downstream stations on the regulated tributary. Mean and minimum temperatures and maximum and mean diurnal ranges for Lower Haddeo are intermediate between those for Upper Haddeo and Pulham. The effects of regulation on diurnal water temperature fluctuations appear to persist relatively weakly, since the mean value of the daily range at Lower Haddeo was almost double that recorded for

Table 4.4 Comparison of water temperature statistics for Upper Haddeo, Lower Haddeo and Pulham in the decade 1981–90[a]

Station	Temperature level			Diurnal fluctuation		
	Mx	Mn	Mi	Mx	Mn	Mi
Upper Haddeo	19.6	10.2	1.6	6.3	1.2	0.1
Lower Haddeo	19.3	10.0	0.2	7.5	2.1	0.1
Pulham	21.3	9.7	−0.9	10.1	2.6	0.2

[a] Temperature values given in °C; Mx, Mn, Mi = maximum, mean and minimum values, respectively

Upper Haddeo, and only 0.5°C below that for the control catchment (Table 4.4). Below the confluence of the River Haddeo with the River Exe, the greater flow volume and increased thermal capacity of the watercourse causes a decrease in diurnal range irrespective of the influence of regulation (Webb and Walling, 1986, 1988b).

Data collected for several sites in the Exe basin show the impact of Wimbleball runoff on the seasonal cycle of water temperature to decline relatively rapidly downstream from the dam. The phase angles of first-order harmonic functions, which were fitted to daily mean water temperature data computed from hourly records in the period 1981–84, indicate that, compared with a lag of approximately one month at Upper Haddeo, the annual march in temperature peaks only six days later at Lower Haddeo than at Pulham on the control catchment (Table 4.5). A small lead of two days in the peak of the seasonal cycle at Upper Exe compared with Pixton, which are located on the mainstream respectively above and below the River Haddeo confluence, can also be attributed to the effects of regulation. Further down the main channel at Stoodleigh and Thorverton, which are located at ca. 14 and 34 km downstream from Pixton and lie below the confluence of the large unregulated River Barle tributary, there is no evidence that the annual cycle of water temperature is affected by runoff from Wimbleball Lake.

Water temperature extremes recorded in the mainstem of the Exe are somewhat more strongly influenced by the Wimbleball Scheme than is the seasonal cycle. Maximum and minimum temperatures observed during the period 1981–84 were almost 2°C lower and more than 1°C higher at Pixton than at Upper Exe above the Haddeo confluence (Table 4.5). A clear impact of Wimbleball runoff on temperature extremes further down the main channel of the Exe is not apparent in these data. However, a comparison of water temperature records for hot spells in the summer of 1976, before regulation, and in the summer of 1983, after regulation, indicates that the influence of Wimbleball Lake can penetrate under hydrologically extreme conditions to the station at Thorverton

Table 4.5 Water temperature statistics for selected stations in the Exe basin for the period 1981–84

Station[a]	Phase angle (deg)[b]	Temperature (°C)	
		Maximum	Minimum
Pulham	32.07	21.3	0.2
Upper Haddeo[*]	49.13	18.0	1.6
Lower Haddeo[*]	35.16	18.4	1.0
Exe	28.07	22.1	−0.3
Pixton[*]	29.28	20.2	0.9
Brushford	25.87	23.6	0.1
Stoodleigh[*]	25.23	24.9	−0.4
Thorverton[*]	24.44	25.2	0.1

[a] Those indicated by an asterisk are influenced by runoff from Wimbleball Lake
[b] Derived from a first-order harmonic function of the form

$$T(x) = a[\sin(bx + c)] + m$$

where $T(x)$ is daily mean water temperature on day x of the calendar year, x is the number of days since 1 January, b is a coefficient, and a, c and m are the amplitude, phase angle and mean of the harmonic curve, respectively

almost 40 km downstream from the dam. Plotting of average values derived from the 100 hours with highest water temperature in July 1976 and July 1983 (Figure 4.9) shows that 1983 values exceeded those in 1976 at all stations *not* influenced by the Wimbleball Scheme. The reverse was true for all the sites receiving runoff from the reservoir, although the difference between 1976 and 1983 values declined with increasing distance downstream (Figure 4.9). It seems that water released from Wimbleball Lake during July 1983 increased flow volume in the Exe mainstream and, in turn, elevated thermal capacity, making it more difficult to achieve the high water temperatures recorded in the main channel during July 1976 when the Exe was unregulated (Webb and Walling, 1988b).

Reservoir releases

Tracing of temperature and flow responses as individual reservoir releases travel through the network of the Exe reveals that the extent of downstream persistence varies with size of the event. Downstream thermal equilibration appears to be quite rapid for the sometimes marked temperature responses that occur at Upper Haddeo as a result of relatively small changes in runoff from the reservoir. For example, a sharp reduction

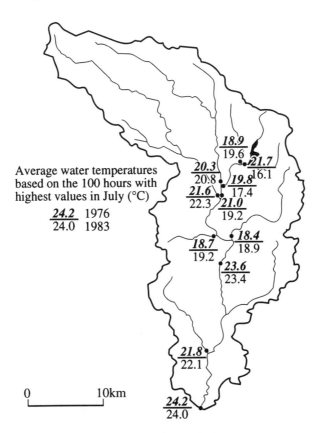

Figure 4.9 Average water temperatures at selected sites in the Exe basin for hot spells in July 1976 and 1983 based on the 100 hours with highest values

and subsequent recovery of water temperature, which was recorded at Upper Haddeo in July 1983 and was associated with alteration of flow from the reservoir, had little or no effect on water temperature further down the River Haddeo or at sites on the mainstream of the Exe, despite the fact that the flow perturbation could be easily traced downstream to Stoodleigh on the mainstream (Figure 4.10(i)–(iii)).

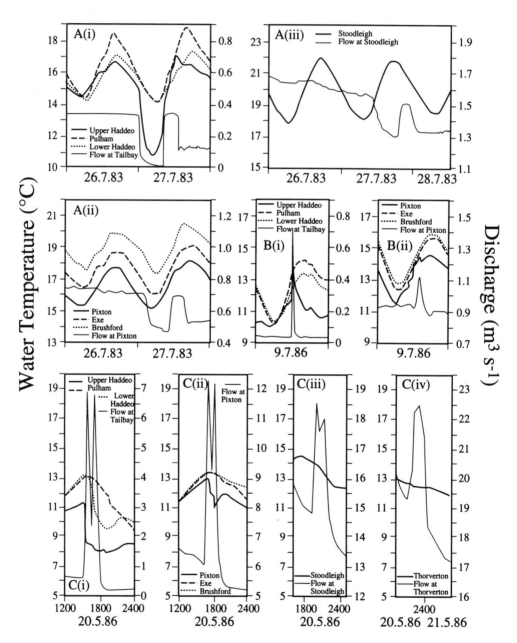

Figure 4.10 Downstream transmission of water temperature and flow changes associated with alterations and releases of flow from Wimbleball Lake. All stations are located on Figure 4.1A

Larger but short-lived releases in the summer months may modify temperatures at farther distances downstream, but with contrasting effect to that recorded close to the dam. The release on 9 July 1986, for example, generated a rapid rise of more than 2°C at Upper Haddeo (Figure 4.10B(i)). In contrast, this release did not cause a marked change in the shape of the diurnal temperature cycle at Lower Haddeo, and its arrival at Pixton was associated with a small *reduction* of water temperature (Figure 4.10B(ii)). The latter response may reflect the fact that solar heating could not keep pace with the increase in discharge and thermal capacity at Pixton caused by the flow release.

The largest releases made from Wimbleball Lake are capable of significantly changing water temperatures in the mainstream of the Exe but rarely at distances of more than 20 km from the dam. A large release on 20 May 1986, for example, caused a reduction of more than 3°C in water temperatures at Lower Haddeo as well as Upper Haddeo, and also depressed values by 2°C at the mainstream site of Pixton (Figure 4.10C(i)–(ii)). However, although this release was clearly transmitted to Stoodleigh and Thorverton, it had little effect on the water temperature responses at these stations (Fig 4.10C(iii)–(iv)).

Recovery processes

An insight into the processes leading to downstream recovery of water temperatures modified by Wimbleball Dam was obtained by modelling the River Haddeo between the stations at Upper and Lower Haddeo. A simplified temperature model developed by Jobson (1981) — which is based on a one-dimensional convective diffusion equation, employs a Lagrangian reference frame and approximates net heat exchange between the river surface and the atmosphere from air temperature and windspeed data — was employed to simulate the change in the temperature of parcels of water as they move through the River Haddeo. The model requires information on flow and temperature at the upstream boundary and for tributaries entering the modelled reach, together with measurements of water surface width and river cross-sectional area at tributary junctions and at the upstream and downstream boundaries. The model was used to predict water temperatures successfully at Lower Haddeo on three occasions during the summer of 1993 when flow conditions were steady in the regulated tributary. Hydraulic and hydrological data for these simulations were collected through field survey, while information on air temperature and windspeed variations was logged using an automatic weather station. Water temperature data were available from the thermographs at Upper Haddeo, Pulham and Lower Haddeo, and from a programme of spot measurements to establish changes in temperature for smaller tributaries entering the study reach.

The model provides information on travel time of parcels of water in the study reach, and attributes changes in temperature of the parcels during downstream transmission to the effects of dispersion, tributary inflows and heat exchange with the atmosphere. Results from the simulations (Figure 4.11) show that dispersion made a very minor contribution to the recovery of water temperatures between Upper and Lower Haddeo, and did not cause changes of more than ±0.1°C. Heat exchange with the atmosphere was seen to be the main process by which the temperature of parcels of water moving downstream from the dam is modified, and was capable of raising or lowering values by *ca.* ±4°C over a reach of no more than 5 km. Heat exchange processes tend to increase the temperature of water moving during the daylight hours

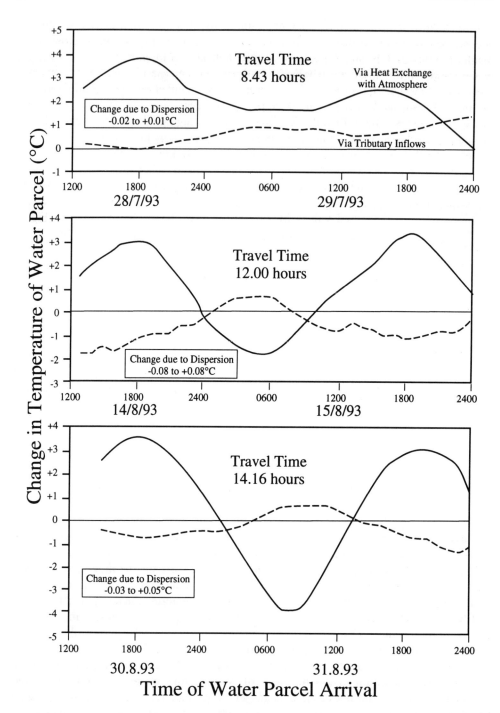

Figure 4.11 The effect of heat exchange with the atmosphere, tributary inflows and dispersion on change in water temperature between the stations at Upper and Lower Haddeo during three periods in summer 1993. Changes are derived from the simplified water temperature model of Jobson (1981)

but to reduce temperature for parcels travelling at night. This phenomenon was accentuated at lower discharge levels and greater travel times, as comparison of the results for 28–29 July and for 30–31 August indicates (Figure 4.11). The changes in water temperature caused by tributary inflows were less marked (-2 to $\pm 1°C$) than those generated by heat exchanges, but also showed clear diurnal variation. However, because the temperature changes in the tributaries tended to be out of phase with those in the mainstream, tributary inflows acted to cool water parcels moving downstream during the day but warm them at night. In consequence, the effects of heat exchange on downstream recovery was somewhat offset by those of tributary inputs.

BIOLOGICAL IMPACTS

The potential impact of the modified thermal regime in the River Haddeo on selected invertebrate and fish species can be evaluated by combining information on daily mean water temperatures at Upper Haddeo, Lower Haddeo and Pulham with published biological models derived from laboratory studies (Webb and Walling, 1993b). This approach makes the simplifying assumptions that the effects of diurnal water temperature variations on freshwater organisms may be ignored, that channel and substratum water temperatures do not differ significantly, and that the influence of temperature on development of aquatic species is independent of season and inter-population variation (Weatherley and Ormerod, 1990). It is also assumed that the effects of other environmental variables are of less importance and do not significantly interact with water temperature, and that relationships developed in laboratory experiments are valid in field situations. Some of these assumptions may not be realistic, but they allow the maximum potential influence of water temperature to be investigated.

Invertebrate development

The mayfly *Baetis rhodani* and four species of stonefly in the genus *Leuctra* were chosen for investigation because they occur in the study rivers, include spring, summer and autumn species, and are known to be variously sensitive to water temperature effects from the results of laboratory experiments. Simulation of egg and larval development, based on water temperature data averaged for the decade 1981–90, reveals that regulation has had a relatively modest impact on the growth and life-cycle of these invertebrates. Incubation periods were a few days slower at Upper Haddeo than at Pulham for eggs of *Leuctra nigra* and *Baetis rhodani* fertilised in spring, while hatching occurred a few days earlier in the regulated compared with the unregulated river for *L. geniculata* and for *B. rhodani* fertilised in late summer and autumn (Table 4.6). This effect can be attributed to the occurrence of lower temperatures in spring and higher temperatures during autumn in the River Haddeo immediately below the dam (Figure 4.3A). The stoneflies *L. hippopus* and *L. moselyi*, which are known to have a relatively insensitive relationship between hatching time and water temperature, were associated with very small differences in simulated egg incubation periods for the regulated and unregulated rivers. In addition, very little effect on average rates of

Table 4.6 Simulated egg development for selected mayfly and stonefly species at the study sites. Incubation period is expressed as number of days to 50% hatch and is based on water temperature data averaged over the decade 1981–90[a]

Species and date of fertilisation	Pulham	Upper Haddeo	Lower Haddeo
B. rhodani			
1 March	41.2	43.0	39.7
1 September	18.9	15.4	17.3
L. nigra			
1 April	41.2	44.0	40.3
1 May	32.8	36.0	32.7
L. geniculata			
1 August	23.9	22.8	23.5
1 September	27.7	23.8	26.0
L. hippopus			
1 March	39.9	40.1	38.8
1 May	30.6	32.3	30.5
L. moselyi			
1 July	24.4	25.4	24.8
1 August	24.7	23.9	24.4

[a]Relationships between incubation period and water temperature are given for *Baetis rhodani* in Elliott (1972) and for *Leuctra* spp. in Elliott (1987a)

invertebrate egg development was apparent at Lower Haddeo, because of downstream recovery of thermal regime (Table 4.6).

Predictions of the time between hatching and emergence of nymphs, again based on water temperature data averaged over the period 1981–90, show that larval development of *B. rhodani* eggs fertilised on 1 March is retarded by a few days in the regulated river because of the slower spring rise of water temperature (Table 4.7). This effect disappears for eggs fertilised on 1 May and is absent for the station at Lower Haddeo. In contrast to *B. rhodani*, which is univoltine, *L. nigra* is semivoltine and has a much longer stage of larval development. Predictions for eggs fertilised in mid-April show that the time between hatching and emergence is reduced by about 12 days at Upper Haddeo and by about 5 days at Lower Haddeo compared with the River Pulham (Table 4.7). This phenomenon is due to more favourable thermal conditions for overwintering stonefly nymphs in the regulated River Haddeo.

Inter-annual variability in the effect of regulation on thermal regime dictates that considerable deviations in the impact on invertebrate development can be expected from year to year in the River Haddeo (Webb and Walling, 1993b).

Trout development

Simulations reveal that thermal modification associated with the Wimbleball Scheme has had a larger impact on the life-cycle and growth of brown trout (*Salmo trutta* L.)

Table 4.7 Simulated period in days between hatching and emergence of invertebrate larvae at the study sites based on water temperature data averaged over the decade 1981–90[a]

Species and date of fertilisation	Pulham	Upper Haddeo	Lower Haddeo
B. rhodani			
1 March	133.4	137.1	133.8
1 May	124.7	123.4	123.4
L. nigra			
15 April	584.6	572.8	579.4

[a] Relationships between larval growth and water temperature are given for *Baetis rhodani* in Elliott *et al.* (1988) and for *Leuctra nigra* in Elliott (1987b). Larval length is assumed to be 0.6 mm on hatching for both species but 9 and 7.5 mm on emergence for *B. rhodani* and for *L. nigra*, respectively

than on invertebrate development. On the basis of water temperature data collected for the decade 1980–91, predictions show that, in the regulated river at Upper Haddeo, incubation period for eggs fertilised on 15 November is shortened by 13 days, rate of alevin development accelerated by 20 days, and weight of underyearling fish at the end of the calendar year following swim-up increased by *ca.* 30% (Table 4.8). Computation of mean instantaneous growth rate per day indicates that underyearling fish do not grow faster at Upper Haddeo than at Pulham. The greater weight attained by 31 December must therefore be a function of earlier emergence and a longer period for growth in the regulated river close to the dam. In the case of larger fish (Growth B,

Table 4.8 Simulated development of brown trout (*Salmo trutta* L.) at the study sites. Predictions are based on water temperature data averaged over the decade 1981–90[a]

Development parameter	Pulhap	Upper Haddeo	Lower Haddeo
Hatching date	16 January	3 January	13 January
Emergence date	11 March	19 February	5 March
Weight of underyearling fish on 31 December (g)	14.4	19.0	16.9
Growth A	0.014 55	0.014 50	0.014 79
Growth B	0.005 08	0.005 44	0.005 34

[a] Fertilisation of eggs is assumed to occur on 15 November, and relationships of incubation period and alevin development with water temperature used in the predictions are given in Crisp (1981, 1988b). An alevin weight of 200 mg on emergence is assumed, and weight of underyearling fish by 31 December is calculated from the relationship given in Elliott (1975) and assumes feeding on maximum rations. Growth A refers to mean instantaneous growth rate per day for underyearling trout during the period between swim-up and 31 December, while Growth B refers to the mean instantaneous growth rate per day during a calendar year for a 'standard trout' weighing 18.6 g on 1 January (Crisp, 1987).

Table 4.8), however, predictions show that growth is faster at Upper Haddeo than at Pulham, and this effect may be related to the impact of regulation in moderating high summer river temperatures (Edwards et al., 1979; Elliott, 1994). Faster egg and alevin development in the early part of the life-cycle may reflect higher winter water temperatures in the regulated stream.

Results of the simulations also show that downstream thermal recovery reduces the impact on trout development, but an effect on incubation period, emergence date and growth of larger fish is still apparent at Lower Haddeo, ca. 5 km downstream of the dam (Table 4.8). Calculation of mean instantaneous rates suggests that underyearling fish grow faster at Lower Haddeo than at the other two study sites. It should be recognised that results averaged over a decade obscure considerable inter-annual variability in the effects of the Wimbleball Scheme on trout development in the River Haddeo. Predictions based on individual years have revealed, for example, that changes in the date of emergence of trout fry at Upper Haddeo can fluctuate between an advance of 57 days and a delay of 14 days, while increases in the weight of underyearling fish in the River Haddeo can vary from less than 15% to more than 60% (Webb and Walling, 1993b).

CONCLUSIONS

The impact of the Wimbleball Scheme on downstream thermal regime is relatively modest compared with larger reservoirs and those in different climatic settings, where temperature extremes have been altered to a much greater extent, annual temperature cycles have been more severely disrupted, and the effects of modification persist for much greater distances downstream (e.g. Lehmkuhl, 1972; Soltero et al., 1973; Ward, 1974; King and Tyler, 1982). However, the thermal modifications associated with the Wimbleball Scheme are consistent with findings reported for other reservoirs in Britain that are relatively shallow and exhibit poorly developed stratification, although some of the biological ramifications, such as the effects on trout growth, will reflect the fine detail of the changed thermal regime, and differ from those reported downstream from comparable UK reservoirs (cf. Crisp, 1987). Some new perspectives on the thermal impact of reservoir construction and river regulation have emerged from the present investigation. In particular, it has been shown that disruption of groundwater circulation may be an important mechanism by which temperature changes below reservoirs are effected, that the rate of downstream recovery varies depending upon which facet of thermal behaviour is considered, and that detailed and long-term records are required to define rigorously the physical and biological consequences of regulation because of considerable inter-annual variability. The Wimbleball study has revealed that the impact on invertebrate and trout development in the regulated river arises largely because of the changes to water temperatures in midsummer, autumn and winter months. In this regard, there is a significant difference between the impact of reservoir construction and that of afforestation in the UK environment (Weatherley and Ormerod, 1990), which has been shown to affect more strongly springtime water temperatures, with different consequences for invertebrate development.

ACKNOWLEDGEMENTS

The author is priviledged to have had Des Walling as an undergraduate teacher, research supervisor and academic colleague throughout a career at the University of Exeter. It is a pleasure to acknowledge that I have benefited enormously at each of these stages from Des's inspiration and leadership, help and guidance, and generosity and friendship.

The success of any long-term investigation such as the Wimbleball study depends on the efforts of several individuals, and the help of Jim Grapes and Andy Bartram in routine maintenance of equipment and field monitoring is gratefully acknowledged. Many staff of South West Water Plc and of NRA South-West Region have kindly provided a wealth of biological, flow and other information relating to the Wimbleball Scheme, and the author is also grateful to Ian Swaffield for data used in connection with modelling of water temperature in the River Haddeo.

Thanks are also due to Rodney Fry and Terry Bacon for drafting the diagrams for this chapter, and I am grateful to Dr Trevor Crisp for constructive comments on the biological sections of this study.

REFERENCES

Battersby, D., Bass, K.T., Reader, R.A. and Evans, K.W. (1979) The promotion, design and construction of Wimbleball. *J. Inst. Water Eng. Scient.* **33**, 399–428.

Bolke, E.L. and Waddell, K.M. (1975) *Chemical Quality and Temperature of Water in Flaming Gorge Reservoir, Wyoming and Utah, and the Effects of the Reservoir on the Green River*, US Geological Survey Water-Supply Paper no. 2039-A.

Boon, P.J. (1987) The influence of Kielder Water on trichopteran (caddisfly) populations in the River North Tyne (Northern England). *Regul. Rivers: Res. Manage.* **1**, 95–109.

Brittain, J.E. and Saltveit, S.J. (1989) A review of the effect of river regulation on mayflies (Ephemeroptera). *Reg. Rivers: Res. Manage.* **3**, 191–204.

Brooker, M.P. (1981) The impact of impoundments on the downstream fisheries and general ecology of rivers. *Adv. Appl. Biol.* **6**, 91–152.

Byren, B.A. and Davies, B.R. (1989) The effect of stream regulation on the physicochemical properties of the Palmiet River, South Africa. *Regul. Rivers: Res. Manage.* **3**, 107–121.

Casado, C., Garcia De Jalon, D., Del Olmo, C.M., Barcelo, E. and Menes, F. (1989) The effect of an irrigation and hydroelectric reservoir on its downstream communities. *Regul. Rivers: Res. Manage.* **4**, 275–284.

Collings, M.R. (1973) *Generalization of Stream Temperature Data in Washington.* US Geological Survey Water-Supply Paper no. 2029-B, pp. B1–45.

Cowx, I.G., Young, W.O. and Booth, J.P. (1987). Thermal characteristics of two regulated rivers in mid-Wales, UK. *Regul. Rivers: Res. Manage.* **1**, 85–91.

Crisp, D.T. (1976) *Northumbrian Water Authority Cow Green Reservoir Releases Experiment, 23rd–28th June 1976*, Directorate of Planning and Scientific Services Report, Northumbrian Water Authority.

Crisp, D.T. (1977) Some physical and chemical effects of the Cow Green (Upper Teesdale) impoundment. *Freshwater Biol.* **7**, 109–120.

Crisp, D.T. (1981) A desk study of the relationship between temperature and hatching time for the eggs of five species of salmonid fishes. *Freshwater Biol.* **11**, 361–368.

Crisp, D.T. (1987) Thermal 'resetting' of streams by reservoir releases with special reference to effects on salmonid fishes. In: Craig, J.F. and Kemper, J.B. (eds), *Regulated Streams, Advances in Ecology*, Plenum Press, New York, pp. 163–182.

Crisp, D.T. (1988a) *Water Temperature Data from Streams and Rivers in North East England*, Freshwater Biological Association, Occasional Publication no. 26.

Crisp, D.T. (1988b) Prediction, from temperature, of eyeing, hatching and 'swim-up' times for salmonid embryos. *Freshwater Biol.* **19**, 41–48.

Delay, W.H. and Seaders, J. (1963) Temperature studies on the Umpqua River, Oregon. In:

Eldridge, E. (ed.), *Water Temperature. Influences, Effects and Control (Proceedings of Twelfth Pacific North West Symposium on Water Pollution Research)*, US Department of Health, Education and Welfare, pp. 57–75.

Edwards, R.W. (1984) Predicting the environmental impact of major reservoir development. In: Roberts, R.D. and Roberts, T.M. (eds), *Planning and Ecology*, Chapman and Hall, London, pp. 55–79.

Edwards, R.W. and Crisp, D.T. (1982) Ecological implications of river regulation in the United Kingdom. In: Hey, R.D., Bathurst, J.C. and Thorne, C.R. (eds), *Gravel-Bed Rivers*, John Wiley & Sons, Chichester, pp. 843–865.

Edwards, R.W., Densem, J.W. and Russell, P.A. (1979) An assessment of the importance of temperature as a factor controlling the growth rate of brown trout in streams. *J. of Animal Ecol.* **48**, 501–507.

Elliott, J.M. (1972) Effect of temperature on the time of hatching Baetis rhodani (Ephemeroptera: Baetidae). *Oecologia* **9**, 47–51.

Elliott, J.M. (1975) The growth of brown trout (*Salmo trutta* L.) fed on maximum rations. *J. Anim. Ecol.* **44**, 805–821.

Elliott, J.M. (1987a) Egg hatching and resource partitioning in stoneflies: the six British *Leuctra* spp. (Plecoptera: Leuctridae). *J. Anim. Ecol.* **56**, 415–426.

Elliott, J.M. (1987b) Temperature-induced changes in the life cycle of *Leuctra nigra* (Plecoptera: Leuctridae) from a Lake District stream. *Freshwater Biol.* **18**, 177–184.

Elliott, J.M. (1994) *Quantitative Ecology and the Brown Trout*, Oxford University Press, Oxford.

Elliott, J.M., Humpesch, U.H. and Macan, T.T. (1988) *Larvae of the British Ephemeroptera, a Key with Ecological Notes*, Freshwater Biological Association, Scientific Publication no. 49.

Gippel, J.C. and Finlayson, B.L. (1993) Downstream environmental impacts of regulation of the Goulburn River, Victoria. In: *Proceedings of Hydrology and Water Resources Symposium (Newcastle, June–July 1993)*, Institution of Engineers Australia National Conference Publication no. 93/14, pp. 33–38.

Jaske, R.T. and Goebel, J.B. (1967) Effects of dam construction on temperatures of Columbia River. *J. Am. Water Works Assoc.* **59**, 935–942.

Jobson, H.E. (1981) *Temperature and Solute-Transport Simulation in Streamflow Using a Lagrangian Reference Frame*, US Geological Survey Water-Resources Investigations Report, no. 81-2.

Jobson, H.E. and Keefer, T.N. (1979) *Modeling Highly Transient Flow, Mass, and Heat Transport in the Chattahoochee River near Atlanta, Georgia*, US Geological Survey Professional Paper no. 1136.

King, R.D. and Tyler, P.A. (1982) Downstream effects of the Gordon River power development, South-west Tasmania. *Aust. J. Marine Freshwater Res.* **33**, 431–442.

Lavis, M.E. and Smith, K. (1972) Reservoir storage and the thermal regime of rivers, with special reference to the River Lune, Yorkshire. *Sci. Total Environ.* **1**, 81–90.

Lehmkuhl, D.M. (1972) Changes in thermal regime as a cause of reduction of benthic fauna downstream of a reservoir. *J. Fisheries Res. Board Can.* **29**, 1329–1332.

Liu, J.K. and Yu, Z.T. (1992) Water quality changes and effects on fish populations in the Hanjiang River, China, following hydroelectric dam construction. *Regul. Rivers: Res. Manage.* **7**, 359–368.

Mackie, G.L., Rooke, J.B. and Gerrath, J.F. (1983) Effects of changes in discharge level on temperature and oxygen regimes in a new reservoir and downstream. *Hydrobiologia* **101**, 179–188.

Marchant, R. (1989) Changes in the benthic invertebrate communities of the Thomson River, Southeastern Australia, after dam construction. *Regul. Rivers: Res. Manage.* **4**, 71–89.

Moore, A.M. (1967) *Correlation and Analysis of Water-Temperature Data for Oregon Streams*, US Geological Survey Water-Supply Paper no. 1819-K.

Neel, J.K. (1963) Impact of reservoirs. In: Frey, D.G. (ed.), *Limnology in North America*, University of Wisconsin Press, pp. 573–593.

Nishizawa, T. and Yambe, K. (1970) Change in downstream temperature caused by the construction of reservoirs, Part I. *Sci. Rep. Tokyo Kyoiku Daigaku* **10**, 27–42.

O'Keeffe, J.H., Palmer, R.W., Byren, B.A. and Davies, B.R. (1990) The effects of impoundment on the physicochemistry of two contrasting South African river systems. *Regul. Rivers: Res. Manage.* **5**, 97–110.

Palmer, R.W. and O'Keeffe, J.H. (1989) Temperature characteristics of an impounded river. *Arch. Hydrobiol.* **116**, 471–485.

Parrish, J.D., Maciolek, J.A., Timbol, A.S., Hathaway, C.B. and Norton, S.E. (1978) *Stream Channel Modification in Hawaii. Part D: Summary Report*, US Department of the Interior, Fish and Wildlife Service, Biological Services Program, FWS/OBS—78/19.

Penaz, M., Kubicek, F., Marvan, P. and Zelinka, M. (1968) Influence of the Vir River Valley reservoir on the hydrobiological and ichthyological conditions in the River Svratka. *Acta Sci. Nat. Acad. Sci. Bohemoslov.–Brno* **2**(1), 1–60.

Petts, G.E. (1984) *Impounded Rivers: Perspectives for Ecological Management*, John Wiley & Sons, Chichester, 329pp.

Petts, G.E. (1986) Water quality characteristics of regulated rivers. *Prog. Phys. Geogr.* **10**, 492–516.

Pitchford, R.J. and Visser, P.S. (1975) The effect of large dams on river water temperature below the dams, with special reference to bilharzia and the Verwoerd Dam. *S. Afr. J. Sci.* **71**, 212–213.

Rader, R.B. and Ward, J.V. (1988) Influence of regulation on environmental conditions and the macroinvertebrate community in the upper Colorado River. *Regul. Rivers: Res. Manage.* **2**, 597–618.

Raney, F.C. (1963) Rice water temperature. *Calif. Agric.* **17**(9), 6–7.

Saltveit, S.J. (1990) Effect of decreased temperature on growth and smoltification of juvenile Atlantic salmon (*Salmo salar*) and brown trout (*Salmo trutta*) in a Norwegian regulated river. *Regul. Rivers: Res. Manage.* **5**, 295–303.

Soltero, R.A., Wright, J.C. and Horpestad, A.A. (1973) Effects of impoundment on the water quality of the Bighorn River. *Water Res.* **7**, 343–354.

Stanford, J.A. and Ward, J.V. (1983) The effects of mainstream dams on the physicochemistry of the Gunnison River, Colorado. In: Adams, V.D. and Lamarra, V.A. (eds), *Aquatic Resources Management of the Colorado River Ecosystem*, Ann Arbor Science Publishers, Michigan, pp. 43–56.

Stevens, H.H., Ficke, J.F. and Smoot, G.F. (1975) Water temperature influential factors, field measurement, and data presentation. *US Geological Survey Techniques of Water-Resources Investigations*, vol. 1, ch. D1.

Sylvester, R.O. (1963) Effects of water uses and impoundments on water temperature. In: Eldridge, E. (ed.), *Water Temperature. Influences, Effects and Control (Proceedings of Twelfth Pacific North West Symposium on Water Pollution Research)*, US Department of Health, Education and Welfare, pp. 6–28.

Troxler, R.W. and Thackston, E.L. (1977) Predicting the rate of warming of rivers below hydroelectric installations. *J. Water Pollut. Control Fed.* August 77, 1902–1912.

Tuch, K. and Gasith, A. (1989) Effects of an upland impoundment on structural and functional properties of a small stream in a basaltic plateau (Golan Heights, Israel). *Regul. Rivers: Res. Manage.* **3**, 153–167.

Tvede, A.M. (1994) Discharge, water temperature and glaciers in the Aurland river basin. *Norsk Geogr. Tidsskr.* **48**, 23–28.

Voeltz, N.J. and Ward, J.V. (1989) Biotic and abiotic gradients in a regulated high elevation Rocky Mountain river. *Regul. Rivers: Res. Manage.* **3**, 143–152.

Walker, K.F., Hillman, T.F. and Williams, W.D. (1978) Effects of impoundments on rivers: an Australian case study. *Verh. Int. Ver. Theor. Angew. Limnol.* **20**, 1695–1701.

Ward, J.V. (1974) A temperature-stressed ecosystem below a hypolimnial release mountain reservoir. *Arch. Hydrobiol.* **74**, 247–275.

Ward, J.V. (1982) Ecological aspects of stream regulation: Responses in downstream lotic reaches. *Water Pollut. Manage. Rev. (New Delhi)* **2**, 1–26.

Ward, J.V. and Stanford, J.A. (1979) Ecological factors controlling stream zoobenthos with emphasis on thermal modification of regulated streams. In: Ward, J.V. and Stanford, J.A. (eds), *The Ecology of Regulated Streams*, Plenum, New York, pp. 35–55.

Ward, J.V. and Stanford, J.A. (1980) Effects of reduced and perturbated flow below dams on fish food organisms in Rocky Mountain trout streams. In: Grover, J.H. (ed.), *Allocation of Fishery Resources*, FAO, Rome, pp. 493–501.

Ward, J.V. and Stanford, J.A. (1983) The serial discontinuity concept of lotic systems. In: Fontaine, T.D. and Bartell, S.M. (eds), *Dynamics of Lotic Ecosystems*, Ann Arbor Science Publishers, Michigan, pp. 29–42.

Weatherley, N.S. and Ormerod, S.J. (1990) Forests and the temperature of upland streams in Wales: a modelling exploration of the biological effects. *Freshwater Biol.* **24**, 109–122.

Webb, B.W. and Walling, D.E. (1986) Spatial variation of water temperature characteristics and behaviour in a Devon river system. *Freshwater Biol.* **16**, 585–608.

Webb, B.W. and Walling, D.E. (1988a) The influence of Wimbleball Lake on river water temperatures. *Rep. Trans. Devon. Assoc. Adv. Sci., Lit. Art* **120**, 45–65.

Webb, B.W. and Walling, D.E. (1988b) Modification of temperature behaviour through regulation of a British river system. *Regul. Rivers: Res. Manage.* **2**, 103–116.

Webb, B.W. and Walling, D.E. (1993a) Longer-term water temperature behaviour in an upland stream. *Hydrol. Processes* **7**, 19–32.

Webb, B.W. and Walling, D.E. (1993b) Temporal variability in the impact of river regulation on thermal regime and some biological implications. *Freshwater Biol.* **29**, 167–182.

Williams, O.O. (1968) *Reservoir Effect on Downstream Water Temperatures in the Upper Delaware River Basin*. U.S. Geological Survey Professional Paper, no. 600-B, pp. B195–199.

Zimmerman, M.J. and Dortch, M.S. (1989) Modelling water quality of a regulated stream below a peaking hydropower dam. *Regul. Rivers: Res. Manage.* **4**, 235–247.

Section II

SEDIMENT DYNAMICS AND YIELDS

5 A Conceptual Model of the Instantaneous Unit Sedimentgraph

KAZIMIERZ BANASIK
Department of Hydraulic Structures, Warsaw Agricultural University, Poland

INTRODUCTION

Estimates of sedimentgraphs (graphs of suspended sediment load associated with hydrographs caused by rainfall) are essential for producing sediment yield estimates, for providing input data for prediction models of sediment deposition in reservoirs, for designing efficient sediment control structures, and for water quality predictions. In these cases, and especially in the frequently considered non-point pollution models in which sediment is a pollutant and transports other pollutants, it is important to estimate sediment transport accurately during individual storms.

The concept of the sedimentgraph model, introduced by Williams (1978), has been used in previous investigations (Banasik and Woodward, 1992; Banasik and Blay, 1994). In these studies, a new definition of the instantaneous unit sedimentgraph (IUSG) is developed. The IUSG procedure is incorporated into a sedimentgraph model, based on a lumped parametric approach.

The sedimentgraph model, which was developed for predicting watershed response to heavy rainfall, consists of two parts: a hydrological submodel and sedimentology submodel (Figure 5.1). The hydrological submodel uses the Soil Conservation Service CN method to estimate effective rainfall, and the instantaneous unit hydrograph (IUH) procedure to transform the effective rainfall into a direct runoff hydrograph. The sedimentology submodel uses a form of the modified universal soil loss equation to estimate the amount of suspended sediment produced during the rainfall–runoff event, and the instantaneous unit sedimentgraph (IUSG) procedure to transform the produced sediment into a sedimentgraph.

IUSG PROCEDURE DESCRIPTION

The IUSG is defined as the time distribution of sediment generated from an instantaneous burst of rainfall producing one unit of sediment. The IUSG presented here is based on the IUH derived by Nash (1957), i.e.

$$u(t) = \frac{1}{k\Gamma(N)} (t/k)^{N-1} \exp(-t/k) \qquad (5.1)$$

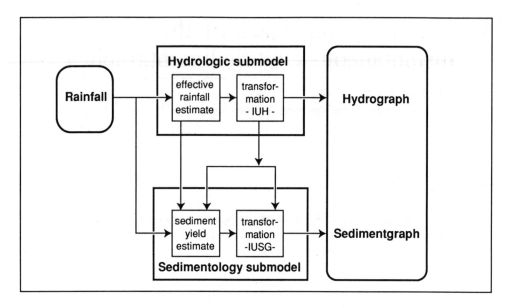

Figure 5.1 Schematic representation of the sedimentgraph model

and the first-order kinetic equation written in dimensionless form and termed the dimensionless sediment concentration distribution (DSCD):

$$c(t) = \exp(-Bt) \qquad (5.2)$$

where $u(t)$ are the ordinates of the IUH (h^{-1}), N and k are the Nash model parameters, N is the number of reservoirs, k is the retention time of reservoir, (h), $c(t)$ are the ordinates of the DSCD, B is the sediment routeing coefficient (h^{-1}) and t is time (h).

The IUSG is calculated from the formula:

$$s(t) = \frac{u(t)c(t)}{\int_0^\infty u(t)c(t)\,\mathrm{d}t} \qquad (5.3)$$

which, after substituting from Eqns (5.1) and (5.2) produces the following formula (Banasik, 1994):

$$s(t) = \frac{Bk+1}{k\Gamma(N)} [t(B+1/k)]^{N-1} \exp[-t(B+1/k)] \qquad \text{for } B > -1/k \qquad (5.4)$$

where $s(t)$ are the IUSG ordinates (h^{-1}).

The IUSG has two parameters, N and k, which are also IUH parameters, and the sediment routeing coefficient, B.

The characteristic values of the IUSG, i.e. time to peak, can be calculated from the formula:

$$t_{ps} = \frac{(N-1)k}{1+Bk} \qquad (5.5)$$

THE INSTANTANEOUS UNIT SEDIMENTGRAPH

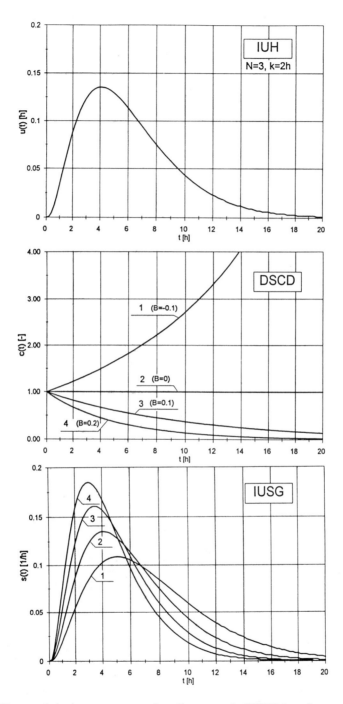

Figure 5.2 Shapes of the instantaneous unit sedimentgraph (IUSG) based on an instantaneous unit hydrograph (IUH) and dimensionless sediment concentration distribution (DSCD) with various sediment routeing coefficients B

and the maximum ordinate of the IUSG can be computed from the equation:

$$s_p = \frac{1+Bk}{k\Gamma(N)} \frac{(N-1)^{N-1}}{\exp(N-1)} \qquad (5.6)$$

where t_{ps} is the time to peak of the IUSG (h) and s_p is the maximum ordinate of IUSG (h^{-1}).

Since the respective values for IUH are calculated from the equations:

$$t_p = (N-1)k \qquad (5.7)$$

and

$$u_p = \frac{1}{k\Gamma(N)} \frac{(N-1)^{N-1}}{\exp(N-1)} \qquad (5.8)$$

where t_p is time to peak of the IUH (h) and u_p is the maximum ordinate of IUH (h^{-1}), then the ratio of the characteristic values of IUSG and IUH are computed from the formulae:

$$\frac{t_{ps}}{t_p} = \frac{1}{(1+Bk)} \qquad (5.9)$$

and

$$\frac{s_p}{u_p} = 1 + Bk \qquad (5.10)$$

It is clear that when B equals zero the characteristic values of IUH and IUSG would be the same, and the right-hand side of Eqn (5.4) assumes the form of the Nash' IUH (Eqn (5.1)). It can also be shown, from Eqn (5.9), that for $B > 0$ the time to peak of the IUSG is shorter than the time to peak of the IUH, and that the peak value of the IUSG is higher than the peak of the IUH (Eqn (5.10)). This is also schematically illustrated in Figure 5.2.

EMPIRICAL ESTIMATION OF SEDIMENT ROUTEING COEFFICIENT

One of the characteristic properties in rainfall–runoff modelling is the retention of the system or lag time, which is defined as the time elapsed between the centroids of effective rainfall and the direct runoff hydrograph. For the IUH derived by Nash (1957), the lag time (LAG) is estimated using the formula:

$$LAG = Nk \qquad (5.11)$$

For the IUSG, the lag time (LAG_s) can be calculated using the equation:

$$LAG_s = \frac{Nk}{1+Bk} \qquad (5.12)$$

Making use of Eqns (5.11) and (5.12), the routeing coefficient B can be computed using the formula:

$$B = (LAG/LAG_s - 1)/k \qquad (5.13)$$

Since the LAG, LAG_s and k can be estimated from rainfall–runoff–suspended sediment data, the routeing coefficient B can be estimated using Eqn (5.13).

Using measured data of rainfall–runoff events, the lag time can be calculated from:

$$LAG = M_{1Q} - M_{1P} \qquad (5.14)$$

where M_{1Q} and M_{1P} are the first statistical moments of the direct runoff hydrograph and the effective rainfall hydrograph (h), respectively.

Many attempts have been made to establish the relationship between the watershed lag time and basin characteristics (e.g. Snyder, 1938; Watt and Chow, 1985; Chang-Xing Jin, 1993). Respectively, based on measured data, the lag time for the sedimentgraph, LAG_s, is defined as the time elapsed between centroids of the sediment production graph (similar to the effective rainfall hyetograph) and sedimentgraph, and can be computed from the formula:

$$LAG_s = M_{1S} - M_{1E} \qquad (5.15)$$

where M_{1S} and M_{1E} are the first statistical moments of the graph of direct suspended sediment rate and the graph of sediment production (h), respectively.

Data from five small Carpathian watersheds were analysed to investigate the relationship between LAG_s and LAG.

Data description and investigation approach

Rainfall, runoff and suspended sediment data collected by the Department of Hydraulic Structures at the Agricultural University of Cracow (Madeyski, 1980) during the summer periods of 1976–79 were used in this study. Data from the five small watersheds, Poniczanka (noted as W-1), Mszanka (W-2), Kasinka (W-3), Lubienka (W-4) and Skawica (W-5), have been described and analysed previously in a joint investigation on estimating rainfall event sediment yields (Banasik and Madeyski, 1994). Some characteristics of the watersheds and measured events are given in Table 5.1.

Direct runoff hydrographs for each of the 64 observed events were obtained from recorded hydrographs after subtracting baseflow. As the baseflow suspended sediment concentrations are negligible, in comparison with the direct discharge of suspended sediment, no correction has been made for the measured sediment discharge. The data set contained incomplete rainfall records, i.e. only rainfall depth was known for most events, except for a few events from catchment W-5. To overcome the need for data on the time distribution of rainfall intensity, it was necessary to estimate the first moments of effective rainfall for each event as well as to estimate graphs of sediment production and their respective moments. These estimates were made using the following assumptions:

- Rainfall started when direct runoff began.
- Rainfall duration was equal to either (a) one time step as used in the computation ($\Delta t = 1$ h) or (b) the time to peak of measured hydrograph, T_p.

As constant values of rainfall intensity, effective rainfall intensity and sediment production during one time step were assumed in the computation procedure, the first moment of the effective rainfall, M_{1P}, in the case of the rainfall duration given in (a),

Table 5.1 Characteristics of the watersheds and data summary

Category	Watershed				
	W-1	W-2	W-3	W-4	W-5
Watershed area (km^2)	33.1	51.0	32.0	48.7	48.6
Crop land ratio	0.40	0.44	0.41	0.43	0.10
Number of events	12	12	13	16	11
Main parameters of the events					
Rainfall depth P_m (mm)					
Avg./event	22.6	17.7	27.3	27.1	16.1
Range	11–53	6–51	13–48	15–44	6–37
Effect. rainfall H (mm)					
Avg./event	7.2	1.2	2.7	3.6	2.1
Range	1.0–39.8	0.1–3.9	0.8–6.0	0.7–12.5	1.1–11.5
Floods					
Peak flow Q_p (m^3 s^{-1})					
Avg./event	5.28	1.86	2.70	3.83	5.08
Range	0.6–23	0.6–4.5	0.8–6.5	0.9–12	1.6–13
Time to peak T_p (h)					
Avg./event	11.0	5.3	9.8	13.4	11.0
Range	8–16	4–8	8–13	10–18	3–23
First statistical moments (h)					
M_1 for hydrographs					
Avg./event	12.9	7.5	10.3	14.7	10.3
Range	7.8–18	4.3–14	8.3–13	11.5–21	7.6–13
M_{1S} for sedimentgraphs					
Avg./event	11.3	6.2	9.8	14.0	7.9
Range	6.9–14	3.4–9.1	7.8–13	11–20	4.3–13

is equal to the first moment of sediment production, M_{1E}, which in turn is equal to half of the time step (i.e. $0.5\Delta t$).

In the case of the rainfall duration given in (b), it has been additionally assumed that the intensity of rainfall is changing in association with its duration according to the scheme of DVWK (1984), which is a simplification of the curve Ia of SCS (USDA-SCS, 1986), and the effective rainfall is distributed according to the Guillot and Duband (1978) formula. Sediment production for each time interval of effective rainfall ΔY_j, which was needed for estimating the first moments, M_{1E}, was computed from the following formula:

$$\Delta Y_j = Y_j - Y_{j-1} \quad \text{for } j = 1, 2, ..., n \quad (5.16)$$

where Y_j is the cumulative sediment production (Mg) computed from the relationship:

$$Y_j = a \left(\sum_{i=1}^{j} \Delta H_i \Delta P_i^{b_1} \right)^b \quad \text{for } j = 1, 2, ..., n \quad (5.17)$$

THE INSTANTANEOUS UNIT SEDIMENTGRAPH 103

where ΔP_i and ΔH_i are incremental rainfall and effective rainfall (mm), and b_1 and b are parameters established using data from 39 events from a small watershed. These were estimated in a previous investigation (Banasik and Walling, 1995) as 0.56 and 0.94, respectively. Finally, a is a fitted parameter and n is number of time intervals of rainfall duration.

Results and concluding remarks

The precision with which the centroids of the direct runoff hydrograph and sedimentgraph can be estimated depends mainly on the accuracy of the measured data (because the separated baseflow and base sediment rate represent only a small part of the storm hydrograph and sedimentgraph), but the precision of estimating the centroids

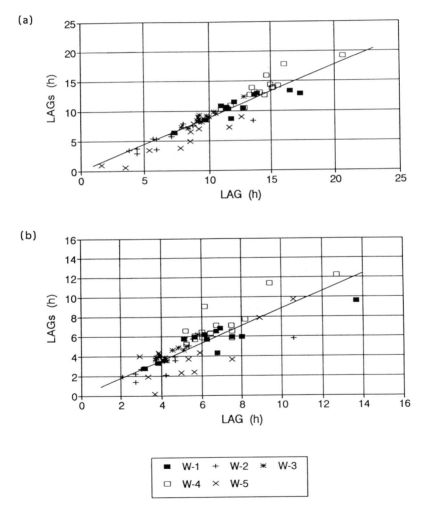

Figure 5.3 Lag time of sedimentgraphs LAG_s versus lag time of hydrographs LAG, for the five watersheds with the assumption of rainfall duration, $D = \Delta t$ (a) and $D = T_p$ (b)

of the effective rainfall, and especially of the sediment production, depends also on assumptions about their distributions. In such cases, the comparison of the first moments of direct runoff hydrographs and sedimentgraphs provides an indication of the difference in centroid location and permits some suggestions regarding the differences between LAG_s and LAG. Mean values and the range of first moments of direct runoff hydrographs (M_{1Q}) and sedimentgraphs (M_{1S}) of the measured events are given in Table 5.1. It can be observed that the average value of M_{1S} for each of the five watersheds is smaller than M_{1Q}.

The regression relationship between the first statistical moments of the sedimentgraphs and hydrographs of the 64 analysed events from the five watersheds was established in the form:

$$M_{1S} = 0.969 M_{1Q} - 0.973 \quad (5.18)$$

with the coefficient of determination $r^2 = 0.87$, standard error of estimate $SEE = 1.30$ and standard error of the coefficient (0.969) $SEC = 0.048$.

If the constant value in Eqn (5.18) is constrained to be zero, the relationship is transformed as follows:

$$M_{1S} = 0.894 M_{1Q} \quad (5.19)$$

with $r^2 = 0.86$, $SEE = 1.32$ and $SEC(0.894) = 0.014$.

The relationship between the lag time of the sedimentgraphs LAG_s and lag time of the hydrographs LAG is are shown in Figure 5.3 under two assumptions for rainfall duration. A regression relationship of the form:

$$LAG_s = a \times LAG \quad (5.20)$$

was computed for the five watersheds and for each of the two assumptions. The results of the computation are presented in Table 5.2. It can be seen from Table 5.2 that, except in one case (W-4(b)), the lag time of the sedimentgraphs LAG_s is smaller than

Table 5.2 Characteristics of the relationship of Eqn (5.20)

Watershed	No. of events	Assumption for rainfall duration	Characteristics[a]			
			a	r^2	SEE	SEC
W-1	12	(a)	0.861	0.727	1.11	0.025
		(b)	0.827	0.632	1.07	0.043
W-2	12	(a)	0.792	0.714	1.09	0.042
		(b)	0.735	0.527	1.02	0.059
W-3	13	(a)	0.929	0.913	0.408	0.015
		(b)	0.993	0.771	0.406	0.025
W-4	16	(a)	0.952	0.822	1.06	0.018
		(b)	1.018	0.690	1.12	0.039
W-5	11	(a)	0.724	0.754	1.32	0.047
		(b)	0.734	0.638	1.62	0.079
All watdesrsheds	64	(a)	0.889	0.880	1.34	0.015
		(b)	0.885	0.672	1.30	0.026

[a] a, parameter of the regression relationship (Eqn (5.20)); r^2, coefficient of determination; SEE, standard error of estimate; **SEC**, standard error of coefficient a

the lag time of the hydrographs LAG (parameter a of Eqn (5.20) is smaller than 1). For the assumption (a), i.e. rainfall duration $D = \Delta t$, the value of the parameter a varied from 0.724 for W-5 to 0.952 for W-4, and for the assumption (b), i.e. rainfall duration $D = T_p$, the value of parameter a varied from 0.735 for W-2 to 1.02 for W-4. The value of the parameter a for all of the events from the five watersheds was estimated as 0.889 and 0.885 for the assumptions of rainfall duration (a) and (b), respectively.

The analysis of the rainfall–runoff–sediment yield data from small watersheds in the Carpathians shows that: (i) a significant linear relationship exists between the lag time of the hydrographs LAG and lag time of the sedimentgraphs LAG_s, (ii) the parameter a of Eqn (5.20) may be assumed to equal 0.89 for the prediction procedure of sedimentgraphs in small ungauged watersheds in this region and (iii) new data, which will also include rainfall records, are needed to verify the relationship in this and other regions.

ACKNOWLEDGEMENT

The author would like to thank Dr M. Madeyski from the Agricultural University of Cracow for his generosity in supplying data from the five Carpathian watersheds.

REFERENCES

Banasik, K. (1994) *Model Sedymentogramu Wezbrania Opadowego w Matej Zlewni Rolniczej (Sedimentgraph Model of Rainfall Event in a Small Agricultural Watershed)* (in Polish with an English summary), Publications of Warsaw Agricultural University — SGGW, 120pp.

Banasik, K. and Blay, D. (1994) An attempt of modelling suspended sediment concentration after storm event in an Alpine torrent. In: *Dynamics and Geomorphology of Mountain Rivers*, Ergenzinger, P. and Schmidt, K.-H. (eds), Springer-Verlag, Berlin, pp. 161–170.

Banasik, K. and Madeyski, M. (1994) Single event sediment yield from small Carpathian watersheds. *Proc. State of the art of River Engineering Methods and Design Philosophies (St. Petersburg)*, 1, pp. 371–380.

Banasik, K. and Walling, D.E. (1995) Predicting sedimentgraphs for a small agricultural catchment. Manuscript submitted for publication to *Nordic Hydrology*.

Banasik, K. and Woodward, D.E. (1992) Prediction of sedimentgraph from a small watershed in Poland in a changing environment. In: *Saving Threatened Resources — in Search of Solution (Proc. of the Irrigation and Drainage Session at Water Forum'92)*, Engman, T. (ed.) ASCE, New York, pp. 493–498.

Chang-Xing Jin (1993) Determination of basin lag time in rainfall–runoff investigations. *Hydrol. Processes* 7, 449–457.

DVWK (1984) *Regeln zur Wasservirtschaft, Arbeitsanleitung zur Anvendung von Niederschlag-Abfluss-Modellen in kleinen Einzugsgebieten*, part II, *Synthese*, Verlag Paul Parey, Hamburg, paper 144.

Guillot, P. and Duband, D. (1978) Function de transfert pluie-debit sur des bassins versante de l'ordre de 1000 km. *Société Hydrotechnique de France (Paris)*, Session of 21–22 November 1978 (personal communication).

Madeyski, M. (1980) Transport of suspended load in little river basins. *Proc. IV Int. Seminar on Transport and Sedimentation of Solid Particles (Wroclaw-Trzebieszowice, Poland)*, vol. II, pp. ES.1–8.

Nash, J.E. (1957) The form of the instantaneous unit hydrograph. *Hydrol. Sci. Bull.*, 3, pp. 114–121.

Snyder, F.F. (1938) Synthetic unitgaph. *Eos*, 19, 447–454.

USDA-Soil Conservation Service (1986) *Urban Hydrology for Small Watersheds* (TR-55), Washington, DC.
Watt, W.E. and Chow, K.C.A. (1985) A general expression for basin lag time. *Can. J. Eng.* **12**, 294–300.
Williams, J.R. (1978) A sediment graph model based on the instantaneous unit sediment graph. *Water Resour. Res.* **14** (4), 659–664.

6 Magnitude and Frequency of Fluvial Sediment Transport Determined From Recent Lake Sediment Cores

R.H.F. CURR
Faculty of Applied Science, Bath College of Higher Education, UK

INTRODUCTION

The increasing length of accurate records of sediment transport in rivers throughout the world provides opportunities to test the approach to the problem of the magnitude and frequency of sediment transport made by Wolman and Miller (1960). Even after several decades of high-quality data collection with estimates of sediment transport based on continuous monitoring of discharge and sediment concentration, the longest accurate records of sediment transport are short when viewed in the context of the possible return intervals of high-magnitude flood events (Webb and Walling, 1982, 1984). Other approaches to the problem have involved the use of lake sediments containing mineralogical, radiometric (lead-210 or caesium-137) or magnetic markers to establish broad patterns of historical sediment transport and deposition in relation to variations in annual rainfall or changes in land use, with further interpretation based on the identification of current sources of sediment established from monitoring programmes (Foster *et al.*, 1985, 1990). Many lakes have high trap efficiencies (Dendy, 1974) and, under certain circumstances, the particle size of sediments deposited in lakes can provide a sequential record of the magnitude of the flood events delivering that sediment. Of particular relevance in the context of this chapter is an example given by Laronne (1990), who identified a number of sedimentation units in two pond-sized reservoirs in the northern Negev, and related those units to high-intensity rainfall inputs, although the absence of runoff records and the rapid drying up of the reservoirs between rainfall events limited the scope of the interpretation. In contrast to the relatively short accurate records of sediment transport, good-quality records of daily precipitation of over 100 years in length are relatively common. The occurrence of long records of precipitation in the vicinity of catchments with established rainfall, runoff and sediment transport records, and which drain to lakes with high trap efficiencies, should provide an alternative approach to the magnitude and frequency debate through the detailed linkage of sedimentation units to flood-generating precipitation.

Sediment and Water Quality in River Catchments. Edited by I.D.L. Foster, A.M. Gurnell and B.W. Webb.
© 1995 John Wiley & Sons Ltd.

THE CORSTON BROOK CATCHMENT AND UPPER LAKE AT NEWTON PARK

The Corston Brook is a small north-east-facing agricultural catchment (ST693641) ultimately draining into the Bristol Avon between Bath and Keynsham, UK (Figure 6.1). The upper 4.1 km^2 of the catchment drains into an artificial lake constructed in the 18th century, and sediment transport from this area was continuously monitored from September 1977 until April 1980. The annual rainfall in the water year 1977–78 was 819 mm, while 860 mm was received in 1978–79 (Curr, 1984). The catchment is underlain by Jurassic strata, which dip gently to the south at about 2°, the lowest strata exposed being the limestones and clays of the White and Blue Lias. These are succeeded by the largely impermeable Lower and Middle Lias Clays, which in turn are overlain by the porous Midford Sands. The latter, with a thin protective cap of Inferior Oolite, dominate the small escarpment that forms the southern boundary to the

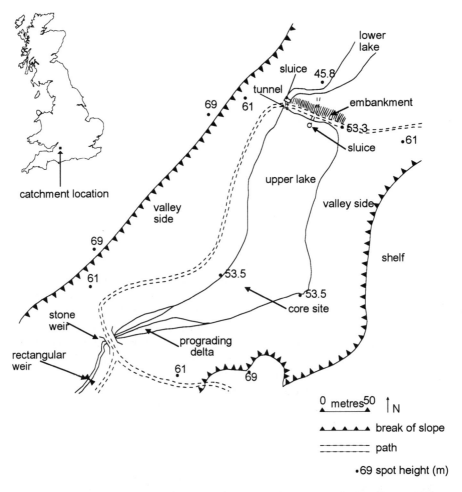

Figure 6.1 The location of the catchment and site of the upper lake at Newton Park

catchment, and has given rise to a spring line at the junction with the Lias Clays. The northern and western watersheds are formed by a series of outliers from the escarpment. The escarpment slopes and the outliers therefore have the best-drained soils, essentially brown earths (Sherborne and Somerton units), with gleyed soils in the valley bottom (Martock and Evesham units) (Findlay et al., 1976). A survey completed in 1977 at the beginning of the study showed that permanent pasture in the valley bottom dominated the land use (51%), with arable cultivation of largely wheat, barley and cattle beet amounting to just over 43% of the catchment area. In addition there were small areas of woodland (3.4%) and the houses, roads and gardens of the village of Stanton Prior (2.4%).

It is of particular significance in the context of this study that the cultivated soils at the base of the southern escarpment (surface water gley soils of the Long Load unit) were often affected by high water tables and free-flowing springs in the winter, when fields were relatively unprotected by vegetation. Winter rainstorms on these wet areas frequently produced saturated overland flow, resulting in important soil erosion from the fields. Such runoff conditions almost invariably resulted in overbank flood events with significant sediment transport, which discharged through a tunnel into an artificial lake of about 0.82 ha, completed between 1760 and 1765, which has never been dredged (Figure 6.1). Originally this lake was over 5 m deep at its distal end but by 1980 it had shallowed to just over 2 m at the deepest point, and a small delta had developed at the proximal end. Based on a tacheometric survey in 1980, which was related to over 100 soundings of sediment and water depth, and estimates of sediment bulk density and the organic content of the deposits, it was estimated that just over 24 000 t of sediment had accumulated in the lake since the construction of the dam (Curr, 1984). The mean annual sediment transport over the (approximate) 218 years of the lake's existence was estimated at 111 t, which compared closely with monitored rates of transport in 1977–79 of 99.7 t yr^{-1} or a sediment yield of about 25 t km^{-2} yr^{-1}. The lake sediments were extremely anaerobic, with no evidence of bioturbation of the sediments.

THE LAKE SEDIMENT CORES

When floodwater and its associated suspended sediment entered the head of the upper lake, it was observed that the storm water advanced across the lake as a discrete front, seemingly as a type of density current. 'Older' lake water was displaced from the upper to the lower lake in advance of the storm water. Monitoring of sediment transport in individual flood events and the analysis of bulk suspended samples from flood events of varying magnitude demonstrated that both total suspended sediment loads and the particle size of the suspended sediment increased with flood magnitude (Curr, 1984). The rate of transfer of water and sediment down-lake depended upon the rate of input of water and sediment into the lake, being faster in large flood events, and relatively slow in small events. Under both conditions, the suspended sediment was projected into an essentially quiescent environment in which sedimentation would bear some relationship to Stokes' law. Shallow excavations in the delta at the head of the lake (Figure 6.1) revealed that the sediment was deposited in discrete layers of varying particle size and thickness. At any given location in the lake, one might therefore expect to find that the sediment column was composed of alternating bands of coarse and fine sediment, whose thickness and particle size were directly related to the magnitude of the transporting flood event.

Cores were extracted from the lake at the site shown in Figure 6.1, by means of a Russian auger. The total depth of the sediment at the core site was 260 cm, lying beneath water a little over 20 cm in depth. The top 60 cm of the sediment was too poorly indurated to prevent disturbance during extraction and, although the particle size of this near-surface material was estimated, it was not included in the analysis discussed below, which therefore considers sediment deposited in at least 80 cm of water and 170 m into the lake. Since the cores were extracted from beneath about 20 cm of water, work was carried out from duckboards and the water surface was used as a datum for estimating 10 cm overlaps between each of the 50 cm auger holes. The extraction sequence was started at the base of the core and successive core sections were removed from separate but immediately adjacent holes. When first withdrawn, the core sections showed buff, grey and black horizons, the position of which was recorded with a waterproof pen on the plastic gutters in which the cores were stored. These enabled tentative correlations between core sections to be made in the field.

In the laboratory, each core section was subdivided into 1 cm increments of sediment. This was regarded as the thinnest section of core that would give a reasonable quantity of sediment to work with. Slices were transferred to numbered, weighed 50 ml conical Pyrex flasks. Any fine organic debris was digested with hydrogen peroxide, and any large pieces of leaf or twig were removed, after washing any attached sediment back into the flask with distilled water. When digestion was complete, the sediment was washed with distilled water and allowed to settle for 24 h, after which the overlying liquid was removed with a fine pipette. The sediment was then dried in an oven for 24 hours at 105°C. After determining the weight of the dried sediment, it was dispersed in about 30 ml of 1% sodium hexametaphosphate and was then wet-sieved through a 45 µm sieve. This sieve size was chosen because it would retain a reasonable quantity of sediment even in those core slices with the finest particulate matter. The sediment retained in the 45 µm sieve was washed into small weighed dishes by means of fine distilled water jets. The water was evaporated from each dish in an oven at 105°C. Each dish was then weighed to determine the weight of sediment from each core slice coarser than 45 µm and the percentage weight calculated.

Figure 6.2 shows the histogram of the particle size analysis of the lake core, which is clearly composed of a sequence of layers in which the coarse fraction of each slice retained in the 45 µm sieve ranged between 5% and 67% of the original weight of sediment in that increment of core. Inspection of these coarse sediments was carried out using using a microscope and graticule, and this revealed a clear spectrum of change. Slices where a high percentage of sediment was retained had poorly sorted sediment with particle sizes of between 45 and 500 µm in diameter. These slices often contained pulverised freshwater mollusc shells. In contrast, in samples where a small percentage of the sediment was retained, it was very well sorted, with particle sizes between 45 and 65 µm, and these sometimes included whole examples of the delicate mollusc shells. These observations supported the notion that the sedimentary layers revealed in the core resulted from variations in the magnitude and turbulence of the transporting flood events.

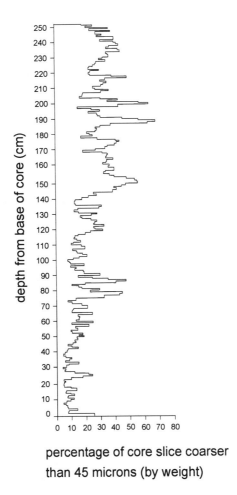

Figure 6.2 The particle size analysis of the complete lake sediment core

Interpretation of the sequence shown in Figure 6.2 presents many problems. For example, the large variation in sediment load transport by flood events of different magnitudes would result in the accumulation of sediment at different rates. Thus very thin layers of fine sediment (perhaps merely fractions of a millimetre) might be deposited by individual small events and, given the dominance of small events in the spectrum of flood responses, accumulations of perhaps 1 cm of fine sediment might represent sedimentation from many sequential small flood events. In contrast, many centimetres of coarse sediment might be deposited by a single high-magnitude event. The chronology of the core was thus not readily apparent, and, while dating using lead-210 and caesium-137 would give useful markers, they would not provide detailed interpretation of the wide variation in particle size within the core. Since only recent runoff records existed for the Corston Brook, and given the long daily rainfall records available at nearby Batheaston, it was clear that the best option for detailed interpretation lay with the meteorological conditions that generated the flood inputs.

THE RAINFALL RECORD

Daily precipitation data have been collected at Batheaston since 1874. This reliable continuous record is now held by Wessex Water Plc. The collection site is in a valley about 9 km to the east of the Corston Brook catchment, and this 103-year record provided an opportunity to establish a chronology for the sediment core. Although very close to Newton Park, the location on the other side of the Avon valley was a slight disadvantage, but as a check, the record from Batheaston was compared with an almost equally long record from Failand House, about 18 km to the west of the Corston Brook. In only two relatively recent cases were significant discrepancies between the records discovered, and these could be corrected using the extensive network of rain gauges maintained by Wessex Water. Although rainfall intensities were not available from Batheaston, in the context of the generally low-intensity rainfall received in the UK this was not considered to be a major disadvantage. Indeed, in an analysis of rainfall and runoff in the Corston Brook catchment over the three-year study period (Curr, 1984), the discharge prior to each flood event (antecedent flood discharge, Q_a) and storm rainfall depth explained 85% of the variation in the maximum discharge generated in the stream, and rainfall intensity was not identified as a significant factor. It is accepted, however, that some high-intensity rainfall events will have occurred in the long Batheaston record, and that both these and snowmelt episodes may provide some sources of error in the modelling of stream response. In addition, some errors were generated because long-duration rainfall inputs could be split between two consecutive days, and details of how these errors were mitigated in the structure of the stream response simulation program are given below.

DEVELOPMENT OF THE STREAM RESPONSE SIMULATION PROGRAM

In order to estimate stream response to daily rainfall inputs over the 103-year period of the Batheaston rainfall record, a physically based simulation model was developed. This was founded on the three years of rainfall and runoff response data collected in the Corston Brook catchment, using an autographic rain gauge, two standard rain gauges and a Munro vertical stage recorder, the latter calibrated to a sharp-crested rectangular weir. Experience gained during this period suggested that estimation of discharge became very difficult when large events overspilled from the channel onto the floodplain. Under these conditions, however, most of the sediment transportation was restricted to the axis of the stream channel, and deposition tended to occur on the vegetated floodplain (Marriott, 1992). For these reasons, stage data were used to calibrate the simulation program.

Statistical analysis of the rainfall and runoff data collected during the three-year study period had established the importance of the antecedent flood discharge as a proxy variable for soil moisture conditions, and it was clear that to be effective the model would require an index of the daily soil moisture status of the catchment. Antecedent precipitation indices (API) provide an indication of how much precipitation has fallen in a preceding period, and can be used to provide an estimate of soil moisture conditions (Gregory and Walling, 1973). Daily exponential decay rates related to season were used in the calculation of the API values, and were similar in magnitude to those quoted by Gregory and Walling (1973), ranging from 0.87 for August to 0.96 for January and February. These seasonal decay rates were applied to all daily rainfall data from the three-year monitoring programme in excess of 2 mm, and a factor F(API, season) was developed and applied to the daily precipitation inputs. This factor summarised evapotranspiration losses and other seasonal factors.

If daily rainfall inputs to the catchment were of sufficient magnitude, it was possible that saturated overland flow would develop. The area contributing such runoff would increase in concert with both the magnitude of the daily rainfall and the soil moisture status, here represented by the value of the API, and it was important that this effect be reflected in the model. Separate regression analyses were therefore carried out on those flood events in which no observed overland flow occurred and those events in which surface runoff was observed within the catchment. This analysis indicated that, with increasing values of the API, smaller daily rainfall inputs were required to initiate

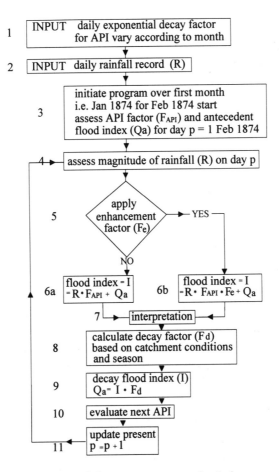

Figure 6.3 The structure of the stream response simulation computer program

overland flow and thus enhance stream response, and a simple enhancement factor F(e) was applied in the simulation model where such runoff conditions were indicated by the rainfall depth and the precipitation history.

Between flood events, the stream stage decayed at rates that reflected the intensity of the flood event, and the season in which it occurred. Using data from the three-year calibration period, stage decay rates over sequential 24 h periods were estimated and used in the calculation of the daily flood index (estimated stage). The flood index was therefore the sum of the antecedent stage of the stream, and the increase in stream stage resulting from the daily rainfall input, including the possible development of overland flow.

Finally, errors resulting from the division of storm inputs between two consecutive days as a result of reading the Batheaston raingauge at a fixed time were also addressed. The worst possible case would occur when a rainfall event was evenly divided between two days. Trial calculations had shown that, given a common API and antecedent flood discharge, the stream response resulting from a single rainfall input was within 10% of that resulting from a split input. This was because as a result of the split input the API and antecedent flood discharge on the second day would be much larger than on day 1, and the flood generated by the second rainfall input would result from interaction with a wetter catchment and from a higher antecedent flood discharge. Details of the structure and calibration of the simulation program can be found in Curr (1984, pp. 361–427).

The stream response simulation program is summarised in Figure 6.3. When checked against the daily rainfall inputs for the three-year study period, the correlation coefficient between 161 simulated stage maxima and the actual daily stage maxima was

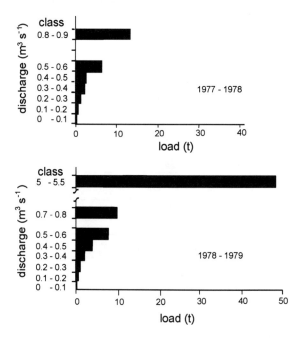

Figure 6.4 Suspended sediment transport in relation to maximum discharge of flood events

0.9335. Daily precipitation data referring to January 1874 were used to initiate the program, giving a reliable API and antecedent flood discharge for the start of February 1874, and the program ran continuously until 1976, identifying each peak flood index resulting from rainfall over 2 mm in depth, as well as the largest flood index in each month and in each year.

The final aspect of modelling the stream response from the daily rainfall record concerned the simulated rate of accumulation of the sediment in the lake as a result of the simulated flood events. The relationship between the actual suspended sediment transport and the maximum discharge of flood events during the three-year monitoring period is shown in Figure 6.4, and it is clear that there is an exponential increase in the transport of suspended sediment with increasing flood magnitude. Consequently, on the basis of all peak flood indices, the program estimated the rate of sediment accumulation in the lake by squaring the flood index (I^2). In this way the effects of large flood events relative to small flood events were emphasised, and the values of I^2 for each year were summed to give a cumulative total representing the annual increment of sediment into the lake. This provided a crude vertical axis to the sequence of flood indices.

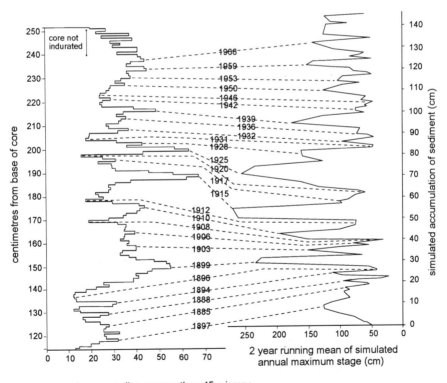

Figure 6.5 Correlation between the lake core particle size histogram and the two-year running mean of the simulated annual maximum flood stage

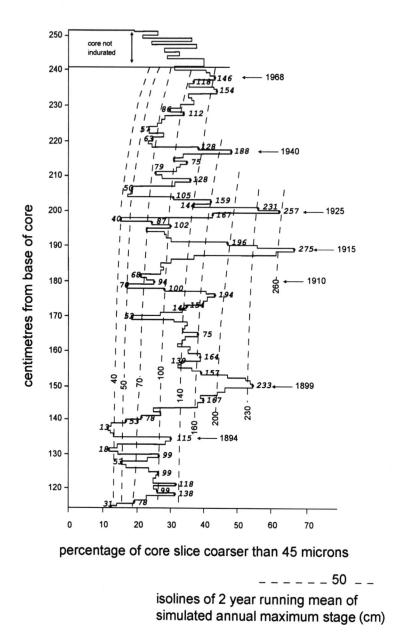

Figure 6.6 Two-year running mean of the simulated annual maximum flood stage superimposed on the lake core histogram

COMPARISON OF THE SIMULATED STREAM RESPONSE DATA WITH THE PARTICLE SIZE ANALYSIS OF THE LAKE CORE

Given that the Batheaston rainfall data referred to the period 1874–1976, it was clear that the pattern of simulated stage data would have to be compared with the upper portion of the core. Evidence based on the sediment transport achieved by the largest floods monitored suggested that the 1 cm incremental slicing of the core had been rather coarse, and consequently a two-year running mean of the simulated stream stage data was compared with the histogram of the particle size analysis of the lake core (Figure 6.5). Not only was there great similarity in the general form of the two records, but even the micro-relief of the peaks, troughs and shoulders were apparent. The two records correlated extremely closely, with 51 peaks identified and a correlation coefficient of 0.9531. Much independent verification of the largest flood events was obtained from the records of local newspapers, although care had to be exercised in eliminating floods on the River Avon, which could have been generated outside the local area. Given the very close correspondence between the core and the simulated flood index or stage, it was therefore possible to date accurately the sedimentary sequence of the core. This opened the way to an examination of the rates of accumulation of sediment in the lake and an analysis of the magnitude and frequency of suspended sediment transport over the 103-year period of the record.

By superimposing the values of the two-year running mean of the simulated stage on the corresponding peaks in the suspended sediment histogram, it was possible to insert a

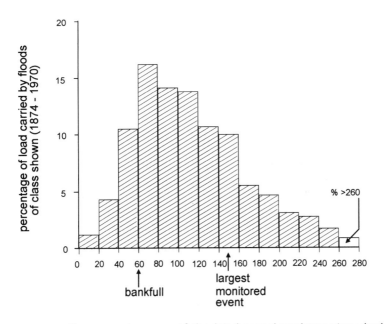

Figure 6.7 Percentage of the sediment load deposited in the lake between 1874 and 1970 by simulated flood events in the classes shown

series of isolines onto the core diagram (Figure 6.6). It was then possible to carry out a simple analysis by counting the total number of core slices to the right of the superimposed isolines. Clearly 100% of the sediment was deposited by events greater than 0 cm of stage, while only 1 cm of sediment was contributed by the largest recorded event, greater than 260 cm of simulated stage. Further confirmation and justification for this approach was provided by the curved nature of the isolines, which clearly indicate the prograding of the lake head delta through time.

Figure 6.7 shows the percentage of the suspended sediment load carried by particular classes of flood event. Over 44% of the suspended sediment was transported by events with simulated stages of between 60 and 120 cm, bankfull at the catchment outlet being estimated at 55 cm of stage. Indeed, 16% of the sediment was transported by events with simulated stages between 60 and 80 cm, and this could be termed the dominant or most effective discharge for suspended sediment transport (Benson and Thomas, 1966).

CONCLUSION

This analysis suggests that events of moderate magnitude and frequency are the most influential in sediment transport in this catchment, and this would seem to support the suggestions made by Wolman and Miller (1960). Simulated stages of between 60 and 80 cm are somewhat above bankfill, and this suggests that the bulk of the sediment is entrained during events which involve overland flow, much of this sediment currently derived from ploughed fields near the spring line at the foot of the escarpment and associated outliers near the head of the catchment. These results compare favourably with those obtained on the River Creedy near Exeter, Devon, by Webb and Walling (1982) where the dominant discharge was identified as being just below bankfull stage. It is not greatly significant that the dominant discharges in these catchments show relatively minor differences in relation to the bankfull stage, for the dominant discharge will reflect not only the proportion of sediment derived from sources such as river banks and field surfaces, but, more particularly, variations in the runoff conditions within the catchments, and the location and extent of the areas suffering soil erosion, as well as the land use and the complexities of sediment delivery. In many agricultural catchments like the Corston Brook, those areas suffering marked soil erosion are generally under arable cultivation, and thus the dominant discharge for sediment transport may also reflect cultivation practice, and will consequently change in response to new farming techniques and changes in land use. In lowland catchments in the UK the dominant discharge for sediment transport will probably lie within the middle range of event size, that is, flood events of moderate magnitude and frequency.

REFERENCES

Benson, M.A. and Thomas, D.M. (1966) A definition of dominant discharge. *Int. Assoc. Sci. Hydrol. Bull.* **11**, 76–80.

Curr, R.H.F. (1984) The sediment dynamics of the Corston Brook. Unpublished PhD thesis, University of Exeter.

Dendy, F.E. (1974) Sediment trap efficiency of small reservoirs. *Trans. ASAE* **17**, PE 5, 898–901

Findlay, D.C., Tomlinson, P.R. and Cope, D.W. (1976) *Soils of the Southern Cotswolds*. Memoirs of the Soil Survey of Great Britain, England and Wales, Rothampsted Experimental Station, Harpenden, Herts.

Foster, I.D.L., Dearing, J.A., Simpson, A.D., Carter, A.D. and Appleby, P.G. (1985) Lake catchment based studies of erosion and denudation in the Merevale catchment, Warwickshire. *Earth Surf. Processes Landforms* **10**(1), 45–68.

Foster, I.D.L., Grew, R. and Dearing, J.A. (1990) Magnitude and frequency of sediment transport in agricultural catchments: a paired lake catchment study in midland England. In: *Soil Erosion on Agricultural Land*, Boardman, J., Foster, I.D.L. and Dearing, J.A. (eds), John Wiley & Sons, Chichester, pp. 153–171.

Gregory, K.J. and Walling, D.E. (1973) *Drainage Basin, Form and Process*, Edward Arnold, London.

Laronne, J.B. (1990) Probability distribution of event sediment yields in the northern Negev, Israel. In: *Soil Erosion on Agricultural Land*. Boardman, J., Foster, I.D.L. and Dearing, J.A. (eds), John Wiley & Sons, Chichester, pp. 481–492.

Marriott, S. (1992) Textural analysis and modelling of a flood deposit, R. Severn, UK. *Earth Surf. Processes Landforms* **17**, 687–697.

Webb, B.W. and Walling, D.E. (1982) The magnitude and frequency characteristics of fluvial transport in a Devon drainage basin and some geomorphological implications. *Catena* **9**, (1/2), 9–23.

Webb, B.W. and Walling, D.E. (1984) Magnitude and frequency characteristics of suspended sediment transport in Devon rivers. In: *Catchment Experiments in Fluvial Geomorphology*, Burt, T.P and Walling, D.E. (eds), Geo Books, Norwich, pp. 399–415.

Wolman, M.G. and Miller, J. (1960) Magnitude and frequency of forces in geomorphic processes. *J. Geol.* **68**, 54–74.

7 Sediment Sources and Their Environmental Controls

COLIN CLARK
Charldon Hill Research Station, Bruton, Somerset, UK

INTRODUCTION

In this review of previous studies and a description of case study material, 'sediment' is taken as that fraction of the load in a river that is suspended by the movement of water in the channel. Bedload is not considered here, although there is inevitably some interchange between suspended load and bedload depending on stream velocity. The environmental controls on sediment sources include both natural factors, such as rainfall depth and intensity, and anthropogenic factors, such as land use, agriculture and construction work. The case study material described here largely derives from the time the author was a research student under the supervision of Des Walling at Exeter. A small catchment in east Devon was instrumented in order to assess sediment yield and sediment source. Subsequently, work by the author on land-use change and flooding has revealed how a knowledge of the rates of change of bankfull discharge can be useful in the rapid determination of the relative importance of land use in controlling sources and rates of surface and bank erosion. Finally, some suggestions are made as to the ways in which over 25 years of scientific study of sediment sources can be applied to be of benefit of humankind.

PREVIOUS STUDIES OF SEDIMENT SOURCES AND THEIR CONTROLS

The published literature on sediment sources and their controls is a rapidly growing one. Walling (1971) gives a useful review of the earlier literature, and this is developed further by Gregory and Walling (1973), Clark (1978), Walling (1988) and Pye (1994). The specific theme of soil erosion is reviewed by Morgan (1979) and by Boardman *et al.* (1990). These reviews and edited volumes point to the importance of the sediment problem in relation to the loss of soil and, while there are some valuable site-specific and single-technique approaches, the use of several parallel and interrelated techniques is the exception rather than the rule.

A central problem in the study of natural systems is to reconcile the apparent slow nature of the processes with their spectacular results. In the brief time-span in which hydrological studies are often made, it would be rash to assume that the measured output of sediment truly represents the state of the system, in both time and space. Hence a study of sediment sources and their controls necessitates a multidisciplinary

---

Sediment and Water Quality in River Catchments. Edited by I.D.L. Foster, A.M. Gurnell and B.W. Webb.
© 1995 John Wiley & Sons Ltd.

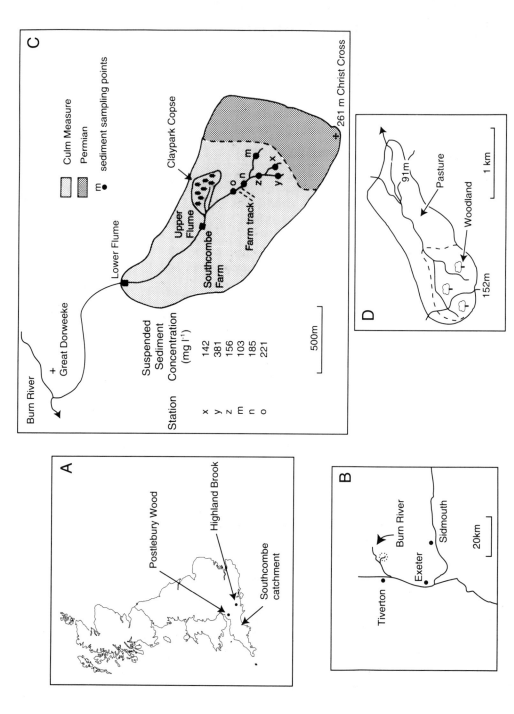

Figure 7.1 Location of catchments referred to in the text (A) with site details of the Southcombe catchment (B and C) and the Postlebury Wood catchment (D)

SEDIMENT SOURCES AND THEIR CONTROLS

approach for three important reasons. First, measurements are needed of both inputs and outputs, and their reliability depends upon good understanding and use of the monitoring equipment. Secondly, analysis and interpretation of the resultant data have to recognise the limitations of accuracy and the spatial and temporal sampling problems that exist. Thirdly, there is the problem of whether a small catchment study represents catchment behaviour on a wider scale.

DESCRIPTION OF THE STUDY SITES

Figure 7.1 shows the location of all study sites and local details of the Southcombe catchment in East Devon whose catchment area is 0.81 km^2. The geology of this site consists of Carboniferous greywackes, sandstones and shaley mudstones of the Culm Measures, and are mapped as the Bude Formation (British Regional Geology, 1975). This formation is overlain by Permian breccias and conglomerates, which weather to give thin, easily erodible soils. The catchment relief is typified by steep slopes of up to 35°, although slopes of 5–10° are more common. The land use consists of sheep and cattle grazing on the better pasture and cereal production in one small area on the shallower slopes. Pigs were housed mainly in sties at Southcombe Farm, although for a brief period were allowed to graze in a field near the farmhouse. Some 95% of the catchment area was grassland, with *Bromus erectus* and *Brachypodium* sp. present. The woodland area covered 4% of the catchment, and included hazel (*Corylus* sp.), alder (*Alnus* sp.) and oak (*Quercus rober*) with some holly (*Ilex* sp.). The field survey began in April 1975 and continued until the end of 1976. During this time England experienced some of the driest weather since records began in 1727 (Grindley, 1976).

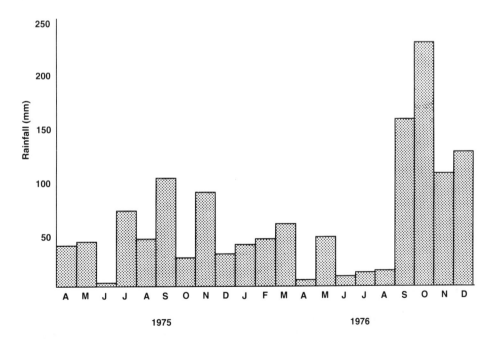

Figure 7.2 Rainfall for the Southcombe catchment, 1975–76

There is no doubt that this was an unusual time for hydrological processes, although towards the end of the study period rainfall increased considerably (Figure 7.2). In 1976, 928 mm were recorded in the catchment, which represents about 93% of the 1941–70 average for this area.

In further studies, two catchments were surveyed for changes in bankfull discharge in relation to land-use change, since this would provide surrogate data on the impact of land use on channel erosion and, therefore, sediment yield. The first is the Postlebury Wood catchment of 1.3 km^2 area on the Oxford Clay in east Somerset, where the change of land use is from deciduous woodland in the headwater area to pasture land downstream (Figure 7.1D). The second is the Highland Brook, a 47 km^2 catchment in the New Forest, where the geology consists of Plateau Gravels, which occur in small areas on the divides and which overly Barton Sand and Barton Clay (Figure 7.1A). Here, the land use is heathland in the upper part of the catchment, which changes to mixed woodland in the lower reaches.

METHODS OF STUDY

A major objective of the Southcombe catchment study was to identify sediment sources and their environmental controls. Full details of the methods and instrumentation are given in Clark (1978) and will only be briefly mentioned here. Two trapezoidal flumes with water-level recorders and sediment samplers were installed in the catchment (Figure 7.1C). Between the flumes were placed 150 erosion pins in order to assess bank erosion. Autographic rainfall was also measured on site. For surface erosion, a Gerlach trough (Gerlach, 1967) was installed. However, owing to the very dry weather, this did not collect any overland flow and eroded sediment during the study period.

With the exception of one or two small storms, all storm events were sampled for both river flow, sediment concentration and storm-period rainfall. The erosion pins were measured at frequencies determined by these events. In addition, water samples were collected from tributary areas and overland flow locations during the winter of 1976, and also bulk sediment samples were collected in order to determine the particle size distribution, organic content and the clay mineralogy of the suspended sediment. Sediment samples from bed, bank and surface soil were also taken and analysed for the same properties.

Bankfull discharge in the other two catchments surveyed was estimated using the slope-area method at suitable sites along the channel reaches. Measurements of the channel were made every 0.1 m and at smaller intervals in the case of small channels. The roughness factors were determined using the methods described in both Barnes (1967) and Chow (1959). The land use at the sites was obtained from 1:25 000 maps and the boundaries checked in the field. The contemporary data were also compared with previous maps of the area in order to establish any major land-use changes of which there were none.

SEDIMENT YIELDS AND BUDGETS

The erosion budget was calculated for 1976, during which both sediment samplers were in operation. Figure 7.3 shows the difference in storm-period sediment yield between the two sampling stations.

SEDIMENT SOURCES AND THEIR CONTROLS

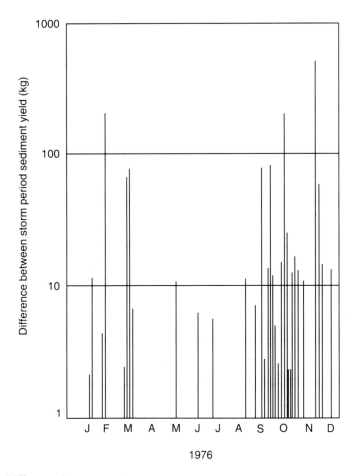

Figure 7.3 Difference in storm-period suspended sediment load between the upper and lower gauging stations of the Southcombe catchment

The sediment budget is:

$$SYi = Bs + Hs \pm St \qquad (7.1)$$

where SYi = sediment yield increase between the gauging stations, Bs = bank erosion, Hs = Hillslope erosion and St = Sediment storage.

Since there was no measured hillslope erosion, the quantities in all components of this budget could not be estimated. Sediment storage changes were assumed to be zero since material is passing through the system all the time and there was no overbank discharge and subsequent floodplain deposition. The hillslope erosion was calculated as the difference between the two sediment yield estimates and bank erosion. A total of 32 storm events were included in the calculation of the erosion budget (Table 7.1). An estimated 1.051 t of bank material was eroded between the two gauging stations. The difference in sediment yield between the two sites was 1.4550 t. This does not include the unmeasured sediment discharge (Colby, 1957), or the coarse material in the banks.

Table 7.1 The sediment transport budget of the Southcombe catchment (tonnes)

Upper sampling station	8.507
Lower gauging station	9.962
Difference	1.455
Bank erosion	1.051
Difference (direct hillslope inputs)	0.404

This material accounts for about 1–2% in rivers and in the sediment-laden overland flow is negligible.

These results can be compared with those of Harvey (1974), who showed that gulley/bank erosion was the sole sediment source in his study area and those of Imeson (1974), who showed that bare ground and gulleys were important sources of sediment. Much material remained stored in Imeson's catchment so that a budget, in percentage terms, could not be calculated, although between a quarter and a third of the suspended solids was suggested to derive from the channel banks. Data collected from a wholly forested area in Luxemborg (Duijsings, 1987) gave a figure of 53% for channel sources and 47% for the valley-side slopes. Clearly local land-use factors will be important.

In order to assess how representative the channel reach budget is of the whole catchment, the results of the present study were extrapolated to include the total channel length within the catchment. The total length was 1640 m and the calculated contribution made by the channel was 5.386 t. Since the recorded yield at the lower sediment sampling station was 9.962 t, then the relative contribution of bank erosion was estimated to be about 54% This is lower than the 72% measured between the two gauging stations, so the contribution from the upper section of the catchment only was then calculated on an areal basis from:

$$Be_t = Be_l + Be_u \tag{7.2}$$

where Be_t = percentage total bank erosion, Be_l = lower catchment area × percentage bank erosion of lower catchment and Be_u = upper catchment area × percentage bank erosion of upper catchment.

The result gave an estimate of 52% bank erosion contribution for the upper catchment area. This result implies that the instrumented-reach erosion-pin data may have overestimated the contribution of the channel to the sediment yield. At the same time the upper catchment is partly forested whereas the land use in the section under detailed study was almost all grassland. Clearly in any sediment budget study there are bound to be certain areas of unmeasured bank erosion: there are errors in the stream load data on account of both discharge measurement errors and water sampling errors. There is no absolute method available to determine the size of these potential errors, but every care was taken to make accurate and frequent measurements. This included an assessment of the intra-storm-period loads and a correction for the slight inaccuracies of the sediment samplers due to fixed-point sampling.

Table 7.2 Sediment erosion budgets of the Southcombe catchment (tonnes)

	Bank source	Hillslope source	Suspended load	Error[a] (%)
Organic	0.04204	0.04848	0.2037	56
Inorganic	1.0089	0.35558	1.2513	9
Total	1.05094	0.40406	1.4550	0

[a] Error (%) = (100 × (bank + hillslope) − total load)/total load

ORGANIC AND INORGANIC CONTRIBUTIONS

Further insights into the erosion budget were made by dividing the suspended solids into organic and inorganic fractions as shown in Table 7.2. Although the inorganic budget is within an acceptable level of accuracy, bearing in mind the measurement errors outlined above, the organic sediment budget needs some explanation. The main source of error was found to be an exceptionally high value of 14% for the organic-matter content of the suspended solids. During high flows the percentage of organic material was lower, so that the load-weighted mean percentage of organic matter was calculated from 33 observations where percentage organic matter, discharge and suspended solids concentration data were available. The result gave a value of about 9%, but in spite of this adjustment the organic budget still remained in error by $ca.$ 40%. In order to balance this budget the organic percentage for surface soil should be around 28%, which suggests that much of the hillslope erosion takes place near the banks where there are considerable quantities of animal manure present. When adjustments for this are made, the discrepancy in the organic budget becomes 20%. Bearing in mind the small quantities of material involved, and the likely measurement errors, these results illustrate some of the problems in calculating a reliable sediment budget under the prevailing catchment conditions.

SEDIMENT SOURCES

In any hydrological study there are always more items to measure than the time allows, so it is important to carry out a general survey that covers a wider area but is, as a result, less detailed. In the present study an attempt was made to assess the relative importance of the tributaries as sources of sediment, and also to isolate the effect of a farm track, which ended by the stream.

Figure 7.1C shows the location of the tributaries, sampling points and the average concentration of suspended solids based on measurements taken during 10 major storm events. The relative contribution of tributaries X and Y cannot be directly compared and interpreted in terms of mechanical erosion unless the total weight of sediment leaving both tributaries can be calculated. This is possible by comparing the mass flow from each tributary and the concentration at points 'x', 'y' and 'z'. Xc, Yc and Zc are the concentrations associated with discharges Xq, Yq and Zq. Considering the mass flows before and after the confluence of tributaries X and Y:

$$XqXc + YqYc = Zc(Xq + Yq) \tag{7.3}$$

it can be shown that:

$$Xq/Yq = (Yc - Zc)/(Zc - Xc) \tag{7.4}$$

Applying Eqn (7.4) to all situations where values of Xc, Yc and Zc are available, the ratio of discharges from the two upper tributaries can be calculated. A mean figure for $Xq:Yq$ was found to be 1.703:1.0. Therefore the adjusted concentrations should be 243 (Xc) and 381 (Yc), where the units are mg l^{-1}. This means that there is 1.567 times more mechanical erosion from tributary Y than tributary X. The difference in erosion may be related to the presence of a gulley on tributary Y just above the confluence, and also to the absence of trees and understorey vegetation in the catchment area of tributary Y. It would be expected that this difference would be most marked on the rising limb of the hydrograph, since sediment transport is greatest there. This could be interpreted in terms of sediment availability, greater bank erosion, or greater sheet erosion during the earlier stages of the storm event.

A comparison can also be made between the sediment concentrations above and below the track area. As there were no significant inputs of water into this area, the concentrations are almost directly comparable. Figure 7.4 shows the ratios above:below this track during parts of 1976. There were hardly any storms during the period April to August, so that the progressive protective effects of the vegetation cover could not be documented. However, there is a clear distinction between the ratios before and after the growth phase. This shows the area to be an important source of suspended sediment.

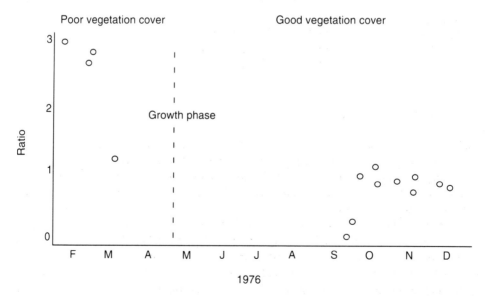

Figure 7.4 Ratio of sediment concentrations above and below the farm track (Figure 7.1C) in relation to time and vegetation cover

These results have shown both the complexity of sediment sources and the inherent difficulty in making realistic assessments of the relative contribution of the banks and valley-side slopes as sediment sources.

FINGERPRINTING SEDIMENT

The problems encountered in trying to assess sediment sources prompted the use of fingerprinting techniques in order to isolate the sources of sediment by the analysis of particle size and mineralogy of suspended sediments. Particle-size analysis was done using the hydrometer method and the sedimentation balance method, having carried out pre-treatment of the samples. Clay mineralogy was determined using an X-ray diffractometer and obtaining the refractive peaks which indicate the clay minerals that are present. These two fingerprinting techniques have several distinct problems. First, if there is a variable source area, then the sediment properties will also vary and the results of the analysis will be difficult to interpret. Secondly, the suspended sediment properties will probably be a mixture of at least two source areas, and direct comparison with samples obtained from any one of these areas could be misleading. Thirdly, selective erosion and transporation of particular size fractions will further confound interpretation of the results.

Particle size

Very little data exist on particle-size characteristics, notable exceptions being those studies undertaken by Walling and Moorehead (1987) and Vaithiyanathan *et al.* (1992). Inferences regarding sediment sources from particle size are hard to determine. From the results of the present analysis, measures of sorting, skewness and kurtosis were employed. However, when these measures were themselves further analysed, the results failed to reveal anything new about sediment sources in the study catchment.

That the source of sediment must be a mixture of both bed and surface soil material is suggested when the average percentages of silt, for five groups of sediment, are compared (Table 7.3). Particle size was investigated by Doty and Carter (1965), who

Table 7.3 Comparison of the average percentages of the silt fraction in possible source sediments with suspended sediment

	Coarse silt (%)	Medium silt (%)	Fine silt (%)	Very fine silt (%)
Surface soil	19	18	15	14
Bed material	15	24	15	15
Bank material	17	10	8	3
Surface runoff	10	10	28	30
Suspended sediment	20	22	22	8

suggested that at high concentrations the particle-size characteristics are similar to those of the source material. At high concentrations of sediment there were higher percentages of silt and lower percentages of clay. The results of the present study did not confirm the findings of Doty and Carter (1965) (Figure 7.5). Furthermore, by dividing up the silt fraction into coarse, medium, fine and very fine silt, the opposite trend was identified (Figure 7.6). These results appear to show that the data consist of two populations: the first tends to behave like clay and decreases at high concentrations, and the second behaves as silt and increases at high concentrations. If this is the case, then samples at high and low concentrations would be expected to behave differently. Figure 7.7 suggests that during storm events the sediment comes from an increasingly wider area. This could mean an expansion of the area of bank covered by the floodwater on the rising limb of the hydrograph, or an increasing contributing area of runoff on the valley-side slopes. When the sediment supply from the land and river banks is exhausted, providing that river discharge is still high, then discharge of suspended bed material greater than 63 µm increases because of the entrainment of these particles. Hence the sediment output is a reflection both of erosional processes

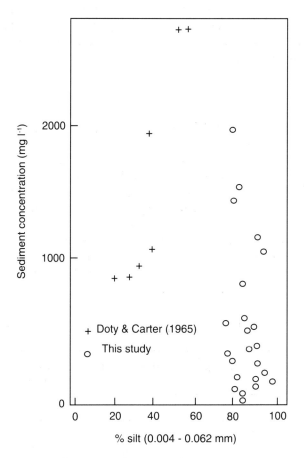

Figure 7.5 The percentage silt content related to sediment concentration from the Southcombe catchment study in comparison with the data of Doty and Carter (1965)

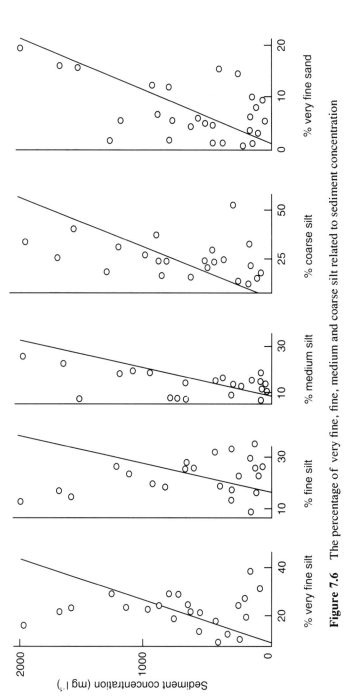

Figure 7.6 The percentage of very fine, fine, medium and coarse silt related to sediment concentration

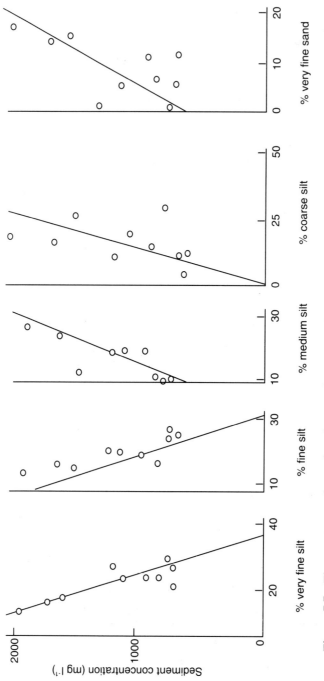

Figure 7.7 The percentage of very fine, fine, medium and coarse silt related to sediment concentration above 600 mg l^{-1}

along the banks and valley-side slopes and also of the hydraulic transport of bed material.

Clay mineralogy

Comparison of sediments from possible source areas (Klages and Hsieh, 1975) can help in the evaluation of sediment sources from within a drainage basin. By using an X-ray diffractometer a trace was produced that can then be compared with reference material (Brown, 1961). Figure 7.8A shows the results for two sediment samples taken from surface runoff. The peak at 8.75° is of special note since it is the dominant peak for the clay mineral illite. The peak at 12.1° is probably kaolinite, although as Cosgrove (1973) pointed out this only amounts to 1% of the clay in Permian strata. Results from surface soil on Permian and Culm Measures were revealing (Figure 7.8B). The peak at 8.75° was absent from the Culm Measures. At the same time this peak was present in bank material on the Culm Measures. Subsequent heating of the samples to 850°C in a furnace destroyed the illite owing to the loss of hydroxyl water from the crystal lattice between 200 and 600°C. In south-west England Cosgrove (1973) showed that illite made up 49% of the total clay present.

These results tend to support and amplify the conclusions drawn from the results of the determination of organic matter present in suspended sediments. The clay mineral illite is present in abundance in Permian strata and it appears to be a marker mineral in the banks. It is absent from soils developed on the Culm measures but it is present on Permian-derived soils, so that it is not possible to separate both channel and land-surface sources using this technique. However, erosion of the Permian soil in the headwater region did not produce sediment that was enriched with illite as compared with sediment passing the lower gauging station.

From an analysis of the erosion budget and sediment properties, it appears that the main source of sediment is the river banks. In areas near the main channel that are free of vegetation, rainsplash erosion must be an important process. Another important source of sediment is the headwater area, which is dominated by grassland.

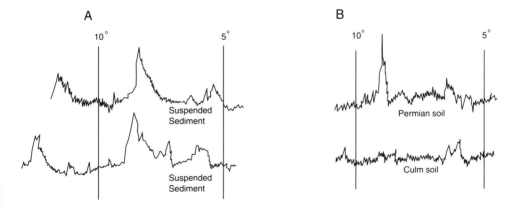

Figure 7.8 X-ray diffractometer traces for suspended sediment (A) and soil on Permian and Culm Measures strata (B)

ENVIRONMENTAL CONTROLS ON SEDIMENT PRODUCTION

The previous section isolated sediment sources and began to consider the background factors such as land use and agricultural practices on erosion and sediment transport. This section will focus on a statistical analysis of physical factors controlling sediment transport, such as rainfall depth and intensity, antecedent conditions and season. These are important because they interact with the catchment and help provide an understanding of how sediment is eroded from certain source areas. Multiple regression analysis was used to sort out the important controlling variables. There are three dependent variables that were predicted in order to identify the importance of a range of environmental factors. These were the storm-period suspended solids load, the discharge-weighted mean concentration for the storm, and the peak suspended solids concentration. The first variable is the most important since it is a direct measure of the amount of sediment lost from the catchment and provides an initial estimate of the rate of denudation provided that both bedload and solution loads are added. The second variable gives the average concentration of sediment, weighted according to the discharge. The value will depend not only on the characteristics of the catchment but also on the sampling interval, because the sample concentration is assumed to remain constant between samples. Generally speaking, the discharge-weighted mean concentration is higher for summer storms when baseflow is low. It is a measure of the intensity of erosion per unit volume of runoff. Independent variables are measures of the rainfall input, catchment condition and catchment runoff, and are listed in Table 7.4 together with the three dependent variables.

Table 7.4 Dependent and independent variables used in the multiple regression analysis

1 Storm period load
2 Discharge weighted mean concentration
3 Peak concentration mg/l
4 Storm period rainfall depth (mm)
5 Rainfall energy in Joules/cm/cm^2 based on 30 minute duration rainfall
6 Rainfall energy calculated as in 5 above, the result of which is multiplied by the maximum 30 minute intensity
7 As 6 except using 1 hour duration rainfall
8 As 6 except using 2 hour duration rainfall
9 Rainfall momentum in kg m^{-2} mm^{-1} s^{-1} calculated using 2 hour duration rainfall
10 As 9 using 1 hour duration rainfall
11 As 9 using 30 minute duration rainfall
12 Sine day index
13 Cosine day index
14 Peak discharge l/s above the pre-storm baseflow level
15 Antecedent precipitation index (using a decay constant of 0.9)
16 Soil moisture deficit for the 75 mm rooting zone
17 Soil moisture expressed as a percentage
18 Storm runoff volume
19 A total runoff peak flow product evaluated using peak $Q.\Sigma Q$
20 Total runoff produced by the storm multiplied by the peak discharge above the preceding level, evaluated using peak $Q.\Sigma(Q-Qn)$ when Qn is the pre-storm discharge

SEDIMENT SOURCES AND THEIR CONTROLS

Table 7.5 The order of entry of variables into the multiple regression equations and the corresponding coefficient of determination

Coefficient of determination (%)	Sediment load (variable 1)	Discharge-weighted mean concentration (variable 2)	Peak concentration (variable 3)
40			14
50			15
60		14	
70		5	12
80	14	12	5
	18	15	
90	5	18	18
>90	12		
	16	13	13
	13	16	15

Multiple regression analysis

As each of the seven selected variables entered the regression equation the multiple correlation coefficient increased. Table 7.5 shows the order of entry of variables into the equation as related to the coefficient of determination, while Table 7.6 gives the multiple regression equations. The coefficient of determination is above 70% in all three cases, with the prediction of storm-period load reaching 79% explained variance. Peak discharge above the preceding level appears to be the most important environmental control on erosion, not only because it is a function of rainfall intensity and duration, but also because it will influence how much channel-side area will be swept clean of sediment. Storm runoff volume is clearly important for similar reasons. The rainfall energy is related to rainfall intensity and was chosen instead of variables 9–11 (Table 7.4), on account of its slightly higher correlation with the discharge-weighted mean concentration. Variable 12 is the sine index, which is a seasonal function in that it roughly follows the annual march of baseflow, being highest at the end of winter and lowest at the beginning of the hydrological year in October. It will also be related to the status of surface vegetation since the biomass of herbs and shrubs and the tree canopy also follow the same pattern. Further details of the interpretation of

Table 7.6 Multiple regression relationships for selected variables

$\log(1) = -0.0185 + 0.6190 \log(14) + 0.8456 \log(18)$
$R = 0.89, r^2 = 79\%$

$\log(2) = 1.7602 + 0.6185 \log(14) + 0.3153 \log(15) + 0.3302 \log(12) + 0.2472 \log(5)$
$R = 0.87, r^2 = 76\%$

$\log(3) = 1.5645 + 0.2586 \log(14) + 0.4515 \log(5) + 0.1663 \log(12)$
$R = 0.84, r^2 = 71\%$

the results for the discharge-weighted mean concentration and the peak sediment concentration can be found in Clark (1978).

If the results reported in the previous sections are now taken together it is clear that both land use and agricultural practices are important in controlling the sources of sediment. The extent to which they lead to widespread erosion is determined by the input variables and the pre-existing state of the catchment. But the interpretation of these results goes much further. First, a change of land use from woodland to either pasture or arable land will have a dramatic effect on erosional processes and sediment production. A change in agricultural practices, such as the timing of ploughing, the presence of tractor wheelings down hillsides, or the overgrazing of fields, will also add to the problem of accelerated erosion. The sediment yield, based only on the suspended solids load for 1976, gives an estimate of surface lowering of 0.011 mm yr^{-1}. For individual storms the rates would be higher if extrapolated over the whole year. For example, for October 1976 the annual rate of erosion, if extrapolated for 12 months, would be 0.60 m^3 ha^{-1}, which is a relatively high value for grassland areas. A change in either the intensity or the seasonal distribution of rainfall could lead to an increase in erosion from agricultural land (Boardman and Favis-Mortlock, 1993). An increase in the area of cereal crops will have similar consequences, and the increase in soil loss will eventually lead to a reduction in crop yield (Boardman, 1986).

LAND-USE CONTROLS ON SEDIMENT PRODUCTION

From the results of the previous investigation, it is clear that land use is an important control on sediment production. However, the isolation of land use as a control is not an easy matter and its effect will depend upon other environmental controls at the time of the investigation. This problem can be avoided by making estimates of bankfull discharge along a river channel where there is a change of land use and comparing both the volume and rate of change of the bankfull discharge. While this can provide data for the on-site effects of land-use change on erosion, it is also possible that the effects of land use extend beyond the confines of the land-use change boundary. Understanding the effects of upstream land-use change on downstream erosion and floods is of vital importance to land-use planning and management. Conventional wisdom would suggest that upstream land-use changes will have an effect beyond the confines of the change, but hard evidence for this effect has not been easily obtained. For example, from a study of flood levels at Iquitos in the Amazon, Gentry and Lopez-Parodi (1980) argued that Andean deforestation was leading to more runoff and, by implication, more erosion and flooding downstream. This was disputed by Nordin and Meade (1982), Sternberg (1987) and Richey et al. (1989). While admitting that land-use change may lead to more runoff, erosion and flooding, Hamilton (1988) dismissed any connection between upstream and downstream channel reaches. On the other hand, the study by Ryan and Grant (1991) showed how river flows must have increased following progressive clear-felling since large reaches of the Elk River basin had enlarged channels over a 24-year period. The effects of these changes must, to a certain extent, be felt further downstream.

By measuring the changes in channel capacity in a drainage basin, wherein the land use changes, it is possible to detect the effects of that change on the runoff characteristics and, therefore, the relative rate of erosion. Furthermore, it will be possible to estimate the effects of the land-use change beyond its boundary. Changes in channel capacity have been assessed under a variety of conditions (Dury, 1973; Gregory and Park, 1974; Richards and Greenhalgh, 1984). However, Ebisemiju (1991) has cast doubt on the extrapolation of observed upstream allometric relations, and has suggested the use of segmented regional relationships between basin area and channel capacity. This caution needs to be extended to the relationship between basin area and bankfull discharge. The efficacy of estimating flood flows from channel size has been demonstrated by Riggs (1976) and Wharton (1992).

In order to test the theory that a change of land use has no effect on surface runoff, and therefore discharge and erosion both on-site and further downstream, the Postlebury Wood stream in east Somerset, underlain by Oxford Clay (Figure 7.1A), was selected for study. Estimates of bankfull discharge were made in forested channel reaches and in pasture reaches further downstream. Figure 7.1D shows the distribution of land use within the Postlebury Wood catchment, while Figure 7.9 shows the changes in bankfull discharge. The main stream channel emerges at a catchment area of 0.025 km^2. The increase in bankfull discharge from the headwater area is very slow over the first 0.2 km^2 increase in catchment area, at a rate of $0.1 \text{ m}^3\text{s}^{-1}\text{km}^2$. This increases to $0.3 \text{ m}^3\text{s}^{-1}\text{km}^2$ from a catchment area of 0.28 km^2 to 0.58 km^2, which is probably the result of the decrease in the percentage area covered by woodland. Yet this rate of increase of bankfull discharge remained uniform until the area increased by a factor of 2 after it had emerged from the wooded headwater area. In other words the protective effect of the forest has extended downstream of its spatial extent. From a catchment area of 0.58–0.83 km^2 there is a very rapid rise in bankfull discharge, a rate of $1.28 \text{ m}^3\text{s}^{-1}\text{km}^2$, but this then decreases to a more modest $0.42 \text{ m}^3\text{s}^{-1}\text{km}^2$

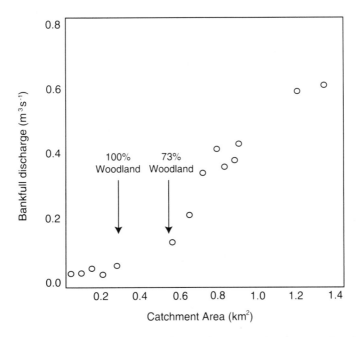

Figure 7.9 Changes in bankfull discharge in relation to catchment area for the Postlebury Wood catchment

Table 7.7 Saturated hydraulic conductivity (mm hr^{-1}) for soils in the New Forest

Geology	Land use		t value	Significance level
	Forest	Heathland		
Barton Sand	13.0	8.8	2.16	0.05
Barton Clay	17.2	9.5	2.38	0.05

(Figure 7.9). The dramatic rise in the rate of increase in bankfull discharge is clearly related to the change of land use.

The controlling effect of runoff from clay soils under woodland and pasture has been investigated by Clark (1987), who measured the saturated hydraulic conductivity of these soils and concluded that woodland soils are about seven times more permeable than those developed on grassland. As the percentage area of woodland increased in the 21 catchments studied, the bankfull discharge decreased. These results clearly have implications for erosion where there is a change of land use.

An example of a change in land use from heathland to forest has also been investigated for the Highland Brook in the New Forest, Hampshire, UK (Figure 7.1A). As expected, there were dramatic decreases in the rate of increase of bankfull discharge after the river entered the forest area; again the lag indicated that upstream land use will affect areas downstream. A detailed field study of the soils developed on the two main rock types — Barton Sand and Barton Clay — and those soils developed on the two land-use types — forest and heathland, showed that land use rather than geology alone has the greater impact on the saturated hydraulic conductivity (Table 7.7). By comparing the results of the survey on the same geology by means of a t test, the significance of the differences between the conductivity values is at once apparent. The low values of conductivity on the heathland areas will give rise to more frequent and higher rates of runoff than areas of woodland. Although the hydrology of this area has been described previously (Gurnell and Gregory, 1987), the effect of a spatial change of land use on bankfull discharge was not reported. These two examples are a clear indication of the effects of land use on surface runoff and, therefore, channel adjustment. In areas where the land use has not changed for decades, then this type of survey can give valuable insights into longer-term effects of land use on runoff and erosion.

DISCUSSION

In view of the increased threat to agricultural land in the future and of the documented examples of soil erosion on arable land, there are surprisingly few examples of soil conservation methods being applied in England. Having given a lecture to a group of farmers and residents in south Somerset, Graham Colbourne (personal communication) stated:

> ...more farmers are now aware that they do have a long-term problem and are trying ameliorative measures. These have included ploughing to different depths, subsoiling more

regularly to break plough pans, keeping the ground surface roughened in winter cereals and part grassing slopes known to erode....Unfortunately it is only the few who try, the rest don't recognise the problem as a long-term soil deterioration, only as a short-term cost loss when their field gets part washed away in an exceptional storm.

While many studies have confirmed the effects of land-use change, crop type and numerous environmental factors on soil erosion, there is a need to apply this knowledge at the farm scale. This will be possible if it can be proved to be worth while on economic as well as ecological grounds. However, there need to be other lines of research, which deserve more attention in the future. First, it is necessary to survey the awareness of farmers to the general and local problem of soil erosion on their land and their knowledge of possible remedial measures. Secondly, we must create the conditions whereby farmers have both the knowledge and technical support needed to implement these measures. This would involve both communication with decision-makers and the need for funds to be made available in order to assist farmers to identify the problem in their own region and then to act accordingly. Thirdly, cost–benefit analysis of the proposed programme of remedial measures is needed so that the value of soil conservation will be better appreciated. Academic research has done much to highlight the problem of soil erosion on arable land and elsewhere. There seems little point in duplicating studies, the outcome of which is often largely predictable. Application of the knowledge of the soil erosion problem should now receive the high priority that it deserves.

CONCLUSIONS

The main points to emerge from this paper are as follows. First, the main source of sediment in a small catchment in east Devon was demonstrated to be the channel banks. Secondly, the main environmental controls on sediment production include rainfall and discharge, as well as the time of year and the antecedent catchment condition. Thirdly, further insights were gained by looking at the distribution of the source of sediment and this clearly showed the importance of land use. Fourthly, measurements of bankfull discharge along selected channel reaches of two catchments showed the crucial importance of land use in controlling channel change and, hence, sediment production. There is a clear need to utilise the results and their implications, which have been described here in order to reduce soil loss from agricultural land.

REFERENCES

Barnes, H.H. (1967) *Roughness Characteristics of Natural Channels*, US Geological Survey Water Supply Paper no. 1849.
Boardman, J. (1986) The context of soil erosion. *Seesoil* 3, 2–13.
Boardman, J. and Favis-Mortlock, J. (1993) Climate change and soil erosion in Britain. *Geogr. J.* 159 (2), 179–183.
Boardman, J., Foster, I.D.L. and Dearing, J.A. (eds) (1990) *Soil Erosion on Agricultural Land*, John Wiley & Sons, Chichester.
Brown, G. (1961) *Identification and Crystal Structures of Clay Minerals*, Mineralogical Society, London.

British Regional Geology (1975) *South-west England*, HMSO, London.
Chow, V.T. (1959) *Open Channel Hydraulics*, McGraw-Hill, New York.
Clark, C. (1978) Sediment dynamics of a small upland catchment. Unpublished PhD Thesis, University of Exeter.
Clark, C. (1987) Deforestation and floods. *Environ. Conserv.* **14** (1), 67–69.
Colby, B.R. (1957) Relationship of unmeasured sediment discharge to mean velocity. *Trans. Am. Geophys. Union* **38** (5), 707–717.
Cosgrove, M.E. (1973) The geochemistry and mineralogy of the Permian Red Beds of southwest England. *Chem. Geol.* **11**, 31–47.
Doty, C.W. and Carter, C.E. (1965) Rates and particle-size distributions of soil erosion from unit source areas. *Trans. ASAE* **8**, 309–311.
Duijsings, J.J. (1987) A sediment budget for a forested catchment in Luxembourg and its implications for channel development. *Earth Surf. Processes Landforms* **12** (2), 173–184.
Dury, G.H. (1973) Magnitude frequency analysis and channel morphometry. In: Morisawa, M. (ed.) *Fluvial Geomorphology*, McGraw-Hill, New York, pp. 91–121.'
Ebisemiju, F.S. (1991) Some comments on the use of spatial interpolation techniques in studies of man-induced river channel changes. *Appl. Geogr.* **11**, 21–34.
Gentry, A.H. and Lopez-Parodi, J. (1980) Deforestation and increased flooding in the Upper Amazon. *Science* **210**, 1354–1356.
Gerlach, T. (1967) Hillslope troughs for measuring sediment movement. *Rev. Geomorphol.* **4**, 173.
Gregory, K.J. and Park, C.C. (1974) Adjustment of river channel capacity downstream from a reservoir. *Water Resour. Res.* **10**, 870–873.
Gregory, K.J. and Walling, D.E. (1973) *Drainage Basin: Form and Process*, Edward Arnold, London.
Grindley, J. (1976) The driest 12-month period recorded in England and Wales since 1727. *The Times* (London), 31 May.
Gurnell, A.M. and Gregory, K.J. (1987) Vegetation characteristics and the prediction of runoff: analysis of an experiment in the New Forest, Hampshire. *Hydrol. Processes* **1**, 125–142.
Hamilton, L.S. (1988) Forestry and watershed management. In: Ives, J. and Pitt, D.C. (eds), *Deforestation: Social Dynamics in Watersheds and Mountain Ecosystems*, Routledge, London, pp. 99–131.
Harvey, A.M. (1974) Gulley erosion and sediment yield in the Howgill Fells, Westmorland. In: Gregory, K.J. and Walling, D.E. (eds), *Fluvial Processes in Instrumented Watersheds*, IBG Special Publication no. 6, pp. 45–58.
Imeson, A.C. (1974) The origin of sediment in a moorland catchment with particular reference to the role of vegetation. In: Gregory, K.J. and Walling, D.E. (eds), *Fluvial Processes in Instrumented Watersheds*, IBG Special Publication no. 6, pp. 59–72.
Klages, M.G. and Hsieh, V.P. (1975) Suspended solids carried by the Gallatin River of southwestern Montana: II. Using mineralogy for inferring sources. *J. Environ. Qual.* **4** (1), 68–75.
Morgan, R.P.C. (1979) *Soil Erosion*, Longman, London.
Nordin, C.F. and Meade, R.H. (1982) Deforestation and increased flooding of the Upper Amazon. *Science* **215**, 426–427.
Pye, K. (ed.) (1994) *Sediment Transport and Depositional Processes*, Blackwell, Oxford.
Richards, K. and Greenhalgh, C. (1984) River channel change: problems of interpretation illustrated by the River Derwent, North Yorkshire. *Earth Surf. Processes Landforms* **9**, 175–180.
Richey, J.E., Nobre, C. and Deser, C. (1989) Amazon River discharge and climate variability 1903–1985. *Science* **246**, 101–103.
Riggs, H.C. (1976) A simplified slope–area method for estimating flood discharges in natural channels. *US Geol. Surv. J . Res.* **4** (3), 285–291.
Ryan, S.E. and Grant, G.E. (1991) Downstream effects of timber harvesting on channel morphology in Elk river basin Oregon. *J. Environ. Qual.* **20**, 60–72.
Sternberg, H.O. (1987) Aggravation of floods in the Amazon as a consequence of deforestation? *Geogr. Ann.* **69A**, 201–219.

Vaithiyanathan, P., Ramanathan, A. and Subramanian, V. (1992) Sediment transport in the Cauvery River basin: sediment characteristics and controlling factors. *J. Hydrol.* **139** (1–4), 197–210.

Walling, D.E. (1971) Instrumented catchments in south-east Devon. Some relationships between catchment characteristics and catchment response. Unpublished PhD thesis, University of Exeter.

Walling, D.E. (1988) Erosion and sediment yield research — some recent perspectives. *J. Hydrol.* **100** (1–3), 113–141.

Walling, D.E. and Moorehead, P.W. (1987) Spatial and temporal variation of the particle size characteristics of fluvial suspended sediment. *Geogr. Ann.* **69A**, 47–59.

Wharton, G. (1992) Flood estimation from channel size: guidelines for using the channel-geometry method. *Appl. Geogr.* **12**, 339–359.

Section III

SEDIMENT QUALITY

8 Sediment Mineralogy and the Environmental Impact of Mining

JOHN R. MEREFIELD

Earth Resources Centre, University of Exeter, UK

INTRODUCTION

The environmental effects associated with mining operations are of particular concern to planning authorities, government and international funding agencies, because they can involve a variety of adverse consequences. Mining may cause environmental damage through a number of routes: land clearance, erosion of spoil tips, hydrological effects, impact on the ecosystem, disruption of natural and human transport systems, pollution of air and water resources, loss of amenity, socio-economic impact, as well as through human health and safety. Most of these routes to pollution will involve water-borne transport at some stage. Watercourses can be polluted by particles of spoil, by discharge of liquid waste and by chemicals used in the treatment and processing of ores, all of which can introduce heavy metals into the surface drainage system.

Stream sediments by their accumulating nature are especially suitable as dynamic monitors of heavy-mineral concentrations. They also retain records of past incidents. Careful examination of sediment profiles can be used to unravel these historical events. Even minor reworking of mine-waste can release small particles of sediment, which can be traced by detailed mineral analysis. Data recorded at reasonable time intervals can thus be used to provide temporal information on the status of a river course and the inferred environmental health status of the drainage basin.

This chapter describes the nature of bed sediments, charts the history of research into modern-day sediment mineralogy, and illustrates the environmental impact of mining on stream and river courses by way of a case study from south-west England. The chapter concludes with a discussion on the needs for future work and current trends in this field of research.

NATURE OF STREAM SEDIMENTS

Streams and rivers are the main routes by which the products of weathering are removed in temperate environments. The mineral composition of bedload deposits, therefore, generally reflects the geology over which a watercourse flows. Significant differences do occur, however, where the introduction of material from naturally eroded ore bodies and/or waste from mining activity causes anomalies that may be displaced over considerable distances. As a result, sediment surveys are now routinely performed in the reconnaissance stages of mineral exploration programmes (Armour-Brown and Nichol, 1970; Levinson, 1974; Watters and Sagala, 1979). Contamination of soils by mining operations is generally localised with identifiable point sources (Alloway, 1990). Tailings from ore processing are inhomogeneous and will vary in particle size from <64 µm (slimes) to >64 µm (sand) (Barbour, 1994). Weathering of ores and spoil, however, will often result in an extensive halo effect, contaminating both soils and streams around the mining area with high levels of heavy metals. In south-west England, correlations between arsenic concentrations in soils and stream sediments, for example, have led to an area of 722 km^2 (7.90% of west Cornwall) to be designated as contaminated land (Abrahams and Thornton, 1987).

The major constituents of stream sediments were originally classified by Hawkes and Bloom (1956) and were more recently restated by Rose *et al.* (1979) as falling into five main components. These are as follows:

- *Lithoclasts*. Detrital rock fragments and mineral fragments including quartz, heavy resistate accessory minerals and ore minerals.
- *Secondary detritals*. Material produced by the chemical weathering of bedrock or mineral veins, including the clays and secondary ore minerals.
- *Precipitates*. Particles precipitated from solution under favourable pH–E_h conditions, including Fe and Mn oxides and hydroxides and associated trace elements.
- *Exchangeable elements*. These are trace elements (often heavy metals) adsorbed onto Fe and Mn oxides, organic matter and clays.
- *Organic matter*. This includes both detrital organic matter and *in situ* organic debris.

In general terms, river bottom sediments may be categorised into three main groups as:

- A detrital fraction.
- A fraction taken out of solution during precipitation and/or adsorption (a reversible process).
- Organic matter.

MINERAL DISPERSION FROM MINING AREAS

In former times, waste debris from mining operations was stored wherever convenient, often on dumps with excessively steep sides. The normal process of gravity-induced surface creep at these sites caused the debris to move downslope to produce fans of contaminated land. Davies (1971) showed that, of the 31% of apparently productive slope soils in the Tamar Valley of Devon and Cornwall, most was contaminated by heavy metals. This was due to the redistribution of mine dump material. Water passing through waste heaps will also cause dispersal, partly by surface wash of colloids and fines and partly by removal of solutes in surface runoff.

The nature and rate of metalliferous mine-waste dispersal from orefields by surface water is influenced by the interaction of both physical and chemical controls. Many researchers have examined the processes involved. Early work focused on the mechanical processes and, subsequently, writers turned their attention to the importance of chemical effects (reviewed by Lewin and Macklin, 1986).

Mechanical

Ore minerals and gangue may be transported in streams and rivers with countryrock, which effectively dilutes the heavy-mineral component and its associated polluting heavy metals. Water sorting of the heavy minerals according to their relative densities will have the opposing effect of concentrating them during transport. There has been little agreement as to the exact mechanisms by which heavy minerals concentrate in streams. Ruby (1933) thought that fall velocity was the most important factor, whereas Rittenhouse (1943) introduced the concept of hydraulic equivalence. Bed configuration and grain density were considered to be more significant than fall velocity by Brady and Jobson (1973). Because a large quartz grain requires the same current velocity to move as a small heavy mineral, when the velocity of a stream is slackened, the first minerals to be deposited are the large heavy minerals. These are followed by the smaller heavy minerals and then the largest of the light minerals such as quartz.

In south-west England, stanniferous gravels weathered from cassiterite-bearing ore bodies in the Pleistocene Period, were often covered with a succession of river terrace deposits. Miners or 'streamers' searching for pebble or stream tin followed these buried river channels. The downstream impact of ore deposit weathering, tin streaming and other mining activity is likely to be the greatest, therefore, at the inside of a meander, where a swift tributary stream meets a slower mainstream and a pay streak has formed, or where projections such as boulders or interbedded rocks serving as natural riffles impede the progress of heavy minerals. These mechanisms gave rise to the concept of entrainment equivalence (Evans, 1980). In west Cornwall, Yim (1976) discovered large concentrations of tin, arsenic, copper, iron, lead, tungsten and zinc in silt particles (<64 µm) of the Hayle Estuary resulting from mining operations inland. These metals were liberated at ore dressing sites located on catchment streams of the estuary, which acts as a highly efficient sediment sink.

Chemical

A number of authors have investigated the reactions that lead to oxidation of the sulphides (Lowson, 1975; Brierley Corale, 1978; Murr, 1980; Harries and Ritchie, 1983). These include physical and chemical processes, which are relatively well understood, and also microbial processes, which are less well researched, as well as the complicating effects of their interaction. Pyrite is a common component of sulphide ores, which, because of its low commercial value, is left to accumulate in the tailings during processing. In a wet environment it oxidises to produce sulphuric acid and iron hydroxide as follows:

$$2FeS_2 + 7O_2 + 2H_2O \rightarrow 2FeSO_4 + 2H_2SO_4 \tag{8.1}$$

This reaction can also be catalysed by bacteria and the ferrous sulphate can be further oxidised to ferric sulphate:

$$4FeSO_4 + O_2 + 2H_2SO_4 \rightarrow 2Fe(SO_4)_3 + 2H_2O \tag{8.2}$$

This process is usually very slow at pH values below 3.5, but it is catalysed by the bacterium *Thiobacillus ferroxidans*, which can speed up the reaction by as much as 10^6 times (Harries and Ritchie, 1983). At pH values between 4 and 7, the ferric ions produced by reaction (8.2) tend to precipitate out as ferric hydroxide or basic ferric sulphates and produce more sulphuric acid. This lowers the pH, which in turn reverses the reaction, consumes the acid and produces ferric sulphate. This buffering effect was investigated by Miller (1980), who concluded that jarosite ($KFe_3(SO_4)_2(OH)_6$) and ferric hydroxide in weathered coal pit spoil buffered in the pH range 3.0 to 3.4, and calculated that it would take 25 years to hydrolyse the jarosite completely in a 0.25 m layer.

Additionally, the sulphuric acid and ferric sulphate produced from pyrite can oxidise other metal sulphides such as chalcopyrite:

$$CuFeS_2 + Fe(SO_4)_3 \rightarrow CuSO_4 + 2FeSO_4 + 2S \qquad (8.3)$$

Usually, pyrite oxidises much faster than chalcopyrite, but when these minerals make intimate contact an electrochemical reaction is set up that modifies reaction times. Similarly, temperature will affect the rate of bacterial action. The activity of *T. ferroxidans* reduces markedly at temperatures above 45°C but *Sulfolobus*-type bacteria, which also derive energy from the oxidation of sulphur, ferrous iron and metal sulphides, have been found active at temperatures above 60°C (Brierley Corale *et al.*, 1980).

Drainage from mine adits and tailings can, therefore, be very acidic, and the stream beds can become coated with ochre (hydrated ferric oxide) slimes (Davies, 1983). Acid oxidising conditions in turn mobilise other heavy minerals/metals in the waste, which will remain in solution to be transported downstream away from the mine source until environmental conditions change. Additional rainfall may reduce the acidity, and under oxidising conditions Fe and Mn may precipitate, causing co-precipitation of other metals such as Pb, Zn, Cu and As (Nichol *et al.*, 1967).

Complex chemical changes occur inside waste heaps, which can be divided into major zones, namely, a surface oxidation zone and a deeper reduction zone (Boorman and Watson, 1976). In Canada, the surface zone of a Pb/Zn sulphide waste dump was shown to have lost 92% of the Zn, because ZnS readily oxidises to $ZnSO_4$. Little Pb was removed, however, as relatively insoluble anglesite ($PbSO_4$) was formed from the galena (PbS). In dry conditions, soluble toxic metal salts may effloresce on the surface to be blown away by wind action (Johnson *et al.*, 1978) to areas where they can then form secondary sources of pollution for subsequent fluvial redistribution.

During fluvial transport the solutes can become complexed and also adsorbed onto suspended sediment. For example, filtered samples of water from the Ottawa River, Canada, were shown to contain complexing agents capable of binding heavy metals (Ramamoorthy and Kushner, 1975). A large proportion of dissolved mercury in some estuarine waters of the USA was additionally shown to be associated with organic matter (Gardiner, 1974). Studies in a former cinnabar (HgS) mining district of Texas proved Hg to be bound strongly to stream alluvia and thus Hg mobility was controlled by the mechanical movement of sediment transport (Blanton *et al.*, 1975).

The transport of toxic materials from mining areas is thus complex, site-dependent and governed by a number of interrelating physical and chemical processes. Geochemical and mechanical factors will cause dissolution of heavy minerals from mining waste and transport by fluvial processes away from the mine. However, environmental conditions such as pH or E_h changes can reverse these processes, switching solute mobilisation to the control of mechanical sediment dispersion and vice versa. Work on iron and manganese coatings on stream detritus by Nichol et al. (1967) drew attention to the role that these elements could play in the co-precipitation of heavy metals and in holding them to the surfaces of sediments of varying particle size, thus flagging both the natural weathering of mineral veins and the environmental impact of mining.

Many sites in the USA contaminated through such mechanisms by non-ferrous mining activity have been designated as Superfund sites. This status demands that some remedial action be taken at these localities in the future (Pierynski et al., 1994).

HISTORICAL PERSPECTIVE

Very little standardisation of procedures has been introduced in stream sediment studies. Most workers have agreed, however, that heavy minerals (representing mining impact) are preferentially associated with the finest fractions in most stream situations. Where Fe/Mn coatings precipitate onto larger particles and trap metals by means of a coating, the coarser grades may then also retain and reflect significant pollution incidents (Nichol et al., 1967; Horsnail, 1968; Merefield, 1987a,b, 1993, 1994). In any study embracing the net contributions of natural and anthropogenic input of heavy metals to the stream channels, the sampling cut-off grade should, therefore, not be too restrictive.

As both physical and chemical processes dominate the interaction between particle size and metal speciation in stream sediments, metals dissolved in acid waters may be deposited later. Work on the Ystwyth river in central Wales has shown that metals in solution continued to be transported from the catchment decades after mining operations had ended (Grimshaw et al., 1976). It has also been demonstrated that barytes ($BaSO_4$) waste from a former mining area has continued to be supplied to the River Teign, Devon, for some 40 years since mining activity ceased (Merefield, 1994).

Most stream sediment surveys have established the need to employ sieving or settling, either in the field or in the laboratory, to recover the appropriate particle-size fraction for analysis. Most particle-size fractions have been examined in bottom sediment surveys. The <250 µm particle-size fraction proved appropriate where high-gradient streams were examined (Merefield, 1974), the <200 µm fraction where geochemical indicators of lithology were assessed (Nichol et al., 1971), the <63 µm grade where surface area and grain-size effects were evaluated (Horowitz and Elrick, 1987), and the <2 µm particles where clay adsorption was examined (Jenne et al., 1980).

Bottom sediments are commonly taken by hand trowel in shallow water or by grab sampler in deeper situations. Corers are used where subsurface material is required. In general, the upper, light brown oxidised zone is different in character from the layers below. It has been suggested that the mainly dark reduced layers beneath it and at 1–3 cm depth represent the pollution events over the last few years (Förstner, 1983). The oxidised zone (0–1 cm) will generally represent most recent sedimentation, however, and should also be sampled either separately or together with anoxic sediments. Where reference is needed to pre-contamination levels, samples must be taken from greater depths. A bullet or piston corer would be used for this.

Any study of bed sediments should take into account variations in the abrasion potential of both natural and artificially introduced components. The engineering properties of shale and limestone lithologies, for example, would suggest these rock types to have low durability. An investigation of the Graveglia river in north-west Italy, a short-headed, high-gradient stream, predictably showed that abrasion of shale and limestone bedrock yielded large amounts of sand. However, the observed downstream decrease in these lithoclasts was due not to less abrasion downstream but to dilution by exotic mineral components introduced from tributaries and slope runoff (McBride and Picard, 1987).

IMPACT OF MINING

Case study: Teign Valley, south-west England

Mineralogy of stream and river sediments in mineralised areas can be complicated by lithology, contamination and secondary environmental changes. The Teign Valley of south-west England provides a valuable location to illustrate these influences, as it contains varied rock types and mineralisation. It is also an area with a long mining history (Hoskins, 1960), and although records are incomplete, tin mining in the Dartmoor catchment dates from 1156 (Worth, 1942). The impact of the tin streamers is still visible today where worked-out areas of cassiterite ore are evident as partly collapsed trenches flanked by low dumps (Beer and Scrivener, 1982). Barytes, galena, zinc blende, haematite, siderite, fluorspar and manganese ore deposits were exploited in the middle Teign Valley (Dines, 1956), with the Bridford Barytes Mine being the last

Table 8.1 Particle-size distribution of Teign Valley stream sediments

Grade	size (μm)	Phi[a]	Amount (wt%)
Granule	>2000		48.4
		−1	
Very coarse sand	<2000–1000		22.6
		0	
Coarse sand	<1000–500		18.7
		1	
Medium sand	<500–250		8.5
		2	
Fine sand	<250–125		1.5
		3	
Very fine sand	<125–64		0.04
		4	
Coarse silt	<64–32		0.00
		5	
Medium-fine silt/clay	<32		0.3
		6	

[a] Phi = $-\log_2$ (diameter/mm)

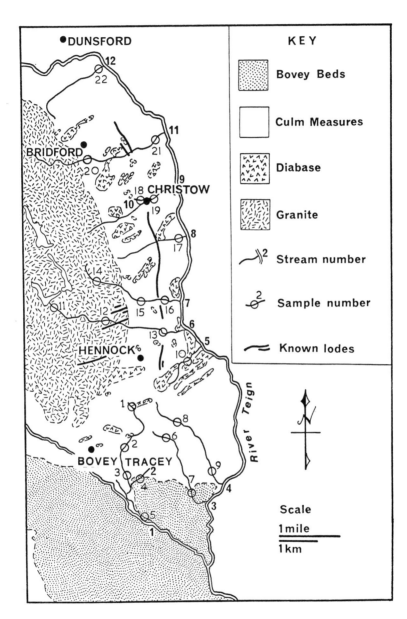

Figure 8.1 Map to show the geology of the middle Teign Valley and stream sediment sample locations (from Merefield, 1974; reproduced by permission of *Journal of Geochemical Exploration*, Elsevier Science, Amsterdam)

Table 8.2 Heavy-mineral analyses of Teign Valley stream sediments[a]

Stream no.[b]	Sample no.	Heavy minerals (%)	Tourmaline (%)	Pyrites (%)	Haematite (%)	Barytes (%)
1	3	18.6	90	–	–	–
6	12	9.1	100	–	–	–
6	13	18.6	10	70	10	5
7	14	4.6	100	–	–	–
7	16	7.4	50	–	–	50

[a] All analyses on fine sand <250 μm (60 mesh)
[b] Sample localities given in Figure 8.1

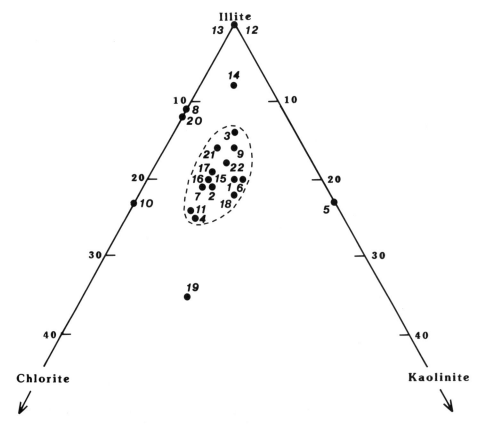

Figure 8.2 Frequency of clay minerals (%) in stream sediments of the Teign Valley (from Merefield, 1974; reproduced by permission of *Journal of Geochemical Exploration*, Elsevier Science, Amsterdam)

to close in 1958 (Schmitz, 1973). Mine dumps and slime ponds remain in the Teign Valley, although most have been lost during its return to agriculture. Inefficient processing and indiscriminate spreading of uneconomic residues over farmland, however, have led to widespread contamination.

The topography is primarily one of high ground reaching elevations of more than 300 m formed by the Dartmoor granite mass, sloping away, often steeply, to the south and east to below 30 m. The west side of the middle Teign Valley includes eight major streams draining the former mining area and its contorted Lower Carboniferous shales, mudstones and cherts. Rainfall there is 40–50 inch. (1016–1270 mm) annually, with a mean annual temperature range of 39–59°F (4–15°C).

Bottom sediments taken by plastic scoop, air-dried, sieved to <250 µm from streams draining the River Teign and from the main river itself were studied during a series of sedimentological and geochemical investigations from 1973 to 1986 (Merefield, 1974, 1976a,b, 1987a,b). Sedimentological studies were made using methods based on the classical techniques of Krumbein and Pettijohn (1938). For heavy-mineral separations, sample splits of bulk (generally <250 µm) grains were boiled with 50% HCl to remove coatings of iron oxide. They were then washed, dried and floated in bromoform. The crop was weighed and examined optically by binocular and scanning electron microscopy. The <2 µm fraction was analysed for clay minerals by X-ray diffraction

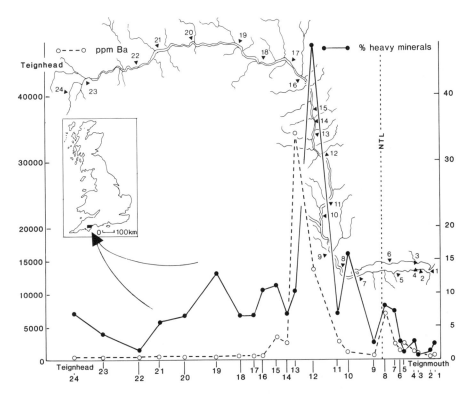

Figure 8.3 Barium and total heavy-metal distribution in River Teign sediments, showing highest values where streams drain the former middle Teign Valley mining area. NTL is the normal tidal limit

(XRD), and semiquantitative clay determinations were carried out employing techniques similar to those of Biscaye (1964).

As might be expected in these high-gradient streams, sieve analysis for waterways draining into the Teign gave a mode of the cumulative weight percentage curve lying in the very coarse fraction (Table 8.1). The major heavy-mineral component of sample 3 (located on Figure 8.1) on Culm (Lower Carboniferous shale) was tourmaline, present in a variety of colours: black, brown, blue, green and colourless (described by Flett and Dewey in Reid *et al.*, 1912). Light minerals comprised mainly rounded quartz, feldspars, biotite and muscovite micas. A few shale fragments were also present. In contrast, sample 13 on the Culm, 1 km downstream from the E–NE haematite lodes, and a few metres downstream from the N–S mineralisation, contained 18.6% heavy minerals. Pyrites was dominant at 70%, with lesser amounts of haematite and tourmaline (10%), barytes (5%) and zircon (Table 8.2). While sample 14 located on Dartmoor granite gave a granitic assemblage with 4.6% heavy minerals, almost entirely tourmaline, sample 16 downstream on the Culm, after the influence of the N–S mineral lodes, demonstrated the mining impact with double the heavy minerals at 7.4%, being barytes and schorlite of equal quantities (Table 8.2).

The clay mineralogy of these streams showed illite as ubiquitous, which is known to result as a weathering product of mica minerals found in the granitic rocks and shales

Figure 8.4 Examples of minerals from the heavy and light fractions of River Teign sediments adjacent to input from weathering mine dumps. The heavy-mineral fraction is dominated by tourmaline (A and inset) from Dartmoor granite and barytes (B and inset) from mine-waste, in equal quantities. The light fraction comprises quartz (C), feldspar (D and inset compared with B) and mica (E) from both granite and carboniferous shales/mudstones

SEDIMENT MINERALOGY AND MINING IMPACT

Figure 8.4 *Continued*

in this area. Fairly equal quantities of kaolinite, from the hydrothermal alteration and weathering of K-feldspars of the Dartmoor granite (Durrance, 1983, 1986), and chlorite, a common component of many sediments especially those having large Fe and Al displacements, also occurred (Figure 8.2).

Samples taken from the River Teign itself, some 15 years after mining operations had ceased, also demonstrated the impact of former mining activities. Gravimetric values of the total heavy minerals appeared less dominated by bedrock lithology, although the values for the granitic section of the upper Teign Valley, which ranged from 1.48 to 7.01 wt%, were generally lower than those from the Lower Carboniferous section downstream (Merefield, 1993). Significantly, a peak value of 46.31% occurred at locality 16, some 2.5 km upstream from the most anomalous barium value recorded during a parallel geochemical survey (Figure 8.3). The heavy minerals then decreased in proportion downstream to the estuary to 2.43%, before rising again at the river/estuary interface. Although the relationship between barium analyses and total mineral content is not straightforward, most Ba was shown to be present as barytes ($BaSO_4$) and derived from weathering of waste dumps introduced by the tributary streams traversing the Teign Valley orefield (Figure 8.4). Point-counting of barytes

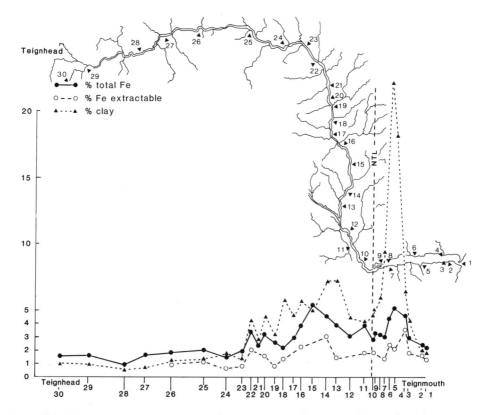

Figure 8.5 Clay, total and extractable iron concentrations in River Teign sediments showing anomalous concentrations adjacent to streams draining the former middle Teign Valley mining area, and at the fresh/salt water interface at the head of the Teign Estuary

grains in these samples showed the highest river value of about 41% barytes to coincide with a total barium peak value of 13 500 parts per million.

Total iron, oxalate-extracted iron and clay concentrations were also compared and showed anomalously high concentrations in the middle Teign Valley, where streams traversing the former mining district drain into the Teign (Figure 8.5). The relationship of clays with precipitated iron is important. The clays provide a host surface area and the precipitated iron a scavenging mechanism for heavy metals weathering from the waste dumps. A second peak also occurred at the salt/fresh water mixing zone at the head of the Teign Estuary where flocculation of suspended sediments takes place. The relative proportions of the dominant clay minerals there gave illite at 41% and kaolinite at 59%, in contrast to clays moving landward from the mouth of the estuary, which are dominated by illite at 89% (Merefield, 1981). Significantly, levels of silt-size grains also increased in the orefield section, signalling mining impact (Figure 8.6).

Traditional methods were also used to obtain panned concentrates both upstream and downstream of the Teign Valley orefield from deeper anoxic sediments. Dominant components of the concentrates above the influence of the middle Teign Valley mining area were ranked as follows:

cassiterite > lead shot > Pb/Zn sulphides

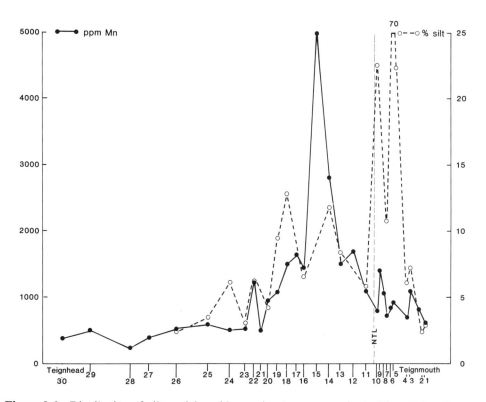

Figure 8.6 Distribution of silt particles with associated manganese in the River Teign. Sample localities numbered on *x*-axis are located in Figure 8.5

and those downstream of the orefield as:

barytes > manganese oxides > lead/zinc/copper sulphides

TRENDS

This chapter has concentrated on the information to be gained from analysis of minerals in surface sediments of stream, river and estuarine deposits. By using active sediments in this way, an overview of the current status of catchments is produced and responses to human activity can be monitored. Some writers have turned their attention to dispersal patterns in valley-floor sediments (Lewin and Macklin, 1986) where heavy-metal input can be dated historically. Heavy-metal-loaded sediments in alluvial deposits are also potential sources of pollution during redevelopment of river valley and estuarine shorelines. Relatively little is known of the extent and magnitude of this potential problem even at reconnaissance level. Heavy-mineral separations of cassiterite (SnO_2) in core material from former tin mining areas, for example, along the lines of the geochemical studies conducted on alluvial deposits of the Lox Yeo (Macklin *et al.*, 1985) and in the Hayle Estuary (Yim, 1976), could produce a good pre-mining baseline datum. Similarly, barytes in cores from past lead mining districts could be used to provide valuable information on both natural and anthropogenic weathering in mineralised areas. This could in turn be used to isolate and to scale any point sources during future investigations of mining impact.

ACKNOWLEDGEMENTS

Grateful thanks are due to staff at the Geology Department of the University of Southampton where the first investigations of the case study used to illustrate this chapter were conducted. Similar acknowledgements are also due to staff at the former Department of Geology and latterly, the Earth Resources Centre, at the University of Exeter, from where the most recent surveys were carried out. Mr Jeff Jones supported with SEM illustrations. One Professor Des Walling gave much valued constructive and critical guidance during the shaping of this long-term progamme into its PhD thesis form.

REFERENCES

Abrahams, P.W. and Thornton, I. (1987) Distribution and extent of land contaminated by arsenic and associated metals in mining regions of southwest England. *Trans. Inst. Min. Metall.* **96**, B1–8.

Alloway, B.J. (1990) Introduction. In: Alloway, B.J. (ed.), *Heavy Metals in Soils*, Blackie, Glasgow, pp. 3–6.

Armour-Brown, A. and Nichol, I. (1970) Regional geochemical reconnaissance and the location of metallogenic provinces. *Econ. Geol.*, **65**, 312–330.

Barbour, A.K. (1994) Mining non-ferrous metals. In: Hester, R.E. and Harrison, R. M. (eds), *Mining and its Environmental Impact*, Royal Society of Chemistry, Letchworth, pp. 1–15.

Beer, K.E. and Scrivener, R.C. (1982) Metalliferous mineralisation. In: Durrance, E.M. and Laming, D.J.C. (eds), *The Geology of Devon*, University of Exeter, pp. 117–147.

Biscaye, P.E. (1964) Distinction between kaolinite and chlorite in recent sediments by x-ray diffraction. *Am. Mineral.* **49**, 1281–1289.

Blanton, C.J., Desforges, C.E., Newland, L.W. and Ehlmann, A.J. (1975) A survey of mercury distributions in the Terlingua area of Texas. *Trace Subst. Environ. Health* **9**, 139–148.
Boorman, R.S. and Watson, D.M. (1976) Chemical processes in abandoned sulphide tailings dumps and environmental implication for northeastern New Brunswick. *CIM Bull.* **69**, 86–96.
Brady, L.L. and Jobson, H.E. (1973) *An Experimental Study of Heavy-Mineral Segregation Under Alluvial Flow Conditions*, US Geological Survey Professional Paper, no. 562-K, Washington, DC.
Brierley Corale, L. (1978) Bacterial leaching. *CRC Crit. Rev. Microbiol.* **6**, 207–262.
Brierley Corale, L., Brierley, J.A., Norris, P.R. and Kelly, D.P. (1980) Metal-tolerant microorganisms of hot, acid environments. In: Gould, G.W. and Corry, J.E.L. (eds), *Microbial Growth and Survival in Extremes of Environment*, Academic Press, London, pp. 39–51.
Davies, B.E. (1971) Trace-element content of soils affected by base metal mining in the west of England. *Oikos* **22**, 366–372.
Davies, B.E. (1983) Heavy metal contamination from base metal mining and smelting: implications for man and his environment. In: Thornton, I. (ed.), *Applied Environmental Geochemistry*, Academic Press, London, pp. 425–462.
Dines, H.G. (1956) *The Metalliferous Mining Region of Southwest England*, 2, Mem. Geol. Surv. Gt. Br., 287pp.
Durrance, E.M. (1983) Radiological indicators of hydrothermal circulation in south east Devon. Unpublished PhD thesis, University of Exeter, 620pp.
Durrance, E.M. (1986) *Introduction to Radioactivity in Geology*, Ellis Horwood, Chichester, 441pp.
Evans, M. (1980) *An Introduction to Ore Geology*, Blackwell, Oxford, 231pp.
Förstner, U. (1983) Assessment of metal pollution in rivers and estuaries. In: Thornton, I. (ed.), *Applied Environmental Geochemistry*, Academic Press, London, pp. 395–423.
Gardiner, J. (1974) Chemistry of cadmium in natural water. I. Study of cadmium complex formation using the cadmium specific-ion electrode. *Water Res.* **8**, 157–164.
Grimshaw, D.L., Lewin, J. and Fuge, R. (1976) Seasonal and short-term variations in the concentration and supply of dissolved zinc to polluted aquatic environments. *Environ. Pollu.* **11**, 1–7.
Harries, J.R and Ritchie, A.I.M. (1983) Pyritic oxidation in mine wastes: its incidence, its impact on water quality and its control. In: O'Loughlin, E.M. and Cullen, P. (eds), *Prediction in Water Quality. (Proc. Symposium on the Prediction in Water Quality*, Australian Academy of Science, Canberra, 1982), pp. 347–377.
Hawkes, H.E. and Bloom, H. (1956) Geochemical prospecting. In: Abelson, P.H. (ed.), *Researches in Geochemistry*, John Wiley & Sons, New York, pp. 62–78.
Horowitz, A J. and Elrick, K.A. (1987) The relation of stream sediment surface area, grain size and composition to trace element chemistry. *Appl. Geochem.* **2**, 437–451.
Horsnail, R.F. (1968) The significance of some regional geochemical patterns in north Wales and south-west England. Unpublished PhD thesis, University of London.
Hoskins, W.G. (1960) *Devon*, Collins, London, 600 pp.
Jenne, E., Kennedy, V., Burchard, J. and Ball, J. (1980) Sediment collection and processing for selective extraction and for total metal analysis. In: Baker, R. (ed.), *Contaminants and Sediments*, vol. 2, Ann Arbor Science, Michigan, pp. 169–189.
Johnson, M., Roberts, D.V. and Firth, N. (1978) Lead and zinc in the terrestrial environment around derelict metalliferous mines in Wales (UK). *Sci. Total Environ.* **10**, 61–78.
Krumbein, W.C. and Pettijohn, F.J. (1938) *Manual of Sedimentary Petrology*, Appleton-Century-Crofts, New York, 549pp.
Levinson, A A. (1974) *Introduction to Exploration Geochemistry*, Applied Publishers, Calgary, 612pp.
Lewin, J. and Macklin, M.G. (1986) Metal mining and floodplain sedimentation in Britain. In: Gardiner, V. (ed.), *International Geomorphology*, John Wiley & Sons, Chichester, pp. 1009–1027.
Lowson, R. (1975) *Bacterial Leaching of Uranium Ores—A Review*, Australian Atomic Energy Commission, AAEC/E356.

Macklin, M.G., Bradley, S.B. and Hunt, C.O. (1985) Early mining in Britain: the stratigraphic implications of metals in alluvial sediments In: Fieller, N.R.J., Gilbertson, D.D. and Ralph, N.G.A. (eds), *Palaeoenvironmental Investigations: Research Design, Methods and Data Analysis*, Oxford, British Archaeological Reports, International Series no. 258, pp. 45–54.

McBride, E.F. and Picard, M.D. (1987) Downstream changes in sand composition, roundness, and gravel size in a short-headed, high-gradient stream, northwestern Italy. *J. Sediment Petrol.* **57**, 1018–1026.

Merefield, J.R. (1974) Major and trace element anomalies in stream sediments ofthe Teign Valley orefield. *J. Geochem. Explor.* **3**, 151–166.

Merefield, J.R. (1976a) Heavy metal accumulations in the Teign Estuary. *Proc. Ussher Soc.* **3**, 390.

Merefield, J.R. (1976b) Barium build-up in the Teign Estuary. *Marine Pollut. Bull.* **7**, 214–216.

Merefield, J.R. (1981) Caesium in the up-estuary transport of sediments. Marine Geol. **39**, M45–55.

Merefield, J.R (1987a) Ten years of barium build-up in the Teign. *Marine. Pollut. Bull.*, **18**, 220–222.

Merefield, J.R. (1987b) Heavy metals in Teign Valley sediments: ten years after. *Proc. Ussher Soc.* **6**, 529–535.

Merefield, J.R. (1993) The use of heavy metals in stream sediments for assessing the environmental impact of mining. Unpublished: PhD thesis, University of Exeter, 368pp.

Merefield, J.R. (1994) Chemical geology in assessing the environmental impact of metalliferous mining. *Trends in Chemical Geology*, Council of Scientific Information, Trivandrum, India (in press).

Miller, S.D. (1980) Sulphur and hydrogen ion buffering in pyritic stripmine spoil. In: Trudinger, P.A., Walter, M.R. and Ralph, B.J. (eds), *Biogeochemistry of Ancient and Modern Environments*, Australian Academy of Science, Canberra, pp. 537–543.

Murr, L.E. (1980) Theory and practise of copper sulphide leaching in dumps and in situ. *Min. Sci. Eng.* **12**, 121–189.

Nichol, I., Horsnail, R.F. and Webb, J.S. (1967) Geochemical patterns related to precipitation of manganese oxides. *Trans. Inst. Min. Metall.* **76**, B113–115.

Nichol, I., Thornton, I., Webb, J.S., Fletcher, W.K., Horsnail, R.F. and Khaleelee, J. (1971) Regional geochemical reconnaissance of part of Devon and North Cornwall. *Rep. Inst. Geol. Sci.* **71** (3), 26pp.

Pierzynski, G.M., Schnoor, J.L., Banks, M.K., Tracy, J.C., Licht, L.A. and Erickson, L.E. (1994) In: Hester, R.E. and Harrison, R.M. (eds), *Mining and its Environmental Impact*, The Royal Society of Chemistry, Letchworth, pp. 49–69.

Ramamoorthy, S. and Kushner, D.J. (1975) Binding of mercuric and other heavy metal ions by microbial growth media. *Microb. Ecol.* **2**, 162–176.

Reid, C., Barrow, G., Sherlock, R.L. MacAlister, D.A, Dewey, H., Bromhead, C.N., Flett, J.S. and Ussher, W.A.E. (1912) *The Geology of Dartmoor*, Mem. Geol. Surv. Eng. Wales, 102pp.

Rittenhouse, G. (1943) Transportation and deposition of heavy minerals. *Geol. Soc. Am. Bull.* **54**, 1725–1780.

Rose, A.W., Hawkes, H.E. and Webb, J.S. (1979) *Geochemistry in Mineral Exploration*, 2nd edn, Academic Press, London, 657pp.

Ruby, W.W. (1933) The size distribution of heavy minerals within a waterlain sandstone. *J. Sediment. Petrol.* **3**, 3–29.

Schmitz, C.J. (1973) *Teign Valley Lead Mines*, NCMRS Occasional Publ. no. 6, 125pp.

Watters, R.A. and Sagala, F.P. (1979) Geochemical reconnaissance for uranium and base metals using heavy mineral separates in central and southern Sumatra. In: Watterson, J. R. and Theobold, P. K. (eds), *Geochemical Exploration*, Proc. 7th. Int. Geochemical Exploration Symposium, Golden, Colorado, 1978, Assn. Exploration Geochemists, Resedale, Ontario, pp. 317–327.

Worth, R.H. (1942) The Dartmoor blowing house. *Proc. Devon. Assoc.* **72**, 201–250.

Yim, W.W.-S. (1976) Heavy metal accumulation in estuarine sediments in a historical mining of Cornwall. *Marine Pollut. Bull.* **7**, 147–150.

9 Long-term Dispersal of Metals in Mineralised Catchments by Fluvial Processes

S. B. BRADLEY
Westlakes Research Institute, Moor Row, Cumbria, UK

INTRODUCTION

Mineral deposits have been exploited in most parts of the world, and great efforts in particular have been made to find and exploit metalliferous deposits. Indeed, the development of civilisations is marked by the processing and usage of metal objects. Some deposits contain metals in their native form, for example, the 'Bridal Chamber' deposit in New Mexico produced 2 500 000 ounces of silver, much of which was horn silver. This was so pure that it was sawed and cut into blocks rather than being blasted (Christiansen, 1974). But in most cases the metals of interest are present in association with the host rock or in other chemical forms.

Where metals have been put to the service of humankind, the benefits they bestow may be countered by the health effects that their extraction and usage may cause. Lead has been widely used in construction, etc., as it is easily worked and intricate components can be fabricated. While these properties were exploited by the Romans in particular, the addition of high concentrations of lead in their diet, by adulteration of wines by the deliberate addition of lead salts together with the lead leached from ceramic glazes, seals and plugs, has been ascribed to the downfall of that civilisation (see Nriagu, 1983). More recently, a chemical plant that was producing plastics discharged methylmercury formed from acetaldehyde and inorganic mercury (used as a catalyst) into a drainage channel leading into Minimata Bay, Japan. A non-infectious neurological illness among inhabitants of the area, especially fishermen and their families, who subsisted mainly on seafoods, was recognised in late 1953. This disease became known as the Minimata disease, as it was revealed in 1959 that the deaths were caused by the consumption of fish and other seafoods contaminated with methylmercury.

While the demand for particular metals has changed over the centuries, especially over the last century, where more exotic alloys have been produced for their very specific properties, the major environmental impact of these metals has been close to the point of extraction. Spoil heaps and tailings lagoons contain materials with elevated concentrations of metals. On exposure these deposits are quickly weathered, and this enhances the release of metals into the environment. Water plays an important role in the working of mineral deposits and, as a result, contamination of river channels away

---

Sediment and Water Quality in River Catchments. Edited by I.D.L. Foster, A.M. Gurnell and B.W. Webb.
© 1995 John Wiley & Sons Ltd.

from mines has been reported for many areas. This impact is not immediately measurable as just the enrichment in metal concentrations in the sediments of the downstream river catchment. The metals released from the mining operations may remain in a stable crystal lattice and will be relatively inert in the river channel. Only a small portion of metals in this form will enter the ecosystem. Conversely, chemical alteration of minerals can result in complexes that are readily available to plants or animals, and smaller numerical concentrations of these metals may have a devastating effect on an ecosystem. This 'bioavailability' of metals has to be considered to determine the full environmental impact of past or current mining operations. Finally, the environmental setting must be considered to determine how, when and where metals may impact the ecosystem. Sediment samples from a river catchment will provide an indication of the levels of contamination at that particular time. Such a snapshot may show the impact of mining operations hundreds of years after the mining activity ceased. If the stream has not reached an equilibrium with the mining-derived sediments, then the levels of contamination of the channel may continue to change.

MINING IMPACTS ON MINERALISED CATCHMENTS

There is a long history of metal mining in the United Kingdom. The principal ore fields are shown in Figure 9.1. Some ore deposits were worked by the Romans, but the most extensive operation of mines occurred in the 19th century. The spoil heaps left after one period of exploitation have often been reworked, especially where developments in technology have enabled metals to be recovered more effectively from the ores and from the discarded tailings from earlier activity. An opportunity is presented to study a range of mining catchments in the United Kingdom to establish the effects of key variables on heavy-metal dispersal and to note the environmental impact of mining operations at a range of times since the mines were in operation (Table 9.1).

The serious effects of mine effluents on water quality have been known for many years; rivers draining the orefields in the uplands of Wales have been grossly affected. One of the first descriptions of such impacts on rivers in the United Kingdom is the fifth report of the 1868 River Pollution Commission, which refers to the severe damage caused by the disposal of toxic metals from lead and zinc mines in mid-Wales (Rivers Pollution Commission, 1874):

> All these streams are turbid, whitened by the wastes of the lead mines in their course; and flood waters in the case of all of them bring down poisonous slimes which, spreading over the adjoining flats, either befoul or destroy the grass, and thus injure cattle and horses grazing the dirtied herbage, or, by killing plants whose roots have held the land together, render the shores more liable to abrasion and destruction on the next occasion of high water. It is owing to the latter cause as well as to the immense quantity of broken rock which every mine sends forth that the small rivers Rheidol and Ystwyth present such surprising widths of bare and stony bed

Early studies on the gross contamination of river channels by mining wastes focused on the loss of the fishery and the reduction in grazing on the floodplain. Carpenter (1924) sampled rivers around Aberystwyth for their fauna and flora, establishing that the relative poverty of certain streams was related to the pollution caused by lead

Figure 9.1 Metal mineralisation in Great Britain and the location of study catchments

Table 9.1 Characteristics of the river catchments used as case studies

River	Catchment area (km^2)	Mineralisation	Peak mining activity	Mining ceased
Ecclesbourne	50.4	Pb, Zn, Cd	1870s	1910
Hamps	58	Cu, Pb, Zn	Unknown	1868
Manifold	148	Cu, later Pb, Zn	1760 Cu, 1840s Zn	ca. 1870
Derwent	689	Mainly Pb	1850s	1950
Ystwyth	116	Pb, Zn, Cu, associated Cd, Ag	1850s	ca. 1920

mining and washing operations. While the mechanical abrasion of fish gills by particulates released from the mines was important, the loss of invertebrate fauna pointed to the impact of dissolved metal salts. The gradual re-colonisation of these channels has been charted in a sequence of publications from the 1920s, and it has become clear that, after a century since active mining in some of the catchments, the biological diversity of the channel is still impaired. Drainage of water through the wastes at these abandoned mines still provides a source of dissolved metals. While it can be anticipated that spoils from lead mines would provide a source for lead salts, the impact of associated metals can be just as important. Spoil heaps in the headwaters of the River Rheidol, besides providing a continuing supply of lead and zinc to the river, provide an important source of low concentrations of silver (Jones, 1986).

Spoil heaps around abandoned mines provide a point source of metals within a river catchment. Fine sediments can be suspended from the weathered surface and be blown to surrounding areas of habitation, constituting an environmental hazard (Davies and White, 1981). The erosion of these spoil materials also supplies metalliferous sediments to the stream channels for transport away from the mine. In general metal concentrations in stream sediments decline with distance away from the mine. Quantitative relationships have been investigated for a number of sites (Lewin and Macklin, 1987), and best-fit curves using power functions, exponential or linear models have been presented. Even in simple catchments within a few tens of kilometres from the mine, the fit between concentration and distance may be poor. Causes of concentration decline are compound. Lewin and Macklin (1987) recognised the operation of several mechanisms:

- Hydraulic sorting according to differential particle density.
- Chemical dispersal through solution or biological uptake.
- Mixing with extra 'clean' sediment, especially as a contribution from a tributary channel that is not mineralised.
- Loss to, or exchange with, stored floodplain sediments.

At any point within the channel the relative importance of each of these mechanisms will influence the observed metal concentrations; in different river systems one of these mechanisms may have a dominant role. Where the sediments contain a suite of metals, the decline in concentration may not be the same for each. Moriarty et al. (1982)

reported for the River Ecclesbourne that: 'In general the concentrations of all three metals decrease downstream, but it is noticeable that the values for zinc and cadmium have very similar patterns and are rather different from that for lead.'

Whilst Moriarty *et al.* (1982) observed a general trend for metal concentrations to decrease downstream, (for lead from over 20 000 mg kg^{-1} near the mine to about 700 mg kg^{-1} near the confluence with the River Derwent), channel sediments collected from the same locations but at different times of the year contained rather different concentrations of metals. They interpreted this temporal pattern as a pulse of sediments with relatively high concentrations that was moving down the river. Consideration of these findings demonstrates the need to consider the dynamics of sediment movement

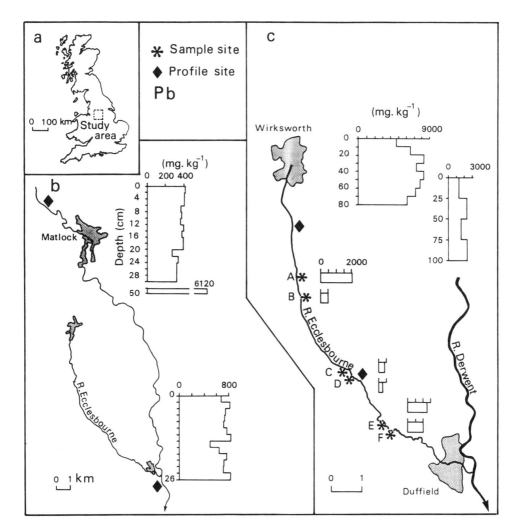

Figure 9.2 (a) Location of the Ecclesbourne catchment in Great Britain. (b) Concentrations of lead in floodplain sequences on the River Derwent, above and below the confluence with the River Ecclesbourne. (c) Concentrations of lead in channel banks and within channel sediments along the River Ecclesbourne

and their supply within the catchment together with those sediments being actively transported in the channel. Channel sediments collected independently, and on a later occasion, showed different concentrations of metals to those reported in the earlier studies, notwithstanding that the samples were not collected from exactly the same locations (Figures 9.2 and 9.3). These sediments were sampled in 1986 where collection of channel sediments was integrated with the sampling of channel bank sediments. The locations chosen had all been undercut by the contemporary channel and there was evidence of slumping of the bank into the river. The bank sediments contained higher concentrations of metals than the active channel sediments. In the upper part of the catchment, but several kilometres downstream of the derelict mines, the sediments

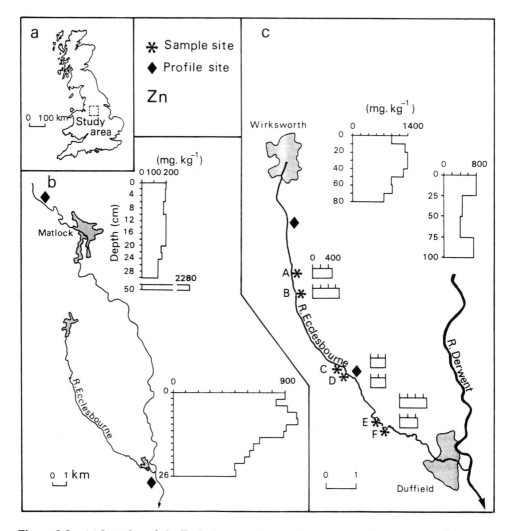

Figure 9.3 (a) Location of the Ecclesbourne catchment in Great Britain. (b) Concentrations of zinc in floodplain sequences on the River Derwent, above and below the confluence with the River Ecclesbourne. (c) Concentrations of zinc in channel banks and within channel sediments along the River Ecclesbourne

contained appreciable concentrations of metals throughout the 1 m sequence that was sampled. Indeed, these sequences contained higher concentrations of lead and zinc than observed from profiles in the floodplain of the River Derwent (the Derwent having received sediments from a large number of mines in its upper catchment). Local bank collapse or slumping would inject contaminated sediments into the channel and these would locally enhance the concentrations in the channel sediments. The pattern of metal concentrations shown in the channel sediments is probably not controlled by the dispersal of sediments from the point source at the mine, but from the local contribution from the channel banks throughout the catchment. It is important then to recognise the storage areas for metal-contaminated sediments within the catchment and to appreciate the timing when these materials are released back into the contemporary river channel.

FLOODPLAIN STORAGE OF METALLIFEROUS SEDIMENTS

Floodplains can act as a sink and a subsequent source of sediments to the river channel. At equilibrium the loss of sediment by channel erosion at the margins of the floodplain or scour from its surface is balanced by accretion over the surface. When the supply of sediment from upstream is changed, naturally by climate change, or artificially by, for example, mining activity, the conditions on the floodplain are affected. Mining in the catchment is accompanied by an enhanced supply of sediments to the channel. The grading of ores and milling operations also result in the supply of sediments with different particle-size distributions to those found naturally. Some early American litigation concerned the impact of mining where the destruction of agricultural land occurred by the deposition of large volumes of inert, sand-sized debris from the hydraulic gold mining in the Sierra Nevada, USA (Kelley, 1959). In this case the vegetation was buried by several feet of sand; flood defences around cities downstream from these operations were progressively raised to keep pace with the elevated channel planform, to the point where cities surrounded by embankments 40 feet high were not unknown. Observations on the sources and the transport of sediments during hydraulic mining in the Sacramento River and its tributaries showed a systematic rise and fall of the channel bed during a period approaching 60 years (Gilbert, 1917), but these effects are relatively short-term compared to the full adjustment of river channels with their floodplain.

The enhanced supply of sediment to the river channel can have repercussions at great distances from the mining centre. Finely milled mine tailings disposed of to the Belle Forche River, South Dakota, USA, were visually distinguishable in floodplain deposits from pre-mining alluvium up to 160 km downstream of the mine (Marron, 1987). Indeed, a rough calculation of the amount of mine tailings that were deposited along the Belle Forche River suggested that at least a third of the mine tailings discharged were stored within the 160 km of floodplain downstream of the mine. Smaller-scale impacts on river floodplains are seen from the more modest mining operations in the United Kingdom.

The impact of mining on floodplains downstream may be recorded in channel characteristics presented in sequences of maps or, for more recent disturbances, in aerial photographs. Where these maps are sufficiently detailed, the impact may be to change the channel pattern from a single to braided regime, for example as recorded for the River Ystwyth between 1800 and 1946 (Lewin *et al.*, 1983) or to indicate the extent of the active sedimentation zone.

An interesting contrast of floodplain sedimentation between two adjacent catchments is presented in the rivers Hamps and Manifold, Staffordshire (Bradley and Cox, 1986; Figure 9.4). In the Manifold catchment contaminated sediments were introduced to the river channel from a number of mines. Stamp mills worked in the catchment and they processed ores from both the Manifold and Hamps catchments and former spoil heaps from both catchments were re-processed. Consequently, the fine sediments from the stamp mills formed a large portion of the anthropogenic supply of sediments to the Manifold catchment, while coarser sediments were added to the channel in the Hamps. The fine sediments added to the River Manifold accumulated in vertical sequences of sediments on the downstream floodplains. In places these fine sediments form 1–2 m high mining terraces. By contrast the coarser sediments added to the River Hamps accumulated laterally as gravel splays and lenses across floodplains.

Floodplain sediments in both the Hamps and Manifold catchments contain elevated concentrations of metals. When the total content of metals is calculated as a combined molarity to enable the sum of lead, copper, zinc and cadmium concentrations to be

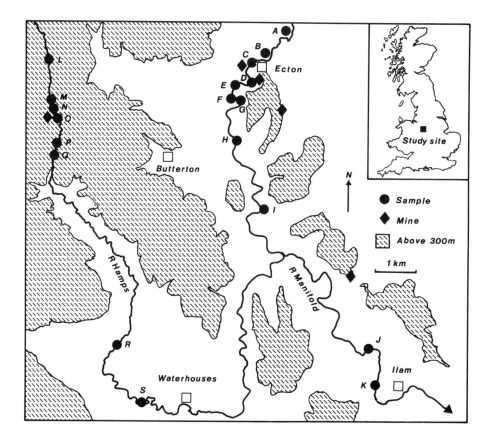

Figure 9.4 Sampling sites and former mines in the Hamps and the Manifold catchments

compared between particle fractions, the fine and medium sands in the floodplains of the Manifold are deficient of metals (Figure 9.5). This pattern is not due to a preferential dispersal of coarser, metalliferous particles, but to the removal of metals from fine sands in the Manifold catchment.

In more complicated situations where the floodplain is receiving sediments as a result of mining in several subcatchments, floodplain aggradation may combine both vertical and lateral components. Many mining ventures were operated irregularly and over long periods in subcatchments of the River Derwent, Derbyshire. As the volume of mining waste contributed to the main channel was a small proportion of the total sediment yield, floodplain sediments for the pre-mining era are not distinguishable from the later sediments. In a study of floodplain sediments in the River Derwent, Bradley and Cox (1990) collected sediment cores on a transect over the floodplain at right angles to the channel (Figure 9.6). There was no clear variation in metal levels across the floodplain or vertically in the sediment profile (Figure 9.7). The highest levels of lead, zinc and cadmium were recorded for sediments from a terrace at the extreme of the floodplain, some 500 m from the present channel. The majority of mining effort in the area ceased by 1913, but spoil heaps throughout the catchment are slowly being eroded and bank collapse along the tributaries provides contemporary channel sediments that contain metals. The study site on the River Derwent is actively aggrading, but the metal content of the sediment does not indicate whether it was added recently or whether it is contemporaneous with the mining activity in the catchment.

The recent (past 30 years) sediment accumulation onto a river floodplain is distinguishable by the excess ^{137}Cs contained in the sediment profile over and above that from fallout alone (see Walling *et al.*, 1986, for methodology). Sediments are not

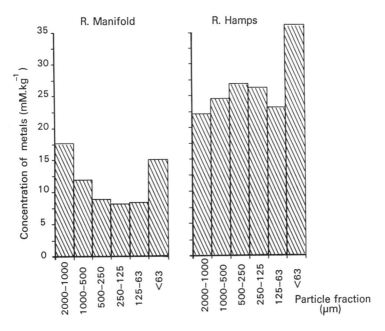

Figure 9.5 Total mass of metals in mine-derived sediments in the Manifold and Hamps catchments

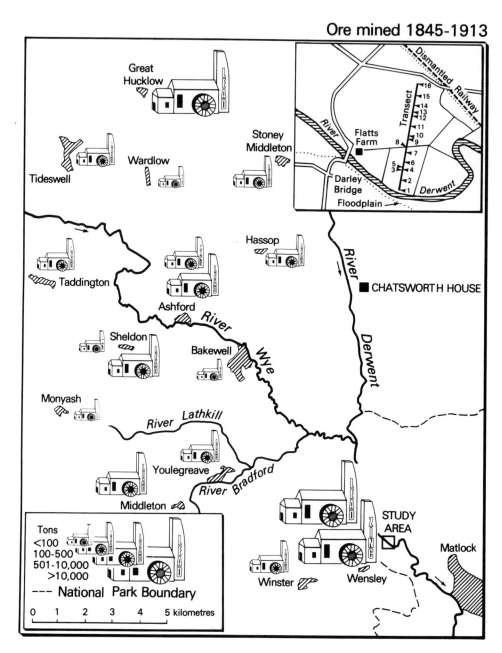

Figure 9.6 Principal mines and their output of lead and zinc ore between 1845 and 1913, and sampling points on the floodplain of the River Derwent

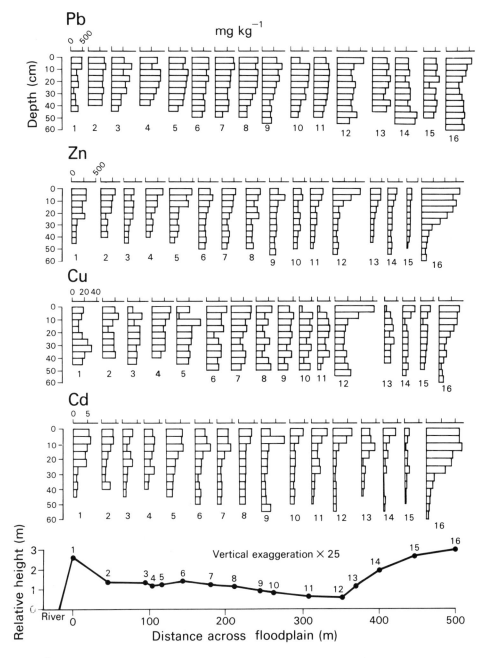

Figure 9.7 Transect across the floodplain, and concentrations of lead, zinc, copper and cadmium in the sediment profiles

added to the floodplain surface as a uniform veneer. The microtopography of the surface controls the dynamics of sediment deposition and re-suspension, and consequently the sediments aggrade at different rates across the floodplain. By determining the excess ^{137}Cs at positions across the floodplain on the Derwent, the recent sediments were distinguishable from earlier materials, while the contemporary transport and deposition of metals onto the floodplain surface could be calculated. It is estimated from this evidence that the contemporary supply of metals to this 0.2 km^2 area of floodplain is in the order of 360 kg lead, 100 kg zinc, 10 kg copper and 2 kg of cadmium each year. There is considerable scope to employ the same methodology to river floodplains elsewhere, particularly to determine the continuing impact of mining operations, even though active mining may have ceased long ago.

CHEMICAL MODIFICATION OF METALS IN SEDIMENTARY DEPOSITS

During the storage of mining-derived sediments in floodplains, the metals in the sediments are exposed to pedogenic processes that may affect the chemical form of the metal. In mining catchments, the metals attached to the sediments are primarily in mineral form. When these sediments are incorporated into floodplains downstream, they may remain in the soil domain for thousands of years before they are re-worked and again become part of the sediment load of the river. While the total metal content is important concerning the polluting effects of the re-mobilised sediments in the river catchment, the *speciation*, or chemical form of the metal, is of equal importance. The release of metals from the sediment to solution may be manifestly different for fresh sediment (or spoil particulates) compared to that which has undergone pedogenic alteration over a number of years.

A number of chemical extractants have been proposed that are deemed to be selective in the chemical form of a metal that they liberate from a sediment sample. Using mild reagents, metals that are least firmly bound to the sediment will be released into the extracting mixture. At the other extreme, the portion of the metal fraction that is only liberated after refluxing with very aggressive acids constitutes that which is unlikely to play an important role in the cycling of metals through the ecosystem. These schemes of chemical extraction have been heavily criticised where the merits of one scheme are pitted against those of another (for example, Martin *et al.* 1987; Nirel and Morel, 1990). Indeed, while there is a multitude of published extraction schemes, each of which reflects the originator's personal preferences, the scientific literature is burdened with opinion of the veracity of one scheme compared with another. If these schemes are used to show only the relative importance of the extractants to remove metals from a sediment, and to provide a means to compare the qualities of samples from several locations, then, within these limits, chemical extractions are useful to untangle the chemical speciation of the metals. Lander (1987) provides a thorough review of the implementation of chemical extractants to provide information on the speciation of metals in water, sediment and soil systems.

A series of chemical extractants proposed by Tessier *et al.* (1979) has been used to investigate the speciation of metals in base-metal mining catchments in Derbyshire and Staffordshire. This scheme, although altered to provide a safer extraction of metals in the least-mobile phase, has been chosen as it has become the closest to a 'standard' method currently available. In the study catchments, metalliferous sediments were dispersed, primarily, during active mining. As the sedimentary units that are preserved on the floodplain were deposited then, and as mining ceased in this area around 1870, pedogenic processes have influenced these sediments for over 100 years. The supply of large amounts of fine-grained sediments from the processing of ores, and later re-processing of spoils in the Manifold catchment, was accompanied by vertical accretion of the floodplains downstream. In places this accretion is represented by uniformly sand-sized particles to over 2 m in depth. The lower sediments have experienced little pedogenesis and may have similar chemical properties to those when they were originally disposed of to the river. These sediments are highly contaminated with lead, zinc, copper and cadmium, with levels declining towards the surface (Figure 9.8). Concentrations of zinc and cadmium show a similar pattern as, geochemically, cadmium is associated with zinc in ores and they are highly correlated ($R^2 = 0.96$). Large portions of cadmium (up to 64%) and zinc (25%) were removed with the mild extractants and were easily exchangeable, and hence 'bioavailable'. Only small amounts of lead and copper were in this phase. For lead the Fe/Mn oxide and the carbonate fraction were important; from 60 cm to the soil surface the organic fraction was also important, and this may reflect the maximum depth of pedogenic alteration on mineral particles.

Sequential extraction on surface sediments from the River Derwent showed that the chemical form of metals varied across the floodplain (Figure 9.9). For lead the sediments from the levee contained a small portion in the exchangeable form, while most was contained in the mineral matrix of the sediment. The organic fraction was most important for copper. This association has been reported for many studies, where the copper has been added with either inorganic sediments or by the application of sewage sludge to fields. Away from the levee, substantial quantities of lead were easily exchangeable, as were copper and cadmium. On the mining terrace at site 16, the pattern of the exchangeable fraction for each of the metals was again different. Investigation of the ^{137}Cs excess in these floodplain sediments had revealed that sediments were not deposited uniformly across the floodplain. Analysis of the sediments showed that the total concentration of metals in the surface was relatively uniform, but inspection of the results of these extractions shows that the chemical forms are more variable.

PROSPECT

The environmental impact of mining operations can be assessed from the results of concurrent monitoring programmes. Where these programmes are planned and implemented before the mining begins, the impact of the mining operation on the catchment can be compared directly with the pre-existing conditions. Such environmental impact assessments were not undertaken for the small-scale base-metal mines in the United Kingdom, albeit such impact assessments would form part of a planning application today. The priorities in earlier centuries were to recover the maximum amount of metals to secure the maximum profit for the mine owners, regardless of the impact on the environment. These mines are now derelict and the spoil

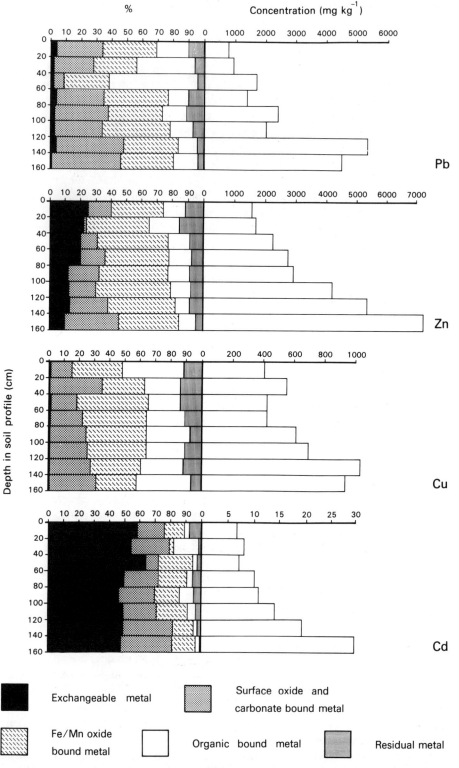

Figure 9.8 Total concentrations and partitioning of lead, zinc, copper and cadmium in a soil profile in the Manifold catchment

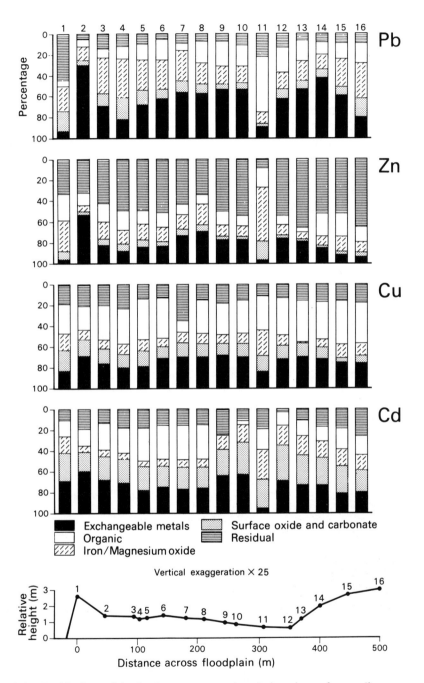

Figure 9.9 Partitioning of lead, zinc, copper and cadmium in surface sediments over the floodplain of the River Derwent

heaps immediately around the installations bear testament to the local impact of these workings. While these point sources are important to supply metals, significant quantities of metals are cycled through the river system by the erosion of floodplain deposits. As the chemical form of the metals may be changed by long-term storage within the floodplain deposits, the contemporary environmental impact of these liberated metalliferous sediments may be severe.

In order to predict the rate and scale of the continuing cycling of metals in these former mining catchments, the interaction between the floodplain and the channel must be understood. Advances in this area have followed the development of new techniques to trap sediments as they are deposited on the floodplain (Lambert and Walling, 1987), to recover fine sediments from the matrix of gravel-bed streams (Lambert and Walling, 1988) and to quantify the rates of sedimentation and erosion (Walling et al., 1986). These techniques can be combined with established methods to characterise the chemical properties of sediments, together they provide a powerful means to untangle the continuing cycling of metals.

REFERENCES

Bradley, S.B. and Cox, J.J. (1986) Heavy metals in the Hamps and Manifold valleys, North Staffordshire, UK: distribution in floodplain soils. *Sci. Total Environ.* **50**, 103–128.

Bradley, S.B. and Cox, J.J. (1990) The significance of the floodplain to the cycling of metals in the River Derwent catchment, UK. *Sci. Total Environ.* **97/98**, 441–454.

Carpenter, K.E. (1924) A study of the fauna of rivers polluted by lead mining in the Aberystwyth district of Cardiganshire. *Ann. Appl. Biol.* **11**(1), 1–23.

Christiansen, P.W. (1974) *The Story of Mining in New Mexico*, Scenic Trips to the Geologic Past no. 12, New Mexico Bureau of Mines and Mineral Resources.

Davies, B.E. and White, H.M. (1981) Environmental pollution by wind blown lead mine waste: a case study in Wales, UK. *Sci. Total Environ.* **20**, 57–74.

Gilbert, G.K. (1917) *Hydraulic-Mining Debris in the Sierra Nevada*, US Geological Survey Professional Paper no. 105, 154pp.

Jones, K.C. (1986) The distribution and partitioning of silver and other heavy metals in sediments associated with an acid mine drainage stream. *Environ. Pollut. B* **12**, 249–263.

Kelley, R.L. (1959) *Gold Vs Grain: The Hydraulic Mining Controversy in California's Sacramento Valley*, A.H. Clark Co., Glendale, CA.

Lambert, C.P. and Walling, D.E. (1987) Floodplain sedimentation: a preliminary investigation of contemporary deposition within the lower reaches of the River Culm, Devon, UK. *Geogr. Ann.* **69A** (3–4), 393–404.

Lambert, C.P. and Walling, D.E. (1988) Measurement of channel storage of suspended sediment in a gravel-bed river. *Catena* **15**, 65–80.

Lander, L. (1987) *Speciation of Metals in Water, Sediment and Soil Systems*, Lecture Notes in Earth Sciences no. 11, Springer-Verlag, Berlin.

Lewin, J. and Macklin, M.G. (1987) Metal mining and floodplain sedimentation in Britain. In: Gardiner, V. (ed.), *International Geomorphology 1986*, part 1) John Wiley & Sons, Chichester, pp. 1009–1027.

Lewin, J., Bradley, S.B. and Macklin, M.G. (1983) Historical valley alluviation in mid-Wales. *Geol. J.* **18**, 331–350.

Marron, D.C. (1987) Floodplain storage of metal contaminated sediments downstream of a gold mine at Lead, South Dakota. In: Averett, R.C. and McKnight, D.M. (eds), *Chemical Quality of Water in the Hydrologic Cycle*, Lewis, Chelsea, MI.

Martin, J.M. Nirel, P. and Thomas, A.J. (1987) Sequential extraction techniques: promises and problems. *Marine Chem.* **22**, 313–341.

Moriarty, F., Bull, K.R., Hanson, H.M. and Freestone, P. (1982) The distribution of lead, zinc and cadmium in sediments of an ore-enriched lotic ecosystem, the River Ecclesbourne, Derbyshire. *Environ. Pollut. B* **4**, 45–68.

Nirel, P.M.V. and Morel, F.M.M. (1990) Pitfalls of sequential extractions. *Water Res.* **24**(8), 1055–1056.

Nriagu, J.O. (1983) *Lead and Lead Poisoning in Antiquity*, John Wiley & Sons, Chichester, 437pp.

River Pollution Commission (1874) *Fifth Report of the Commissioners Appointed in 1868 to Inquire into the Best Means of Presenting the Pollution of Rivers.*

Tessier, A., Campbell, P.G.C. and Bisson, M. (1979) Sequential extraction procedure for the speciation of particulate trace metals. *Anal. Chem.* **51**, 844–851.

Walling, D.E., Bradley, S.B. and Lambert, C.P. (1986) Conveyance losses of suspended sediment within a flood plain system. In: *Drainage Basin Sediment Delivery*, IAHS Publication no. 159, pp. 119–132.

10 Fingerprinting Sediment Sources: An Example From Hong Kong

MERVYN R. PEART

Department of Geography and Geology, University of Hong Kong, Hong Kong

INTRODUCTION

Deposition of sediment by rivers can generate many problems (e.g. Sundborg and Rapp, 1986) and Hong Kong is no exception. Berry (1955) comments upon the threat posed to the Kowloon, Jubilee and Tai Lam Chung reservoirs by sedimentation resulting in the loss of storage. Davis (1949) reports that the old Aberdeen reservoir lost 20% of its storage due to sedimentation between 1890 and 1930. Owing to remedial action, such as afforestation, it is now believed that sedimentation does not seriously affect water supply operations. However, it does affect the use of weirs constructed to supply water for irrigation to agriculture. The Agriculture and Fisheries Department operates a maintenance programme for sediment removal and, for example, in the years 1990–91 and 1991–92 removed 919 and 1017 t of sediment at a cost per year of HK$450 000 and HK$517 000 respectively. They also need to remove sediment from the irrigation reservoirs. The Port Works Division (1988) examined sedimentation in a number of man-made river channels. They found that over 300 000 m^3 of material, or about 400 000 t, had accumulated. These sediments are often heavily polluted, and the Port Works Division (1988) estimated that it would cost HK$20 million per annum for dredging in order to control the problem and remove the polluted sediments and improve the aesthetics of the sedimented reaches. The impact of sediment deposition in river channels as a contributor to flooding has been documented by a number of studies in Hong Kong (e.g. Territory Development Office, 1989) and has been identified by the Drainage Services Department as a causal factor. Sediment transport and siltation also impacts upon stream fauna. Dudgeon (1994) reports that, after roadworks had caused modification of sediment transport in a section of the Lam Tsuen river, the zoobenthic taxa decreased from 52 to 24.

Hong Kong has a sedimentation problem, and currently the remedial action consists of removal of the deposited sediment, afforestation and protection of eroding sites, and the control of animal waste pollution in streams, which, in some areas, has contributed significantly to the sedimentation problem (e.g. Port Works Division, 1988). It might prove possible to take more effective remedial and preventive action if the sources of sediment were known. Apart from the Port Works Division (1988) study and Peart (1993), little attempt has been made to identify sediment sources in Hong Kong.

A number of methods can be used to identify sediment sources and they have been reviewed by Peart (1989). One of the most attractive ways to elucidate sediment source

---

Sediment and Water Quality in River Catchments. Edited by I.D.L. Foster, A.M. Gurnell and B.W. Webb.
© 1995 John Wiley & Sons Ltd.

is to use the properties of the sediment to identify its origin. By comparing the sediment properties from the river to those of potential source material, such as channel bank or topsoil, it may be possible to identify their origin. As noted by Walling *et al.* (1993) the use of sediment properties to trace the source of the material is an alternative method to the more traditional monitoring techniques (discussed by Peart, 1989) and it may avoid some of the problems associated with them. Early examples of this fingerprinting approach are offered by Wall and Wilding (1976) and Oldfield *et al.* (1979). More recent developments, especially in the quantification of the contribution made by various sources to the sediment output, are exemplified by, for example, Walling *et al.* (1993), while the study of Caitcheon (1993) indicates that the method may be used in a more spatial context. This chapter reports upon an attempt to use sediment properties in order to try to determine their source in a river in the New Territories of Hong Kong.

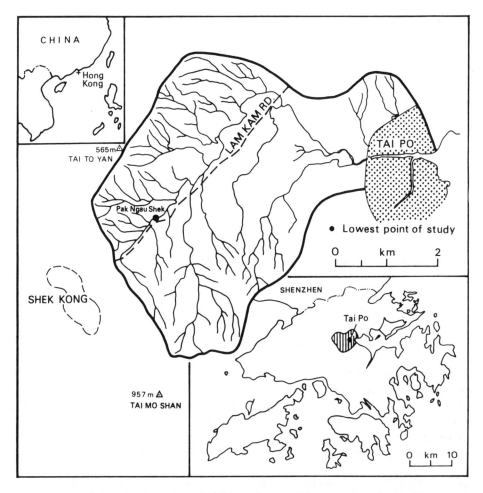

Figure 10.1 Location of the Lam Tsuen study basin, Hong Kong

STUDY AREA

The basis of this study area is the Lam Tsuen river, which has a length of about 9 km and drains an area of around 20 km² in the central New Territories of Hong Kong (Figure 10.1). It has its origins on the slopes of Tai Mo Shan, which at 957 m is Hong Kong's highest peak. One of the western tributaries near Pak Ngau Shek, which is about 7 km from the mouth of the river, flows in a valley, which provides a major road link between Tai Po, Shek Kong and Yuen Long. The road is being improved near Pak Ngau Shek, and here work impinges upon the river and it is this part of the basin that is being studied. Most of the basin is covered with grassland, shrubland and woodland. Acid volcanics and acid intrusive igneous rocks form the bedrock. Slopes are generally steep, 30° to 40° being common, and a number of landslides occurred during heavy rainfall in May 1989.

MATERIALS AND METHODS

Two major categories of material were recognised as potential sources of sediment in the Lam Tsuen river basin, namely surface soil and substrate material. These were selected in order to provide a basis for distinguishing between sediment sources associated with shallow topsoil erosion and those based upon the deep-seated erosion of the subsoil or substrate exposed by human impact such as construction activity. Surface soil samples were collected from the top few centimetres of material exposed in landslide scars that occurred in the basin in May 1989. The topsoil had a darker appearance than the lower substrate owing to organic matter incorporation and this aided delimitation. Additional topsoil samples were collected from areas of the basin that did not experience landslips. Substrate material was classified into that from cut-and-fill slopes associated with roadwork construction and that from the slopes of the basin. Use was made of exposed landslide scars to obtain samples of substrate but, in areas lacking natural exposures associated with landslips, soil pits were excavated or other exposures such as cuts for paths utilised. The cut-and-fill slopes were sampled on a number of occasions because the engineers have modified the exposed material as work progressed. Bulk samples were taken so as to provide a spatially integrated coverage of the exposed cut-and-fill slopes.

Samples of suspended and bottom sediment were collected for analysis downstream of the roadworks. The bottom sediments were air-dried and sieved, and the <2 mm fraction was kept for analysis. Suspended sediment samples were filtered through pre-weighed Whatman GF/C filter papers, which were retained for determination of sediment concentrations and loss-on-ignition. For a limited number of samples, the suspended matter was obtained by means of settlement and decanting of the clarified water. The levels of suspended sediment were also determined in the Lam Tsuen river upstream of the roadworks or at other control sites, such as the tributary stream near Pak Ngau Shek in Figure 10.1, for comparison with those observed at the site impacted by construction. Suspended sediment samples were collected under baseflow and stormflow conditions. Only stormflow samples contained sufficient sediment for tracing source and only these have been reported. Samples were obtained over a wide range of water levels, from little more than baseflow to near bankfull. Both rising and falling limbs were sampled and depth-integrated using 500 ml sampling bottles.

The source material and sediment were analysed for organic carbon and total nitrogen using Walkley–Black and Kjeldahl methods. In order to ensure complete digestion of the sample, a pestle and mortar was used to grind the samples after sieving. Loss-on-ignition treatment was carried out at 400°C for 6 h. Bottom sediment samples were collected during 1991, 1992, 1993 and 1994 while those for suspended matter were obtained in 1991, 1992 and 1993.

RESULTS

Mean values of the sediment properties used to fingerprint sediment source are presented in Table 10.1. It can be seen that organic carbon, loss on ignition and carbon/nitrogen ratios are all much higher for surface soil in comparison to substrate material. In contrast, there is little difference between values of total nitrogen for soil and substrate. Consequently, it is not possible to use this determinand to differentiate sediment source. Furthermore, none of the sediment properties are able to differentiate between the two substrate sources, that is to say material associated with roadwork construction and that derived from substrate on the slopes.

Table 10.1 also presents mean values of the same properties for bottom sediment from the Lam Tsuen river. In the case of the bottom sediments, the organic carbon and loss-on-ignition values most closely approximate those of the substrate material and suggest a substrate origin for the <2 mm size fraction of the bottom sediments. The C/N ratio, however, is closer to that of soil materials and implies that surface sources may be important.

Values for loss on ignition, organic carbon and total nitrogen are presented in Table 10.1 for suspended sediment. The loss-on-ignition values (based upon ignition of sediment collected on filter papers) most closely resemble those of surface soil. In contrast, the organic carbon content of the suspended matter is intermediate between that of soil and subsurface materials, and suggests that substrate sources may make a significant contribution.

Table 10.1 Mean values and standard deviations of source and sediment properties

Material	Organic carbon (%)	Total nitrogen (%)	C/N ratio	Loss on ignition (%)
Soil	2.56 ± 0.83	0.19 ± 0.07	15.00 ± 5.67	5.05 ± 1.31
Substrate	0.34 ± 0.11	0.10 ± 0.07	4.03 ± 2.30	1.65 ± 0.46
Cut-and-fill substrate	0.21 ± 0.09	0.17 ± 0.13	5.47 ± 8.38	1.14 ± 0.48
Bottom sediment[b]	0.35 ± 0.22	0.07 ± 0.11	10.43 ± 8 27	1.77 ± 1.18
Suspended sediment	1.12 ± 0.92	0.14 ± 0.09	NA[a]	6.38 ± 2.51
Number of samples, n^c	13	8	NA	24

[a] Not available   [b] 26 Samples analysed   [c] Suspended sediment

DISCUSSION

Use has been made in this study of bottom sediment properties to identify their origin. They were collected at various times since observations in the river began. The sample standard deviations reported in Table 10.1 reveal that the properties of the bottom sediments are quite variable but showed no clear seasonal pattern. This suggests that the bottom sediments in this river channel may not act to integrate properties through time, and the collection of one sample with which to characterise the bottom material is not possible. Ongley (1982) has indicated that bottom sediments in rivers with mobile beds will not provide a time-integrated record of nutrient and contaminant levels. Phillips (1985) has also recognised the problem of bottom sediment mobility and time integration. The bed material in the study river is quite mobile, with large storms moving bed sediment through the study reach.

The mobility of the channel bottom sediments may be of benefit to the fingerprinting methods using the tracers employed in the present study. If the bed were stable, organic matter from aquatic plants and zoobentha could accumulate. This would certainly enrich the organic carbon and loss-on-ignition values. In consequence, these determinands along with the C/N ratio would no longer reflect those of the source materials. Contamination and alteration of bottom sediments so that they no longer reflect those of the parent material remain a serious difficulty for the fingerprinting technique, especially when using bed material. In the present study, the high mean value for the C/N ratio results from five samples with very low nitrogen levels. It is possible that during storage, however short the duration, nitrogen was lost from the sediments. Such a transformation would mean that nitrogen and hence the C/N ratio would not reflect the original source material and may falsely point to surface soil as a significant source. Information on the storage of sediment, especially fines, in the channel system, such as that provided by Lambert and Walling (1986), would be of benefit in assessing the potential for contamination and alteration.

The different results regarding the origin of suspended sediment provided by organic carbon and loss on ignition is somewhat problematic. Further work is clearly needed to explore this situation, and the use of a wide range of tracers as advocated by Peart and Walling (1986) might resolve the situation. Unfortunately, insufficient material was available to expand the database, although the recent acquisition of equipment to examine the magnetic properties of sediment should help to achieve this in the future. Simple process measurements, however, can help to elucidate the origin of the suspended matter. In the Lam Tsuen basin, the fingerprinting results can be confirmed by observations of erosion and suspended sediment concentrations. Overland flow troughs were installed on slopes covered by fern and grass vegetation, and under woodland in adjoining drainage basins. The overland flow troughs were around 1.0 m wide under the grass and fern vegetation and 1.5 m in width in the woodland. They were all unbounded, giving contributing slope lengths in some cases of over 30 m on the grass and fern and 50 m for the woodland sites. In 1993, on only three visits out of 37 following rainfall were significant volumes of water found in the collectors. The sediment content was also low, being at most a few kilograms. This suggests that, with an undisturbed vegetation cover, little material that may become bedload and suspended matter is likely to be produced from topsoil erosion. This conclusion is in part confirmed by the work of Lam (1974), who reports much lower suspended sediment

Table 10.2 Comparison of suspended sediment concentrations (mg l^{-1}) in the Lam Tsuen river and the control sites

Year		Lam Tsuen river	Control sites
1991	Max.	4 400	408
	Min.	209	12
	Median	1 378	59
	16 observations		
1992	Max.	18 222	152
	Min.	114	2
	Median	866	15
	40 observations		
1993	Max.	6 426	267
	Min.	12	1
	Median	711	20
	62 observations		

yields originating from a small basin with a complete vegetation cover in comparison with two other basins that had 24% and 40% of their slopes devoid of vegetation.

Observations of suspended sediment concentration made in the Lam Tsuen river and paired with samples collected in undisturbed tributaries or control sites in adjacent basins are presented in Table 10.2. Suspended sediment concentrations below the roadworks are clearly much higher than for undisturbed sites and this was the case for every paired set of samples. The data suggest that the substrate of the cut-and-fill slopes associated with the roadworks provide a ready supply of sediment. These observations also clarify the fingerprinting results, which for suspended matter were inconsistent. The process observations are indicative of the importance of substrate exposed in cut-and-fill slopes as a source of suspended sediment.

CONCLUSION

Four properties of sediment have been investigated in terms of elucidating sediment source. One of them, total nitrogen, was not sufficiently different between surface soil and substrate material to be used as a tracer. Values of organic carbon and loss on ignition, when determined for the <2 mm fraction of bottom sediment in the Lam Tsuen river, indicated the dominance of substrate material. They could not, however, distinguish between sediment derived from roadwork construction sites and that from natural erosion of slopes by, for example, gullying or reworking of material exposed through landslides. The results of the C/N ratio were possibly influenced by the transformation of nitrogen during storage.

The two suspended sediment properties used to identify source gave conflicting results, one indicating a dominant surface origin and the other suggesting that subsurface sources might be important. However, the paired observations of suspended sediment concentrations revealed substrate derived from roadworks to be the major contributor of suspended sediment. Process observations may be useful in explaining and clarifying fingerprinting results, especially where only a limited number of properties are available with which to evaluate source.

This study indicates that the fingerprinting approach outlined so well by Meier (1977) can help to identify sediment source. However, as indicated, it is not without its problems, but it remains an attractive methodology, in part, because of the inherent simplicity of the concept.

REFERENCES

Berry, L. (1955) Effects of erosion in Hong Kong. *Far East. Econ. Rev.* **19**, 98–100.

Caitcheon, G.G. (1993) Applying environmental magnetism to sediment tracing. In: *Tracers in Hydrology*, Peters, N.E., Hoehn, E., Leibundgut, Ch., Tase, N. and Walling, D. E. (eds), International Association of Hydrological Sciences, Wallingford, pp. 285–302.

Davis, S.G. (1949) *Hong Kong in Its Geographical Setting*, Collins, London.

Dudgeon, D. (1994) Functional assessment of the effects of increased sediment loads resulting from riparian-zone modification of a Hong Kong stream. *Verh. Internat. Verein. Limnol.* **25**, 1790–1792.

Lam, K.C. (1974). Some aspects of fluvial erosion in three small catchments, New Territories, Hong Kong. Unpublished MPhil thesis, Department of Geography and Geology, University of Hong Kong.

Lambert, C.P. and Walling, D.E. (1986). Suspended sediment storage in river channels: a case study of the River Exe, Devon, UK. In: *Drainage Basin Sediment Delivery*, Hadley, R.F. (ed.), International Association of Hydrological Sciences, Wallingford, pp. 263–276.

Meier, M.C. (1977). Research needs in erosion and sediment control. In: *Soil Erosion: Prediction and Control (Proceedings of a National Conference on Soil Erosion, May 1976, Purdue University, Indiana)*, Soil Conservation Society of America, pp. 85–89.

Oldfield, F., Rummery, T.A., Thompson, R. and Walling, D.E. (1979) Identification of suspended sediment sources by means of magnetic measurements: some preliminary results. *Water Resour. Res.* **15**, 211–218.

Ongley, E.D. (1982) Influence of season, source and distance on physical and chemical properties of suspended sediment. In: *Recent Developments in the Explanation and Prediction of Erosion and Sediment Yield*, (ed.), Walling, D.E. International Association of Hydrological Sciences, Wallingford, pp. 371–384.

Peart, M.R. (1989) Methodologies currently available for the determination of suspended sediment source: a critical review. In: *River Sedimentation (Proceedings of the Fourth International Symposium)*, China Ocean Press, Beijing, China, vol. 1, pp. 150–157.

Peart, M.R. (1993) Using sediment properties as natural tracers for sediment source: two case studies from Hong Kong. In: *Tracers in Hydrology*, Peters, N.E., Hoehn, E., Leibundgut, Ch., Tase, N. and Walling, D.E. (eds), International Association of Hydrological Sciences, Wallingford, pp. 313–318.

Peart, M.R., and Walling, D.E. (1986) Fingerprinting sediment sources: the example of a drainage basin in Devon, UK. In: *Drainage Basin Sediment Delivery*, (ed.), Hadley, R.F. International Association of Hydrological Sciences, Wallingford, pp. 41–55.

Phillips, D.J.H. (1985) Monitoring and control of coastal water quality. In: *Pollution in The Urban Environment: POLMET 85*, Hong Kong Institution of Engineers Publications Division, Hong Kong, pp. 559–572.

Port Works Division (1988) *The Sedimentation Study of Various Man-Made River Channels in the Territory*, Civil Engineering Services Department, Hong Kong Government, Hong Kong, 2 vols.

Sundborg, A. and Rapp, A. (1986) Erosion and sedimentation by water: problems and prospects. *Ambio* **15**(4), 215–225.

Territory Development Office (1989) *River Indus Study*, Territory Development Office, New Territories North-East, Hong Kong Government, Hong Kong, 5 vols.

Wall, G.J. and Wilding, L.P. (1976) Mineralogy and related parameters of fluvial suspended sediments in Northwestern Ohio. *J. Environ. Qual.* **5**, 168–173.

Walling, D.E., Woodward, J.C. and Nicholas, A.P. (1993) A multi-parameter approach to fingerprinting suspended-sediment sources. In: *Tracers in Hydrology*, Peters, N.E., Hoehn, E., Leibundgut, Ch., Tase, N. and Walling, D.E. (eds), International Association of Hydrological Sciences, Wallingford, pp. 329–338.

Section IV

SEDIMENT SOURCES AND SINKS

11 The Identification of Catchment Sediment Sources

ROBERT J. LOUGHRAN[1] AND BRYAN L. CAMPBELL[2]
[1]*Department of Geography, University of Newcastle, NSW, Australia; and*
[2]*Department of Geography, University of Newcastle, NSW, Australia (formerly: ANSTO, Lucas Heights, Menai, NSW, Australia)*

INTRODUCTION

A major research need in the study of drainage basins is an improved knowledge of sediment sources to the stream system (Walling, 1983). 'In most of the world's drainage basins the principal sediment source areas have not been identified and there is no accurate base on which to ascertain either from where or when the sediment load is being derived' (Campbell, 1992, p. 462). Problems of spatial lumping of potential sediment sources arise because of the diversity of topographic, land-use and soil conditions, which can be expected to produce variations in sediment delivery response (Walling, 1983). Furthermore, temporal variations in the magnitude of sediment sources and storages within basins will also complicate relationships between source and sediment output from the catchment. There appear to be two approaches that have been adopted in sediment source identification: one attempts to measure or infer the magnitude and location of sources without direct reference to the stream transport system (e.g. soil erosion studies); and the other attempts to link the source to the stream by way of tracer techniques. Reviews of methods used in the measurement of soil erosion have appeared, for example, in the texts of Zachar (1982) and Morgan (1986), while Walling (1990) has written on sediment delivery linkages between agricultural fields and the river. This chapter reviews and re-examines both of these areas of sediment source research in the context of how drainage basin sediment yield may be explained.

Sources of sediment within drainage basins may be subdivided into slopes and channels, and be activated by raindrop impact, sheet runoff, concentrated slope runoff (producing rills, gullies and tunnels), mass movements and stream runoff. Methods used for measuring soil and stream channel erosion can be employed to quantify the effects of more than one process.

Tracers make it possible to follow the dynamic behaviour of a substance, provided the tracer acts 'in every respect like the population being traced' (McHenry, 1969, p. 280). There are several properties of sediments that can be used to trace their movements from source to channel, including molecular identity, orbital-electron properties and nuclear properties.

Sediment and Water Quality in River Catchments. Edited by I.D.L. Foster, A.M. Gurnell and B.W. Webb.
© 1995 John Wiley & Sons Ltd.

SURVEYING METHODS FOR THE IDENTIFICATION OF SEDIMENT SOURCES

Visual appraisal

Visual appraisal of potential sources on photographs and in the field can provide semiquantitative information, particularly if a time series of vertical air photographs is available. This method is used mostly for the mapping of bare ground that may be subject to sheet and rill erosion (Kirkbride and Reeves, 1993; Wilson et al., 1993), the occurrence and advance of gully and stream bank erosion (Hooke, 1980) and mass movement features such as landslides (e.g. Temple and Rapp, 1972). In their study of 1000 landslides in Tanzania, Temple and Rapp (1972) used ground reconnaissance and field sketching from high vantage points, aerial reconnaissance and oblique photography, because vertical aerial photographs were not available.

Percentage ground cover, gully length (m km^{-2}), proportion of the basin given over to particular land uses, and area of landslides (m^2 km^2) can be used as indicators of sediment source, particularly in comparative catchment studies of sediment yield explanation. Disadvantages with this method of appraisal are: the degraded surfaces may not be active sources, perhaps owing to the historic removal of erodible horizons, exposing less-erodible subsoils; the subjectivity of the assessment; and the uncertainty of linkages between the apparent source and the river.

Some of the disadvantages listed above may be overcome with the use of quantitative surveying techniques, whereby a datum is set against which changes in surface elevation can be measured.

Surveying with level, tape and staff

Cross-profiles of rills and gullies may be surveyed to determine cross-sectional area from shoulder to shoulder, assuming that an original pre-erosion surface can be identified. Adjacent cross-sectional areas, multiplied by the distance between them, can be used to estimate the volume of soil lost over a known time-span determined by field observations and/or aerial photographs (Boardman and Robinson, 1985; Duck and McManus, 1987; Gabriels et al., 1977; Govers, 1987). This method is useful for measuring rills produced by runoff on agricultural land shortly after a seed-bed has been prepared. The rate of erosion is usually underestimated by this method, however, because sheet and inter-rill erosion are not able to be included (Verhaegen, 1987). Associated deposition in the form of fans has been quantified by surveying surface areas and estimating depths by soil augering (Boardman and Robinson, 1985). Soil and sediment bulk density measurements can be used to convert volumes to masses.

The exposure of tree roots may also be used to ascertain the elevation of original surfaces, enabling estimates of the degree of sheet erosion to be made. Dunne et al. (1978), for example, used this technique to determine erosion rates in semi-arid Kenya.

Photogrammetry

Aerial photographs for the rapid assessment of erosion have been used in a variety of ways, including a national survey of Zimbabwe (Whitlow, 1986), the determination of the annual growth rate of gullies over a 20-year period (Majorov et al., 1986), river channel erosion in California (Kondolf and Curry, 1986) and the estimation of erosion and deposition in the 1980 crater of Mount St Helens, Washington, USA (Mills, 1992). In the latter, a series of digitised topographic maps was produced from aerial photographs taken in 1980, 1981, 1983, 1986 and 1988, and analysed using a geographical information system (GIS).

Close-range photogrammetry to a vertical accuracy of 2 mm was used by Sneddon et al. (1984) to determine surface elevation changes in forest soils in southern New South Wales, Australia, while large-scale aerial photographs keyed to the cropping cycle were used to determine the volume of soil lost to concentrated flow (Thomas et al., 1986). In the latter case, plotted contours with an accuracy of ±55 mm, equal to those obtained by ground survey, were produced. In addition, measurements of channel and gully cross-sections with an airborne laser altimeter have provided quick, accurate and readily obtained data (Ritchie et al., 1994).

Erosion pins and profilometers

Ground 'advance' and 'retreat' can be measured with nails or steel rods (erosion pins) inserted into a surface. The length of pin protruding from the ground surface is measured periodically with a gauge or ruler. Sometimes a washer has been used over the pin to estimate erosion (beneath the washer) and sedimentation (above the washer). A review of the technique by Haigh (1977) suggested that pins should be thin, and should not be set flush with the soil surface. Best results are obtained when pins are used in short rows or clusters. Pins have been used for measuring sheet, gully sidewall and river bank erosion (Mackay et al., 1985; Sutherland and Bryan, 1991; Crouch, 1987; Neller, 1988; Wolman, 1959; Murgatroyd and Ternan, 1983; Davis and Gregory, 1994). Erosion pins are a cheap and potentially effective method for obtaining erosion data, but they suffer a number of disadvantages. For example, their presence may affect the flow of water and sediment, and washers may protect the soil surface from raindrop impact. Animal disturbance, frost heave, rejection by soil swelling, operator disturbance and loss can also reduce the quality of measurements.

A novel approach to erosion-pin technology has been developed by Lawler (1992): the photoelectronic erosion pin (PEEP). Here, an array of photosensitive cells within a transparent tube are inserted like an erosion pin into the erosional/depositional surface. The retreat of the face exposes the cells to light, which increases the PEEP voltage output, whereas deposition decreases sensor outputs. The PEEP is connected by cable to a data-logger.

Profilometers, consisting of frames with vertical sliding rods that are lowered to the soil surface, are mounted on permanent benchmarks. The length of rod extended from the horizontal portion of the frame is either measured in the field, or photographed so that soil surface change can be determined from a photographic print (McCool et al., 1981). This method has the advantage of not disturbing the soil surface, nor interfering with erosion and transport processes (Lam, 1977). A similar approach was used by Toy (1983) in the development of a linear erosion/elevation measuring instrument (LEMI). Two carpenter's levels mounted on a series of support rods implanted on the slope allow the vertical distance between the LEMI and the soil surface to be recorded to the nearest 0.5 mm.

Soil profile properties

Soil profile reconstruction, such as the matching of soil horizons, assumes that an uneroded soil pedon can be used as a reference for the assessment of an eroded site (Lewis and Lepele, 1982; Olson and Beavers, 1987). However, Daniels *et al.* (1988) consider this assumption to be without foundation, because it ignores the natural variability of soils in the landscape.

EROSION PLOTS

The measurement of soil loss by sheet erosion and rilling can be undertaken on plots (Morgan, 1986). These are sections of the landscape from which runoff and transported soil materials are collected in a catching trough and led to a container. The plots may have constructed boundaries, the dimensions of which have varied from study to study (Loughran, 1989). Plot boundaries will interrupt the flow of water and sediment onto the plot from upslope, and may channel flow along margins. Alternatively, a plot may be without boundaries (Gerlach, 1967), but it could then be difficult to determine the area contributing to the trough so that soil loss per unit area can be calculated.

Plots were first used by the US Forest Service in 1915 (Moldenhauer and Foster, 1981), and have been employed in a variety of environments to determine the factors that control soil erosion. From these experiments, models to predict soil loss have been developed (Foster, 1988). Plots have also been used directly to indicate soil losses within drainage basins in an effort to quantify sediment sources in relation to land use (Loughran *et al.*, 1986, 1992).

Quite apart from the difficulty of attempting to scale up plot observations to catchment scale, other problems have been identified in the use of plots. For example, there was a relatively large amount of unexplained variability in soil loss from 40 apparently uniform plots when 25 rainfall–runoff events were compared (Wendt *et al.*, 1986). Lang (1992) pointed out the occurrence of errors in estimating the sediment concentration in catching devices on plots, where mixtures of known sediment concentration were tested. After vigorous agitation of the soil–water mixture to suspend the sediment, dip samples were collected to determine the sediment concentration, a common, standard procedure (Lang, 1992). It was found that this produced gross underestimation of the concentrations, because of the rapid settling of the solid particles and the inability of the samplers to trap a representative sample (Lang, 1992).

THE IDENTIFICATION OF CATCHMENT SEDIMENT SOURCES USING CAESIUM-137

The fallout isotope caesium-137 (^{137}Cs), a product of thermonuclear weapons tests, has been distributed globally and may be used as an indicator of soil erosion status. On reaching the Earth's surface, ^{137}Cs has become firmly adsorbed to surface soils (Ritchie and McHenry, 1990). Fallout from atmospheric weapons testing during the 1950s to 1970s has labelled soil materials, so that sites with no net soil loss should have ^{137}Cs inventories that reflect the amount of ^{137}Cs fallout, less the loss due to radioactive decay, (^{137}Cs has a half-life of 30 years). The level of ^{137}Cs at an undisturbed site should provide a reference value for assessing the degree of erosion and deposition within the catchment, where sites with less than the reference value can be considered eroded, and sites with more than the reference value can be regarded as depositional (Loughran, 1989; Walling and Quine, 1990a, 1991a,b, 1992). Mapping the spatial distribution of ^{137}Cs within drainage basins may therefore provide an indication of sediment sources (e.g. McCallan et al., 1980; Loughran et al., 1992; Zhang et al., 1994).

There are several assumptions that are commonly made when the ^{137}Cs technique is used to map soil erosion. Because ^{137}Cs fallout commenced in the mid-1950s, erosional losses are time-averaged from that period. It is essential that the reference site provides an accurate assessment of the ^{137}Cs fallout inventory, because this value is used as the baseline for all erosion/deposition determinations. There have been several studies that have examined the variability of ^{137}Cs, and Sutherland (1991) recommends that between 12 and 20 soil samples are required for the estimation of the reference value at the 95% confidence level, with an allowable error of ±10%. Also, it is assumed that fallout has been uniform over the landscape being studied, and that the isotope is irreversibly bound to soil materials.

Because the depth distribution of ^{137}Cs within the soil profile may vary from site to site, it is advisable to sample the soil at depth increments (Campbell et al., 1988). In this way judgements can be made on the likely total depth of ^{137}Cs penetration when bulk-core samples are being collected for ^{137}Cs mapping (Morris and Loughran, 1994). The spatial frequency of soil sampling should take into consideration the potential variability of ^{137}Cs. For example, the cultivation of soils by ploughing has probably reduced the microvariability of ^{137}Cs, compared with sites vegetated by grass tussocks, where rainsplash, localised runoff and stock trampling can locally redistribute soil materials on rough surfaces (Loughran et al., 1993). Grid and transect sampling at intervals of 100 m (Longmore et al., 1983), 10 m (Campbell et al., 1986), 20–25 m (Walling and Quine, 1991b), 15 × 20 m (Morris and Loughran, 1994), 4 × 6 m and 4 m (Zhang et al., 1994) and 10 m and 25 m (Sutherland, 1994) have all been employed to determine the pattern of ^{137}Cs distribution in basins and on slopes.

In preparation for ^{137}Cs analysis, soil samples are oven-dried, disaggregated and sieved to make them comparable with the standard soil that has been used for calibration. The sieved fraction of the soil is placed on a gamma-detector to measure the area under the 661.6 keV photopeak to determine the ^{137}Cs concentration in millibecquerels per gram (mBq g^{-1}) (Campbell et al., 1988). The concentration of ^{137}Cs is multiplied by the total mass of the sieved sample and divided by the area of the sampling device to calculate the areal ^{137}Cs activity (mBq cm^{-2}) (Campbell et al., 1988).

The frequency of grid sampling on a site on the Darling Downs, Queensland, Australia, was sufficient for McCallan et al. (1980) to draw isolines of ^{137}Cs content (isocaes), and to indicate zones of net ^{137}Cs loss and gain and, therefore, soil. The same technique has been successfully used by Walling and Quine (1990a) on arable land in the UK. To determine the presence of erosion or deposition at a site, Walling (1990) took ^{137}Cs values that were at least −20% and +20% of the reference value, respectively. In later studies, limits of ±10% were used (Walling and Quine, 1991b).

These margins represented the limits of precision of ^{137}Cs measurement and the natural sampling variability observed in the soil cores.

The quantification of net soil loss from measurements of caesium-137

Soil loss–^{137}Cs loss from erosion plots, accounting procedures and theoretical models have been used to calculate the magnitude of soil erosion from ^{137}Cs measurements (Walling and Quine, 1990b).

Erosion plot calibrations

Ritchie et al. (1974), taking plot soil-loss data from other workers using ^{137}Cs, and other applied radionuclides, and their own ^{137}Cs and the Universal Soil Loss Equation (USLE) measurements, found a relationship between radionuclide loss and soil loss:

$$Y = 1.6X^{0.68} \qquad (r = 0.95) \qquad (11.1)$$

where Y is radionuclide loss (as a percentage of radionuclide input) and X is soil loss (t ha^{-1} yr^{-1}).

Over a 10-year period (1965–76), losses of ^{137}Cs were correlated with measured soil losses from erosion plots in Canada ($r = 0.8$, $n = 10$), and there was found to be almost a direct proportionality between the two measures (Kachanoski, 1987).

In Australia, Loughran et al. (1988) used soil-loss data from long-term erosion plots to construct a soil loss – ^{137}Cs loss calibration equation for black earth soils:

$$Y = 4.35X^{1.526} \qquad (r = 0.85) \qquad (11.2)$$

where Y is net soil loss (kg ha^{-1} yr^{-1}) and X is the loss of ^{137}Cs, expressed as a percentage of the reference value. Additional data from Australian soil-loss plots were utilised to construct separate calibration equations for conditions of grazing and cultivation (Elliott et al., 1990). For grazed and cultivated sites, respectively, the relationships are:

$$Y = 7.74/1.09^x \qquad (n = 31, r = 0.82) \qquad (11.3)$$

$$Y = 80.6/1.07^x \qquad (n = 60, r = 0.85) \qquad (11.4)$$

where Y is net soil loss (kg ha^{-1} yr^{-1}) and x is percentage ^{137}Cs loss, compared with the reference value.

Because the relationships of Elliott et al. (1990) are based largely on data collected from runoff–erosion plots of the NSW Soil Conservation Service, some revision of Eqns (11.3) and (11.4) will be required in the light of the errors identified by Lang (1992). Revision is in progress, and it seems that, for ^{137}Cs losses between 1% and 50%, soil losses are underestimated by a factor of 2. Above 50% ^{137}Cs loss, underestimates diminish, and the revised and original curves coincide at approximately 85% ^{137}Cs loss.

Some of the scatter in the statistical relationships of Eqns (11.3) and (11.4) (Elliott et al., 1990) may be explained by soil properties and the variable depth of ^{137}Cs labelling. As part of the Maluna Creek drainage basin study (Loughran et al., 1992), soil loss and ^{137}Cs loss from three plots (2 m^2 in area) were monitored over a period of 2.2 yr

(Figure 11.1). The plots were prepared by removing all vegetation cover (largely grass) by hand and trowel, and were maintained in that condition for the study period. There was no deep cultivation of the plots, and all events were from natural rainfall. The decision to conduct the experiment on bare soil plots was based on a requirement for erosion to produce sufficient soil for ^{137}Cs analysis. On five occasions, sediment caught in the catching troughs of the three plots was collected, dried, weighed and analysed for ^{137}Cs content. Sediment delivered to the catching container in runoff was not analysed for ^{137}Cs. The ^{137}Cs content of the plots before the experiment commenced was estimated from samples collected adjacent to the plots, because soil sampling of the plots would have greatly disturbed them. Details of the plots are given in Table 11.1. Plot C had the lowest estimated ^{137}Cs content before the experiment began, but had the highest total ^{137}Cs loss (11.3 mBq cm^{-2}) for the 803-day study period (Table 11.1). Plot D experienced almost the same rate of soil erosion as plot C, but its rate of ^{137}Cs loss was approximately one-third, despite having a 39% higher estimated original ^{137}Cs inventory. Plot E had the lowest soil and ^{137}Cs losses. It appears that the original (pre-plot experiment) ^{137}Cs values, as indicators of soil erosion status, are correlated with soil erodibility, and that the most erodible soils have lost ^{137}Cs at a greater rate. For plots D and E, the rate of soil loss versus rate of ^{137}Cs loss shows little or no scatter (Figure 11.2). Plot C, however, shows greater variability between soil and ^{137}Cs loss, but there is nevertheless a strong trend. In all cases, trends indicate a proportionality between soil loss and ^{137}Cs, confirming the experiments of Kachanoski (1987) in Canada.

The depth distribution of ^{137}Cs adjacent to the three plots indicated that ^{137}Cs was confined to the 0–4 cm layer in the earthy sand of plot C, whereas it was found to at

Figure 11.1 Runoff–erosion plot C, before the removal of vegetation cover, Maluna Creek catchment, NSW, Australia

Table 11.1 Runoff–erosion plot details: Maluna Creek drainage basin, Hunter Valley, New South Wales, Australia[a]

Plot	Soil type	Soil ^{137}Cs at time of installation (mBq cm^{-2})	Total soil loss (kg)	Total ^{137}Cs loss (mBq cm^{-2})
C	Earthy sand	52.9	9.93	11.29
D	Brown podzolic	73.6	9.17	4.58
E	Black earth	80.0	5.38	1.27

[a] Study period: 4 October 1983–17 December 1985 (803 days). Plot area: 2 m^2

least 12 cm in the brown podzolic soil of plot D and to 16 cm in the black earth soil of plot E. In the last case, the soil cracks deeply when dry and surface soil falls down the cracks, thus labelling the profile with ^{137}Cs to greater depth. Southard and Graham (1992) noted that ^{137}Cs labelling could occur to a depth of 72 cm, the average depth of soil cracking within a vertisol. It would appear, therefore, that the rate of ^{137}Cs loss is partly controlled by the depth distribution, or concentration, of ^{137}Cs within the soil profile. Plots C, D and E, with their different soil properties, show different relationships between soil and ^{137}Cs losses (Figure 11.2).

Given that the same reference value can be applied to all plots in the Maluna Creek basin, (100 mBq cm^{-2}; Loughran *et al.*, 1992), the ^{137}Cs loss per unit soil loss varies as a ratio from 1 (plot E), to 2 (plot D) to nearly 5 (plot C) (Figure 11.2). Rates of ^{137}Cs loss are, therefore, directly related to the concentration of the isotope in the profile. This effect is reflected in Eqns (11.3) and (11.4) for cultivated and uncultivated

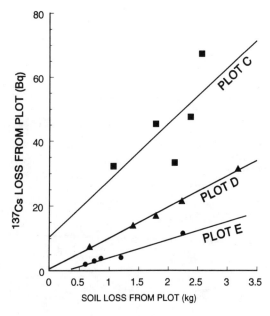

Figure 11.2 Relationship between soil loss and ^{137}Cs loss from three runoff–erosion plots, Maluna Creek catchment, NSW, Australia

(grazed) soils, respectively, where, for the same percentage ^{137}Cs loss, there is a substantially greater soil loss from cultivated soils because of the deeper distribution of ^{137}Cs.

Because the plot calibration models are designed to calculate soil loss, soil sampling points that show ^{137}Cs gain cannot be accommodated. No equations exist to relate soil gain to ^{137}Cs gain. If depth profiling is used to determine sedimentation depths, there should be sufficient of them for the calculation of depositional volumes (Loughran et al., 1987).

Other models to predict soil loss from caesium-137 measurements

For cultivated soils, a theoretical proportional model has been used for estimating soil erosion, where soil loss from the profile has been assumed to be proportional to the loss of ^{137}Cs content from the plough layer, compared with the reference value (Mitchell et al., 1981). The proportional model assumes that ^{137}Cs is uniformly mixed within the cultivated layer, and the net rate of soil erosion is divided by the length of time between the commencement of ^{137}Cs fallout and the time of survey. A measure of soil bulk density is necessary to convert soil depth loss into soil mass loss (Walling and Quine, 1990b). This approach has been used to measure soil erosion at sites in Canada (e.g. de Jong et al., 1983; Martz and de Jong, 1987) and Australia (Elliott and Cole-Clark, 1993).

A model for predicting the temporal relationship between soil ^{137}Cs and the erosion rate has been devised by Kachanoski and de Jong (1984). The model predicts the amount of ^{137}Cs remaining in the soil as a function of time and the erosion rate. The deposition of atmospheric ^{137}Cs, radioactive decay, tillage dilution and erosional transport of ^{137}Cs are accounted for. Walling and Quine (1990b) have developed a similar model, which has been calibrated using long-term sediment yield measurements from catchments in Devon. A family of curves has been produced to predict soil loss from ^{137}Cs loss for various depths of cultivation (Walling and Quine, 1990b).

Two models were employed by Brown et al. (1981) in estimating soil erosion in nested drainage basins in Oregon, USA. A volumetric model, involving the estimation of the volume and mass of sediment deposited in sedimentation zones, was based on a survey of the areal extent and depth of ^{137}Cs. It was assumed that the sediment was derived from sources on sideslopes and ridgetops since fallout commenced. A second, gravimetric, model involved algebraic manipulations of the areal concentration of ^{137}Cs in both depositional and erosional zones to calculate erosional losses of ^{137}Cs from upland sites.

THE USE OF TRACERS TO IDENTIFY CATCHMENT SEDIMENT SOURCES

The methods for mapping and measuring sediment sources, described above, do not link the source to the stream system. The effect of the source on catchment sediment output can only be inferred, therefore. Furthermore, potential variability in sediment storages in space and time would need to be considered.

The so-called 'fingerprinting' of sediments may allow the tracing of eroded material from source to stream, and Walling et al. (1993) have provided an example of a multiparameter approach using fallout radionuclides, mineral magnetics and organic carbon and nitrogen levels in two drainage basins in Devon, England. Furthermore, radioisotopes such as ^{59}Fe may be labelled onto sediments (Wooldridge, 1965), but it is necessary to ensure that there is no environmental hazard associated with the experiment. This problem may be overcome with the application of neutron activation to samples after their collection from the field, as in the study of Kiernan et al. (1993), which employed gold.

The use of radioisotopes to trace sediment

As indicators of sediment source, ^{137}Cs, ^{134}Cs, ^{210}Pb, ^{7}Be, ^{226}Ra and ^{232}Th have been employed, but to be effective the tracer needs to be able to be linked with certainty to the potential source. For example, a simple division of potential sediment sources into two land-use categories enabled an estimate of the relative magnitude of contributions to be made (Loughran et al., 1986). Here, ^{137}Cs concentrations in the surface layers (0–2 cm) of uncultivated soils (forest and grassland) and cultivated vineyard soils (0–20 cm) were contrasted. The ^{137}Cs concentration on uncultivated soils was in every sampled case higher than on vineyard soils and, on average, 93% of the ^{137}Cs would have to have been contributed from the cultivated source to produce the composite stream sediment ^{137}Cs concentration in Maluna Creek (Loughran et al., 1986). Allowance was made for the enrichment of clay materials during sediment transport to the basin outlet stream sampling point. Walling and Bradley (1990), using a similar approach in the Jackmoor Brook catchment, were able to suggest that arable fields were more likely to provide the bulk of suspended sediment in transport, compared with stream bank material and pasture topsoil. Furthermore, any uncertainties in this approach can be overcome with the use of more than one diagnostic property (Peart and Walling, 1988). Unsupported ^{210}Pb and ^{7}Be, as well as ^{137}Cs, were therefore utilised to identify the relative magnitude of four potential sediment sources in two basins in Devon: cultivated fields, permanent pasture, woodland and channel banks (Walling and Woodward, 1992). The preferential adsorption of ^{137}Cs onto the finer fraction of soils and sediments was also taken into account.

In the Whiteheads Creek catchment near Goulburn, NSW, Australia, Olley et al. (1993) were able to distinguish sediment sources with unique ratios of ^{226}Ra to ^{232}Th, produced as a result of rock type and environmental factors. The ratio of ^{226}Ra to ^{232}Th in transported sediment indicated that a gully was the source of sediment, an observation confirmed by measurements of the ratio of ^{137}Cs to ^{7}Be. The latter ratio can indicate the part of the soil profile that has been eroded because of the differential penetration of the two isotopes into the soil. Cosmogenic ^{7}Be was found on average to penetrate between 0.7 and 10 mm, whereas 90% of the total ^{137}Cs activity was detected in the top 10 to 20 cm of the soil profile. Therefore, sediments with high concentrations of both ^{137}Cs and ^{7}Be suggest shallow sheet erosion as a likely source, with low concentrations of both indicating material which had never been exposed to atmospheric fallout, such as gully walls. High values of ^{137}Cs, coupled with low amounts of ^{7}Be, might indicate sediments derived from rill erosion.

The use of gold as a tracer in sediment studies

The artificial labelling of sediments and soils with radionuclides, such as ^{59}Fe, may lead to environmental hazards, and it may therefore be necessary to fence off experimental sites. Isotopes with a short half-life may be preferred under such circumstances, but this may conflict with the need for the isotope dosage to remain detectable for the duration of the study. The post-sampling activation technique overcomes problems of this kind. An experiment to trace sediment eroded from a roadside batter devoid of vegetation utilised gold and the nuclear activation technique, where it was important to know if roadside erosion in an urban area was contributing to sedimentation in Lake Macquarie, near Newcastle, NSW, Australia.

The batter, at Fennell Bay, Lake Macquarie, was sampled to determine if gold was naturally present in the environment prior to the application of a solution of gold chloride. The solution was applied from an aerosol container onto three marked areas (100 × 50 cm^2) of the eroded surface. Two years later, sediments in the roadside gutter downslope from the labelled site were sampled. Approximately 50 m from the site, drainage was conveyed to Lake Macquarie in an underground a culvert system. Samples were also collected from sedimentary bars where the culvert entered the lake (Table 11.2).

Analysis of samples was by neutron activation, a non-destructive technique capable of detecting many elements at a level of 10^{-10} g. Samples are exposed to measured fluxes of thermal neutrons to produce radioactive ^{198}Au, and gamma-ray spectroscopy is used to measure concentrations. Results are given in Table 11.2. Detectable levels of gold were found in sediment samples collected from the roadside gutter to a distance of 51.3 m downslope from the labelled area. However, in sediments at the outfall into Lake Macquarie, approximately 1000 m from the labelled site, there was no gold present. This may mean that gold concentrations were diluted below the level of

Table 11.2 Sediment tracing experiment using gold: eroded roadside batter, Fennell Bay, Lake Macquarie, NSW, Australia

Site	Au ($\mu g\ g^{-1}$)	Distance from labelled site (m)	
FB1	0.40 ± 0.05	1.3	Road gutter
FB2	0.29 ± 0.02	11.3	Road gutter
FB3	0.23 ± 0.02	21.3	Road gutter
FB4	0.03 ± 0.01	31.3	Road gutter
FB5	0.38 ± 0.04	41.3	Road gutter
FB6	0.10 ± 0.04	51.3	Road gutter
Drainage enters underground culvert to Lake Macquarie			
FB7	<0.04	ca. 1000	Sedimentary bar at
FB8	<0.04	ca. 1000	culvert mouth at
FB9	<0.04	ca. 1000	Lake Macquarie
FB10	<0.04	ca. 1000	
FB11	<0.04	ca. 1000	
FB12	<0.05	ca. 1000	

detection, or that labelled sediments had not yet reached the lake. Further field sampling is possible because the gold label is not subject to decay.

Magnetic measurements for tracing sediments

'The ubiquitous nature of iron oxides in soils makes their study significant from a number of viewpoints', including the tracing of soil materials and sediments between environmental systems (Dearing et al., 1985, p. 245). There are several magnetic parameters that may be used to describe sediments, including magnetic susceptibility, isothermal remanent magnetisation and coercivity. Environmental factors and parent material control these soil magnetic properties, which, together, may produce a magnetic fingerprint allowing the identification of sediment source. For example, in the Jackmoor Brook catchment, Devon, England, the magnetic properties of suspended sediments delivered from the basin closely matched those of cultivated surface soils, and that source increased in importance during major runoff events (Oldfield et al., 1979). Further work in the same catchment matched source with sediment yielded to stream in relation to the hydrograph, but it was noted that more information on sediment particle-size and organic-matter effects was required if the technique was to become more effective (Walling et al., 1979).

The source of sediments deposited in stream channels, and in lakes and reservoirs has also been determined from magnetic measurements (e.g. Dearing et al., 1986; Dearing, 1992; Caitcheon, 1993; Yu and Oldfield, 1993). In a study conducted in several Australian environments, Caitcheon (1993) used a nested subcatchment approach to sediment sourcing by collecting samples at tributary junctions for magnetic measurement. The method was found to be successful where two subcatchments had delivered magnetically distinguishable sediments to a trunk stream. Relative contributions to the channel sediments below the confluence were then able to be determined.

CONCLUSIONS

The location and magnitude of sediment sources within catchments may only be inferred when methods to measure soil erosion are employed, although these may prove reliable where sediment budgets can be constructed (Loughran et al., 1992). The erosion status of soils throughout a drainage basin may be determined by ^{137}Cs, but the quality of results will depend on the reliability of the reference, or background, value of ^{137}Cs. However, this method is probably the most suitable for mapping sediment sources, because it provides information about sheet and rill erosion that is time-averaged since the mid-1950s (Walling and Quine, 1990a). Furthermore, it does not interfere with erosional processes and only one visit to the site may be required. It is, therefore, a cost-effective method for providing quantifiable data on net soil loss.

Tracing the source of sediment delivered in stream runoff can be accomplished by several means, provided such difficulties as sediment enrichment, particle-size effects and acquisition of sufficient sample for analysis can be overcome. Choice of methods will often depend on research budgets and the availability of laboratory equipment to measure isotopes and magnetic properties, for example.

The most successful studies of sediment sources have employed a variety of techniques, an example of which is that of Walling et al. (1993) in the catchments of Jackmoor Brook and the River Dart in Devon, England. They identified some fundamental requirements for fingerprinting or tracer properties. First, the properties should reflect different environmental controls and should be capable of discriminating potential sediment sources. The application of statistical tests and other methods to confirm individual sediment sources and their magnitude should also be possible. In their study, Walling et al. (1993) used organic carbon and nitrogen concentrations, four mass-specific mineral magnetic parameters and two fallout radionuclides as tracers. They conclude that such an approach 'offers a powerful tool for sediment budget investigations', but suggest there is scope for further refinement (Walling et al., 1993, p. 337). This included the identification of other tracer properties, the inclusion of further source types, a refinement of the mixing model, improvement in procedures for taking into account the effects of particle size and organic matter, and the application of the approach to detail the sediment sources and dynamics during and between storm runoff events.

ACKNOWLEDGEMENTS

Funds for the gold tracing experiment were provided by the Lake Macquarie City Council, NSW, Australia. Field and laboratory assistance was provided by Andrew Jenkinson and John Fardy of ANSTO, Lucas Heights, and Allan Williams, Darren Shelly, Neil Gardner, Andrew Philippa and Chris Dever of the Department of Geography, The University of Newcastle. Contribution no. 3, Geomorphology and Quaternary Science Research Unit, Department of Geography, The University of Newcastle, Australia.

REFERENCES

Boardman, J. and Robinson, D.A. (1985) Soil erosion, climatic vagary and agricultural change on the Downs around Lewes and Brighton, autumn 1985. *Appl. Geog.* **5**, 243–258.
Brown, R.B., Kling, G.F. and Cutshall, N.H. (1981) Agricultural erosion indicated by ^{137}Cs redistribution: II Estimation of erosion rates. *Soil Sci. Soc. Am. J.* **45**, 1191–1197.
Caitcheon, G.G. (1993) Sediment source tracing using environmental magnetism: a new approach with examples from Australia. *Hydrol. Processes* **7**, 349–358.
Campbell, B.L., Loughran, R.J., Elliott, G.L. and Shelly, D.J. (1986) Mapping drainage basin sediment sources using caesium-137. In: *Drainage Basin Sediment Delivery*, IAHS Publication, no. 159, pp. 437–446.
Campbell, B.L., Loughran, R.J. and Elliott, G.L. (1988) A method for determining sediment budgets using caesium-137. In: *Sediment Budgets*, IAHS Publication no. 174, pp. 171–177.
Campbell, I.A. (1992) Spatial and temporal variations in erosion and sediment yield. In: *Erosion and Sediment Transport Monitoring Programmes in River Basins*, IAHS Publication no. 210, pp. 455–465.
Crouch, R.J. (1987). The relationship of gully sidewall shape to sediment production. *Aust. J. Soil Res.* **25**, 531–539.
Daniels, R.B., Gilliam, J.W. and Cassel, D.K. (1988) Comments on 'A method to estimate soil loss from erosion'. *Soil Sci. Soc. Am. J.* **52**, 542–543.
Davis, R.J. and Gregory, K.J. (1994) A new distinct mechanism of river bank erosion in a forested catchment. *J. Hydrol.* **157**, 1–11.
Dearing, J.A. (1992) Sediment yields and sources in a Welsh upland lake-catchment during the past 800 years. *Earth Surf. Processes Landforms* **17**, 1–22.

Dearing, J.A., Maher, B.A. and Oldfield, F. (1985) Geomorphological linkages between soils and sediments: the role of magnetic measurements. In: Richards, K.S., Arnett, R.R. and Ellis, S. (eds), *Geomorphology and Soils*, Allen and Unwin, London, pp. 245–266.

Dearing, J.A., Morton, R.I., Price, T.W. and Foster, I.D.L. (1986) Tracing movements of topsoil by magnetic measurements: two case studies. *Phys. Earth Planet. Interiors* **42**, 93–104.

de Jong, E., Begg, C.B.M. and Kachanoski, R.G. (1983) Estimates of soil erosion and deposition for some Saskatchewan soils. *Can. J. Soil Sci.* **63**, 607–617.

Duck, R.W. and McManus, J. (1987) Soil erosion near Barry, Angus. *Scot. Geogr. Mag.* **103**, 44–46.

Dunne, T., Dietrich, W.E. and Brunengo, M.J. (1978) Recent and past erosion rates in semi-arid Kenya. *Z. Geomorphol.* **29**, 130–140.

Elliott, G.L. and Cole-Clark, B.E. (1993) Estimates of erosion on potato lands on krasnozems at Dorrigo, NSW, using the caesium-137 technique. *Aust. J. Soil Res.* **31**, 209–223.

Elliott, G.L., Campbell, G.L. and Loughran, R.J. (1990) Correlation of erosion measurements and soil caesium-137 content. *Int. J. Radiat. Appl. Instrum. (A) Appl. Radiat. Isotopes* **41**, 713–717.

Foster, G.R. (1988) Modelling soil erosion and sediment yield. In: Lal, R. (ed.), *Soil Erosion Research Methods*, Soil and Water Conservation Society, Ankeny, Iowa, pp. 97–117.

Gabriels, D., Pauwels, J.M. and De Boodt, M. (1977) A quantitative rill erosion study on a loamy sand in the hilly region of Flanders. *Earth Surf. Processes* **2**, 257–260.

Gerlach, T. (1967) Hillslope troughs for measuring sediment movement. *Rev. Geomorphol. Dyn.* **4**, 173.

Govers, G. (1987) Spatial and temporal variability in rill development processes at the Huldenberg experimental site. *Catena* **8**, Suppl., 17–34.

Haigh, M.J. (1977) The use of erosion pins in the study of slope evolution. In: *Shorter Technical Methods (II)*, Technical Bulletin no. 18, British Geomorphological Research Group, pp. 31–49.

Hooke, J.M. (1980) Magnitude and distribution of rates of river bank erosion. *Earth Surf. Processes* **5**, 143–157.

Kachanoski, R.G. (1987) Comparison of measured soil cesium-137 losses and erosion rates. *Can. J. Soil Sci.* **67**, 199–203.

Kachanoski, R.G. and de Jong, E. (1984) Predicting the temporal relationship between soil cesium-137 and erosion rate. *J. Environ. Qual.* **13**, 301–304.

Kiernan, K., Campbell, B.L. and Elliott, G.L. (1993) *A Reconnaissance of the Erosion of Dolerite Soils in the Eastern Tiers, Tasmania*, Report to the Tasmanian Forest Research Council, 55pp.

Kirkbride, M.P. and Reeves, A.D. (1993) Soil erosion caused by low-intensity rainfall in Angus, Scotland. *Appl. Geogr.* **13**, 299–311.

Kondolf, G.M. and Curry, R.R. (1986) Channel erosion along the Carmel River, Montery County, California. *Earth Surf. Processes Landforms* **11**, 307–319.

Lam, K.-C. (1977) Patterns and rates of slopewash on the badlands of Hong Kong. *Earth Surf. Processes* **2**, 319–332.

Lang, R.D. (1992) Accuracy of two sampling methods used to estimate sediment concentrations in runoff from soil-loss plots. *Earth Surf. Processes Landforms* **17**, 841–844.

Lawler, D.M. (1992) Design and installation of a novel automatic erosion monitoring system. *Earth Surf. Processes Landforms* **17**, 455–463.

Lewis, D.T. and Lepele, M.J. (1982) Quantification of soil loss and sediment produced from eroded land. *Soil Sci. Soc. Am. J.* **46**, 369–372.

Longmore, M.E., O'Leary, B.M., Rose, C.W. and Chandica, A.L. (1983) Mapping soil erosion and accumulation with the fallout isotope caesium-137, *Aust. J. Soil Res.* **21**, 373–385.

Loughran, R.J. (1989) The measurement of soil erosion. *Prog. Phys. Geogr.* **13**, 216–233.

Loughran, R.J., Campbell, B.L. and Elliott, G.L. (1986) Sediment dynamics in a partially cultivated catchment in New South Wales, Australia. *J. Hydrol.* **83**, 285–297.

Loughran, R.J., Campbell, B.L. and Walling, D.E. (1987) Soil erosion and sedimentation indicated by caesium-137, Jackmoor Brook catchment, Devon, England. *Catena* **14**, 201–212.

Loughran, R.J., Elliott, G.L., Campbell, B.L. and Shelly, D.J. (1988) Estimation of soil erosion from caesium-137 measurements in a small, cultivated catchment in Australia. *Int. J. Radiat. Appl. Instrum.* (A) *Appl. Radiat. Isotopes* **39**, 1153–1157.

Loughran, R.J., Campbell, B.L., Shelly, D.J. and Elliott, G.L. (1992) Developing a sediment budget for a small drainage basin in Australia, *Hydrol. Processes* **6**, 145–158.

Loughran, R.J., Elliott, G.L., Campbell, B.L., Curtis, S.J., Cummings, D. and Shelly, D.J. (1993) Estimation of erosion using the radionuclide caesium-137 in three diverse areas in eastern Australia. *Appl. Geogr.* **13**, 169–188.

Mackay, S.M., Long, A.C. and Chalmers, R.W. (1985) Erosion pin estimates of soil movement after intensive logging and wildfire. In: Loughran, R.J. (ed.), *Drainage Basin Erosion and Sedimentation, Conference and Review Papers*, vol. 2, University of Newcastle, NSW, pp. 15–22.

Majorov, Y.I., Borovkov, S.N., Soloshenko, V.M. and Shevtsov, A.E. (1986) Gully growth estimated for economic appraisal of fluvial erosion and its control. *Geomorfologiya* **3**, 31–35.

Martz, L.W. and de Jong, E. (1987) Using cesium-137 to assess the variability of net soil erosion and its association with topography in a Canadian prairie landscape. *Catena* **14**, 439–451.

McCallan, M.E., O'Leary, B.M. and Rose, C.W. (1980) Redistribution of caesium-137 by erosion and deposition on an Australian soil. *Aust. J. Soil Sci.* **18**, 119–128.

McCool, D.K., Dossett, M.G. and Yecha, S.J. (1981) A portable rill meter for field measurement of soil loss. In: *Erosion and Sediment Transport Measurement*, IAHS Publication no. 133, pp. 479–484.

McHenry, J.R. (1969) Use of tracer technique in erosion research. *Isotopes Radiat. Technol.* **6**, 280–287.

Mills, H.H. (1992) Post-eruption erosion and deposition in the 1980 crater of Mt St Helens, Washington, determined from digital maps. *Earth Surf. Processes Landforms* **17**, 739–754.

Mitchell, J.K., Bubenzer, G.D., McHenry, J.R. and Ritchie, J.C. (1981) Soil loss estimation from fallout cesium-137 measurements. In: De Boodt, M. and Gabriels, D. (eds), *Assessment of Erosion*, John Wiley & Sons, Chichester, pp. 393–401.

Moldenhauer, W.C. and Foster, G.R. (1981) Empirical studies of soil conservation techniques and design procedures. In: Morgan, R.P.C. (ed.), *Soil Conservation: Problems and Prospects*, John Wiley & Sons, Chichester, pp. 13–29.

Morgan, R.P.C. (1986) *Soil Erosion and Conservation*, Longmans, London.

Morris, C.D. and Loughran, R.J. (1994) The distribution of caesium-137 in soils across a hillslope hollow. *Hydrol. Processes* **8**, 531–541.

Murgatroyd, A.L. and Ternan, J.L. (1983) The impact of afforestation on stream bank erosion and channel form. *Earth Surf. Processes Landforms* **8**, 357–369.

Neller, R.J. (1988) A comparison of channel erosion in small urban and rural catchments, Armidale, New South Wales. *Earth Surf. Processes Landforms* **13**, 1–7.

Oldfield, F., Rummery, T.A., Thompson, R. and Walling, D.E. (1979) Identification of suspended sediment sources by means of magnetic measurements: some preliminary results. *Water Resour. Res.* **15**, 211–218.

Olley, J.M., Murray, A.S., Mackenzie, D.H. and Edwards, K. (1993) Identifying sediment sources in a gullied catchment using natural and anthropogenic radioactivity. *Water Resour. Res.* **29**, 1037–1043.

Olson, K.R. and Beavers, A.H. (1987) A method to estimate soil loss from erosion. *Soil Sci. Soc. Am. J.* **51**, 441–445.

Peart, M.R. and Walling, D.E. (1988) Techniques for establishing suspended sediment sources in two drainage basins in Devon, UK: a comparative assessment. In: *Sediment Budgets*, IAHS Publication no. 174, pp. 269–279.

Ritchie, J.C. and McHenry, J.R. (1990) Application of radioactive fallout cesium-137 for measuring soil erosion and sediment accumulation rates and patterns: a review. *J. Environ. Qual.* **19**, 215–233.

Ritchie, J.C., Spraberry, J.A. and McHenry, J.R. (1974) Estimating soil erosion from the redistribution of cesium-137. *Soil Sci. Soc. Am. Proc.* **38**, 137–139.

Ritchie, J.C., Grissinger, E.H., Murphey, J.B. and Garbrecht, J.D. (1994) Measuring channel and gully cross-sections with an airborne laser altimeter. *Hydrol. Processes* **8**, 237–243.

Sneddon, J., Olive, L.J., Rieger, W.A. and Lutze, T.A. (1984) Erosion measurement using close-range photogrammetry. In: Loughran, R.J. (ed.), *Drainage Basin Erosion and Sedimentation*, NSW Soil Conservation Service and The University of Newcastle, pp. 153–159.

Southard, R.J. and Graham, R.C. (1992) Cesium-137 distribution in a California pelloxerert: evidence of pedoturbation. *Soil Sci. Soc. Am. J.* **56**, 202–207.

Sutherland, R.A. (1991) Examination of caesium-137 areal activities in control (uneroded) locations. *Soil Technol.* **4**, 33–50.

Sutherland, R.A. (1994) Spatial variability of ^{137}Cs and the influence of sampling on estimates of sediment redistribution. *Catena* **21**, 57–71.

Sutherland, R.A. and Bryan, R.B. (1991) Sediment budgeting: a case study in the Katiorin drainage basin, Kenya. *Earth Surf. Processes Landforms* **16**, 383–398.

Temple, P.H. and Rapp, A. (1972) Landslides in the Mgeta area, western Uluguru Mountains, Tanzania. *Geogr. Ann.* **54A**, 157–193.

Thomas, A.W., Welch, R. and Jordan, T.R. (1986) Quantifying concentrated-flow erosion on cropland with aerial photogrammetry. *J. Soil Water Conserv.* **41**, 249–252.

Toy, T.J. (1983) A linear erosion/elevation measuring instrument (LEMI). *Earth Surf. Processes Landforms* **8**, 313–322.

Verhaegen, T. (1987). The use of small flumes for the determination of soil erodibility. *Earth Surf. Processes Landforms* **12**, 185–194.

Walling, D.E. (1983) The sediment delivery problem. *J. Hydrol.* **65**, 209–237.

Walling, D.E. (1990) Linking the field to the river: sediment delivery from agricultural land. In: Boardman, J., Foster, I.D.L. and Dearing, J.A. (eds), *Soil Erosion on Agricultural Land*, John Wiley & Sons, Chichester, pp. 129–152.

Walling, D.E. and Bradley, S.B. (1990) Some applications of caesium-137 measurements in the study of erosion, transport and deposition. In: *Erosion, Transport and Deposition Processes*, IAHS Publication no. 189, pp. 179–203.

Walling, D.E. and Quine, T.A. (1990a) Use of caesium-137 to investigate patterns and rates of soil erosion on arable fields. In: Boardman, J., Foster, I.D.L. and Dearing, J.A. (eds), *Soil Erosion on Agricultural Land*, John Wiley & Sons, Chichester, pp. 33–53.

Walling, D.E. and Quine, T.A. (1990b) Calibration of caesium-137 measurements to provide quantitative erosion rate data. *Land Degrad. Rehab.* **2**, 161–175.

Walling, D.E. and Quine, T.A. (1991a). Use of ^{137}Cs measurements to investigate soil erosion on arable fields in UK: potential applications and limitations. *J. Soil Sci.* **42**, 147–165.

Walling, D.E. and Quine, T.A. (1991b) Recent rates of soil loss from areas of arable cultivation in the UK. In: *Sediment and Water Quality in a Changing Environment: Trends and Explanation*, IAHS Publicaiton no. 203, pp. 123–131.

Walling, D.E. and Quine, T.A. (1992) The use of caesium-137 measurements in soil erosion surveys. In: *Erosion and Sediment Transport Monitoring Programmes in River Basins*, IAHS Publication no. 210, pp. 143–152.

Walling, D.E. and Woodward, J.C. (1992) Use of radiometric fingerprints to derive information on suspended sediment sources. In: *Erosion and Sediment Transport Monitoring Programmes in River Basins*. IAHS Publication no. 210, pp. 153–164.

Walling, D.E., Peart, M.R., Oldfield, F. and Thompson, R. (1979) Suspended sediment sources identified by magnetic measurements. *Nature* **281** (5727), 110–113.

Walling, D.E., Woodward, J.C. and Nicholas, A.P. (1993) A multi-parameter approach to fingerprinting suspended-sediment sources. In: *Tracers in Hydrology*, IAHS Publication no. 215, pp. 329–338.

Wendt, R.C., Alberts, E.E. and Hjelmfelt, A.T. (1986) Variability of runoff and soil loss from fallow experimental plots. *Soil Sci. Soc. Am. J.* **50**, 730–736.

Whitlow, R. (1986) Mapping erosion in Zimbabwe: a methodology for rapid survey using aerial photographs. *Appl. Geogr.* **6**, 149–162.

Wilson, P., Clark, R., McAdam, J.H. and Cooper, E.A. (1993) Soil erosion in the Falkland Islands: an assessment. *Appl. Geogr.* **13**, 329–352.

Wolman, M.G. (1959) Factors influencing erosion of a cohesive river bank *Am. J. Sci.* **257**, 204–216.

Wooldridge, D.D. (1965) Tracing soil particle movement with Fe-59. *Soil Sci. Soc. Am. Proc.* **29**, 469–472.

Yu, L. and Oldfield, F. (1993) Quantitative sediment source ascription using magnetic measurements in a reservoir–catchment system near Nijar, SE Spain. *Earth Surf. Processes Landforms* **18**, 441–454.

Zachar, D. (1982) *Soil Erosion*, Developments in Soil Science no. 10, Elsevier, Amsterdam.

Zhang, X., Quine, T.A., Walling, D.E. and Li, Z. (1994) Application of the caesium-137 technique in a study of soil erosion on gully slopes in a yuan area of the Loess Plateau, Xifeng, Gansu Province, China. *Geogr. Ann.* **76A**, 103–120.

12 Determination of Suspended Sediment Provenance Using Caesium-137, Unsupported Lead-210 and Radium-226: A Numerical Mixing Model Approach

QINGPING HE AND PHIL OWENS
Geography Department, University of Exeter, UK

INTRODUCTION

In most situations the sediment transported in suspension in river channels is a mixture of materials derived from different sources within the drainage basin. Consequently, much has been written about the need to determine the source areas of fine river sediment (cf. Wolman, 1977; Peart and Walling, 1986; Walling, 1989; Walling and He, 1993) and a variety of different approaches have been used in provenance studies over the last few decades. These can be broadly grouped into two main types. First, the contribution from various *selected* source areas or types can be directly measured using traditional field monitoring techniques (e.g. Lewin *et al.*, 1974; Balteanu *et al.*, 1984) or can be indirectly estimated using models or physical reasoning (e.g. Dietrich and Dunne, 1978; Trimble and Lund, 1982; Jordan and Slaymaker, 1991). Secondly, source areas can be identified by examining one or more of the diagnostic properties of suspended sediment and potential source materials, an approach commonly known as 'fingerprinting' (Walling and Kane, 1984; Peart and Walling, 1988). Recent research has tended to focus on the use of diagnostic properties, and examples of this approach include the use of sediment mineralogy (Klages and Hsieh, 1975; Wall and Wilding, 1976; Johnson and Kelley, 1984; Garrad and Hey, 1989; Phillips, 1992; Woodward *et al.*, 1992), sediment colour (Grimshaw and Lewin, 1980; Imeson *et al.*, 1984), sediment chemistry (Peart and Walling, 1986, 1988) mineral magnetics (Oldfield *et al.*, 1979; Walling *et al.*, 1979, 1993), heavy-metal content (Lewin and Wolfenden, 1978; Passmore and Macklin, 1994) and radionuclide activity (Loughran *et al.*, 1982; Peart and Walling, 1986, 1988; Walling and Woodward, 1992; Walling *et al.*, 1993).

To date, many of the sediment provenance studies have made qualitative statements regarding likely source areas and the relative proportions that these different sources contribute to the transported fluvial sediment. Few studies have attempted to determine the absolute amounts that are derived from each of the sources. One approach that can be used to provide quantitative information on suspended sediment sources is to use several diagnostic properties (cf. Peart and Walling, 1986; Walling and Woodward,

---

Sediment and Water Quality in River Catchments. Edited by I.D.L. Foster, A.M. Gurnell and B.W. Webb.
© 1995 John Wiley & Sons Ltd.

1992) combined with a mixing model (cf. Stott, 1986; Yu and Oldfield, 1989; Walling et al., 1993). In this chapter we report on the development of a simple numerical mixing model using three different radionuclides to identify sediment sources in a drainage basin in Devon, UK.

RADIONUCLIDES AS SOURCE DISCRIMINATORS

Radionuclides in source materials

The use of radionuclides for suspended sediment source discrimination is now reasonably well established (cf. Peart and Walling, 1986; Wasson et al., 1987; Walling and Woodward, 1992). Generally most studies have tended to use caesium-137 (^{137}Cs) measurements alone, although Walling and Woodward (1992) also used unsupported lead-210 (^{210}Pb) and beryllium-7 (^{7}Be) concentrations for source identification in the River Exe basin, Devon, UK, while Wallbrink and Murray (1993) also used these three radionuclides to determine the source of artificially-generated suspended sediment in New South Wales, Australia. The use of several different diagnostic sediment properties is to be recommended as it provides a means of ensuring that the results obtained are consistent and reliable (Peart and Walling, 1986). Here, we use the radionuclides ^{137}Cs, unsupported ^{210}Pb and ^{226}Ra (radium-226).

Fallout ^{137}Cs is an artificial radionuclide (half-life 30.2 yr) that was produced during the 'atom bomb' tests in the 1950s and 1960s. It was ejected into the stratosphere where it circulated globally. Deposition to the biosphere was mainly associated with precipitation. The temporal pattern of ^{137}Cs fallout is well documented in many regions around the world. Since the mid-1980s the amount of ^{137}Cs deposition has been negligible, except in areas affected by the Chernobyl incident in 1986 (Cambray et al., 1989). Over relatively small areas (i.e. at the field or small basin scale) the total amount of atmospheric fallout can often be assumed to be uniform, although it is recommended that some assessment be made of the spatial variability of the amount of ^{137}Cs fallout within a study area (cf. Sutherland, 1991, 1994; Owens, 1994; Foster et al., 1994). Once the ^{137}Cs reaches the soil surface, it is generally sorbed tightly to soil material within the top few centimetres, although there are exceptions, such as free-draining soils and acid organic soils and peats (cf. Oldfield et al., 1979). Experimental and empirical studies have shown that, with time, ^{137}Cs may migrate slightly down the profile owing to the combined effect of diffusive and convective processes and bioturbation (Owens, 1994). However, in many cases most of the ^{137}Cs is still retained in the top few centimetres of the soil profile. Figure 12.1A illustrates ^{137}Cs depth distributions for uncultivated soil profiles in two drainage basins in Devon. In contrast, in soils that have been cultivated, the ^{137}Cs is mixed to the plough depth (Figure 12.1B); consequently, the surface concentrations of ^{137}Cs in cultivated soils are usually significantly lower than those in uncultivated soils.

The depth distribution of ^{137}Cs in woodland or forest soils is often similar to that in uncultivated soils, with high concentrations at the surface and a rapid decline in concentration with depth. However, in woodland soils *surface* concentrations of ^{137}Cs are often higher than for uncultivated soils, probably reflecting the high organic matter content of the surface layer of woodland sites (Walling and Woodward, 1992; Owens,

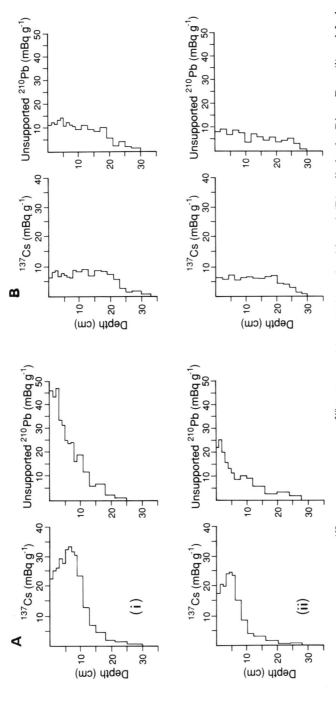

Figure 12.1 Depth distributions of ^{137}Cs and unsupported ^{210}Pb in uncultivated (A) and cultivated (B) soils in the River Dart (i) and Jackmoor Brook (ii) drainage basins, Devon. Both soils are sandy silt brown earths. See Figure 12.2 for basin location and characteristics (based on Walling and Woodward, 1992)

1994). The ^{137}Cs content of channel bank material tends to be significantly lower than surface material in cultivated and uncultivated fields and woodland. This is because of the near-vertical surface geometry of channel banks and because, in many cases, much of the material delivered to the channel from bank erosion is likely to be below the depth to which fallout ^{137}Cs penetrates. Even in cases where the height of channel banks is similar to the depth to which the ^{137}Cs extends in the soil, the average ^{137}Cs content of channel bank material will be closer to that of cultivated topsoil and considerably lower than that of uncultivated and woodland topsoil.

Thus, the nature of ^{137}Cs input to the soil surface and its strong association with soil particles enables ^{137}Cs measurements to distinguish between cultivated and uncultivated soils, woodland and channel bank material in the same region. It is independent of geology and also soil type in most cases.

Another radionuclide that can be employed for sediment source ascription is ^{210}Pb. However, unlike ^{137}Cs, ^{210}Pb reaches surficial materials via two pathways: *in situ* production and atmospheric deposition, both of which can be considered to be essentially constant through time. It is, therefore, important that the total ^{210}Pb contained within a soil or sediment sample is separated into these two components. Lead-210 is a natural radionuclide with a half-life of 22.3 yr and is one of the end-products of the uranium-238 (^{238}U) decay series. Uranium-238 is present in the Earth's crust and decays to ^{226}Ra (half-life 1622 yr), and this, in turn, decays to its daughter radon-222 (^{222}Rn, half-life 3.8 day). Radon-222 is an inert gas, part of which diffuses out of the Earth's crust into the atmosphere, troposphere and stratosphere, where it circulates, finally decaying to ^{210}Pb through a series of short-lived processes (cf. Robbins, 1978). Deposition of the atmospherically derived ^{210}Pb to the Earth's surface is associated with wet deposition and dry fallout. Fallout ^{210}Pb is termed 'unsupported' ^{210}Pb in order to distinguish it from 'supported' ^{210}Pb, which is produced by the *in situ* decay of ^{226}Ra. As supported ^{210}Pb will be in equilibrium with ^{226}Ra, the amount of unsupported ^{210}Pb in a sample is calculated by measuring both total ^{210}Pb and ^{226}Ra, and subtracting the supported component. As with ^{137}Cs, as unsupported ^{210}Pb accumulates in surface horizons of soils and sediments, cultivated, uncultivated and woodland soils and channel bank material will have different surface concentrations and depth distributions of ^{210}Pb (see Figure 12.1).

Radium-226 may also be used for identifying suspended sediment sources. Although the ratio of ^{226}Ra to ^{232}Th has been used to trace sediment from different geological formations (cf. Stanton *et al.*, 1992), the results obtained in this investigation indicate that ^{226}Ra concentrations may also be used to discriminate sediment from different sources. Unlike ^{137}Cs and unsupported ^{210}Pb, which are derived from atmospheric fallout, the depth distribution of ^{226}Ra in uncultivated soils usually shows a depletion in the surface layers (cf. Nozaki *et al.*, 1978). The depth distribution of ^{226}Ra in cultivated soils, on the other hand, is relatively uniform to the plough depth. Thus, as the ^{226}Ra content of uncultivated topsoil is likely to be lower than that of cultivated topsoil, in certain situations it may be possible to use this radionuclide to discriminate source materials.

Identifying suspended sediment sources

Having demonstrated how the chosen radionuclides can be used to discriminate between different types of source material, it is important to consider how they can be employed to determine suspended sediment provenance. Perhaps the simplest way to identify suspended sediment sources using radionuclides is to compare the radionuclide concentrations of suspended sediment qualitatively with that of selected source materials, and to make judgements concerning the likely proportions that the different sources make to the sediment load. To illustrate this approach Table 12.1 presents information concerning the ^{137}Cs content of suspended sediment samples, collected during the period November 1990 to August 1992 over a range of discharge conditions, compared with potential source materials from the Jackmoor Brook and River Dart drainage basins in Devon (Figure 12.2). Detailed information concerning the study sites and the methodology used to collect and analyse suspended sediment and source samples is contained in Walling and Woodward (1992) and Owens (1994). In brief, the River Dart drainage basin (46 km^2 at Bickleigh) is underlain by sandstones and shales of Upper Carboniferous age, has a deeply dissected, incised terrain, and is dominated by a land use of permanent pasture. The nearby Jackmoor Brook basin (9.3 km^2 at Pynes Cottage) is underlain by sandstones, breccias and conglomerates of Permian age, has a gentler relief, and is dominated by mixed arable land use. In this study, because of the expected differences in the particle size composition of suspended sediment and source material, source samples were screened through a 63 μm sieve in order to make them comparable with suspended sediment (cf. Walling and Woodward, 1992). It is important to stress that no attempt has been made to correct for differences in the particle size characteristics of the <63 μm fraction of the source and suspended sediment material, as advocated by Walling and Woodward (1992) and Owens (1994). Thus, the comparisons made in this section are only of a general nature. The development and application of corrective measures for differences in particle size composition are discussed in more detail below.

The average ^{137}Cs content of the different source types in Table 12.1 agrees with expectations based on the behaviour of ^{137}Cs in soils as described earlier. The concentrations of ^{137}Cs in source materials are higher in the Dart basin than in the Jackmoor Brook basin. This reflects greater mean annual rainfall (see Figure 12.2) and thus atmospheric deposition of ^{137}Cs in the former (the ^{137}Cs reference inventories for the Dart and Jackmoor Brook basins are 341 ± 13 and 217 ± 13 mBq cm^{-2},

Table 12.1 Mean value of the ^{137}Cs content of fluvial suspended sediment and the <63 μm fraction of potential source materials in the Jackmoor Brook and River Dart catchments, Devon

Soil source	Jackmoor Brook		River Dart	
	n	(mBq g^{-1})	n	(mBq g^{-1})
Bank	10	1.4 ± 0.8[a]	16	0.6 ± 1.0
Cultivated	21	8.3 ± 0.5	11	11.3 ± 0.5
Uncultivated	7	14.3 ± 0.7	10	21.6 ± 0.8
Woodland	2	45.0 ± 1.4	3	62.2 ± 1.7
Sediment	44	13.0 ± 1.4	43	12.6 ± 1.2

[a] The ± values represent the uncertainties associated with the ^{137}Cs measurement process (cf. Owens, 1994)

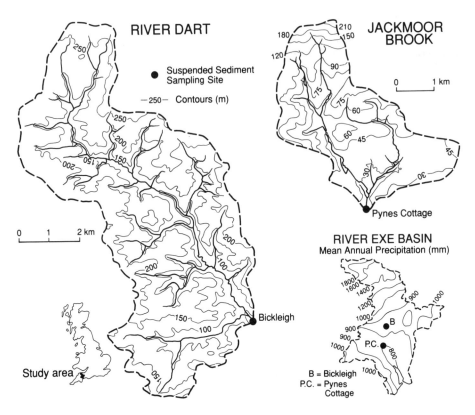

Figure 12.2 Location and characteristics of the River Dart and Jackmoor Brook drainage basins, Devon (based on Walling and Woodward, 1992)

respectively). The ^{137}Cs concentrations of channel bank and woodland sources are considerably different from suspended sediment. In the case of woodland, given the fact that there is relatively little woodland in the study basins, it is unlikely that the erosion of topsoil from wooded areas makes a significant contribution to the sediment load of these two streams. Similarly, the ^{137}Cs content of eroding channel bank material is very low, almost negligible, and is noticeably different from that of the suspended sediment. An initial inspection of the data, therefore, would seem to suggest that channel banks are not a major sediment source.

The ^{137}Cs content of topsoil from both cultivated and uncultivated fields is similar to that of the suspended sediment in both drainage basins. From the data presented in Table 12.1 it appears that, generally, most of the suspended sediment transported in these streams is derived from cultivated and uncultivated fields. The slightly lower mean ^{137}Cs content of suspended sediment for the River Dart, compared with Jackmoor Brook, may be due to a higher contribution from cultivated fields and/or bank material in the former.

While this example helps to illustrate the potential of using fallout radionuclide measurements, it also illustrates the main problem associated with using one discriminating property to identify sediment sources. For example, although an initial inspection of the data presented in Table 12.1 suggests that cultivated and uncultivated topsoils are the main sediment sources in both drainage basins, it is possible that the suspended sediment transported in these streams is instead derived from a mixture of woodland and channel bank material. Indeed, there are several possible source combinations that would give the same ^{137}Cs concentrations for suspended sediment. Clearly, in order to identify sediment sources with some confidence, a more rigorous approach is required. One such approach is to use several different diagnostic properties combined with a numerical mixing model.

SUSPENDED SEDIMENT SOURCE ASCRIPTION USING A NUMERICAL MIXING MODEL

Owing to the selectivity of soil erosion and transport processes, suspended sediment properties are usually different from those of its source materials. Figure 12.3 illustrates the typical particle size composition of the <63 μm fraction of suspended sediment and soil source materials in the Jackmoor Brook basin. Each distribution represents the amalgamation of several samples. It is clear that the suspended sediment collected from the Jackmoor Brook is enriched in particles <8 μm and depleted in particles >8 μm compared with catchment source soils. Because fallout ^{137}Cs is primarily associated with fine soil material, an increase in the fine fraction of suspended sediment relative to its source material will lead to an increase in its ^{137}Cs concentration. Any attempt aimed at using this radionuclide for quantitative source discrimination will need to take this into consideration. Although less is known about the behaviour of fallout ^{210}Pb and ^{222}Ra in soils and sediments, a preliminary study by the authors has shown that unsupported ^{210}Pb is also primarily associated with fine-grained material, while ^{226}Ra is partly associated with fine-grained material. The following section considers in more detail the relation between particle size and radionuclide concentration.

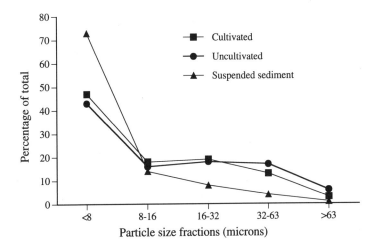

Figure 12.3 The particle size composition of the <63 μm fraction of suspended sediment and soil source material in the Jackmoor Brook basin

The grain size behaviour of radionuclides in soils

If it is assumed that the sorption of fallout ^{137}Cs and ^{210}Pb by soil particles is associated with cation-exchange or similar processes (Sawhney, 1972), it can be expected that the specific surface area (or size) of the particles determines the amount of ^{137}Cs and ^{210}Pb sorbed (cf. Hillel, 1982). Furthermore, assuming that soil particles are spheres, then the specific surface area of soil particles is inversely proportional to their radius (cf. Marshall and Holmes, 1979). Concentrations of ^{137}Cs or unsupported ^{210}Pb, $C_{s,r}(\bar{S})$ (mBq g^{-1}) ($s =$ u (uncultivated) or c (cultivated), and $r = ^{137}$Cs or ^{210}Pb), in a sample with weighted-average specific surface area $\bar{S}_s$ (m^2 g^{-1}) from surface soils may be expressed as:

$$C_{s,r}(\bar{S}_s) = v_{s,r}\bar{S}_s v_{1s,r} \tag{12.1}$$

where $v_{s,r}$ (mBq m^{-2}) is a constant scaling factor reflecting the atmospheric input of the radionuclide and the parameter $v_{1s,r}$ is introduced to reflect the deviation of the soil particles from perfect spheres. The parameter $v_{1s,r}$ depends on the type of radionuclide and may be independent of the atmospheric fallout input flux. The weighted-average specific surface area $\bar{S}_s$ of the sample is related to the frequency density $\rho(x)$ (μm^{-1}) and specific surface area S_{sp} (m^2 g^{-1}) of size x (μm) according to:

$$\bar{S}_s = \int_0^\infty \rho(x) S_{sp}(x) \, dx \tag{12.2}$$

It is clear from Eqn (12.1) that samples composed of coarser particles will exhibit lower fallout radionuclide concentrations than samples composed of finer particles, because the former generally have smaller specific surface areas than the latter. In order to test the above assumption, surface (top 1 cm) samples were collected from uncultivated and cultivated fields in the vicinity of Exeter, Devon. These were then air-dried, gently ground and chemically treated (with H_2O_2) to destroy organic matter. This treatment did not alter the radionuclide concentrations of the samples (i.e. there was no radionuclide loss in solution). Subsamples with different grain size distributions were obtained by repeatedly settling soil particles in solution. The samples were then analysed for their grain size distributions using a Malvern MasterSizer, and for fallout ^{137}Cs, ^{210}Pb and ^{226}Ra concentrations using gamma spectrometry (more detailed information on these techniques is contained in He (1993) and Owens (1994)). Figure 12.4 illustrates the relationships between measured radionuclide concentration and specific surface area. Eqn (12.1) has been used to simulate the measured results and a least-squares fit gives the following relationships for uncultivated topsoil:

$$C_{u,r}(\bar{S}_u) = \begin{cases} 54.6\bar{S}_u^{0.75} & r^2 = 0.93 \quad \text{for } ^{137}\text{Cs} \\ 69.5\bar{S}_u^{0.71} & r^2 = 0.98 \quad \text{for unsupported } ^{210}\text{Pb} \end{cases} \tag{12.3}$$

For cultivated soil, the relationships are:

$$C_{c,r}(\bar{S}_c) = \begin{cases} 14.5\bar{S}_c^{0.65} & r^2 = 0.85 \quad \text{for } ^{137}\text{Cs} \\ 14.4\bar{S}_c^{0.76} & r^2 = 0.89 \quad \text{for unsupported } ^{210}\text{Pb} \end{cases} \tag{12.4}$$

The particle size behaviour of ^{226}Ra will be different from that of fallout ^{137}Cs or unsupported ^{210}Pb. This is because the mechanisms controlling its association with soil

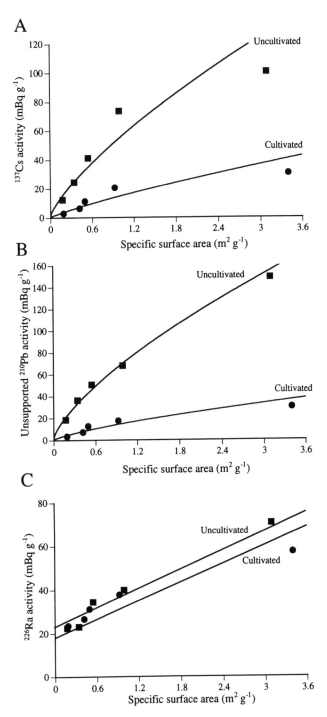

Figure 12.4 The relationship between radionuclide concentration and specific surface area for (A) ^{137}Cs, (B) ^{210}Pb and (C) ^{226}Ra. The smoothed curves are least-squares fitted results

particles are different from that of the association of fallout ^{137}Cs or ^{210}Pb with soil particles (the former has existed in soils since the formation of the Earth). He (1993) found that ^{226}Ra concentrations $C_{u,Ra}(\bar{S}_u)$ (or $C_{c,Ra}(\bar{S}_c)$) in soil samples collected in the vicinity of Exeter are linearly related to sample specific surface areas with a non-zero regression constant for both uncultivated and cultivated soils. For uncultivated soil:

$$C_{u,Ra}(\bar{S}_u) = 22.0 + 14.4\bar{S}_u \qquad (12.5)$$

where $r^2 = 0.92$. For cultivated soil:

$$C_{c,Ra}(\bar{S}_c) = 19.0 + 13.1\bar{S}_c \qquad (12.6)$$

where $r^2 = 0.94$. Eqns (12.5) and (12.6) suggest that the ^{226}Ra concentration of soils is partly particle size-independent (characterised by the first constant term on the right-hand side of the equations) and partly particle size-dependent (characterised by the second term on the right-hand side of the equations). Eqns (12.5) and (12.6) can be written in a more common form that is not site-specific:

$$C_{s,Ra} = \nu_{s,Ra}(1 + \nu_{1s,Ra}\bar{S}_s) \qquad (12.7)$$

where the subscript s denotes either uncultivated (u) or cultivated (c) soil, and $\nu_{1s,Ra}$ (0.65 for uncultivated soil and 0.70 for cultivated soil) and $\nu_{s,Ra}$ are constants.

The radionuclide concentrations of source materials

The radionuclides chosen in this investigation can be used to discriminate uncultivated, cultivated, woodland and channel bank source material. In most drainage basins in *lowland* Britain, topsoil from woodland (and forested land) is unlikely to be a major sediment source, and the examples from Devon presented earlier tend to support this argument. For this reason, woodland soil is ignored as a potential sediment source in this chapter. Furthermore, for simplicity, it has been assumed that the radionuclides are primarily associated with the mineral component, although we recognise that a small proportion of the ^{137}Cs and unsupported ^{210}Pb contained in soils is associated with organic matter (cf. He, 1993), and that differences in organic matter content between suspended sediment and source materials will therefore influence their concentrations in suspended sediment.

In the case of uncultivated and cultivated surface soils, the relationships between radionuclide concentrations and soil specific surface areas can be described by Eqns (12.1) and (12.7). To estimate the radionuclide content of channel bank materials, it has been assumed that channel banks act like uncultivated soils in receiving fallout radionuclides, and that the subsequent movement of the radionuclides in the soil profile can be described by the same mechanisms as for uncultivated soils. However, the differences in the radionuclide content of sediment derived from channel bank erosion and that generated by surface erosion must be emphasised. Surface erosion involves removal of a thin layer of soil and associated radionuclides from the soil surface, whereas bank erosion may involve the collapse of a whole soil profile into the channel or erosion of all or specific parts of the vertical bank surface (cf. Thorne, 1992). All the radionuclides contained in the soil profile can, therefore, be mobilised to the channel provided that the depth of the eroding profile is greater than the depth to which the radionuclides are distributed. Here, it has been assumed that bank collapse is the main mechanism of bank erosion. The relationships between the radionuclide concentrations and the grain size distributions may therefore be assumed to be the same as those for uncultivated soils. Generally, radionuclide concentrations in source materials can be expressed as:

$$C_{s,r}(\bar{S}_s) = v_{s,r} f_{s,r}(v_{ls,r}, \bar{S}_s) \qquad (12.8)$$

where the subscript s denotes u, c, b (bank) and the subscript r denotes ^{137}Cs, unsupported ^{210}Pb and ^{226}Ra.

The sediment mixing model

For the purpose of modelling source mixtures, it has been assumed that the particle size distribution of fluvial suspended sediment is representative of sediment mobilised from all possible sources, and that uncultivated topsoil, cultivated topsoil and channel banks are the main sediment sources. Under such assumptions, the radionuclide concentrations of fluvial suspended sediment $C_{r,\text{sd}}(S)$ (mBq g^{-1}) can be estimated by the following sediment-mixing model:

$$C_{r,\text{sd}}(S) = a_u C_{u,r}(S) + a_c C_{c,r}(S) + a_b C_{b,r}(S) \qquad (12.9)$$

where S is the mean specific surface area of suspended sediment and $C_{s,r}(S)$ (mBq g^{-1}) represents the radionuclide concentrations in source materials with the *same* specific surface area (S). The area S can be written as:

$$S = \int_0^\infty \rho'(S_{\text{sp}}) S_{\text{sp}} \, dS_{\text{sp}} \qquad (12.10)$$

where $\rho'(S_{\text{sp}})$ is the frequency density for suspended sediment with specific surface area S_{sp}. In Eqn (12.9), a_u, a_c and a_b are the mass proportions of sediment (or the *relative contributions*) from uncultivated topsoil, cultivated topsoil and channel banks, respectively. The relative contribution from each source is a function of, among other things, rates of soil erosion in cultivated and uncultivated fields, the efficiency of the sediment delivery system, the percentage land use within the drainage basin and the erosivity of the channel network. As the total mass of a sample is unity, then:

$$a_u + a_c + a_b = 1 \qquad (12.11)$$

and the contribution from each source can range from zero to one, i.e.:

$$0 \leq a_s \leq 1 \quad \text{for } s = u, c, b \qquad (12.12)$$

In Eqn (12.9), the radionuclide concentration $C_{s,r}(S)$ can be related to that of source material $C_{s,r}(\bar{S}_s)$ if it is assumed that both the suspended sediment and mobilised source materials have similar particle size distributions. The frequency density of suspended sediment $\rho'(S_{\text{sp}})$ is generally not the same as the frequency density $\rho_s(S_{\text{sp}})$ for source material because of the difference in grain size distribution between the suspended sediment and the original source. As the radionuclide concentrations associated with suspended sediment particles with specific surface area S are the same as those of source soil particles with the same specific surface area, the radionuclide concentrations in sediment can also be represented as (see Eqn (12.8)):

$$C_{s,r}(S) = v_{s,r} f_{s,r}(v_{1s,r}, S) \tag{12.13}$$

Because the function $f_{s,r}$ takes the fixed form, from Eqns (12.8) and (12.13) there is:

$$C_{s,r}(S) = R_{en,s,r}(\bar{S}_s, S) C_{s,r}(\bar{S}_s) \tag{12.14}$$

where $R_{en,s,r}(\bar{S}_s, S)$ may be termed the radionuclide *enrichment* (or depletion) ratio associated with sediment. This is defined as the ratio of the radionuclide concentration in sediment to that in the original source materials:

$$R_{en,s,r}(\bar{S}_s, S) = \frac{C_{s,r}(S)}{C_{s,r}(\bar{S}_s)} = \frac{f_{s,r}(v_{1s,r}, S)}{f_{s,r}(v_{1s,r}, \bar{S}_s)} \tag{12.15}$$

$R_{en,s,r}(\bar{S}_s, S)$ depends on radionuclide type and the difference in grain size distribution between the sediment and the source materials. From Eqn (12.15) it is apparent that $R_{en,s,r}$ reflects enrichment or depletion of finer particles (larger specific surface area) in sediment. It is not, however, linearly related to the enrichment of the percentage of fines as proposed by Peart and Walling (1988) and Walling and Woodward (1992). It should also be recognised that the enrichment ratio is unlikely to be a constant and will probably vary between sediment samples. This is because different sediment samples usually have different particle size distributions $p'(S_{sp})$ or different average specific surface areas S, even though they may be collected from a single storm event. The enrichment ratio $R_{en,s,r}(\bar{S}_s, S)$ can be related to the specific surface areas of the sediment and soil samples according to:

$$R_{en,s,r}(\bar{S}_s, S) = \begin{cases} (S/\bar{S}_s)^{v_{1s,r}} & \text{for } ^{137}\text{Cs and unsupported } ^{210}\text{Pb} \\ (1 + v_{1s,Ra}S)/(1 + v_{1s,Ra}\bar{S}_s) & \text{for } ^{226}\text{Ra} \end{cases} \tag{12.16}$$

The radionuclide concentrations in suspended sediment derived from each source can be estimated from Eqn (12.15). The corresponding enrichment ratio has been assumed to be independent of input fluxes and can be calculated from Eqn (12.16). Eqn (12.9) can now be represented as:

$$C_{r,sd}(S) = a_u R_{en,u,r}(\bar{S}_u, S) C_{u,r}(\bar{S}_u) + a_c R_{en,c,r}(\bar{S}_c, S) C_{c,r}(\bar{S}_c) + a_b R_{en,b,r}(\bar{S}_b, S) C_{b,r}(\bar{S}_b) \tag{12.17}$$

In the above equation all the parameters can be measured or estimated except for a_u, a_c and a_b. It is this property that provides a means of using radionuclides to fingerprint suspended sediment sources (i.e. the relative weight percentage of sediment from each possible source), provided that the number of *independent* radionuclides is greater than or equal to the number of sources and that the concentration of each radionuclide is significantly different between source materials (cf. Walling and Woodward, 1992; also see next section). Thus, using the independent radionuclides fallout ^{137}Cs, unsupported ^{210}Pb and ^{226}Ra, the relative contributions from different sources (a_u, a_c and a_b) in each sediment sample can be obtained by solving the following set of independent linear equations with a_u, a_c and a_b as unknown variables and the radionuclide concentrations in source materials as coefficients:

$$R_{en,u,Cs}\bar{C}_{u,Cs}(\bar{S}_u)a_u + R_{en,c,Cs}\bar{C}_{c,Cs}(\bar{S}_c)a_c + R_{en,b,Cs}\bar{C}_{b,Cs}(\bar{S}_b)a_b = C_{Cs,sd}(S)$$
$$R_{en,u,Pb}\bar{C}_{u,Pb}(\bar{S}_u)a_u + R_{en,c,Pb}\bar{C}_{c,Pb}(\bar{S}_c)a_c + R_{en,b,Pb}\bar{C}_{b,Pb}(\bar{S}_b)a_b = C_{Pb,sd}(S) \quad (12.18)$$
$$R_{en,u,Ra}\bar{C}_{u,Ra}(\bar{S}_u)a_u + R_{en,c,Ra}\bar{C}_{c,Ra}(\bar{S}_c)a_c + R_{en,b,Ra}\bar{C}_{b,Ra}(\bar{S}_b)a_b = C_{Ra,sd}(S)$$

where $\bar{C}_{s,r}(\bar{S}_s)$ (mBq g^{-1}) is the mean concentration for source s.

Two important facts should be borne in mind when interpreting the results obtained from the above equation set. First, it is likely that these equations are almost always mathematically independent for all values of the coefficients derived from environmental samples, even though they may not be statistically different. Thus, solutions to Eqn (12.18) always exist. A statistical significance test must, therefore, be used to see if they can be used as source discriminators. Secondly, the restrictions of Eqns (12.11) and (12.12) might not always be met (see next section).

THE RIVER CULM: A CASE STUDY

To illustrate the application of the numerical mixing model for suspended sediment source identification, the River Culm drainage basin (a tributary of the River Exe) in Devon, UK, has been chosen. Figure 12.5 illustrates the drainage network and relief of the basin and the suspended sediment sampling site at Rewe, near the outlet of the river. Further information concerning the drainage basin and the properties of the suspended sediment transported by this river is contained in Walling and Kane (1984), Lambert (1986), Walling and Bradley (1989) and Walling and Moorehead (1989). In brief, the total basin area is 276 km², and is mainly underlain by Permian and Triassic marls, breccias, conglomerates and sandstones, and Cretaceous greensand. Land use is a mixture of permanent pasture and arable farming, with a small amount of woodland. The suspended sediment samples ($n = 13$) reported here were collected during flood events that occurred during the winter of 1992–93. Source samples to represent uncultivated topsoil ($n = 17$), cultivated topsoil ($n = 12$) and channel bank material ($n = 9$) were also collected. These samples are assumed to represent the main sources of the suspended sediment load transported by this river. All the source and suspended sediment samples were prepared in the normal way and subsamples of the <2 mm fractions were analysed for ^{137}Cs, unsupported ^{210}Pb and ^{226}Ra concentration, organic matter content and grain size distribution.

The mean radionuclide concentrations and specific surface areas of the source materials are given in Table 12.2. along with values for v_{ls}. In this table the bank materials were treated as uncultivated soils (with $v_{lb} = v_{lu}$), and it has been assumed that

Table 12.2 Properties of surface materials from different sources

	$\bar{C}_{s,Cs}$ (mBq g^{-1})	$\bar{C}_{s,Pb}$ (mBq g^{-1})	$\bar{C}_{s,Ra}$ (mBq g^{-1})	$\bar{S}$ (m² g^{-1})	$v_{ls,Cs}$	$v_{ls,Pb}$	$v_{ls,Ra}$
Uncultivated soils	22.0	40.0	23.0	0.390	0.75	0.71	0.65
Cultivated soils	9.4	20.0	29.0	0.444	0.65	0.76	0.70
Channel banks	2.8	8.5	30.0	0.380	0.75	0.71	0.65

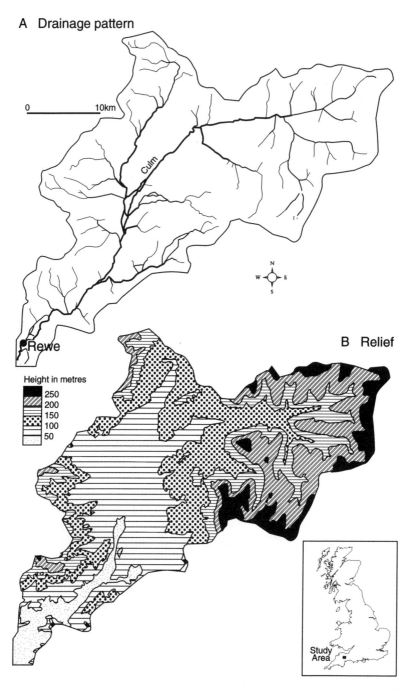

Figure 12.5 The drainage pattern (A) and relief (B) of the River Culm, Devon, and the location of the sampling site at Rewe

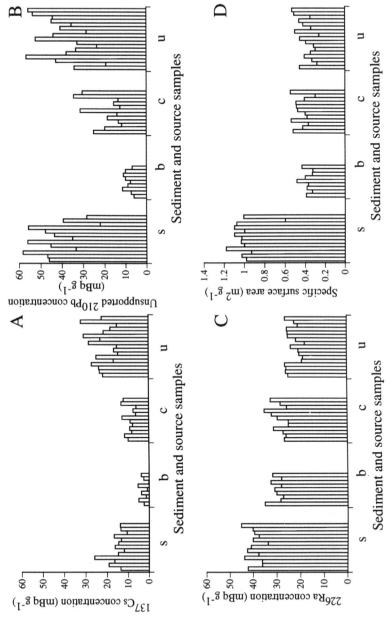

Figure 12.6 Distribution of the concentrations of ^{137}Cs (A), unsupported ^{210}Pb (B) and ^{226}Ra (C), and of specific surface area (D) in suspended sediment and source samples in the River Culm basin. Groups s, b, c and u represent suspended sediment, channel bank, cultivated and uncultivated samples respectively

v_{lu} and v_{lc} can be calculated from Eqns (12.3), (12.4) and (12.7). Figure 12.6 shows the distribution of the radionuclide concentrations of suspended sediment and source materials. The concentrations of ^{137}Cs, unsupported ^{210}Pb and ^{226}Ra vary between the three different source materials. The ^{137}Cs concentrations range from 15.0 to 33.0 mBq g^{-1} in uncultivated topsoil samples, and from 7.0 to 14.0 mBq g^{-1} in cultivated topsoil samples. The unsupported ^{210}Pb concentrations range from 19.0 to 57.0 mBq g^{-1} in uncultivated topsoils, and from 11.0 to 34.0 mBq g^{-1} in cultivated topsoils. In bank materials, both the ^{137}Cs and unsupported ^{210}Pb concentrations are relatively low, with mean concentrations of 2.8 mBq g^{-1} for ^{137}Cs and 8.5 mBq g^{-1} for ^{210}Pb, respectively. However, concentrations of ^{226}Ra are highest in bank materials and lowest in uncultivated topsoil samples, with mean concentrations of 30.0 mBq g^{-1} for bank material, 29.0 mBq g^{-1} for cultivated topsoil and 23.0 mBq g^{-1} for uncultivated topsoil.

It is apparent from Figure 12.6 that there is some overlap in the radionuclide concentrations of the samples collected to represent cultivated and uncultivated soils. This is because it is difficult to determine the precise land use history of a field based on a single site visit. Several of the fields thought to have been uncultivated at the time of sampling may have been ploughed decades before and then left as temporary grass. If there is sufficient evidence to support this view, and if land use has changed since the start of radionuclide fallout (and this varies between radionuclides), temporary grass samples could be reclassified as cultivated. Here, we have used the original data sets in order to discriminate sources. Furthermore, it is important to stress that in this example, for simplicity, the radionuclide concentrations of the sediment sources are represented by mean values. As described above, in reality there is noticeable variation in concentrations for each source type, and ideally this should be incorporated into the numerical mixing model.

In order to remove the complicating effects associated with the organic matter component of the samples (He, 1993), values of radionuclide concentrations are represented as activities per gram of mineral soil. Based on the assumption that the mean concentrations are normally distributed, a significance test of the differences of the mean concentrations for each radionuclide between the sources was carried out. The results show that differences between these sources are significant (at a significance level of $\alpha < 0.01$ for ^{137}Cs and unsupported ^{210}Pb, and $\alpha < 0.05$ for ^{226}Ra), except for the case of ^{226}Ra concentrations in bank materials and cultivated surface soils (with $\alpha > 0.05$). Therefore, the three equations in Eqn (12.18) may be viewed as statistically independent. The differences in ^{226}Ra concentrations between uncultivated topsoils and bank material or cultivated topsoils probably reflect the slow downward migration of this radionuclide in soils. The ^{226}Ra concentration profile in soils usually shows a depletion in surface layers at uncultivated sites (cf. Nozaki *et al.*, 1978). The relatively higher concentrations of ^{226}Ra in cultivated surface soils and channel bank materials may be associated with soil mixing due to cultivation and the collection of samples from middle or lower parts of vertical bank surfaces.

The grain size distributions and concentrations of ^{137}Cs, unsupported ^{210}Pb and ^{226}Ra are different between the suspended sediment samples and the source materials. Figure 12.6D shows the distributions of the specific surface area of the suspended sediment and source samples. Compared with source materials, suspended sediment is enriched in fine soil particles, with specific surface areas generally over two times higher than the average values for source samples. The ^{137}Cs and unsupported ^{210}Pb concentrations in the suspended sediment samples range from 12.0 to 27.0 mBq g^{-1} and from 22.0 to 58.0 mBq g^{-1}, respectively. These values are generally lower than the radionuclide concentrations in uncultivated soil source materials and higher than the radionuclide concentrations in cultivated soil source materials and channel bank materials. The ^{226}Ra concentrations, on the other hand, exhibit a relatively lower variation and are higher than those in all source samples, demonstrating that suspended sediment is significantly enriched in this radionuclide.

In order to obtain the mass percentage of each sediment sample from the three sources, enrichment ratios were estimated using Eqn (12.16) and substituted into Eqn (12.18). The radionuclide concentrations of the suspended sediment samples were also represented as activities per gram of mineral soil. In order to meet the physical requirements of Eqns (12.11) and (12.12), instead of directly solving Eqn (12.18), the least-squares method was adopted to estimate the appropriate variables. This method is commonly used to solve similar problems (cf. Thompson and Edwards, 1981; Wadden et al., 1986; Mackas et al., 1987; Yu and Oldfield, 1989). The values of a_s, (s = u, c and b) in Eqn (12.18) with the constraints of Eqns (12.11) and (12.12) can be estimated by minimising the sum of the squares of the residual R_{es} of all variables involved:

$$R_{es} = \left[C_{Cs,sd} - \left(\sum_s a_s R_{en,s,Cs} \bar{C}_{s,Cs}\right)\right]^2 W_{Cs} + \left[C_{Pb,sd} - \left(\sum_s a_s R_{en,s,Pb} \bar{C}_{s,Pb}\right)\right]^2 W_{Pb}$$

$$+ \left[C_{Ra,sd} - \left(\sum_s a_s R_{en,s,Ra} \bar{C}_{s,Ra}\right)\right]^2 W_{Ra} \qquad (12.19)$$

where W_{Cs}, W_{Pb} and W_{Ra} are weight factors associated with the magnitude and errors of the corresponding radionuclide concentrations. The minimisation of the sum of the weighted squares of the residuals was used for each sample because the magnitude of the concentrations and the errors associated with the measurements of ^{137}Cs, ^{210}Pb and

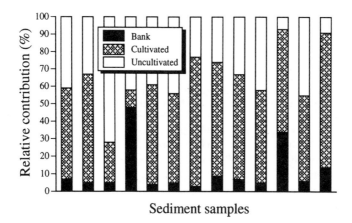

Figure 12.7 The relative contributions of different sources to the suspended sediment load transported by the River Culm during storm events in the winter 1992–93 period

^{226}Ra are different. Values of W_{Cs}, W_{Pb} and W_{Ra} were set to the inverses of the squares of the variances of the related radionuclide concentrations (cf. Mackas *et al.*, 1987).

Figure 12.7 illustrates the relative contributions from the different sources to each of the 13 suspended sediment samples, derived by the radionuclide fingerprinting method developed in this paper. The relative importance of the different sources varies between the samples. For example, the proportion of sediment derived from channel banks varies from 3 to 48%, while the contribution from cultivated fields varies from 10 to 77%. This variation may be explained by changes in sediment source within and between different storm events (cf. Walling and Woodward, 1992) caused by changes in the mobilisation and delivery of source materials to the channel network. During the winter period of 1992–93, average mass proportions of suspended sediment transported by this river derived from cultivated topsoil, uncultivated topsoil and channel banks are estimated to be 53%, 35% and 12%, respectively.

CONCLUSION

In certain situations, radionuclides can be used to identify suspended sediment sources. However, as has been illustrated earlier, even though one radionuclide may be able to distinguish between source types, it may not be sufficient for reliably identifying the sources of the sediment load transported by a stream. This is because of the large number of possible source combinations. In this paper three different radionuclides have been used along with a numerical mixing model to identify the sources of the suspended sediment transported in the River Culm in Devon. While this approach was not able to distinguish source area on the basis of geology, it was able to distinguish confidently between channel bank material, cultivated topsoil and uncultivated topsoil. Information relating to source type (as opposed to the precise spatial location) is often more important for understanding the processes involved in landscape evolution and for making land management decisions. In the case of the River Culm drainage basin, the suspended sediment load transported during the winter of 1992–93 was derived mainly from the erosion of topsoil in cultivated fields.

ACKNOWLEDGEMENTS

The work reported in this paper was carried out under the supervision of Des Walling. Financial support was provided by an ORS Award and a University of Exeter Postgraduate Scholarship to Qingping He and a NERC Postgraduate Studentship to Phil Owens. We would like to thank Jamie Woodward for assistance in sample collection, Terry Bacon and Andrew Teed for producing some of the diagrams, and Alison Foskett, Ian Foster and anonymous referees for comments on earlier versions of this chapter.

REFERENCES

Balteanu, D., Mihaiu, G., Negut, N. and Caplescu, L. (1984) Sources of sediment and channel changes in small catchments of Romania's hilly regions. In: Burt, T.P. and Walling, D.E.

(eds), *Catchment Experiments in Fluvial Geomorphology*, Geo Books, Norwich, pp. 277–288.

Cambray, R.S., Playford, K., Lewis, G.N.J. and Carpenter, R.C. (1989) *Radioactive Fallout in Air and Rain: Results to the End of 1988*, United Kingdom Atomic Energy Authority Report AERE-R 13575, HMSO, London.

Dietrich, W.E. and Dunne, T. (1978) Sediment budget for a small catchment in mountainous terrain. *Z. Geomorphol.* **29**, 191–206.

Foster, I.D.L., Dalgleish, H., Dearing, J.A. and Jones, E.D. (1994) Quantifying soil erosion and sediment transport in drainage basins; some observations on the use of Cs-137. In: Olive, L., Loughran, R.J. and Kesby, J.A. (eds), *Variability in Stream Erosion and Sediment Transport*, IAHS Publication no. 224, IAHS Press, Wallingford, pp. 55–64.

Garrad, P.N. and Hey, R.D. (1989) Sources of suspended and deposited sediment in a broadland river. *Earth Surf. Processes Landforms* **14**, 41–62.

Grimshaw, D.L. and Lewin, J. (1980) Source identification for suspended sediments. *J. Hydrol.* **47**, 151–162.

He, Q. (1993) Interpretation of fallout radionuclide profiles in sediments from lake and floodplain environments. Unpublished PhD thesis, University of Exeter.

Hillel, D. (1982) *Introduction to Soil Physics*, Academic Press, Orlando, FL.

Imeson, A.C., Vis, M. and Duysings, J.J.H.M. (1984) Surface and subsurface sources of suspended solids in forested drainage basins in the Keuper region of Luxembourg. In: Burt, T.P. and Walling, D.E. (eds), *Catchment Experiments in Fluvial Geomorphology*, Geo Books, Norwich, pp. 219–235.

Johnson, A.G. and Kelley, J.T. (1984) Temporal, spatial, and textural variation in the mineralogy of Mississippi River suspended sediment. *J. Sediment. Petrol.* **54**, 67–72.

Jordan, P. and Slaymaker, O. (1991) Holocene sediment production in Lillooet River basin, British Columbia: a sediment budget approach. *Geogr. Phys. Quatern.* **45**, 45–57.

Klages, M.G. and Hsieh, Y.P. (1975) Suspended solids carried by the Gallatin River of southwestern Montana: II. Using mineralogy for inferring sources. *J. Environ. Qual.* **4**, 68–73.

Lambert, C.P. (1986) The suspended sediment delivery dynamics of river channels in the Exe basin. Unpublished PhD thesis, University of Exeter.

Lewin, J. and Wolfenden, P.J. (1978) The assessment of sediment sources: a field experiment. *Earth Surf. Processes* **3**, 171–178.

Lewin, J., Cryer, R. and Harrison, D.I. (1974) Sources for sediments and solutes in mid-Wales. In: Gregory, K.J. and Walling, D.E. (eds), *Fluvial Processes in Instrumented Watersheds*, Institute of British Geographers Special Publication no. 6, London, pp. 73–85.

Loughran, R.J., Campbell, B.L. and Elliott, G.L. (1982) The identification and quantification of sediment sources using ^{137}Cs. In: Walling, D.E. (ed.), *Recent Developments in the Explanation and Prediction of Erosion and Sediment Yield*, IAHS Publication no. 137, IAHS Press, Wallingford, pp. 361–369.

Mackas, D.L., Denman, K.L. and Bennett, A.F. (1987) Least squares multiple analysis of water mass composition. *J. Geophys. Res.* **92**, C3, 2907–2918.

Marshall, T.J. and Holmes, J.W. (1979) *Soil Physics*, Cambridge University Press, Cambridge.

Nozaki, Y., DeMaster, D.J., Lewis, D.M. and Turekian, K.K. (1978) Atmospheric ^{210}Pb fluxes determined from soil profiles. *J. Geophys. Res.* **83**, 4047–4051.

Oldfield, F., Appleby, P.G., Cambray, R.S., Eakins, J.D., Barber, K.E., Battarbee, R.W., Pearson, G.R. and Williams, J.M. (1979) ^{210}Pb, ^{137}Cs and ^{239}Pu profiles in ombrotrophic peat. *Oikos* **33**, 40–45.

Owens, P.N. (1994) Towards improved interpretation of caesium-137 measurements in soil erosion studies. Unpublished PhD thesis, University of Exeter.

Passmore, D.G. and Macklin, M.G. (1994) Provenance of fine-grained alluvium and late Holocene land-use change in the Tyne basin, northern England. *Geomorphology* **9**, 127–142.

Peart, M.R. and Walling, D.E. (1986) Fingerprinting sediment source: the example of a drainage basin in Devon, UK. In: Hadley, R.F. (ed.), *Drainage Basin Sediment Delivery*, IAHS Publication no. 159, IAHS Press, Wallingford, pp. 41–55.

Peart, M.R. and Walling, D.E. (1988) Techniques for establishing suspended sediment sources in two drainage basins in Devon, UK: a comparative assessment. In: Bordas, M.P. and

Walling, D.E. (eds), *Sediment Budgets*, IAHS Publication no. 174, IAHS Press, Wallingford, pp. 269–279.

Phillips, J.D. (1992) Delivery of upper-basin sediment to the Lower Neuse River, North Carolina, USA. *Earth Surf. Processes Landforms* **17**, 699–709.

Robbins, R. A. (1978) Geochemical and geophysical application of radioactive lead. In: Nriagu, J.O. (ed.), *The Biochemistry of Lead in the Environment*, Elsevier Science, Amsterdam, pp. 286–383.

Sawhney, B.L. (1972) Selective sorption and fixation of cations by clay minerals: a review. *Clays Clay Miner.* **20**, 93–100.

Stanton, R.K., Murray, A.S. and Olley, J. M. (1992) Tracing the source of recent sediment using environmental magnetism and radionuclides in the karst of the Jenolan Caves, Australia. In: Bogen, J., Walling, D. E. and Day, T. (eds), *Erosion and Sediment Transport Monitoring Programmes in River Basins*, IAHS Publication no. 210, IAHS Press, Wallingford, pp. 125–133.

Stott, A.P. (1986) Sediment tracing in a reservoir-catchment system using a magnetic mixing model. *Phys. Earth Planet. Interiors* **42**, 105–112.

Sutherland, R.A. (1991) Examination of caesium-137 areal activities in control (uneroded) locations. *Soil Technol.* **4**, 33–50.

Sutherland, R.A. (1994) Spatial variability of ^{137}Cs and the influence of sampling on estimates of sediment redistribution. *Catena* **21**, 57–71.

Thompson, R. and Edwards, R.J. (1981) Mixing and water-mass formation in the Australian subantarctic. *J. Phys. Oceanogr.* **11**, 1399–1406.

Thorne, C.R. (1992) Bend scour and bank erosion on the meandering Red River, Louisiana. In: Carling, P.A. and Petts, G.E. (eds), *Lowland Floodplain Rivers: Geomorphological Perspectives*, John Wiley & Sons, Chichester, pp. 95–115.

Trimble, S.W. and Lund, S.W. (1982) *Soil Conservation and the Reduction of Erosion and Sedimentation in the Coon Creek Basin, Wisconsin*, US Geological Survey Professional Paper no. 1234.

Wadden, R.A., Uno, I. and Wakamatsu, S. (1986) Source discrimination of short-term hydrocarbon samples measured aloft. *Environ. Sci. Technol.* **20**, 473–483.

Wall, G.J. and Wilding, L.P. (1976) Mineralogy and related parameters of fluvial suspended sediments in northwestern Ohio. *J. Environ. Qual.* **5**, 168–173.

Wallbrink, P.J. and Murray, A.S. (1993) Use of fallout radionuclides as indicators of erosion processes. *Hydrol. Processes* **7**, 297–304.

Walling, D.E. (1989) Physical and chemical properties of sediment: the quality dimension. *Int. J. Sediment Res.* **4**, 27–39.

Walling, D.E. and Bradley, S.B. (1989) Rates and patterns of contemporary floodplain sedimentation: a case study of the River Culm, Devon, UK. *GeoJournal* **19**, 53–62.

Walling, D.E. and He, Q. (1993) Towards improved interpretation of ^{137}Cs profiles in lake sediments. In: McManus, J. and Duck, R.W. (eds), *Geomorphology and Sedimentology of Lakes and Reservoirs*, John Wiley and Sons, Chichester, pp. 31–53.

Walling, D.E. and Kane, P. (1984) Suspended sediment properties and their geomorphological significance. In: Burt, T.P. and Walling, D.E. (eds), *Catchment Experiments in Fluvial Geomorphology*, Geo Books, Norwich, pp. 311–334.

Walling, D.E. and Moorehead, P.W. (1989) The particle size characteristics of fluvial suspended sediment: an overview. *Hydrobiologia* **176/177**, 125–149.

Walling, D.E. and Woodward, J.C. (1992) Use of radiometric fingerprints to derive information on suspended sediment sources. In: Bogen, J., Walling, D.E. and Day, T. (eds), *Erosion and Sediment Transport Monitoring Programmes in River Basins*, IAHS Publication no. 210, IAHS Press, Wallingford, pp. 153–164.

Walling, D.E., Peart, M.R., Oldfield, F. and Thompson, R. (1979) Suspended sediment sources identified by magnetic measurements. *Nature* **281**, 110–113.

Walling, D.E., Woodward, J.C. and Nicholas, A.P. (1993) A multi-parameter approach to fingerprinting suspended-sediment sources. In: Peters, N.E., Hoehn, E., Leibundgut, Ch., Tase, N. and Walling, D.E. (eds), *Tracers in Hydrology*, IAHS Publication no. 215, IAHS Press, Wallingford, pp. 329–338.

Wasson, R.J., Clark, R.L., Nanninga, P.M. and Waters, J. (1987) ^{210}Pb as a chronometer and tracer, Burrinjuck Reservoir, Australia. *Earth Surf. Processes Landforms* **12**, 399–414.

Wolman, M.G. (1977) Changing needs and opportunities in the sediment field. *Water Resour. Res.* **13**, 50–54.

Woodward, J.C., Lewin, J. and Macklin, M.G. (1992) Alluvial sediment sources in a glaciated catchment: the Voidamatis basin, northwest Greece. *Earth Surf. Processes Landforms* **17**, 205–216.

Yu, L. and Oldfield, F. (1989) A multivariate mixing model for identifying sediment sources from magnetic measurements. *Quatern. Res.* **32**, 168–181.

# 13 Processes of River Bank Erosion and Their Contribution to the Suspended Sediment Load of the River Culm, Devon

DAVID ASHBRIDGE
Stamford Bridge, York, UK (formerly: University of Exeter, UK)

INTRODUCTION

Early studies of bank erosion have typically viewed the processes involved from one of a number of perspectives. These include studies of bank erosion in the context of channel change (e.g. Wolman and Leopold, 1957; Hooke, 1977), from an engineering perspective (e.g. Burgi and Karaki, 1971; Thorne, 1982) and as an agent of fluvial geomorphology (e.g. Knighton, 1973; Brunsden and Kesel, 1982). Only limited attention has been paid, however, to those aspects of bank erosion that relate to the supply of sediment to the river channel and the incorporation of suitably sized sediment in the suspended load of the river.

In examining catchment sediment budgets, several authors have developed estimates of the potential contribution made by bank erosion to the suspended sediment load of rivers. These include Walling (1971), Grimshaw and Lewin (1980), Collins (1981), Walling *et al.* (1979), Walling and Kane (1982) and Peart and Walling (1982). However, relatively few studies have attempted to examine processes of river bank erosion in the specific context of their contribution to the suspended sediment load of rivers. Neill and Mollard (1982), for example, in a study of bank erosion and sediment transport in the upper Oldman River basin, described the ways in which banks retreated, but did not address the underlying processes or attempt to estimate the significance of bank erosion as a suspended sediment source.

This chapter presents the result of research work undertaken to determine the processes by which river banks are eroded and to assess the contribution made by those processes to the sediment budget of the River Culm, Devon.

Background information on the study catchment and methods of data collection is provided, followed by an analysis of the processes responsible for bank retreat. Observed patterns of retreat through time are described, and the potential significance of bank erosion as a source of suspended sediment in the Culm is examined. A fuller description of this study can be found in Ashbridge (1990).

Sediment and Water Quality in River Catchments. Edited by I.D.L. Foster, A.M. Gurnell and B.W. Webb.
© 1995 John Wiley & Sons Ltd.

THE STUDY CATCHMENT

The River Culm, with a catchment area of 276 km², rises from a spring source in the Blackdown Hills, and flows for a distance of 55 km to its confluence with the Exe, near Stoke Canon (Figure 13.1). In this distance the river descends from an altitude of 270 m to 20 m OD (Ordnance Datum).

Much of the course of the Culm is sinuous and frequently meandering, with the river flowing between interlocking spurs in the upper part of its catchment and then, further downstream, within an alluvial floodplain, the width of which increases to over 1 km at Stoke Canon. Although much of the river channel is cut into alluvium, particularly in the lower reaches, the underlying solid geology comprises mostly Permo-Triassic marls and sandstones in the headwaters, passing into breccias of similar age further downstream.

Land use adjacent to the channel, both on the floodplain and also further upstream in the headwaters, is almost entirely pasturage. This is interspersed with occasional areas of woodland in the upper part of the catchment.

The climate experienced by the River Culm catchment is largely controlled by its position within the south-west peninsula. This gives rise to a mild and moderately moist

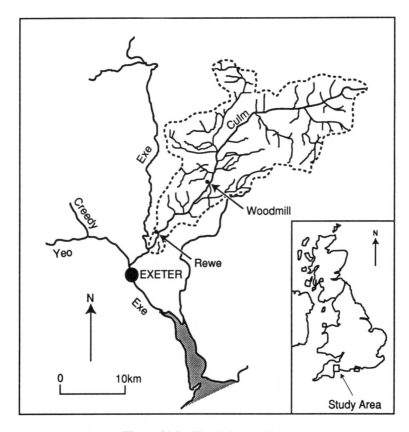

Figure 13.1 The Culm catchment

climatic regime, with most of the annual precipitation occurring in late autumn and early winter.

DATA COLLECTION

The rates and spatial distribution of bank erosion in the Culm, together with streamflow and suspended sediment transport in the catchment, were monitored over a two-year period in 1980 and 1981. Data on a range of environmental variables were collected throughout the same period. The data collection programme focused on a study reach situated between the South West Water Authority's hydrometric stations at Woodmill and Rewe (Figure 13.1), where bank erosion was most prevalent.

Bank erosion

Bank erosion processes were observed and rates of retreat measured at 57 actively eroding banks on the River Culm between Woodmill and Rewe. A total of 379 erosion pins were used to monitor retreat. Measurements were taken on 30 occasions during the study period, with more frequent measurements being taken at two selected sites at Rewe. This provided a total of over 12 000 individual measurements. In order to be able to extend the findings of the study to the catchment as a whole, bank retreat was also monitored, in less detail, at 12 sites in the upper part of the catchment. Additionally, in Spring 1981, a reconnaissance of all actively eroding banks in the Culm basin was undertaken.

Streamflow discharge and suspended sediment discharge

A continuous discharge record for the Culm at Woodmill has been maintained by the South West Water Authority since January 1962. The annual average daily flows measured for 1980 and 1981 were, at 3.89 and 4.41 $m^3 s^{-1}$ respectively, statistically similar to those measured for the preceding 17 years (1963–79). Additional recording equipment was installed at Rewe in order to allow continuous streamflow discharge measurement during the study period.

Dual photometric turbidity probes were used at Rewe and Woodmill to provide continuous records of suspended sediment transport. In addition, water samples were taken from the Culm at Rewe to allow particle-size analysis of the sediment. During the study period, 31 samples each of 125 litres were taken from the Culm at Rewe during storm events when sediment concentrations were high. The sediment was extracted from the water by continuous flow centrifuges and then analysed for particle-size distribution.

Environmental parameters

Data were collected on a range of environmental variables through a programme of field measurement and laboratory analysis.

Bank material characteristics

All 57 erosion sites between Rewe and Woodmill were sampled for bulk density measurements. Two samples were taken at each site so that some account could be taken of vertical variations in the bulk density of the bank material. Subsamples taken from the bulk density samples were further analysed for their particle-size distribution. Eight soil samples were taken from each of the two banks at Rewe for shear strength testing. Finally, in order to combine data on soil matrix potentials and soil shear strength at varying water contents, the moisture retentivity characteristics of soil samples taken from study sites at Rewe were determined using a Haines apparatus constructed for this purpose.

Soil moisture conditions

A continuous record of local soil moisture conditions was obtained from automatically recording systems employing porous cup tensiometers installed at the Rewe sites. Seasonal fluctuations in soil moisture conditions were indicated by longer-term variations in matrix potentials. These could be discerned clearly, even though shorter-term changes were often masked.

Climate

With the exception of air and soil temperatures, which were monitored at Rewe using purpose-built equipment, all other climatic data were obtained from the South West Water Authority.

PROCESSES OF BANK EROSION

Observation revealed that erosion at most retreating banks took place through a cyclic process involving both corrasion and collapse. This cycle comprised corrasive undercutting that was followed by collapse of the resulting overhang into the channel; this material was then removed through corrasion and entrainment by the river, so that the cycle might recommence. Each element of the cycle is examined in the following sections.

Corrasion and undercutting

Corrasion occurred, to a variable extent, at all of the sites of bank retreat that were studied on the River Culm. At a minority of sites, corrasion occurred in isolation and there was no evidence of slumping. These banks typically comprised cohesive material that was relatively resistant to particle entrainment, and measured rates of erosion at these sites were low. Corrasion, therefore, did not appear to represent an effective erosion agent, other than when the cohesiveness of the bank material had been reduced by sub-aerial weathering or frost action in the period preceding the storm event. Weathering was much more important than frost action as a means of preparing the bank for corrasion, at least partly because of the low incidence of sub-zero temperatures during the study period. Data from the air-and-soil-temperature probes used at Rewe also revealed that freezing was confined to the outer 10–15 mm of the bank, and could therefore weaken only a relatively thin outer layer of material prior to corrasion by the river. Consequently, although corrasion acting alone clearly did cause erosion, the amounts were small in comparison to those produced by corrasion and slumping acting together.

At the majority of eroding sites, slumping was preceded by active undercutting of the bank by corrasive activity. The undercutting resulted not only from streamflow conditions within the channel and the more frequent action of corrasive forces on the lower parts of the bank, but often also from the composition of the banks themselves. In the study area, many alluvial banks comprise a cohesive silt/clay alluvium underlain by a layer of relatively cohesionless, coarser sand/gravel-sized material that represents the remains of an old point bar upon which the upper finer material was later deposited. The cohesionless lower layer is particularly susceptible to corrasion and it therefore undergoes relatively rapid erosion, leading to the development of overhangs.

Bank collapse

Shear failure and, to a lesser extent, tensile failure were found to be the main mechanisms responsible for bank collapse on the River Culm (Figure 13.2). Shear failure occurs when the weight of an overhang exceeds the maximum that can be supported by the upward shear force acting vertically along the plane of potential failure. The mechanism of tensile failure closely resembles that of shear failure, except that tensile forces at the top of the block must also be considered in addition to the shear forces acting at the side of the block. However, the tensile strength of most soils is low, usually not greater than one-tenth of the compressive strength and may approach zero under conditions of high soil water content. Under such circumstances, the factors governing the tensile stability of a cantilever are essentially the same as those controlling its shear stability.

Observations revealed that bank collapse was frequently, though not always, associated with the falling stage of a storm hydrograph. The presence of many failed

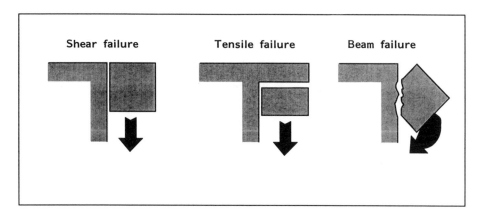

Figure 13.2 Mechanisms of bank collapse

blocks at the foot of banks was powerful evidence to support this. Had the blocks fallen before or at the peak of the event they would have been removed, or at least displaced, by the streamflow, something that frequently occurred during a subsequent storm of similar magnitude.

It is suggested that failure of a stable overhang is usually brought about by one or more of the following changes in the condition of the overhang and its constituent material.

Extent of undercutting

Any increase in the width of the overhang as a result of further erosion of the underlying cohesionless stratum will increase the weight of the overhang. If this then exceeds the maximum upward shear force that can be exerted by the soil along the vertical plane of potential failure, shearing will occur along this plane and the block will fall into the channel. In this way, progressive undercutting will increase the weight of an overhang, reducing its stability and leading to its eventual collapse into the channel.

Soil water content

The weight of the overhang will also be affected by changes in the water content of the bank material. Hydraulic conditions within the river bank fluctuate considerably, both on a seasonal basis and also on a much shorter time-scale. Field measurements at Rewe and laboratory testing of bank material samples showed that the moisture content of these banks typically fluctuated between 20% and 35% by weight. An increase from 20% to 35% would appreciably increase the weight of an overhang, possibly to such an extent that shear failure would occur.

Soil shear strength

If the mass of the overhang remains constant but the shear strength of the bank material is progressively decreased, shear failure will again occur once the critical point is passed and the maximum shear force available to support the overhanging block becomes less than the weight of that block. Tests carried out on soil samples taken from both sites at Rewe revealed that the shear strength of the bank material was inversely related to its moisture content, as had been suggested by several earlier authors, for example Hooke (1979). Thus by increasing the water content of bank material, failure may be induced as a result of both decreased shear strength and increased weight.

River stage

Changes in river stage may contribute to shear failure. During a storm event, rapid erosion of the underlying coarse layer may increase the width of an overhang to a point where its weight could not normally be supported by the bank material. However, if river stage is high and the overhang partially or wholly submerged, its immersed weight will be less than its own weight as a consequence of it displacing river water. As stage declines after the peak of the storm, the effective weight of the overhang will increase progressively as less of it remains submerged. Once the effective weight, that is, the true weight minus the weight of the water displaced, exceeds the maximum supportable mass, failure will occur. Consider, for example, a cantilever of overall bulk density 1625 $kg\,m^{-3}$ (that is, a dry bulk density of 1250 $kg\,m^{-3}$ with a moisture content of 30%); when immersed, its apparent bulk density falls to 625 $kg\,m^{-3}$. Thus when the block re-emerges from the water its apparent mass increases by 160% and the shear strength needed to support it will increase by the same amount. It is because of this that many river bank failures can be seen to occur after the passage of peak discharge.

Soil structure

The development of cracks and minor failures within the structure of the overhang will further contribute to its eventual collapse into the channel. The existence of near-vertical cracks coinciding with the shear failure plane has been noted by many authors, and may be the result of one or more of three causes. Tension cracks result from the tendency for an overhang to rotate downwards towards the channel and characteristically develop from the floodplain downwards. They may extend over a large proportion of the total height of the bank (Turnbull et al., 1966; Brunsden and Kesel, 1973) and thus significantly reduce the weight of overhang that can be supported by shear forces within the bank. Pressure release cracks may develop in a bank as a consequence of the removal of bank material at the channel boundary. Desiccation of the overhang material will also lead to the growth of cracks, frequently upwards from the base of the cantilever. This may be accompanied by a loss of soil particles at the base of the overhang as these dry out and fall into the channel. Pressure release and desiccation cracks, like tension cracks, will lead to a reduction in the overall effective shear strength of the material supporting the overhang. Tension, pressure release and desiccation thus combine to promote the development of a soil structure that is in turn conducive to the collapse of sizable blocks of material from the banks into the channel.

In the Culm, it is typically through a combination of these mechanisms that the majority of bank collapses occur. A simple mathematical model of the process of overhang failure was developed in order to allow theoretical verification of the combined operation of these mechanisms. Application of the model to empirical data obtained from the Culm supported the preceding analysis of the bank collapse process.

In addition to the above, several other factors may be expected to exert an influence on the process of bank collapse. However, it is believed that these will have a modifying, rather than controlling, influence. For example, fluid drag and shear forces act upon the submerged overhang. These are, however, estimated to rarely exceed 5% of the cantilever weight (Thorne, 1978) and are unlikely to have a serious effect on the stability of the overhang.

It might also be expected that the passage of a storm peak would lead to an increase in soil moisture content through rainfall percolation and seepage from the channel, with consequent implications for shear strength and bulk density. However, the extent to which this would be significant is debatable, and analysis of the soil moisture tension records for Rewe revealed that in many cases there was no significant reduction in tension as a result of a storm event. It would appear, rather, that the high moisture contents that were measured in the winter months were the product of a more general rise in soil moisture levels, reflecting the combined effects of increased rainfall and reduced evapotranspiration rates at this time of year.

Water movement across the channel boundary may, however, have some effect on the stability of eroding banks. At both Rewe sites, networks of tensiometers provided information on hydraulic potentials within the floodplain adjacent to the channel. The general pattern of seepage was one of gradual water movement from bank to channel at almost all times, except during conditions of high discharge when the direction might be reversed. Although it was not possible to determine the effect of this on the banks at Rewe, the potential destabilising effect on such seepage was indicated by Burgi and Karaki (1971). Water movement from the lower coarser stratum to the channel at times of low river discharge may help to prepare this material for effective corrasion by reducing the cohesiveness of the matrix. It is also possible that, while not directly affecting the parameters controlling the bank collapse model, increased seepage towards the channel may be a minor influence on the precise timing of block collapse from the banks. Overall, however, it is considered unlikely that seepage could significantly affect the stability analysis presented in the model of bank collapse, given the low rate of water movement and the potentially much stronger influence of other factors that affect bank stability.

Removal of eroded material

Removal of collapsed bank material is an essential element of the overall retreat process, in allowing the cycle of corrasive undercutting and bank collapse to recommence. The debris from bank collapse may comprise intact blocks, or a block may have been broken up as a result of the fall to the channel bed. Blocks are then abraded and broken up by corrasion and translocation, and eventually disperse to become part of the sediment load of the river.

As most cantilevers fail after peak flow, the debris often remains at the foot of the bank until it is removed by subsequent high flows. Blocks that fall into the river at the base of the bank will be affected by slaking and a rise in water content to beyond the liquid limit of the material. They will consequently begin to disintegrate, becoming highly liable to processes of particle and aggregate entrainment.

In the case of blocks that fall from the bank but are only partly, if at all, below water level, stability will in general increase through time, often augmented by the development of vegetation on the block. If the river is unable to remove the block, it may cease to have an identity of its own and become part of the lower river bank. Fallen blocks resulting from storms late in the winter, or even from early summer events, are particularly liable to become stabilised in this way, as high flows are rare during the summer. Such a block may remain at the foot of the bank throughout the summer, becoming vegetated and highly stabilised during this time. It then proves relatively difficult for early winter storms to remove this material, and a few blocks may remain throughout the winter and into the following summer. This process of stabilisation has important implications for the operation of the corrasion–collapse–removal cycle of erosion. Unless the failed material is removed from the base of the bank, it will protect the non-cohesive layer from corrasion and thus prevent the cycle of corrasion–collapse–removal from being repeated.

It is therefore important to the continuing operation of the cycle that debris from a collapsed overhang be removed from the base of the bank relatively soon after collapse. This will occur through corrasion and translocation by the river, aided by the gradual break-up of the block when immersed for a long period. The dominant influences on this part of the erosion process are therefore the magnitude and frequency of succeeding storm events.

PATTERNS OF EROSION THROUGH TIME

It is difficult to model effectively a cyclic process that advances in intermittent steps over an extended time period. Thus, although the mechanisms by which bank retreat occurs have been identified, it is impossible to include in such an analysis any consideration of the precise timing of bank collapse at a specific site. As a result of both this and the effects of local variations in conditions, the occurrence or otherwise of erosive activity cannot be predicted with any accuracy for a given site and storm event. Instead, erosion will take place in an essentially stochastic manner throughout the catchment, reflecting the general, rather than specific, operation of the processes indicated previously.

Confirmation of the stochastic nature of the bank retreat process was provided by statistical analysis of the incidence of erosion events at the sites between Rewe and Woodmill during the two-year measurement period. Although many sites showed similar amounts of retreat over the two-year period, considerable inter-site variation occurred when shorter time periods were considered.

Despite this variability, some broad patterns in the incidence of erosion were detected. A general increase in erosional activity in the winter months was evident, and a small number of large storm events each led to erosion occurring at a majority of sites. Consideration of the processes responsible for bank retreat can explain most, if not all, of this observed seasonal pattern in erosion in terms of two important factors.

Firstly, the frequency and magnitude of storm events reach a maximum from October to March. These factors have been demonstrated to be major determinants of the precise timing and scale of bank retreat.

Secondly, seasonal variations in soil moisture conditions result in the river banks being most susceptible to general erosive activity in the winter months, when moisture levels are at a maximum. More specifically, observations confirm that bank collapse, which is responsible for a large part of overall retreat, occurs predominantly in winter, with erosion at most banks being restricted to more gradual corrasion during the summer months when the banks are relatively dry. This reflects the influence of soil moisture conditions on the collapse process: high moisture content levels reduce bank strength, while at the same time increasing the weight of overhanging banks and hence the likelihood of collapse.

BANK EROSION AS A SOURCE OF SUSPENDED SEDIMENT

The concept of the suspended load being essentially a non-capacity load, where the capacity of the river to transport material exceeds the available supply of suitable material, has been proposed by a number of authors. In the case of the Culm, the capacity of the river to transport sediment delivered to the channel is never approached under normal conditions. It is therefore the rate of supply of suitable material to the river that controls sediment output from the catchment during the course of a storm event.

Sediment entering the Culm as a result of bank erosion processes normally does so in one of two ways: the corrasion of submerged banks, and the collapse of parts of undercut banks into the channel. The silt/clay fraction of material removed from the banks by corrasion will normally be incorporated in the suspended load almost immediately. Collapsed blocks will be further eroded by corrasion and abrasion, and suitable sediment particles entrained, over a period of time that can encompass several separate storm events. Consequently, the supply of sediment to the suspended load through bank collapse will rarely coincide with bank retreat, but will depend more on the availability of collapsed blocks within the channel for particle entrainment.

Nevertheless, the absence of long-term and augmenting sediment deposits in the channel indicates a long-term equilibrium between sediment supply from bank erosion and particle entrainment and transport by the river. It is therefore reasonable to use bank erosion as an indicator of the quantities of sediment entering the suspended load of the river over seasonal or longer time periods.

Quantities of sediment eroded from channel banks

In the spring of 1981, the location, length and height of all eroding banks in the Culm were determined through an extensive programme of field surveying. Sample measurements were also made of eroding banks in several tributaries as part of this exercise.

The validity of only a single survey of eroding sites was carefully assessed, as there was the possibility that some sites that would have been eroded in the winter prior to the survey had not been active for a number of years, and conversely some sites that were normally active might have been excluded from the survey. However, experience of monitoring erosion over a two-year period (three winters) suggests that the majority of sites remain active for many seasons at a time. Reference to earlier studies carried out on the Culm (e.g. Hooke, 1977) indicates that most sites of bank erosion remain in an active state for many years and that the total number of eroding sites is relatively constant. The small number of ephemeral sites included in the survey would, on average, remain nearly constant from year to year. Therefore, the single sampling exercise represented by this survey was considered to provide a realistic estimate of the number, size and location of eroding river banks within the River Culm.

The survey revealed that there is a progressive decrease in the size of eroding banks in an upstream direction. Observations also suggested that, upstream of a point beyond which reduced bank height precludes active bank retreat through collapse, and, in particular, where the river is no longer free to select its own course, bank retreat rates become insignificant. Because of this, a 'cut-off' point is proposed, above which no significant bank retreat is assumed to occur. Field observations also revealed that bank retreat in tributary streams is sufficiently limited in extent to be excluded from calculations of sediment production. The total area of actively eroding bank within the catchment above Rewe is thus calculated to be 10 302 m^2 from the measurements taken of each eroding site in early 1981.

In the course of 1980 and 1981, the mean measured rates of bank retreat at eroding sites in the study reach were 227 and 329 mm respectively, total mean retreat being 556 mm over the two-year period. Measurements taken at additional sites in the upper part of the catchment demonstrated that bank retreat at similar rates could be expected to occur throughout the catchment where the river was laterally mobile within an alluvial valley floor.

The bulk density of soil samples taken from sites in the study reach was determined, from which a mean bank material bulk density of 1232 kg m^3 was calculated. The coefficient of variation of this mean was sufficiently low (approximately 12%) for this value to be considered representative of all eroding alluvial banks in the lower part of the catchment.

From consideration of the mean particle-size distribution of sampled bank material, the breakdown of overall sediment production is obtained (Table 13.1).

Input to the suspended load

Consideration of the factors affecting the rate of particle entrainment within the channel reveals streamflow discharge as the dominant influence. Streamflow discharge acts as a surrogate indicator of streamflow velocity and river stage within the channel. Bank corrasion rates are influenced by streamflow velocity, and the area of bank affected is strongly dependent on the depth of the water at the bank face. Similarly, while sediment supply from collapsed blocks is clearly dependent on the number and size of such blocks within the channel, the rate of particle entrainment from those blocks is primarily a function of water velocity past the block and, in the case of larger blocks, the depth of that water.

Sediment production from channel sources will also be influenced by a number of other factors. Bank erodibility will be affected by the length of time available prior to the storm event for sub-aerial weathering of the bank face. Collapsed blocks will be subject to similar influences and, if at least partly submerged in the river, soil strength and cohesion will be reduced through the absorption of water by the block. However, these factors have a modifying, rather than controlling, effect, and it is streamflow discharge that remains the major influence on sediment production from channel bank sources. As streamflow discharge increases, an accompanying increase in the rate of sediment supply may therefore be expected, as increased water velocity and depth result in more rapid particle entrainment from banks and collapsed blocks.

There are, however, two potential limits to this process, beyond which an increase in discharge will not result in a corresponding increase in particle entrainment. First, the supply of readily removable sediment within the channel may become exhausted.

Table 13.1 Sediment production from channel bank erosion

	1980	1981	1980 + 1981	Percentage of total
Sand (t)	1412	2047	3459	49
Silt (t)	774	1121	1895	27
Clay (t)	695	1008	1703	24
Total (t)	2881	4176	7057	100

Secondly, a point is reached when storm event discharge increases to the extent that the channel can no longer contain the river. Overbank flow then develops, with the river spilling out of the channel and across its floodplain. Once bankfull discharge has been reached, further increases in discharge will not result in any increase in the areas of bank or block material subject to corrasive forces. In addition to this, water velocity in the channel will increase only gradually with further increases in discharge. Sediment entrainment ability thus reaches a threshold at bankfull discharge, beyond which it will increase more slowly as river discharge rises.

Sediment losses from the channel

The second main effect of overbank flow on sediment transport is the large-scale inundation of the Culm's floodplain. In large storm events, sediment-laden river water spills out over the floodplain, slows and eventually becomes almost stationary over large areas. This reduction in velocity leads to widespread deposition, as the water is no longer able to maintain a sediment load in suspension. Available data suggest that overbank storm events occur about seven times per year, predominantly in the winter months.

Work by Walling (Walling et al., 1986; Lambert and Walling, 1987) on the extent of such losses in the Culm demonstrated that virtually all of the measured sediment losses could be attributed to overbank flow and deposition, with only very small amounts of sediment otherwise being lost at times of stable within-channel flow. Average annual suspended sediment output at Rewe was estimated in this study to be approximately 4250 t, broadly comparable to the 3850 t measured in the present study. Moreover, the reach of river in which conveyance losses were monitored corresponds closely to the study reach in which erosion was monitored in detail. Some of the estimates of conveyance loss developed may therefore be applied to the findings of the present study, to obtain more realistic estimates of the potential contribution made by bank erosion to the suspended sediment load.

Walling (Walling et al., 1986; Lambert and Walling, 1987) concluded that the annual total loss of suspended material in the reach between Woodmill and Rewe was equivalent to the sum of all sediment inputs to the river within the reach plus 1750 t of the suspended load entering the reach at Woodmill. There are two main sediment inputs within the reach, tributary streams and bank erosion; no significant sheet erosion of the floodplain soils is envisaged. Walling suggests that sediment input from tributaries will be of the order of 600 t per annum, assuming an average erosion rate of 40 t km^2 per annum. Data obtained in the present study indicate that average annual sediment production from retreating banks between Woodmill and Rewe was approximately 1960 t, out of a catchment total (above Rewe) of 3530 t per annum, in the years 1980–81. As indicated earlier, the annual catchment sediment output in the period studied by Walling (November 1982 to May 1984) was some 10% higher than that in 1980–81. If it is assumed that the overall balance between different sediment sources remained fairly constant, then this suggests that the average sediment production from bank erosion in the later period would have been of the order of 2150 t per annum.

Combining these estimates of conveyance loss, tributary input and bank-derived sediment input yields an estimated approximate annual loss of 4500 t of sediment. Comparison with the average catchment output at Rewe, over the same period, of

PROCESSES OF RIVER BANK EROSION   241

4250 t per annum indicates that, in total, of the order of 50% of suspended sediment entering this reach may be lost from the river each year.

It is suggested that sediment losses in the catchment above Woodmill will also be significant. While opportunities for, and the frequency of, large-scale overland flow and widespread deposition are reduced, a sizeable floodplain does exist in many places, particularly just north of Woodmill, and the total valley length involved is much greater. Consequently, it is proposed that overall suspended sediment losses in the catchment above Woodmill could be of the order of 40%. Bank-derived sediment in the Culm will thus be subject to a range of losses, both within the study reach and also further upstream. In particular, sediment originating in the upper part of the catchment will be subject to high combined losses.

From the above discussion, it is possible to develop a generalised model of the annual movement of bank-derived sediment in the Culm. It is accepted that the

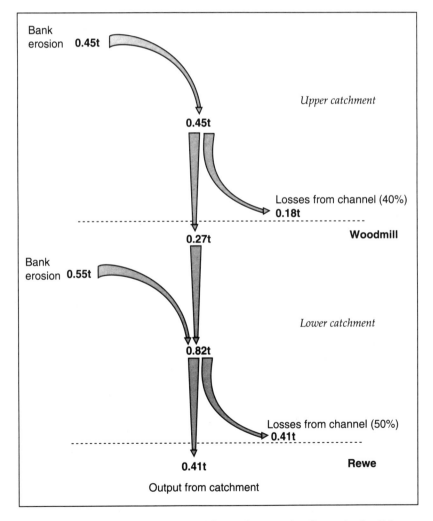

Figure 13.3 Sediment loss model for bank-sourced sediment in the Culm

estimates of sediment losses are necessarily somewhat crude, but it is believed that this model can nevertheless contribute to an understanding of the role of bank erosion in the sediment budget of the Culm. The sediment loss model, which is presented in Figure 13.3, assumes that proportional losses are the same for sediment from all sources.

Of each tonne of sediment lost from the banks in the catchment as a whole, 0.55 t is contributed, on average, by the 76 actively retreating sites in the study reach, with the remainder of the catchment providing the other 0.45 t of material. Consideration of the fractional losses involved in different parts of the catchment, as shown in Figure 13.2, reveals that, at most, only approximately 40% of the sediment eroded annually from the banks can be expected to contribute to the catchment suspended sediment output measured at Rewe.

CONTRIBUTION TO THE SUSPENDED SEDIMENT LOAD

Suspended sediment output from the Culm

Monitoring of suspended sediment concentrations, combined with analysis of the particle-size distribution of the suspended load, allowed determination of the quantities of sediment output from the catchment (Table 13.2). No significant quantities of material larger than 63 μm, i.e. sand-sized material, were found in the samples of suspended sediment extracted from the Culm. This indicated that only a fraction of the sediment entering the river system from bank erosion was of sufficiently fine calibre for incorporation into the suspended part of the load.

Contribution to the suspended load of the Culm

It can be seen from the estimates of sediment output from the catchment and sediment production, described in the preceding sections, that some year-to-year variability occurs in both estimates. The figures show that, while slightly more sediment was output from the catchment in 1980 than in the following year, 4090 and 3570 t respectively, the situation regarding sediment production from bank sources was reversed, with 2881 t being input to the channel in 1980, in comparison to 4176 t in 1981.

This reversal may reflect the operation of different controlling factors over the two variables. Analysis of the sediment output record of the Culm indicates that sediment

Table 13.2 Suspended sediment output of the Culm at Rewe

	1980	1981	1980 + 1981	Percentage of total
Sand (t)	0	0	0	0
Silt (t)	1779	1571	3370	44
Clay (t)	2291	1999	4290	564
Total (t)	4090	3570	7660	100

output is strongly influenced by the passage of particularly large storm events, with the majority of sediment being moved by a relatively small number of large events. The annual sediment output from the catchment may thus be disproportionately influenced by the number of large storm events in that year. Bank retreat processes operate through a cycle of events, which will usually encompass a number of storm events, and sediment production from banks in the catchment is therefore less liable to short-term fluctuation.

Data from individual years cannot therefore provide an adequate basis for comparison between sediment production and output from the catchment, as these respond in different ways to short-term fluctuations in environmental conditions. In order to permit a more realistic appraisal of the role of bank erosion as a sediment source in the Culm, data from an extended time period are required, and comparisons between sediment production and output have been restricted to data from the two years, 1980 and 1981, combined.

In these two years, bank retreat generated a total of 7057 t of sediment. However, almost all no material coarser than 63 μm equivalent spherical diameter will be incorporated in the suspended part of the sediment load. Instead, sand-sized particles will either be almost immediately redeposited close to the eroding bank (frequently on the floodplain adjacent to the bank in the case of overbank floods) or be moved downstream as channel bedload, through saltation processes. It is, for practical purposes, only the silt- and clay-sized particles that will be entrained by the river and which can therefore contribute to the suspended load.

On this basis, bank retreat produced a total of 3598 t of sediment suitable for inclusion as suspended load. This comprised 1895 t of silt and 1703 t of clay-sized material.

By incorporating the transport losses described earlier, more realistic estimates can be developed of the absolute contribution made by bank erosion processes to the suspended sediment output of the catchment. It has been demonstrated how, on an annual basis, fractional losses totalling some 60% of the sediment derived from bank sources can be expected. Inclusion of these losses in the estimates of sediment quantities provided from bank sources significantly reduces both the absolute level of contribution and also the magnitude of the relative contribution made to the measured suspended sediment output of the catchment.

Estimates of the absolute contribution made by bank retreat to the measured suspended sediment output are presented in Table 13.3, together with the calculated relative contribution, i.e. the fraction of the suspended sediment output that appears to

Table 13.3 Estimated contribution from bank sources to the suspended sediment output of the Culm at Rewe, 1980–81

	Net amount from banks (t)	Total catchment output (t)	Relative contribution (%)
Silt	758	3370	22
Clay	681	4290	16
Silt + clay	2439	7660	19

Table 13.4 Confidence limits (95%) of estimated relative contribution from bank sources to the suspended sediment output of the Culm at Rewe, 1980–81

	Lower limit	Mean	Upper limit
Silt (%)	14	22	35
Clay (%)	10	16	25
Silt + clay (%)	13	19	30

be derived from bank erosion processes. Estimates are provided for the suspended load as a whole and also for the silt and clay fractions separately.

It is estimated that 19% of the total sediment output in 1980 and 1981 is attributable to bank retreat processes, the remaining 81% being the product of other sources within the catchment. Considering only the coarser, silt-sized, fraction of the load, bank processes are capable of generating 22% of this material, whereas only 16% of the clay fraction can be provided by channel sources.

These estimates are based on the calculated most probable level of sediment production from bank erosion and sediment output from the catchment. Consideration of the statistical sampling error involved in these quantities allows the 95% confidence intervals to be established for the revised estimates of relative contribution. Mean estimates and upper and lower confidence limits are summarised in Table 13.4.

From this it may be inferred that, at a 95% confidence level, between 13% and 30% of the suspended sediment load of the Culm at Rewe may be derived from bank sources. The equivalent ranges for the silt and clay fractions of the load are from 14% to 35% and from 10% to 25% respectively.

CONCLUSION

Estimates of the relative contribution made to the suspended load of rivers by bank erosion vary considerably among previous authors. The overall level of 19% derived for the Culm lies towards the upper end of the scale of published values for rivers in comparable environments. It is believed that this is caused primarily by low inputs from non-channel sources rather than by any fundamental difference in the quantities of sediment produced by channel bank erosion. As a result, bank erosion is able to contribute a significant proportion of the total sediment load of the Culm.

These findings suggest that channel bank erosion may also be a significant sediment source in some other rivers characterised by low sediment yields. Any attempt to model the sediment budget of such rivers must therefore take into account those amounts of sediment that may be generated by bank erosion processes. In addition, this study has shown that the supply of material to the channel from the collapse of eroding banks takes place in a mainly episodic manner, and any model developed should include suitable functions for both this and the delay in this material being incorporated into the suspended load.

REFERENCES

Ashbridge, D. (1990) Processes of riverbank erosion and their contribution to the suspended sediment load of the River Culm. PhD thesis, University of Exeter.

Brunsden, D. and Kesel, B.H. (1973) Slope development on a Mississippi River bluff in historic time. *J. Geol.* **81**, 576–597.

Burgi, P.H. and Karaki, S.M. (1971) Seepage effect on channel bank stability. *J. Irrig. Drain. Div., Proc. ASCE* **97**, 59–72.

Collins, M.B. (1981) Sediment yield studies of headwater catchments in Sussex, SE England. *Earth Surf. Processes Landforms* **6**, 517–539.

Grimshaw, D.L. and Lewin, J. (1980) Source identification for suspended sediments. *J. Hydrol.* **47**, 151–162.

Hooke, J.M. (1977) The distribution and nature of changes in river channel patterns: the example of Devon. In: Gregory, K.J. (ed.), *River Channel Changes*, John Wiley & Sons, Chichester, pp. 265–280.

Hooke, J.M. (1979) An analysis of the processes of river bank erosion. *J. Hydrol.* **42**, 39–62.

Knighton, A.D. (1973) River bank erosion in relation to streamflow conditions, river Bollin-Dean Cheshire. *E. Midl. Geogr.* **5**, 416–426.

Lambert, C.P. and Walling, D.E. (1987) Floodplain sedimentation: a preliminary investigation of contemporary deposition within the lower reaches of the River Culm, Devon, UK. *Geogr. Ann.* **69A** (3–4), 393–404.

Neill, C.R. and Mollard, J.D. (1982) Erosional processes and sediment yield in the upper Oldman River basin, Alberta, Canada. In: *Recent Developments in the Explanation and Prediction of Erosion and Sediment Yield*, IAHS Publication no. 137, pp. 183–192.

Peart, M.R. and Walling, D.E. (1982) Particle size characteristics of fluvial suspended sediment. In: *Recent Developments in the Explanation and Prediction of Erosion and Sediment Yield*, IAHS Publication no. 137, pp. 397–408.

Thorne, C.R. (1978) Processes of bank erosion in river channels. PhD thesis, University of East Anglia.

Thorne C.R. (1982) Processes and mechanics of river bank erosion. In: Hey, R.D., Bathurst, J.C. and Thorne C.R. (eds), *Gravel Bed Rivers*, John Wiley & Sons, Chichester, pp. 227–259.

Turnbull, W.J.F., Krinitzsky, E.L.M. and Weaver, F.J. (1966) Bank erosion in soils of the lower Mississippi valley. *J. Soil Mech. Found. Div.* **92**(1), 121–136.

Walling, D.E. (1971) Sediment dynamics of small instrumented catchments in South-East Devon. *Trans. Devon. Assoc.* **103**, 147–165.

Walling, D.E. and Kane, P. (1982) Temporal variation of suspended sediment properties. In: *Recent Developments in the Explanation and Prediction of Erosion and Sediment Yield*, IAHS Publication no. 137, pp. 409–420.

Walling, D.E., Peart, M.R., Oldfield, F. and Thompson R. (1979) Suspended sediment sources identified by magnetic measurements. *Nature* **281**, 110–113.

Walling, D.E., Bradley, S.B. and Lambert, C.P. (1989) Conveyance losses of suspended sediment within a floodplain system. In: Hadley, R.F. (ed.), *Drainage Basin Sediment Delivery*, IAHS Publication no. 159, pp. 119–131.

Wolman, M.G. and Leopold, L.B. (1957) *River Flood Plains: Some Observations on Their Formation*, USGS Professional Paper no. 282.C.

14 The Rates and Patterns of Overbank Deposition on a Lowland Floodplain

D.J. SIMM
Department of Science, St Mary's University College, Strawberry Hill, UK

INTRODUCTION

Many floodplains of lowland rivers undergo regular flooding, and consequently, during overbank inundation, experience enhanced deposition of fine material from suspension. Despite the infrequency and irregularity of flooding, floodplain environments experience slow cumulative vertical aggradation and there is often no major discernible hiatus in the sedimentary record (Alexander and Prior, 1971; Nanson and Young, 1981). The misconception often persists that overbank deposition is of lesser importance compared to lateral accretion. However, floodplain alluvium is areally and volumetrically significant and, when a river with a high suspension load shifts laterally at only a slow rate, overbank deposition becomes the dominant process (Ritter *et al.*, 1973). Rumsby (1991) observes that, in Britain, overbank deposition is more prominent than in-channel deposition in catchments underlain by shale or clay lithologies.

The study of overbank deposition rates and patterns has, until recently, received little attention. First, this is because of the difficulties of accurately measuring small quantities of spatially and temporally variable sediment accretion in an environment that experiences only transitory and infrequent inundation. Secondly, both equipment and fieldworkers may not be able to measure what is desirable under potentially hazardous flood conditions (Lewin, 1989). Although the sedimentology of contemporary flood deposits is well documented (Coleman, 1969; Singh, 1972; Hughes and Lewin, 1982; Farrell, 1987; Marriott, 1992), studies are hindered by low rates and often imperceptible variability of deposition, the absence of discernible storm couplets (cf. Singh, 1972) and thus the homogeneous appearance of sediment due to bioturbation and pedogenic processes. Early studies relied on visually distinct characteristics such as colour, particle size or mineralogy (cf. Carlson and Runnels, 1952), and even sun desiccation cracks (cf. Klimek, 1974), to distinguish flood layers. As a result of the lack of a clear structure in floodplain deposits, much early work on floodplain accretion rates was restricted to the study of freshly deposited sediment, in particular crevasse splaying and levee deposits from rare, high-magnitude storm events (cf. McPherson and Rannie, 1969; Williams and Guy, 1973; Kesel *et al.*, 1974; Alexander and Prior, 1971; Brown, 1983) (Table 14.1). However, deposits from such events are not typical of the floodplain as a whole, since overbank sedimentation mainly proceeds by the settling out of fines during pondage.

Table 14.1 Selected examples of contemporary and recent overbank deposition rates

Source	Location	Temporal scale	Deposition rate
Contemporary timescale			
Alexander and Prior (1971)	Black Bottom, Lower Ohio R., S Illinois	1964 spring flood (50 yr event)	3.2 mm on ridges at distance from channel; 0.46 mm on levee; 1.0–1.6 cm yr^{-1} on levee more typical
Brice (1966)	Medicine Creek, Nebraska	Single flood	Trace–15.2 cm yr^{-1}
Brown (1983)	R. Stour, Blandford Forum, Dorset	27–30 Dec 1979 (100 yr event)	Sandy veneer, $\bar{x} = 16$ mm; bush obstruction, $\bar{x} = 50$ mm, max. = 102 mm; sandy mounds in lee of grass tussocks, $\bar{x} = 2.1$ mm
Carlson and Runnels (1952)	Kansas R., Lawrence and Topeka, Kansas	July 1951 flood	Avg. 5–6 mm; 1–19 mm range
Gretener and Strömquist (1987)	Lower R. Fryrisan, Sweden		8–240 g m^{-2}; approx. 0.08–2.40 mm yr^{-1}
Jahns (1947)	Connecticut R., Massachusetts	March 1936 flood	Avg. 3.5 cm on floodplain; avg. 25.9 cm on river banks; avg. 9.1 cm on artificial levee lee
		1938 flood	Avg. 2.2 cm on floodplain; avg. 10.7 cm on tributary banks; avg. 4.3 cm on artificial levee lee
Kesel *et al.* (1974)	Mississippi R.	1936 + 1938 floods combined April–June 1973 (30 yr event) (77 days of inundation)	Up to 2 m deposited; avg. 9 mm across floodplain Backswamp, 1.1 cm avg., 0.5–2.5 cm range; sloughs, 6 cm avg., 4–9 cm range; natural levee, 53 cm avg., 10–84 cm range; levee back, 12.5 cm avg., 4–9 cm range
Lambert (1986)	R. Culm, Devon, SW England		Avg. 0.42 mm yr^{-1}
Macklin and Newson (unpubl.) (cited by Rumsby, 1991)	R. Swale (550 km^2)	1986 flood	Max. 13 cm
Mansfield (1939)	Ohio R.	Jan–Feb 1937	Avg. 7.0 cm, 0.36–56 cm range, 0.246 cm yr^{-1}
Marriott (1992)	R. Severn	Jan–Feb 1990	Up to 150 mm near channel; up to 5 mm in ponded areas
McKee *et al.* (1967)	Bijou Creek	1965 flood	Near channel, avg. 61–91 cm, 0–366 cm range
McPherson and Rannie (1969)	Graburn watershed, Cypress Hills, Alberta	May 1967 flood (50 yr event)	2.77 cm yr^{-1}
Mitchell and Gerrard (1987)	R. Wye	Dec 1960 flood	0.134 mm
		Dec 1979 flood	0.010 mm
Nelson (1966)	Chemung R. (Susquehanna R.), New York	1960 flood (10 yr event)	Avg. 13 mm, 4.6 mm yr^{-1}
Sigafoos (1964)	Potomac R., Maryland and Virginia	7 + 23 Mar 1963 floods	5.1–10.2 cm
Simm (1993)	R. Culm		Avg. 0.40 mm yr^{-1}

Reference	Location	Time period	Rate/Description
Velikanova and Yarmykh (1970)	Ob'	1969 flood	Point bar, up to 150 cm; crevasse splay, up to 60 cm; flood basin, 0.2–3.0 cm
Walling et al. (1992)	R. Leira, S Norway		Up to 10 cm per flood
Williams and Guy (1973)	James R. basin, Virginia	Hurricane Camille, 1969 (130 yr event)	6.10–91.44 cm
Wolman and Eiler (1958)	Farmington	1955 flood	Avg. <1.5 cm, 0–94 cm range
Wolman and Leopold (1957)	Missouri	1881 flood	122–183 cm on levee
	Kansas	1951 flood	Avg. 3.0 cm
Recent timescale			
Brice (1966)	Upper Dry Creek, Nebraska	33 years	0.46–0.55 cm yr^{-1}
	Medicine Creek, Nebraska	22 years	8.3 cm yr^{-1}
	Well Canyon	40 years	1.5–2.0 cm yr^{-1}
Cooper et al. (1987)	N Carolina		Floodplain swamp, <0.26 cm yr^{-1}
Hupp and Bazemore (1993)			Floodplain wetland, 0.24–0.28 cm yr^{-1}
Hupp and Morris (1990)	Cache R., Arkansas		Forested wetland, floodplain: sloughs, $\bar{x} = 0.34$ cm yr^{-1}; rises $\bar{x} = 0.07$ cm yr^{-1}
Johnston et al. (1984)	Wisconsin		Natural levee, 2.62 cm yr^{-1}
Leopold (1973)	Watts Branch, near Washington, DC	13 years	1.17 cm yr^{-1}
McHenry et al. (1976)	Upper Mississippi	20 years	2.5–3.5 cm yr^{-1} in artificial backwater lakes and ponds
Mitchell and Gerrard (1987)	R. Wye	Since 1963	6.67 mm yr^{-1} in ox-bow lake; 3.67 mm yr^{-1} on floodplain
Mitsch et al. (1979)	S Illinois		Cypress tupelo swamp, floodplain, <0.2 cm yr^{-1}
Ritchie et al. (1975)	Little Tallahatchie	8 years	Natural levee, 4.7–6.5 cm yr^{-1}; crevasse splay, 2.8–3.4 cm yr^{-1}; backswamp and abandoned channels, 0.9–2.8 cm yr^{-1}
Rowan (1990)	R. Severn, Welshpool	Since 1963	Approx. 1.5 cm yr^{-1}
	R. Exe, Devon	Since 1963	Approx. 0.9 cm yr^{-1}
Schumm and Lichty (1963)	Cimarron R., SW Kansas	12 years	5.2 cm yr^{-1}
Simm (1993)	Rewe, R. Culm, Devon		Average 0.3 ± 0.1 mm yr^{-1}
	Paddleford, R. Culm		Average 0.28 ± 0.1 mm yr^{-1}
	Kensham, R. Culm		Average 0.12 ± 0.06 mm yr^{-1}
Walling et al. (1992)	Silverton Mill, R. Culm, Devon		Inside 'S' bend, >7 mm yr^{-1}; large depression, <1 mm yr^{-1}; small depression, 3 mm yr^{-1}; levee, 4–6 mm yr^{-1}
			Up to 4 mm yr^{-1}
Wolman and Leopold (1957)	R. Severn, Tewkesbury		General floodplain, avg. 1.5 mm yr^{-1}

A variety of methods have been tried to study contemporary overbank deposition, including augering and surveying the pre- and post-flood surface (Williams and Costa, 1988). Attempts have been made to estimate contemporary deposition rates from conveyance losses (e.g. Walling *et al.*, 1986), but these can be considered as tentative estimates because of uncertainties in the precision of suspension load measurements (Walling and Webb, 1989). Furthermore, overbank deposition rates are not uniformly distributed across the area of the floodplain, sediment may not accumulate at a constant rate, and sediment sinks on the floodplain may become abandoned or switch position. However, such spatial and temporal complexities are difficult to characterise and so time-averaged estimates of vertical accretion have proved more practicable to obtain. Much research has switched towards documentation of the floodplain as a sediment

Figure 14.1 Locality map of the lower Culm floodplain

sink over a time-scale of a few hundred to several thousand years (Table 14.1). A wide range of methods have been used, including heavy-metal concentrations (cf. Magilligan, 1985; Knox, 1987; Macklin and Dowsett, 1989; Macklin and Klimek, 1992), radiocarbon dating (e.g. Baker *et al.*, 1985), historical sources (Wolman, 1971) and archaeological (e.g. Miller and Werdorf, 1968; Robinson and Lambrick, 1984), sedimentological (e.g. Helley and LaMarche, 1973), vegetational (Sigafoos, 1964; Bedinger, 1971; Fonda, 1974), lichenometric (e.g. Rumsby, 1991), dendrological (Everitt, 1968; Hupp, 1988; Hupp and Morris, 1990) and biological evidence (Daniels *et al.*, 1962; Robinson and Lambrick, 1984).

Recently, renewed interest in the spatial patterns and temporal dynamics of overbank deposition has been prompted by an increasing awareness that floodplains act as both a sink and a storage compartment for pollutants (Marron, 1992). Floodplains can also cause secondary contamination by the re-introduction of these pollutants, such as heavy metals, to the channel as alluvial deposits are reworked (Bradley, 1989). The distribution of such pollutants, therefore, provides a basis for explaining the evolution of the floodplain in some detail. In recent years most study has centred on heavy metals (Macklin and Dowsett, 1989) and radionuclides, namely caesium-137 and caesium-134 (Walling and Bradley, 1989, 1990; Walling *et al.* 1992). Although the caesium method, by allowing numerous samples to be taken from a site, promotes the detailed study of the spatial patterns of recent overbank deposition, the use of floodplain sediment traps provides an accurate and practicable method of measuring overbank deposition rates for contemporary individual flood events (Lambert, 1986). The sediment trap approach has been adopted by Masikkaniemi (1985), by Gretener and Stromquist (1987) using plywood board, by Mitsch *et al.* (1979), and by Lambert (1986), Lambert and Walling (1987) and Simm (1993) using Astroturf mats (Table 14.1). However, little comparison has been undertaken between the sediment trap and caesium methods. In addition, although much mapping of lowland floodplain sites has been undertaken (cf. Walling and Bradley, 1989; Rowan, 1990; Walling *et al.*, 1992), little attention has been given to understanding how site characteristics, such as microtopography and vegetation, determine flooding behaviour and the resultant pattern of deposition. The theme of this chapter is to document the spatial and temporal rates and patterns in overbank deposition on a local scale, and to relate the resultant patterns to the microtopography and the character of the hydrological conditions operating.

LOCALITY

Because of the problems of measuring small quantities of overbank deposition and the spatial and temporal infrequency of depositional behaviour, a suspension load-dominant river that floods regularly was chosen for this study to ensure that deposition in measurable quantities would accumulate over a relatively short time period. The lower reaches of the River Culm, a tributary of the River Exe, south-east Devon, was chosen (Figure 14.1). Flooding along the Culm valley is common downstream of Cullompton (ST 023066) and is caused principally by the configuration of tributaries in the upper valley and the physiography of the floodplain. The bulk of the transported sediment is derived from the highly dissected Cretaceous marl plateau of the Blackdown Hills in east Devon, with smaller contributions from Greensand, flinty Head deposits and New Red Sandstone. Consequently, the Culm is a suspension load-dominant river, typically 95% <63 µm and 70% <2 µm (Walling and Bradley, 1989). Most suspended sediment is mobilised during storm events, with only 4% being associated with stable low flows (Lambert, 1986), but concentrations rarely exceed 1000 mg l^{-1}. Typical deposition values for the lower Culm floodplain are 0.4 mm yr^{-1} (based on sediment traps) and 0.50 ± 0.3 mm yr^{-1} (caesium studies) (Simm, 1993).

The study site at Columbjohn (SX 959997) lies 4 km upstream of the confluence with the Exe (Figure 14.1). Inundation of the floodplain occurs, on average, on seven occasions per year, with an average floodwater depth of 20–30 cm during an annual flood event (58 m^3 s^{-1} at the upstream gauging station at Woodmill (ST 021059)). The study site (see Figures 14.3 and 14.4A) is centred on a meander bend immediately upstream of the road bridge at Columbjohn (SX 957997), adjacent to a high terrace on the eastern flank of the floodplain. The gravel-bed channel is 8–10 m wide. The right-hand bank has a well-developed levee and several depressions. To the west, the floodplain extends for about 400 m and is traversed by several sinuous ditches flanked by hedgerows. On the left bank, inside the northern meander bend, there is a near-circular boggy depression that promotes persistent retention pondage, and during winter is generally too boggy for sheep and cattle grazing. Immediately downstream, the road forms a physical barrier to flow, but the road is susceptible to severe flooding, up to 40 cm deep. Note that the topography map (see Figure 14.4A) is based on a horizontal datum rather than the water gradient.

METHODOLOGY

Two methods, covering two time-scales and with different attributes and spatial and temporal resolutions, were adopted to investigate depositional patterns within the study area. First, a contemporary time-scale was adopted by studying individual flood events through the field observation of flooding behaviour and the use of sediment traps on the floodplain to monitor deposition rates. Secondly, a recent time-scale, covering the past 35 years or so within the temporal range of caesium-137 studies, provided a general perspective of deposition trends. Caesium data can reveal the longer-term trends, whether net accumulation or scour and their rates, whereas the sediment trap method records the potential for deposition but, because of removal of material after each event, fails to indicate the occurrence of inter-storm scour remobilisation. Although not directly comparable in what they measure, the comparison of these two approaches and time-scales is a valuable exercise.

Plastic Astroturf mats (20 cm × 35 cm), comprising 1.5 cm tufts attached to a pliable base, were used to assess localised rates of deposition during individual flood events. These replicate, to a certain degree, the natural surface, are robust, can withstand submersion and, because deposited sediment can be readily recovered, allow individual storm events to be sampled (Lambert, 1986; Simm, 1993). The traps were positioned on representative geomorphic units, such as the levee crest, levee backslope, open and closed topographic depressions and breaches (Figure 14.2B). Two mats were installed adjacent to each other in order to gauge the variability of deposition rates crudely over the small scale of decimetres. The installation procedure for Astroturf sediment traps is fully documented in Lambert (1986), Lambert and Walling (1987)

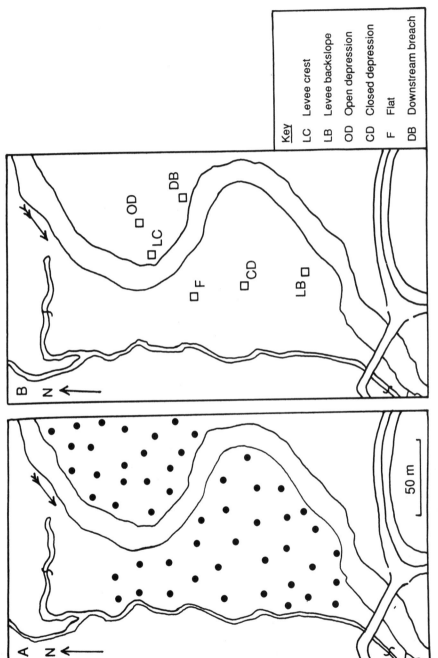

Figure 14.2 Sampling distribution of (A) caesium cores and (B) sediment traps

and Simm (1993). The mats were retrieved following a flood event, as soon as they emerged from the receding floodwaters. After careful transport to the laboratory in individual polythene bags, the mats were thoroughly sprayed with distilled water and scrubbed clean. The collected sediment–water mixture was centrifuged, decanted and freeze-dried for subsequent analysis for their bulk density and (ultimate) particle-size distribution.

The longer time-scale covered by the caesium method overcomes the inherent difficulties of quantifying slow vertical accretion rates (Campbell et al., 1988; Ritchie and McHenry, 1990; Walling et al., 1992). At each of 52 sample points (Figure 14.2A), a single core (42 cm^2) to a depth of at least 50 cm was obtained using a Cobra percussion corer. Sediment was carefully exhumed from the cylinder and, before bulking, a basal slice, 1.5 cm thick, was removed to be assayed independently to ensure that the core had penetrated the full depth of caesium-contaminated alluvium. A surface sample, from the topmost 0.5 cm, was also taken from each sample point. A core was also obtained from a local control site to act as reference datum because of inter-regional spatial and temporal variability due to local precipitation patterns and rates (Ritchie and McHenry, 1990). The prerequisite site characteristics are outlined in Campbell et al. (1988).

The samples were air-dried in preference to oven drying at 105°C (as suggested by Campbell et al. (1988)) because of the high clay content. Following disaggregation and the separation of the <2 mm fraction by dry sieving to achieve a consistent packing geometry, the samples were placed in 1500 g Marinelli pots similar to that used in calibration and the samples were assayed for their ^{137}Cs content using a Canberra series 35 multichannel analyser linked to a germanium detector housed in a 10 cm thick lead shield. Regular monitoring of detector efficiency and a counting time of 20000 s provided a typical analytical precision of ±6% (two standard deviations) (Walling and Bradley, 1989). By comparison with a reference datum of 250 mBq g^{-1}, derived from the control site, isocaesium plots reveal areas of enriched ('excess') and depleted ('deficit') inventories across the floodplain, which are indicative of deposition and scour, respectively. Estimates of vertical accumulation rates may be made using the mean bulk density of the sample (cf. Walling et al., 1992).

Deposition is often site-specific (Simm, 1993), and it was therefore necessary to monitor floodwater behaviour at the site. This was achieved through the current metering, sequential photographs of inundation and the mapping of flood marks, such as trash lines and aligned grass blades, following floodwater recession.

RESULTS AND DISCUSSION

Sequence of inundation and the relationship with deposition

During the early stages of a flood event (Figure 14.3A), passive back-ponding occurs prior to overbank spillage, for instance via downstream breaches (B1, B2) into a semicircular, boggy depression (DEP1) or via a ditch-fed depression (DEP2), on the right-hand bank, immediately upstream of Columbjohn bridge (SX 958993). The latter case induces overbank spillage towards the channel, while inundation directly from the channel is restricted by a high levee (L2). Floodwater afflux effects are produced by the

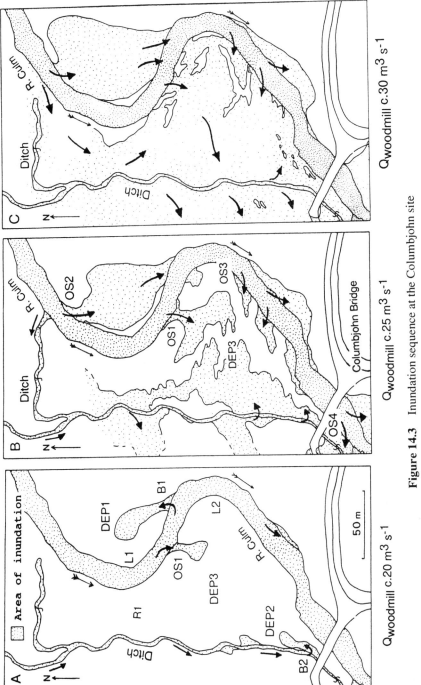

Figure 14.3 Inundation sequence at the Columbjohn site

bridge, there being only one main archway for the main channel and a small leat tunnel for the ditch. This causes the floodwaters to spread back on to the right bank. Consequently, velocities are reduced to about 0.1 m s^{-1} despite water depths in excess of 0.3 m. Lengthy retention storage is promoted in breaches (B1, B2) and depressions (DEP1, DEP2). As stage rises (Figure 14.3B), a chute develops using a shallow breach (OS2) on the left bank. However, inundation extent is limited on the left bank by a high terrace. Overbank spillage first occurs where the thalweg intercepts the concave bank on the bend (Figure 14.3B, OS1). The shallow levee slopes away from the channel into a wide depression (DEP3), which is flanked by a shallow rise (R1) to the west and a prominent levee (L2) to the east (see Figure 14.3A for feature locations). Once out-of-bank (Figure 14.3B), floodwaters are inhibited from returning to the channel by the levees, are diverted towards the ditch and laterally displaced into the adjacent field to the west. In and immediately adjacent to the ditch, elevated velocities of up to 0.6 m s^{-1} are recorded. Fed primarily by overbank spillage from ditches, the floodwaters subsequently extend across the fields on the right-hand bank. However, velocities are typically lower, about 0.2 m s^{-1}, with variable floodwater depths, although depths generally increase with distance from the channel. The road downstream becomes inundated via a low breach (OS4) and overbank spillage from a diversion channel immediately downstream of Columbjohn bridge (Figure 14.3C). Water depths in excess of 30 cm are common.

A plot of the caesium inventories (Figure 14.4B), combined with the mean bulk density of the sediment, allows the spatial trends in deposition rate (Figure 14.4C) to be constructed and permits qualitative assessment of the physical factors determining deposition rates and pattern. The mean for the ^{137}Cs inventories for this site is an 'excess' of 198 mBq cm^{-2}, equating to 2343 g m^{-2} yr^{-1} (2.9 mm yr^{-1}). Only eight out of 52 cores obtained record a 'deficit' value, found mainly at points of overbank spillage and throughflow. Overbank deposition is typically non-uniform across the floodplain and appears to be related primarily to microtopography (see Table 14.2).

High deposition rates principally occur on levees on the inside of meander bends (e.g. L2), with a steep depositional gradient away from the channel margin (Figure 14.4C). High deposition rates, typically of the order of 4400 g m^{-2} yr^{-1} (5.5 mm yr^{-1}), occur in depressions fed by breaches (DEP1) or ditches (DEP2). Most deposition occurs during the back-ponding phase of small localised floods and the retention pondage phase of larger flood events (Simm, 1993). During the back-ponding phase of small localised floods and the retention pondage phase of larger flood events, depressions fed by breaches or ditches undergo enhanced deposition. However, the timing of the suspended sediment peak relative to overbank spillage is important in determining the amount of deposition (Bradley, 1988). Where the suspended sediment peak typically precedes the discharge peak, most deposition will occur during back-ponding during the early phase of inundation (cf. Simm, 1993). Deposition is often promoted in the lee of upstanding vegetation, for instance sedge (*Carex*) tussocks in DEP1 (Figure 14.3A), or by dense vegetation, particularly on the levee, reducing floodwater velocities and trapping sediment. The downstream breach B1 undergoes high (approximately 6353 g m^{-2} yr^{-1} or 7.9 mm yr^{-1}) but variable deposition (Table 14.2) because, as an open depression, freshly deposited material may be susceptible to scour remobilisation by return flows. DEP1 is often poached by cattle trampling (cf. Trimble, 1994), thereby liberating sediment for scour remobilisation during flooding.

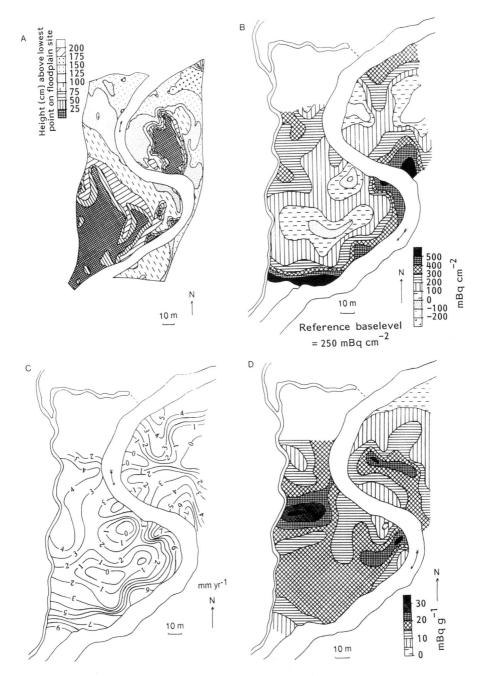

Figure 14.4 (A) Summary map of topography. (B) 'Excess' caesium-137 inventories (mBq cm^2) of whole cores. (C) Vertical accumulation estimates (mm yr^{-1}) (based on 35-year period). (D) Caesium-137 activities (mBq g^{-1}) of surface samples

The highest 'excess' inventory of 616 mBq cm^{-2} (equating to 7301 g m^{-2} yr^{-1} or 9.1 mm yr^{-1}), occurs immediately upstream of Columbjohn bridge associated with DEP2 and L2. This enhanced deposition is mainly caused by the afflux effects on floodwaters produced by the obstruction, but also by the 'trapping' of sediment by dense vegetation, mainly brambles and nettles.

Depleted caesium inventories can be found at points of overbank spillage (e.g. OS1) or of throughflow (DEP3). For instance, the right bank receives scour at the initial point of overbank spillage (OS1) at the inflection point of the meander. Depleted activities of 112 mBq cm^{-2} are recorded from the levee backslope at OS1, flanked on either side by higher inventories on the levee (L2). Floodwater velocities, induced by displacement by topographic restrictions to flow, appear to be sufficient to limit the amount of deposition in DEP3 (Figure 14.4B). A possible 'scour' path, undergoing floodwater velocities in excess of 0.3 m s^{-1}, can be identified linking DEP3 to the ditch.

Comparison with surface concentrations

Comparison of the pattern of caesium inventories (Figure 14.4B) with the surface concentrations (Figure 14.4D) reveals apparently contradictory trends in deposition. Deposition appears to be consistently high in B1 (Figures 14.4B, 14.4D) and on the rise (R1). However, the high surface concentrations in DEP1, DEP3 and OS3 (Figure 14.4D) are not mirrored by the inventories in these depressions (Figure 14.4B). Similarly, an inverse relationship occurs in DEP2 and on L2 (Figures 14.4B, 14.4D). This suggests that either there has been a significant change in depositional patterns over the past few years, or that, in recent years, frequent small floods have been important for deposition during pondage while more infrequent flood events may be responsible for removing material. A core from DEP1 (SX 959999) comprises predominantly homogeneous alluvium, averaging about 70% silt and 20% clay, and mean grain size and sorting remain consistent down-profile (at roughly 1.9 and 6.3 phi respectively). However, a layer of up to 30% sand occurs at a depth of 25 cm and probably represents a crevasse splay or scour deposit (cf. Gomez and Sims, 1981). The core contains occasional, well-preserved organic detritus (mainly wood), with *Phragmites* stem remnants at 65 cm depth, which indicates that this depression has existed for in excess of a century with some scour activity occurring during large floods.

The variability of deposition

Deposition averages, based on 14 flood events gauged by sediment trap point samples on a representative point on each topographic unit, are shown in Table 14.2. Means of 1625 g m^{-2} per flood (8.1 mm yr^{-1}) and 627 g m^{-2} per flood (3.1 mm yr^{-1}) are recorded on the levee backslope and in closed depressions, respectively. In contrast, the lowest means occur on the levee crest (187 g m^{-2} per flood, equating to 0.9 mm yr^{-1}) and on flat sections of floodplain (15 g m^{-2} per flood, 0.1 mm yr^{-1}). Some of these estimates, in particular that for the levee crest, appear to contradict the caesium data, but they have large standard deviations, highlighting the variability in deposition.

Temporal variability in overbank deposition may be shown using data collected from sediment traps. In general, the study site displays lower coefficients of variation for

Table 14.2 Overbank deposition rates (g m^{-2} per flood) for various sediment trap point samples (based on 14 flood events)

Topographic unit	Mean $\bar{x}$	SE of $\bar{x}$	SD s	SE of s	Min.	Max.	Median	LQ	UQ	IQ range	Deposition rate (mm yr^{-1})
Levee crest	187	54.6	181.9	38.6	0	398	161	6.3	383.2	376.9	0.9
Levee backslope	1625	307.0	651.3	217.0	710	2177	1986	1348.3	2081.8	733.6	8.1
Open depression	276	28.5	76.6	20.2	176	347	303	198.4	347.3	148.9	1.4
Closed depression	627	73.1	155.0	51.7	518	846	518	517.5	681.9	164.5	3.1
Flat	15	1.3	3.3	0.9	9	20	15	15.3	15.3	0.0	0.1
Breach	392	138.5	440.7	97.9	50	1576	356.5	63.6	359.5	295.9	2.0

deposition rates on specific topographic units on the floodplain compared to other sites (Simm, 1993). This is most probably because the site is regularly inundated and undergoes lengthy retention pondage following floodwater recession. Other localities, both upstream and downstream, that experience more transient inundation experience a coefficient of variation for deposition of up to 130 cv% (Lambert and Walling, 1987; Simm, 1993). At this site, the downstream breach B1 (112 cv%) shows the greatest variability because, as a preferential flow route, it undergoes notable scour remobilisation. Generally, the geomorphic units that undergo high deposition rates, such as the levee crest and backslope (L1), undergo the greatest variability (97 and 41 cv% respectively), whereas areas of low deposition, such as flat areas (23 cv%), remain consistently low. An exception to this is DEP2, where lengthy retention pondage promotes relatively low variability (25 cv%).

In addition, the comparison of deposition rates on adjacent traps can provide crude information on the small-scale variability in deposition for a flood event. Not only does overbank deposition vary temporally between floods, but this variability varies over a small scale during a particular flood event. Highly fluctuating ranges in deposition rates for adjacent traps between different floods are recorded on the levee crest (2–19%) and backslope (8–17%). There is a possible association with flood magnitude, with the greatest variability occurring during high-discharge events. DEP1 displays the greatest consistency in deposition rates, with an average difference in deposition between adjacent mats of 6%, while the downstream breach (B1) records the largest range of 1–23%, with no clear association with discharge magnitude. This small-scale variability in deposition rates demonstrates the difficulty in assessing the standard error.

General pattern of particle size

The readily inundated and ponded nature of the site promotes the deposition of finer material than other sites along the Culm floodplain (Simm, 1993). The material is predominantly <63 μm, characteristically >60% silt with a mean of around 6.5 phi (Table 14.3). Sand fractions are present on elevated parts of the floodplain, namely on the levee crest and backslope, usually associated with deposition during overbank spillage, and also on rises. The finest material is consistently deposited in depressions undergoing lengthy retention pondage, whereas breaches experience variable deposition, with some sand component (about 8%). Deposition is typically site-specific and depends upon factors such as microtopography and the mode of inundation.

Table 14.3 General particle size (phi)[a] of sediment deposited on floodplain sediment traps

Topographic unit	Mean $\bar{x}$	SE of $\bar{x}$	Var.	SD s	SE of s	Particle size (% by weight)			
						Clay	Silt	Sand	Gravel
Levee crest	6.404	0.194	0.272	0.520	0.137	23.67	62.46	13.87	0
Levee backslope	6.225	0.030	0.035	0.187	0.021	22.57	58.48	18.96	0
Rise	5.768					15.71	61.97	22.32	0
Depression	6.937					26.53	69.41	4.07	0
Breach	6.509	0.115	0.071	0.267	0.016	23.13	67.94	8.93	0

[a] Phi = $\log_2$ (diameter/mm)

CONCLUSIONS

The spatial patterns of overbank deposition are successfully documented using the caesium technique, whereas the point samples of the sediment trap method provide detail on contemporary deposition rates and the particle size for individual flood events. The comparison of these methods allows contrast of the potential for deposition and the general trends. Deposition principally occurs on the levee, in breaches and depressions. However, deposition is highly variable, particularly on the levee and in breaches.

Different modes of flooding, including passive back-ponding and overbank spillage, can be observed at this site and become operative at different stages of the flood event. These often determine the rates and pattern of deposition and the calibre of the material deposited. Other interactive variables include: (i) hydrological factors, such as the relative timing and magnitude of the suspended sediment and discharge peaks (Bradley, 1988), and floodwater velocity (Lambert, 1986); (ii) geomorphological factors, namely the role of microtopography, which determines the inundation sequence and the conductivity of the floodplain, i.e. the interconnectability either of internal breaches (Hughes, 1976) or ditches (Simm, 1993); (iii) sedimentological factors, including the particle-size distribution of the suspended load and the role of flocculation (Droppo and Ongley, 1982); and (iv) the vegetation type and density, which play an important role in trapping sediment and reducing floodwater velocities.

The caesium studies show some areas with negative inventories, suggesting the occurrence of scour erosion. High velocities are needed for erosion, but remobilisation and redeposition possibly occur during the same storm, otherwise the sediment is stabilised by vegetation. However, material can be poached by cattle and subsequently exposed to remobilisation.

ACKNOWLEDGEMENTS

This work was undertaken while the author was in receipt of an NERC award (no. GT4/89/AAPS/23). Thanks go to Professor Des Walling for his supervision, to the technical staff in the Department of Geography, University of Exeter, in particular Mr A. Ames and Mr J. Grapes, and also to the referees of this chapter for their valued comments.

REFERENCES

Alexander, C.S. and Prior, J.C. (1971) Holocene sedimentation rates in overbank deposits in the Black Bottom of the Lower Ohio River, southern Illinois. *Am. J. Sci.* **270**, 361–372.

Baker, V.C., Pickup, G. and Polach, H.A. (1985) Radiocarbon dating of flood deposits, Katherine Gorge, Northern Territory, Australia. *Geology* **13**, 344–347.

Bedinger, M.S. (1971) *Forest Species as Indicators of Flooding in the Lower White River Valley, Arkansas.* US Geological Survey Professional Paper no. 750-C, pp. 248–254.

Bradley, S.B. (1988) Sediment-water interactions: the physical transport of heavy metals in the fluvial system. In: *Metals and Metalloids in the Hydrosphere: Impact Through Mining and Industry and Prevention Technology*, manuscript.

Bradley, S.B. (1989) Incorporation of metalliferous sediments from historic mining into river floodplains. *GeoJournal* **19**(1), 5–14.

Brice, J.C. (1966) Erosion and deposition in the loess-mantled Great Plains Medicine Creek drainage basin, Nebraska. US Geological Survey Professional Paper no. 352-H, pp. 255–339.

Brown, A.G. (1983) An analysis of overbank deposits of a flood at Blandford Forum, Dorset, England. *Rev. Geomorphol.* **32**(3), 95–99.

Campbell, B.L., Loughran, R.J., Elliott, G.L and Shelly, D.J. (1988) Mapping drainage basin sediment sources using caesium-137. In: Hadley, R.F. (ed.), *Drainage Basin Sediment Delivery (Proceedings of the Albuquerque Symposium, August 1986))*, IAHS Publication no. 159, pp. 437–446.

Carlson, W A. and Runnels, R.T. (1952) A study of silt deposited by the July 1951 flood, Central Kansas River Valley. *Trans. Kansas Acad. Sci.* **55**(2), 209–213.

Coleman, J.M. (1969) Brahmaputa River: channel processes and sedimentation. *Sediment. Geol.* **3**, 129–239.

Cooper, J.R., Gilliam, J.W., Daniels, R.B. and Robarge, W.P. (1987) Riparian areas as filters for agricultural sediment. *Soil Sci. Soc. Am. J.* **51**, 416–420.

Daniels, R.B., Rubin, M. and Simonson, G.H. (1962) Alluvial chronology of the Thompson Creek Watershed, Harrison County, Iowa. *Am. J. Sci.* **261**, 473–487.

Droppo, I.G. and Ongley, E.D. (1982) Flocculation of suspended sediment solids in southern Ontario rivers. In: *Sediment and the Environment (Proceedings of the Baltimore Symposium, May 1989)*, IAHS Publication no. 184, pp. 95–103.

Everitt, B.L. (1968) Use of cottonwood in an investigation of the recent history of a floodplain. *Am. J. Sci.* **226**, 417–439.

Farrell, K.M. (1987) Sedimentology and facies architecture of overbank deposits of the Mississippi River, False River region, Louisiana. *J. Sediment. Petrol.* **57**, 111–120.

Fonda, R.W. (1974) Forest succession in relation to river terrace development in Olympic National Park, Washington. *Ecology*, **55**, 927–942.

Gretener,B. and Strömquist, L. (1987) Overbank sedimentation rates of fine grained sediments: a study of recent deposition in the Lower River Fryisan. *Geogr, Ann.* **69A**, 139–146.

Gomez, B. and Sims, P.C. (1981) Overbank deposits of the Narrator Brook, Dartmoor, England. *Geol. Mag.* **118**(1), 77–82.

Helley, E.J. and LaMarche, V.C. (1973) *Historic Flood Information for Northern California Streams from Geological and Botanical Evidence*, US Geological Survey Professional Paper no. 485-E, 16 pp.

Hughes, D.A. (1976) Flooding and floodplain inundation, Unpublished PhD thesis, University College of Wales, Aberystwyth.

Hughes, D.A. and Lewin, J. (1982) A small scale floodplain. *Sedimentology*, **29**(6), 891–895.

Hupp, C.R (1988) Plant ecological aspects of flood geomorphology and palaeoflood history. In: Baker, V.R., Kochel, R.C. and Patton, P.C. (eds), *Flood Geomorphology*, Wiley-Interscience, New York, ch. 20, pp. 335–356.

Hupp, C.R. and Bazemore, D.E. (1993) Temporal and spatial patterns of wetlands sedimentation, West Tennessee. *J. Hydrol.* **141**, 179–196.

Hupp, C.R and Morris, E.E. (1990) A dendrogeomorphic approach to measurement of sedimentation in a forested wetland, Black Swamp, Arkansas. *Wetlands*, **10**, 107–124.

Jahns, R.H. (1947) *Geologic Features of the Connecticut Valley, Massachusetts, as Related to Recent Floods*, US Geological Survey Water Supply Paper no. 996, 158 pp.

Johnston, C.A., Bubenzer, G.D., Lee, G.B., Madison, F.W. and McHenry, J.R. (1984) Nutrient trapping by sediment deposition in a seasonally flooded lakeside wetland. *J. Environ. Qual.* **13**(2), 283–290.

Kesel, R.H., Dunne, K.C., McDonald, R.C., Allison, K.R. and Spicer, B.E. (1974) Lateral erosion and overbank deposition on the Mississippi River in Louisiana caused by 1973 flooding. *Geology* **2**(9), 461–464.

Klimek, K. (1974) The structure and mode of sedimentation of floodplain deposits in the Wisloka Valley (South Poland). *Stud. Geomorphol. Carpatho-Balcanica* **8**, 135–151.

Knox, J.C. (1987) Historical valley floor sedimentation in the Upper Mississippi Valley. *Ann. Assoc. Am. Geogr.* **77**(2), 224–244.

Lambert, C.P. (1986) The suspended sediment delivery dynamics of river channels in the Exe basin. Unpublished PhD thesis, University of Exeter, 302 pp.

Lambert, C.P and Walling, D.E (1987) Floodplain sedimentation: a preliminary investigation of contemporary deposition within the lower reaches of the River Culm, Devon, UK. *Geogr. Ann.* **69A**, 393–404.

Leopold, L.B. (1973) River channel change with time: an example. *Geol. Soc. Am. Bull.* **84**, 1845–1860.

Lewin, J. (1989) Floods in fluvial geomorphology. In: Beven, K and Carling, P. (eds), *Floods: Hydrological, Sedimentological and Geomorphological Implications*, BGRG Symposia Series, John Wiley & Sons, Chichester.

Macklin, M.G. and Dowsett, R.B. (1989) The chemical and physical speciation of trace metals in fine-grained overbank flood sediments in the Tyne Basin, north-east England. *Catena* **16**, 135–151.

Macklin, M.G. and Klimek, K. (1992) Dispersal, storage and transformation of metal-contaminated alluvium in the Upper Vistula basin, southwest Poland. *Appl. Geogr.* **12**, 7–30.

Magilligan, F.J. (1985) Historical floodplain sedimentation in the Galena River basin, Wisconsin and Illinois. *Ann. Assoc. Am. Geogr.* **75**(4), 583–594.

Mansfield, G.S. (1939) *Flood Deposits of the Ohio River, January–February, 1937–38—A Study of Sedimentation.* US Geological Survey Water Supply Papers no. 838, pp. 693–733.

Marriott, S. (1992) Textural analysis and modelling of a flood deposit: River Severn, UK. *Earth Surface Processes and Landforms* **17**(7), 687–697.

Marron, D.C. (1992) Physical and chemical characteristics of a metal-contaminated overbank deposit, west–central South Dakota, USA. *Earth Surface Processes and Landforms* **14**, 419–432.

McHenry, J.R., Ritchie, J.C. and Verdon, J.(1976) Sedimentation rates in the Upper Mississippi River. In: *Rivers '76. (Symposium on Inland Waterways or Navigation, Flood Control and Water Diversions)*, ASCE, New York, vol. II, pp. 1339–1349.

McKee, E.D., Crosby, E.J. and Berryhill, H.L. (1967) Flood deposits, Bijou Creek, Colorado. *J. Sediment. Petrol.* **37**, 829–851.

McPherson, H.J. and Rannie, W.F. (1969) Geomorphic effects of the May 1967 flood in Graburn Watershed, Cypress Hills, Alberta, Canada. *J. Hydrol.* **9**, 307–321.

Masikkaniemi, H. (1985) Sedimentation and water quality in the flood basin of the River Kyronjoki in Finland. *Fennia* **163**, 155–194.

Miller, J.P. and Werdorf, F. (1968) Alluvial chronology of the Tesque Valley, New Mexico. *J. Geol.* **66**, 177–194.

Mitchell, D.J. and Gerrard, A.J. (1987) Morphological responses and sediment patterns. In: Gregory, K.J., Lewin, J. and Thornes, J.B. et al. (eds), *Palaeohydrology in Practice: A River Basin Analysis*, John Wiley & Sons, Chichester, ch. 9, pp. 177–200.

Mitsch, W.J., Dorge, G.L. and Weimhoff, J.R. (1979) Ecosystem dynamics and a phosphorous budget of an alluvial swamp in southern Illinois. *Ecology* **60**, 1116–1124.

Nanson, G.C. and Young, R.W. (1981) Overbank deposition and floodplain formation on small coastal streams of New South Wales. *Z. Geomorphol.* **25**(3), 332–347.

Nelson, J.G. (1966) Man and geomorphic process in the Chemung River Valley, New York and Pennsylvania. *Ann. Assoc. Am. Geogr.* **56**, 24–32.

Ritchie, J.C. and McHenry, J.R (1990) Application of radioactive fallout caesium-137 for measuring erosion and sediment accumulation rates and patterns: a review. *J. Environ. Qual.* **19**(2), 215–233.

Ritchie, J.C., Hawks, P.H. and McHenry, J.R. (1975) Deposition rates in valleys determined using fallout cesium-137. *Geol. Soc. Am. Bull.* **86**, 1128–1130.

Ritter, D.F., Kinsey, W.F. and Kauffman, M.E. (1973) Overbank sedimentation in the Delaware River valley during the last 6000 years. *Science* **179**, 374–375.

Robinson, M.A. and Lambrick, G.H. (1984) Holocene alluviation and hydrology in the upper Thames basin. *Nature* **308**, 809–814.

Rowan, J.S. (1990) The sediment-associated transport and redistribution of Chernobyl-derived radiocaesium in fluvial systems. Unpublished PhD thesis, University of Exeter, 358 pp.

Rumsby, B. (1991) Flood frequency and magnitude estimates based on valley floor morphology and floodplain sedimentary sequences: the Tyne Basin, NE England. Unpublished PhD thesis, University of Newcastle-upon-Tyne, 208 pp.

Schumm, S.A. and Lichty, R.W. (1963) *Channel Widening and Floodplain Construction, Cimarron River, Kansas.* US Geological Survey Professional Paper no. 352-D, 71–88.

Sigafoos, R.S. (1964) *Botanical Evidence of Floods and Floodplain Deposition*, US Geological Survey Professional Paper no. 485-A.

Simm, D.J. (1993) The deposition and storage of suspended sediment in contemporary floodplain systems: a case study of the River Culm, Devon. Unpublished PhD, University of Exeter, 347 pp.

Singh, I.B. (1972) On the bedding in the natural-levee and point bar deposits of the Gomti River, Uttar Pradesh, India. *Sediment. Petrol.* **7**, 309–317.

Trimble, S.W. (1994) Erosional effects of cattle on streambanks in Tennesses, USA. *Earth Surface Processes and Landforms* **19**(5), 451–464.

Velikanova, Z.M. and Yarnykh, N.A. (1970) Field investigations of the hydraulics of a floodplain during a high flood. *Sov. Hydrol.* **5**, 426–440.

Walling, D.E. and Bradley, S.B. (1989) Rates and patterns of contemporary floodplain sedimentation: a case study of the River Culm, Devon, UK. *Geo Journal* **19**(1), 53–62.

Walling, D.E. and Bradley, S.B. (1990) Some applications of caesium-137 measurements in the study of erosion, transport and deposition. In: *Erosion, transport and Deposition Processes (Proceedings of the Jerusalem Workshop, March–April 1987)*, IAHS Publication no. 189, pp. 179–203.

Walling, D.E and Webb, B.W. (1989) The reliability of rating curve estimates of sediment yield: some further comments. In: *Sediment Budgets (Proceedings of the Porto Alegre Symposium)*, IAHS Publication no. 174, pp. 337–350.

Walling, D.E, Bradley, S.B. and Lambert, C.P. (1986) Conveyance losses of suspended sediment within a floodplain system. In: *Drainage Basin Sediment Delivery (Proceedings of the Albuquerque Symposium, August 1986)*, IAHS Publication no. 159, pp. 119–131.

Walling, D.E, Quine, T.A. and He, Q. (1992) Investigating contemporary rates of floodplain sedimentation. In: Carling, P.A. and Petts, G.E. (eds), *Lowland Floodplain Rivers: Geomorphological Perspectives*, BGRG Symposia Series, John Wiley & Sons, Chichester, ch. 7, pp. 165–184.

Williams, G.P. and Costa, J.E (1988) Geomorphic measurements after a flood. In: Baker, V.R., Kochel, R.C. and Patton, P.C. (eds), *Flood Geomorphology*, Wiley-Interscience, New York, ch. 4, pp. 65–77.

Williams, G.P. and Guy, H.P. (1973) *Erosional and depositional aspects of Hurricane Camille in Virginia, 1969*, US Geological Survey Professional Paper no. 804, 80 pp.

Wolman, M.G. (1971) Evaluating alternative techniques of floodplain mapping. *Water Resour. Res.* **7**, 1383–1391.

Wolman, M.G. and Eiler, J.P. (1958) Reconnaissance study of erosion and deposition produced by the flood of August, 1955, in Connecticut. *Trans. Am. Geophys. Union* **39**, 1–14.

Wolman, M.G. and Leopold, L B. (1957) River floodplains: some observations on their formation. US Geological Survey Professional Paper no. 282-C, pp. 87–107.

15 Lake and Reservoir Bottom Sediments as a Source of Soil Erosion and Sediment Transport Data in the UK

IAN D.L. FOSTER
Centre for Environmental Research and Consultancy, Coventry University, UK

INTRODUCTION

> Although the importance of monitoring suspended sediment loads in rivers has been clearly recognised in many countries ... there has been no attempt to establish a national monitoring programme in Britain.
>
> Walling, (1990, p. 130)

The paucity of long-term suspended sediment data and the different monitoring strategies adopted in various research programmes in the UK make it difficult to provide both reliable and directly comparable estimates of suspended sediment yields for the country as a whole. Available data suggest that suspended sediment transport rates lie typically in the range 50–100 t km^{-2} yr^{-1}, which are low by world standards (Walling and Webb, 1987). Highest values occur in upland regions, and lowland agricultural systems have rarely been found to exceed sediment yields of 50 t km^{-2} yr^{-1}, although these can be significantly higher than geological rates in adjacent undisturbed catchments (Foster *et al.*, 1986, 1990a). Indeed, water erosion of lowland agricultural soils is now seen as a major cause for concern in terms of areal extent, frequency of occurrence and local impact (Arden-Clarke and Evans, 1993).

One of the major problems in interpreting sediment yield data is the disparity that is often found between river-based estimates of yield and hillslope monitoring-based estimates of soil erosion, which in part relates to the problem of linking the field to the river through the sediment delivery process (Walling, 1983, 1990). The difference between soil erosion and sediment yield estimates for a region also partly reflects the fact that the units used to report both measurements, i.e. mass per unit area per unit time, are the same (e.g. tonnes per square kilometre per year (t km^{-2} yr^{-1})) and do not consider the relative area, at either the plot or catchment scale, from which sediment is derived. The time-scale is essentially arbitrary, since most sediment in UK rivers is transported in 5–10% of the time. Furthermore, the length of time that sediment is stored in intermediate locations within the catchment (e.g. behind hedgerows and in floodplains) will vary from location to location, and sediment may be remobilised over different time-scales from the same store. It is perhaps not surprising, therefore, that relationships between the annual amount of sediment transported and the rainfall and/or runoff in the catchment often produce a poor correlation (Walling and Webb, 1983).

Sediment and Water Quality in River Catchments. Edited by I.D.L. Foster, A.M. Gurnell and B.W. Webb.
© 1995 John Wiley & Sons Ltd.

Sediment yields obtained from a single point in the drainage basin fluctuate in response to annual variations in runoff (Walling, 1978), and many years of monitoring and data processing may be required before reliable regional estimates are obtained. Particular problems occur in the UK, where high-frequency monitoring is often associated with three-year PhD training programmes. Furthermore, the UK river monitoring agencies do not routinely collect sediment concentration data at sufficient frequency to enable reliable sediment yield estimates to be made from rating curves (cf. Walling, 1978; Walling and Webb, 1981). Lack of long-term monitoring does not allow the identification of trends in sediment yields in response to changes in climate or land management practices, and individual estimates are thus difficult to set in a regional and/or historical context.

Alternative sources of proxy hydrological data have been sought in recent years in order to overcome some of the spatial and temporal shortcomings in river monitoring programmes. This chapter is particularly concerned with the interpretation of suspended sediment yield information derived from reservoir and lake sediment-based reconstructions over short to intermediate time-scales (150–200 years), and with the integration of reconstructed sediment yields with source tracing and sediment delivery processes from the drainage basin over similar time-scales. Before considering this approach in detail, however, it is appropriate to examine controls on soil erosion, sediment transport and delivery processes that operate over these same time-scales.

SOIL EROSION AND SEDIMENT TRANSPORT

Soil erosion and sediment transport in drainage basins are, in part, a function of intrinsic factors that, first, control the balance between soil erodibility and the erosivity of rainfall, overland flow and other sediment transport processes and, secondly, provide opportunities for sediment transfer between the hillslope, floodplain and river channel (Boardman et al., 1990). Extrinsic controls can also affect sediment movement, and modifications of catchment land use may be significant in changing rates of sediment transport in river basins in three ways. First, they may act by increasing the amount of sediment available for erosion by landscape disturbance and subsequent management (e.g. deforestation and cultivation). Secondly, the nature of the internal linkages may be modified through processes such as hedgerow removal and artificial drainage. Thirdly, hydrological pathways may be altered by, for example, decreases in infiltration rates under high animal stocking densities (Arden-Clark and Evans, 1993). Changes in climate are also significant. In the short term, increasing the energy available for sediment detachment and transport may occur through changes in the amount and/or timing of rainfall (Walling, 1990; Arden-Clark and Evans, 1993). In the long term, shifts in average climate conditions may occur over decades and centuries (Lockwood, 1983; Starkel, 1983). Furthermore, climate may significantly influence the type and density of natural vegetation, which has a significant impact on soil erosion (cf. Thornes, 1990), although the change in vegetation may lag behind the change in climate over time (Knox, 1972).

Identifying the relative significance of intrinsic and extrinsic controls on sediment transport in the whole drainage basin is complex (Schumm, 1979) and is inherently linked to both the sediment delivery problem (Walling, 1983, 1988) and the time-scales over which sediment transport occurs (cf. Trimble, 1981; Foster et al., 1988). Specific sediment delivery, the mass of sediment passing through a river cross-section, has been related to upstream catchment area (e.g. Livesey, 1972) and to other physiographic and catchment factors (Maner, 1958; Foster et al., 1990a). A decrease in sediment delivery with increasing catchment size is often interpreted as resulting from a decrease in slope and channel gradient and an increase in the opportunities for sediment storage, particularly in the floodplain, although the controls are spatially complex (Walling, 1983, 1990). The conventional view of decreasing specific sediment yield with increasing catchment size is not universally applicable, and many studies have shown that sediment delivery can increase with increasing basin size, often as a result of the reworking of sediment stored in the lower valley, which may have been derived from earlier erosional and depositional episodes (Meade, 1982; Meade et al., 1985; Church and Slaymaker, 1989).

From the above discussion, it is evident that sediment yields from a single drainage basin are a function of both intrinsic and extrinsic factors, which may operate at different time-scales, and it is therefore not surprising that attempts to model sediment yields in relation to single controls, such as climate, have proved problematic, although regional analysis of data, based on homogeneous lithological grouping of data for example, appears to be more valid (Walling and Webb, 1983).

Given the lack of long-term data in the UK, surrogate measures of soil erosion and sediment transport are required in order to provide reliable estimates of average regional rates and to identify land-use and other controls upon contemporary and historical sediment yields. The following section briefly reviews various approaches to estimating sediment yield based upon the analysis of lake and reservoir sediments before examining case studies of reconstruction in detail.

RESERVOIR AND RECENT LAKE SEDIMENTS

Various opportunities exist to utilise the bottom sediments of lakes and reservoirs for providing proxy hydrological data. First, sub-bottom profiling or ground surveys after drawdown can be used to estimate the total mass of sediment accumulated since construction, and sediment yield estimates can then be made (Duck and McManus, 1987; Butcher et al., 1993; Neil and Mazari, 1993). The advantage of this approach is that it can provide rapid estimates of regional sediment yields and pinpoint areas of concern. However, a major limitation is that, without an established chronology, the estimate of sediment yield obtained may cover time-scales in excess of 100 years and will not allow temporal trends to be identified.

Secondly, where varved sediments are deposited, a chronology may be provided simply by counting the annual layers. The mass of sediment for each annual increment can be calculated from varve thickness and sediment density. Such an approach appears to be particularly applicable in cold temperate environments where a highly resolved time-frame can be developed over hundreds of years (cf. Saarnisto, 1986; Grönlund, 1991; Desloges and Gilbert, 1991). More recently, attempts have been made by Desloges and Gilbert (1994) to correlate a sediment accumulation index with long-term hydrological data from the inflowing rivers to Lillooet Lake, British Columbia, Canada. In arid and semi-arid environments, the magnitude and frequency of sediment delivery are irregular in both time and space, and it is less relevant to consider the annual timescale for sediment yield interpretation. Individual events may be identified in the deposited sequence as sedimentary couplets, the thickness of which may be directly related to the mass of sediment moved during each event (Laronne, 1987, 1990). More recently, attempts have been made to identify similar sedimentary sequences in humid temperate environments in south Limburg, The Netherlands (Laronne, J. pers. comm. 1994).

Although laminated sediments may have been identified in the UK (O'Sullivan, 1990), and even some high-magnitude events have been recorded in reservoir sediments (e.g. Smith *et al.*, 1983; Foster and Walling, 1994), preservation of individual events or laminations appears to be poor, and attempts to subdivide the sediment column into time periods shorter than the life of the reservoir have used radiometric dating (caesium-137 (^{137}Cs) and lead-210 (^{210}Pb) and core correlation techniques developed largely from palaeolimnological research (Ritchie *et al.*, 1973; Pennington *et al.*, 1973; Dearing, 1986). Dating and core correlation techniques provide an opportunity to subdivide the sediment record into time zones covering between *ca.* 5 and 50 years (Foster *et al.*, 1990a). Dating is usually performed on one or two sediment cores and the chronology transferred to other sediment cores by means of core correlation techniques (cf. Dearing, 1986). The total mass of sediment in each synchronous zone is multiplied by sedimentation area to provide an estimate of sediment yield (cf. Foster *et al.*, 1985). In rapidly sedimenting basins, the resolution of the chronology becomes poor because of dilution effects (cf. Flower *et al.*, 1989), a problem that may be overcome partially by careful selection of reservoir catchments (Dearing and Foster, 1993).

Comparison of estimated sediment yields based on both bottom sediment reconstruction and river monitoring is problematic owing to the assumptions made about reservoir trap efficiency, the relative proportion of atmospheric, autochthonous and allochthonous sediment accumulating at the reservoir bed, and the different timescales and errors associated with the two methods of yield estimation (cf. Rausch and Heinemann, 1984; Foster *et al.*, 1985, 1990a; Owens and Slaymaker, 1993; Butcher *et al.*, 1993). Few comparative studies exist, but the data of Table 15.1 suggest that

Table 15.1 Comparative estimates of suspended sediment yield from river monitoring and lake or reservoir bottom sediment-based estimations (t km^{-2} yr^{-1})

Site[a]	River monitoring estimate	Bottom sediment estimate
Merevale Lake, N Warwickshire[b]	10.3	10.3
Seeswood Pool, N Warwickshire[c]	51.5	36.2
Slapton Ley, Devon[d]	17.4	13.4

[a] Site locations are given in Figure 15.1
[b] From Foster *et al.* (1985)
[c] From Foster *et al.* (1986)
[d] From O'Sullivan *et al.* (1991)

+ Atmospheric Cs-137 record

Figure 15.1 Location of selected UK lake and reservoir sites of sediment and ^{137}Cs surveys. Shaded area covers 22 southern Pennine reservoirs surveyed by Butcher et al. (1993).

A–H are reservoirs surveyed by Duck and McManus (1990): A, Lambieletham; B, Harperleas; C, Holl; D, Drumain; E, Glenfarg; F, Glenquey; G, Cullaloe; H, Pinmacher.

1–19 are selected UK sites where sediment yields and/or ^{137}Cs profiles have been obtained from analysis of bottom sediments: 1, Slapton Ley (O'Sullivan et al., 1991); 2, Old Mill Reservoir (Foster and Walling, 1994); 3, Chard Reservoir (Walling and He, 1993); 4, Wadhurst Park Lake (Wise, 1980; Walling and He, 1993); 5, Wyken Slough (Charlesworth and Foster, 1993); 6, Coombe Pool (Foster, 1985); 7, Seeswood Pool (Foster et al., 1986); 8, Merevale Lake (Foster et al., 1985); 9, Holmer Lake (Gaskell, 1992); 10, Llyn Lygard Rheidol (Bonnett and Appleby, 1991); 11, Lake Vyrnwy (Rowan et al., 1993); 12, Rostherne Mere (Livingstone and Cambray, 1978); 13, Windermere (Pennington et al., 1973); 14, Blelham Tarn (Pennington et al., 1973); 15, Esthwaite Water (Pennington et al., 1973); 16, Wastwater (Pennington et al., 1973); 17, Ennerdale Water (Pennington et al., 1973); 18, Brotherswater (Eakins et al., 1981); 19, Ponsonby Tarn (Bonnett and Cambray, 1991)

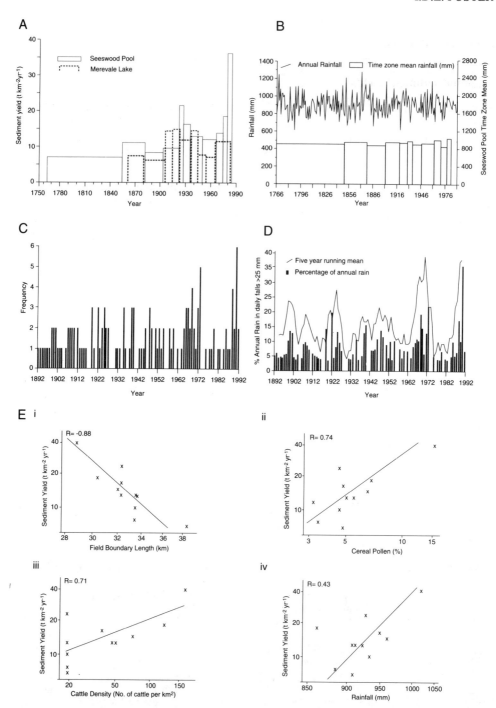

Figure 15.2 Sediment yields, climate and land-use controls on long-term sediment yields in the Midlands. (A) Sediment yields reconstructed from the Merevale Lake and Seeswood Pool catchments (after Foster *et al.*, 1985, 1986). (B) Trends in the homogenised rainfall series for England and Wales and the variations in rainfall by sediment yield time zones in Seeswood Pool

reconstructed sediment yields are reasonably consistent with river-monitored estimates in the same catchments.

Regional sediment yields

Average post-1954 sediment yields for five Midland reservoir catchments are given in Table 15.2 (locations of these sites and other sites referenced in the text are given in Figure 15.1). Sediment yields vary from *ca.* 10 t km^{-2} yr^{-1} in undisturbed and urban catchments to 36 t km^{-2} yr^{-1} in lowland agricultural catchments. On average, however, the yields lie within the range estimated by Walling and Webb (1987) for lowland systems, are generally consistent with estimates for inorganic sediment yields of lowland reservoir catchments in Scotland, which range from 1.8 to 61.7 t km^{-2} yr^{-1} (Duck and McManus, 1990), but are considerably lower than the maximum sediment yields for upland southern Pennine reservoirs reported by Butcher *et al.*, (1993). From a survey of 22 reservoirs in this latter study, sediment yields exceeded 200, 100 and 50 t km^{-2} yr^{-1} in four, seven and six catchments respectively.

Temporal trends

Examples of the temporal patterns of sediment yield reconstructed using radiometric dating and core correlation for three lowland UK reservoir catchments are given in Figures 15.2A and 15.3A. Figure 15.2A shows the sediment yields for two rural catchments in north Warwickshire. The forested catchment of Merevale Lake has had low yields for the last 130 years, although the underlying trends are similar to those of the agricultural catchment of Seeswood Pool up until *ca.* 1970 and may have a similar regional control such as climate over this period (Dearing and Foster, 1987). Sediment yields in the Seeswood Pool catchment rose rapidly in the most recent two time zones in

Table 15.2 Post-1954 estimates of sediment yield from Midland lakes and reservoirs

Site	Catchment area (km^2)	Lake area (ha)	Land use	Sediment yield (t km^{-2} yr^{-1})
Merevale Lake	1.95	6.5	Mixed forest	10.0
Wyken Slough	4.50	2.3	Urban	10.0
Holmer Lake	18.37	3.8	Urban/mining	20.1[a]
Coombe Pool	22.25	35.0	Lowland Agriculture	36.0[b]
Seeswood Pool	2.21	6.7	Lowland Agriculture	17.0

[a] Since 1969 (data from Gaskell, 1992)
[b] Since 1946

Figure 15.2 *Continued* (based in part on data in Wigley *et al.*, 1982). (C) Annual frequency of daily rainfalls of greater than 25 mm in the Coventry rainfall record (based on data in Sheard 1994). (D) Percentage and smoothed (five-year running mean) percentage contribution of daily rainfalls of greater than 25 mm to total annual rainfall in the Coventry rainfall series, 1892–1992 (based on data in Sheard, 1994). (E) Relationship between reconstructed sediment yield at Seeswood Pool and land-use and climatic variables: (i) field boundary length; (ii) percentage cereal pollen; (iii) cattle stocking density; (iv) annual average rainfall by time zone (see part B)

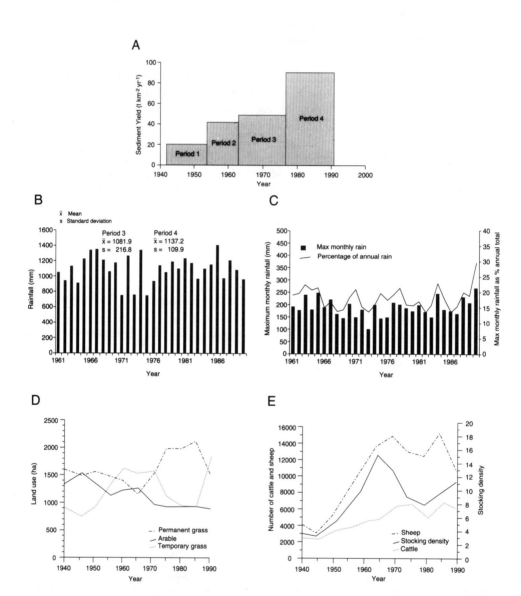

Figure 15.3 Sediment yields, climate and land-use controls on sediment yield in the Old Mill Reservoir catchment. (A) Reconstructed sediment yields, 1941–91 (based on Foster and Walling, 1994). (B) Annual rainfall for the Old Mill catchment (based on the homogenised rainfall series at Dartmouth and Slapton). (C) Maximum monthly rainfall as a proportion of annual rainfall for the Old Mill catchment with the five-year running mean. (D) Trends in land utilisation in the Slapton catchments (based on data in Heathwaite, 1993). (E) Trends in stocking densities in the Slapton catchments (based on data in Heathwaite, 1993)

the 1970s and 1980s. A similar pattern to that at Seeswood Pool is revealed in the reconstructed suspended sediment yield record of the Old Mill Reservoir catchment (Foster and Walling, 1994) (Figure 15.3A). The earliest period shows yields of *ca.* 20 t km^{-2} yr^{-1}, rising to *ca.* 90 t km^{-2} yr^{-1} since the late 1970s.

The explanation for the increase in sediment yield in the Seeswood Pool and Old Mill catchments is complex and has involved consideration of both climate and land-use controls. For example, long-term trends in annual Midland rainfall over the last 220 years have been compiled and analysed (based on data in Wigley *et al.*, 1982) and extrapolated since 1980 from the 130-year Coventry rainfall record. The data show no sustained increase or decrease in rainfall through time and are best characterised by the mean (921.1 mm) and standard deviation (115.6 mm). The most recent period covered by the lake sediments in Seeswood Pool (Figure 15.2A) is associated with slightly higher average annual rainfall over the same time period (Figure 15.2B). The average annual rainfall for each time period for which sediment yields have been estimated is significantly correlated with the sediment yield estimates at Seeswood Pool (Figure 15.2E(iv)), although the r^2 value of 18.5% is low. Since over 90% of the sediment in Midland rivers is transported in *ca.* 5–10% of the time, it is probably more appropriate to examine shorter-term events in the rainfall record. Analysis of extreme daily rainfalls (>25 mm day^{-1}) in Coventry over the last 100 years suggests that there has recently been an increase both in the annual frequency of events exceeding 25 mm day^{-1} (Figure 15.2C) and in the percentage of annual rainfall delivered in daily rainfalls exceeding 25 mm (Figure 15.2D). The last 30 years of record is notable for an increase in the number of extreme events per year and in their contribution to total annual rainfall. The five-point moving average of Figure 15.2D shows some periodicity in the record, with increasing amplitude over the last three decades. However, the period of highest sediment yield in the Seeswood Pool catchment is associated with a reduction both in the number of extreme events and in the proportion of annual rainfall delivered by extreme events.

A similarly complex picture has emerged from an analysis of the rainfall record for the Old Mill catchment, where there is no evidence of increasing or decreasing annual rainfall (Foster and Walling, 1994) (Figure 15.3B). Furthermore, despite the presence of a coarse sedimentary layer in the post-1980 sediments, analysis of available daily rainfall data indicates a number of possible events spanning a five-year period that might have triggered extreme erosion. Daily rainfall records for the area between 1961 and 1990 are incomplete, and no attempt has been made to identify the existence of a change in the frequency of daily rainfall above a critical threshold. Monthly data have been analysed in order to identify the existence of trends by plotting maximum monthly rainfall in each year and the percentage of the annual totals delivered in these months (Figure 15.3C). The data indicate some variability but no significant trend.

It would seem likely that controls other than climate have a profound effect on sediment delivery and transport in these lowland agricultural catchments. For example, at Seeswood Pool, a range of secondary data sources have been investigated, which include: historical maps, from which data on field boundary length has been extracted (Owen, 1985); the agricultural census and other local historical sources, from which cattle population data have been obtained (Talbot, 1992); and pollen analysis of a lake sediment core, where an increase in arable cultivation for each time zone has been estimated from the percentage cereal pollen content (Rennie, 1988). Figure 15.2E(i)–(iii) shows the three statistically significant relationships between land use and sediment yield that were obtained from these analyses. The highest correlation is with field boundary length where, over the last 200 years, some 9–10 km of field boundary have been removed from the catchment. Sediment yield is also significantly correlated with an increase in cereal cultivation and with increasing cattle stocking density. All three land-use controls have higher correlations with sediment runoff than the rainfall index of Figure 15.2E(iv).

A similar pattern emerges from the the Old Mill catchment. The data of Figures 15.3D and 15.3E, derived from a land-use analysis of the nearby Slapton Ley catchments (Heathwaite, 1993), shows post-1940 increases in sheep and cattle stocking densities and, since 1970, an increase in the area of permanent grassland. Furthermore, recent analysis of historical data from the Old Mill catchment (Holt, 1994) has shown that, between 1900 and 1973, no field boundaries had been removed. Between 1973 and 1993, however, some 4.2 km of field boundaries were removed, thereby increasing average field size from 1.43 to 2.38 ha at a time when sediment yields increased to *ca.* 90 t km^{-2} yr^{-1}.

One of the most significant impacts of land use in both Seeswood Pool and the Old Mill Reservoir is the effect on hydrological response. For example, Table 15.3 provides summary unit hydrograph data for the Merevale catchment and the two sub-catchments of Seeswood Pool dominated by arable cultivation and permanent pasture respectively (Foster *et al.*, 1990b). Although the percentage runoff is lower at Seeswood Pool, the short lag times of the pasture catchment and the high peak flows generated are a direct function of the impact of cattle trampling which, in winter months, reduces soil infiltration rates to below 5 mm h^{-1} in the riparian zones immediately adjacent to the main stream channel.

Hydrological and infiltration records are not available for the Old Mill catchment, but direct comparison with infiltration and bulk density data for the nearby Slapton catchments (Heathwaite *et al.*, 1990) would suggest a similar pattern in hydrological behaviour to Seeswood Pool. Infiltration capacities on permanent grass grazed by cattle at Merrifield (Slapton) in March were less than 20 mm h^{-1} in comparison with temporary grass at the same time of *ca.* 80 mm h^{-1} and kale grazed by sheep of *ca.* 100 mm h^{-1}. The infiltration characteristics show a marked seasonal pattern in all land uses examined by Heathwaite *et al.* (1990), which would suggest that antecedent conditions have a significant impact in regulating the rainfall input and, in consequence, the hydrological and sediment transport pathways.

Table 15.3 Summary of catchment response; unit hydrograph derivation (from Foster *et al.*, 1990b)

Site	Lag (h)	Peak flow (m^3 s^{-1} per 50 km^2)	Percentage runoff
Merevale (oak woodland)	13.1	9.8	44.3
Seeswood (arable)	6.1	15.4	24.3
Seeswood (pasture)	3.9	23.5	29.8

The historical climate and land-use data from these two lowland agricultural catchments have provided some insight into the major factors controlling the delivery of sediment from the hillslopes to the river. On balance it would appear that increasing field size and, in consequence, reduction in the opportunity for sediment storage behind lateral hedge boundaries, may have had a significant impact on sediment delivery to the transporting rivers and receiving reservoirs, even if the rate of detachment on hillslopes had remained constant through time. Furthermore, the significant hydrological effect of cattle trampling on both infiltration capacities and storm hydrograph response may have a direct impact on soil erosion from the fields and on channel erosion consequent upon increasing flow magnitudes.

SEDIMENT SOURCES AND SEDIMENT DELIVERY

The above section has highlighted some of the difficulties in identifying controls on soil erosion and sediment transport over the time-scales for which short- to medium-term sediment yields can be reconstructed from an analysis of reservoir sediments. A further significant problem associated with reservoir sediment-based reconstructions of sediment yield is whether the change in yield represents an increase in erosion rate from the same source, a change in sediment source, a change in the sediment delivery ratio or a combination of all three factors (cf. Boardman *et al.*, 1990; Walling, 1990).

A variety of geochemical, mineral magnetic and radiometric fingerprinting techniques have been applied to the reservoir and source sediments in the catchments of the Old Mill Reservoir and Seeswood Pool (Dearing *et al.*, 1986; Dearing and Foster, 1987; Foster *et al.*, 1990b; Grew, 1990; Foster and Walling, 1994; Foster *et al.*, 1994). The following section concentrates upon the application of the ^{137}Cs technique.

The fallout radionuclide ^{137}Cs has been used for the last two decades to date lake and reservoir sediments (cf. Pennington *et al.*, 1973; Eakins *et al.*, 1981) (site locations are given in Figure 15.1). More recently, this radionuclide has been used in the UK for soil erosion and sediment transport research, including studies of sediment delivery, soil redistribution on arable fields and floodplain sediment accumulation, and to investigate the controls imposed by sediment delivery from the catchment on the form of the ^{137}Cs profile in reservoirs (Walling *et al.*, 1986; Walling and Bradley, 1990; Walling and Quine, 1990, 1992; Walling and Woodward, 1992; Walling and He, 1993).

The atmospheric influx of ^{137}Cs in the English Midlands is shown in Figure 15.4A. This diagram includes the original fallout data uncorrected for radioactive decay and the decayed record (to 1992) in order to illustrate the decline in the contribution that this early period would make to the ^{137}Cs activity of lake sediments collected at the present time. Some 74% of the total fallout, and 68% of the decayed fallout, occurred before 1965. With the exception of Chernobyl-derived ^{137}Cs, the influx has shown a marked post-1965 decline in activity through time. Incorporation and fixation of ^{137}Cs into stable (non-eroding or depositing) catchment soils (Figure 15.4B) typically shows high activities in the near-surface layers with an exponential decline in activity with depth, usually ascribed to a combination of diffusion and bioturbation processes. In arable fields (Figure 15.4C) the ^{137}Cs is usually well mixed and the differences between individual soil core sections lie within the counting errors on the ^{137}Cs estimate. The depth to which ^{137}Cs is recorded in the soil profile may vary. On accumulating sites,

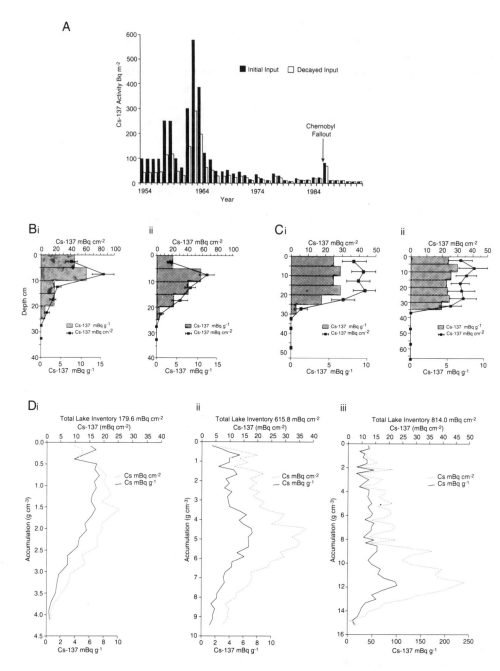

Figure 15.4 Caesium-137 in soils and lake sediments. (A) Trends in ^{137}Cs atmospheric influx, 1954–92, Chilton, Oxfordshire (see Figure 15.1 for site location); decayed input calculated to 1992 (data courtesy of AERE Harwell). (B) Profiles of ^{137}Cs in pasture soils of the Seeswood Pool catchment (based on data in Foster *et al.*, 1994). (C) Profiles of ^{137}Cs in cultivated soils of the Seeswood Pool catchment (based on data in Foster *et al.*, 1994). (D) The ^{137}Cs activities plotted against sediment accumulation for Merevale Lake (i), Seeswood Pool (ii) and the Old Mill Reservoir (iii) sediments

^{137}Cs may be detected below the current plough depth of 30 cm (Figure 15.4C(ii)), whereas on eroding and/or stable sites ^{137}Cs would be found at or above the plough layer provided no significant bioturbation, eluviation or chemical diffusion occurred (Figure 15.4C(i)).

The ^{137}Cs influx to the UK is not uniform and broadly reflects the distribution of average annual rainfall (Cawse and Horrill, 1986). For example, the background inventory in stable soils for the Seeswood Pool catchment is around 168 mBq cm^{-2} (corrected to 1992) with a mean annual rainfall of ca. 670 mm, whereas for the Old Mill Reservoir it lies between 250 and 300 mBq cm^{-2} (corrected to 1991; Foster and Walling, 1994) with a mean annual rainfall of ca. 1200 mm.

Caesium-137 activities (mBq cm^{-2}) and concentrations (mBq g^{-1}) in reservoir sediments have been plotted against sediment accumulation, rather than depth, in order to overcome the effects of sediment compaction through time. Data for Merevale Lake, Seeswood Pool and the Old Mill Reservoir are shown in Figure 15.4D. The calculated ^{137}Cs inventories for individual or small numbers of lake sediment cores should be treated with some caution and cannot be compared directly with fallout inventories in soils because of factors such as reservoir trap efficiency, sediment focusing and the possibility of losing atmospheric ^{137}Cs reaching the lake surface to the outflow before it is incorporated into bottom sediments (Owens, 1994). The calculated total ^{137}Cs inventory for the three sites of Figure 15.4D is based on the average of two cores for Merevale Lake and Seeswood Pool but only one core in the Old Mill Reservoir, and the estimates have been adjusted for focusing by comparing the ratio of the sedimentation area in each lake to the lake surface area (cf. Foster and Walling, 1994). The calculated influx to Merevale Lake, where sediment yields and lake sediment accumulation rates are low, is only marginally higher than soil inventories at nearby stable soil sites, which would suggest that the adjustment for sediment focusing in this lake provides a reasonable approximation of the atmospheric fallout record. In contrast, the ^{137}Cs inventory in Seeswood Pool is some 450 mBq cm^{-2} higher than the background atmospheric influx despite adjusting for sediment focusing, suggesting that ca. 2.5 times as much ^{137}Cs has entered the lake with eroded soil than from atmospheric sources. The sedimentation rate since ^{137}Cs was first detected in this reservoir is about double that of Merevale Lake. The Old Mill Reservoir sediments contain 546 mBq cm^{-2} more ^{137}Cs than can be attributed to the atmospheric influx for the region. Sedimentation rates are 1.5 times faster than for Seeswood Pool over the same time period.

In the absence of post-depositional mobility, the form of the ^{137}Cs profile in a lake sediment core will reflect the relative importance of the contributions to the total inventory provided by fallout to the lake surface and sediment input from the surrounding drainage basin (Walling and He, 1993). In the absence of a catchment component, the lake sediment profile should mirror that of Figure 15.4A. The lake sediment ^{137}Cs profile of Merevale Lake shows little resemblance to the atmospheric influx record, which may suggest that bioturbation may significantly redistribute ^{137}Cs throughout the profile in this slowly sedimenting basin.

By contrast, the ^{137}Cs profile of Seeswood Pool, with an estimated atmospheric component of ca. 30% (assuming no direct loss of atmospheric input over the spillweir) of the total ^{137}Cs influx, shows evidence of sustained high ^{137}Cs activities upcore of the 1963 peak, suggesting that soil erosion is maintaining the supply of

^{137}Cs to the reservoir as the relative significance of the atmospheric influx declines (Figure 15.4A). However, the most recent period appears to be associated with a significant decline in total concentration, which would suggest either a change in source to ^{137}Cs-deficient catchment materials or an exhaustion of ^{137}Cs-bearing sediments from the same source area. Recent catchment monitoring, and geochemical and mineral magnetic reconstruction (Dearing and Foster, 1987; Foster and Dearing, 1987; Foster et al., 1987, 1990b; Grew, 1990) have suggested that the most recent time period of high sediment yields in the Seeswood Pool catchment is associated with destruction of the riparian zone by cattle poaching. The heavily poached areas and channel bank sediments have negligible ^{137}Cs activities and would contribute to a significant reduction in ^{137}Cs activity in the sediment core as evidenced at the top of the lake sediment sequence (Figure 15.4D(ii)). The use of ^{137}Cs to investigate soil redistribution on a steep arable field at Seeswood Pool suggests that some limited reworking of soil occurs (Foster et al., 1994). However, two further factors point to the fact that cultivated fields make little contribution to the amount of sediment entering Seeswood Pool. First, low sediment yields are associated with the arable subcatchment of Seeswood Pool (ca. 10 t km^{-2} yr^{-1}) and, secondly, the ^{137}Cs budget for the arable field indicates no net soil loss from the slopes to the channel (Foster et al., 1994).

The various sources of information collected with regard to the dynamics of sediment transport in the Seeswood Pool catchment consistently suggest that the pasture dominates sediment production mainly due to the destruction of the riparian zone by cattle poaching.

The ^{137}Cs profile of the Old Mill Reservoir would suggest that high activity and influx is sustained up-core of the 1963 peak (Figure 15.4D). However, the most recent time periods are associated with highly variable activities, the lowest of which would suggest a switch in source from grassland topsoils to channel bank and/or subsoil sources, as confirmed by recent mineral magnetic modelling of sediment sources (Foster and Walling, 1994). On the basis of mineral magnetic signatures, it has been demonstrated that woodland soils are unlikely to make a major contribution to the sediments in the Old Mill Reservoir (Holt, 1994). The relatively high concentrations of ^{137}Cs of ca. 40 mBq g^{-1} in the upper reservoir sediments are consistent with activities in surface soils from fields of permanent pasture and rough grazing (Table 15.4). The evidence presented for the Old Mill Reservoir catchment would therefore suggest a

Table 15.4 Representative mean values for the current ^{137}Cs activities of potential source materials within the catchment of the Old Mill Reservoir (after Foster and Walling, 1994)

Potential source material (<63 μm fraction)	Mean ^{137}Cs concentration (mBq g^{-1})
Woodland	65
Permanent pasture and rough grazing	36
Improved and ley Pasture	24
Arable fields	20
Channel banks	1.5

similar pattern of erosion, sediment transport and sediment sources to that of Seeswood Pool, where grassland has been shown to be the most susceptible to erosion.

DISCUSSION

Lake and reservoir bottom sediments offer an important and, as yet, not fully evaluated source of proxy hydrological data. Research over the last two decades has witnessed a significant growth in bottom sediment-based studies of sediment yield and ^{137}Cs in a range of physical environments in the UK (Figure 15.1), allowing a number of ideas concerning sediment transport and its control to be tested. The case studies presented above have demonstrated that meaningful sediment yield data can be obtained from an analysis of bottom sediments and point to common controls on erosion in two contrasting lowland agricultural catchments in Warwickshire and Devon. However, interpretation of the reconstructed sediment yield is fraught with difficulties largely as a result of the complexity of drainage basin response and the problem of linking the sediment yield record to climate, land-use and hydrological controls at the appropriate time-scale.

In addition to providing information on sediment yield history, a range of radionuclide, mineral magnetic and geochemical properties of deposited sediments have been compared with the same properties of catchment soils and sediments in order to identify changes in the source of eroded sediments at time-scales that match the reconstructed sediment yield record. In utilising ^{137}Cs, there are a number of difficulties in interpreting the form of the bottom sediment profile and in comparing the lake inventory with inventories from stable soils of the same region in order to calculate the influx of eroded ^{137}Cs from the catchment. Nevertheless, analysis of ^{137}Cs will allow the sedimentary record to be subdivided into time periods shorter than the life of the lake or reservoir, and the shape of the profile may enable trends in the erosional history of the catchment to be inferred. The case studies presented above suggest that increasing rates of erosion in two contrasting agricultural environments over the last 30 years are associated with an increase in the amount of sediment delivered from the same area of the catchment as a result of increasing pressures imposed by high stocking densities.

The hydrological value of bottom sediment reconstruction needs to be considered in the light of research requirements and the conceptual limitations of the approach. The use of palaeoenvironmental reconstruction will provide important information on temporal trends in sediment yields, for which few long-term data are available in the UK, and may also indicate changes in sediment source. However, the approach will provide limited information on process dynamics and sediment transport controls. For both approaches, there is an urgent need to consider the appropriateness of the units in which soil erosion and sediment transport rates are measured and reported, and to develop a greater understanding of the fundamental links that exist between hydrological and sediment-contributing areas of the catchment and of the time-scales at which sediment stores operate and interact within the catchment. Despite these limitations, a combination of bottom-sediment analysis and catchment monitoring appears to provide a powerful conceptual and methodological framework for improved understanding of drainage basin sediment dynamics.

ACKNOWLEDGEMENTS

This chapter is dedicated to Des Walling, who, as PhD supervisor and mentor, has provided considerable stimulus to my own thoughts on catchment processes, soil erosion and sediment transport problems. I am indebted to NERC, Coventry University and Coventry City Council for funding the Midland research programmes, and to numerous colleagues and research students at Coventry University and elsewhere for providing stimulating discussions and ideas over the last 15 years of hydrological and sedimentological research, including Toff Berry, Steve Bradley, Andre Carter, Sue Charlesworth, Heather Dalgleish, John Dearing, Rob Duck, Roger Flower, Rob Grew, David Jones, David Keen, Joan Lees, John McManus, Phil Owens, Tim Quine and Steven Wade. Special thanks go to Phil Owens and Bruce Webb for making valuable comments on an earlier draft of this chapter. Final thanks go to Ruth Gaskell for the diagrams.

REFERENCES

Arden-Clarke, C. and Evans, R. (1993) Soil erosion and conservation in the United Kingdom. In: Pimental, D. (ed.) *World Erosion and Conservation*, Cambridge University Press, Cambridge, pp. 193–217.

Boardman, J., Dearing, J.A. and Foster, I.D.L. (1990) Soil erosion studies, some assessments. In: Boardman, J., Foster, I.D.L. and Dearing, J.A. (eds), *Soil Erosion on Agricultural Land*, John Wiley & Sons, Chichester, pp. 659–672.

Bonnett, P.J.P. and Appleby, P.G. (1991) Deposition and transport of radionuclides within an upland drainage basin in mid-Wales. *Hydrobiologia* **214**, 71–76.

Bonnett, P.J.P. and Cambray, R.S. (1991) The record of deposition of radionuclides in the sediments of Ponsonby Tarn, Cumbria. *Hydrobiologia* **214**, 63–70.

Butcher, D.P., Labadz, J.C., Potter, A.W.R. and White, P. (1993) Reservoir sedimentation rates in the southern Pennine region, UK. In: McManus, J. and Duck, R.W. (eds), *Geomorphology and Sedimentology of Lakes and Reservoirs*, John Wiley & Sons, Chichester, pp. 73–92.

Cawse, P.A. and Horrill, A.D. (1986) *A Survey of Cs-137 and Plutonium in British Soils in 1977*, AERE Harwell Report R-10155, HMSO, London.

Charlesworth, S.M. and Foster, I.D.L. (1993) Effects of urbanisation on lake sedimentation: the history of two lakes in Coventry, UK—Preliminary results. In: McManus, J. and Duck, R.W. (eds), *Geomorphology and Sedimentology of Lakes and Reservoirs*, John Wiley & Sons, Chichester, pp. 15–29.

Church, M. and Slaymaker, O. (1989) Disequilibrium of Holocene sediment yield in glaciated British Columbia. *Nature* **337**, 452–454.

Dearing, J.A. (1986) Core correlation and total sediment influx. In: Berglund, B. (ed.), *Handbook of Holocene Palaeoecology and Palaeohydrology*, John Wiley & Sons, Chichester, pp. 247–270.

Dearing, J.A. and Foster, I.D.L. (1987) Limnic sediments used to reconstruct sediment yields and sources in the English Midlands since 1765. In: Gardiner, V. (ed.), *International Geomorphology*, John Wiley & Sons, Chichester, pp. 853–868.

Dearing, J.A. and Foster, I.D.L. (1993) Lake sediments and geomorphological processes: some thoughts. In McManus, J. and Duck, R.W. (eds), *Geomorphology and Sedimentology of Lakes and Reservoirs*, John Wiley & Sons, Chichester, pp. 5–14.

Dearing, J.A., Morton, R.I., Price, T.W. and Foster, I.D.L. (1986) Tracing movements of topsoil by magnetic measurements; two case studies. *Phys. Earth Planet. Interiors* **42**, 93–104.

Desloges, J.R. and Gilbert, R. (1991) Sedimentary record of Harrison Lake: implications for deglaciation in southwestern British Columbia. *Can. J. Earth Sci.* **28**, 800–815.

Desloges, J.R. and Gilbert, R. (1994) Sediment source and hydroclimatic inferences from glacial lake sediments: the postglacial and sedimentary record of Lilloet lake, British Columbia. *J. Hydrol.* **159**, 375–393.

Duck, R.W. and McManus, J. (1987) Sediment yields in lowland Scotland derived from reservoir surveys. *Trans. R. Soc. Edinb., Earth Sci.* **78**, 369–377.

Duck, R.W. and McManus, J. (1990) Relationship between catchment characteristics, land use and sediment yield in the Midland Valley of Scotland. In: Boardman, J., Foster, I.D.L. and Dearing, J.A. (eds), *Soil Erosion on Agricultural Land*, John Wiley & Sons, Chichester, pp. 285–299.

Eakins, J.D., Cambray, R.S., Chambers, K.C. and Lally, A.E. (1981) *The Transfer of Natural and Artificial Radionuclides to the Brotherswater Lake from its Catchment*, AERE Harwell Report R-10375, HMSO, London.

Flower, R.J., Stevenson, A.C., Dearing, J.A., Foster, I.D.L., Airey, A., Rippey, B., Wilson, J.P.F. and Appleby, P.G. (1989) Catchment disturbance inferred from studies of three contrasted sub-humid environments in Morocco. *J. Palaeolimnol.* **1**, 293–322.

Foster, I.D.L. (1985) *Sedimentation in Coombe Pool: A Preliminary Appraisal*, Sediments and Water Group, Geography Department, Coventry University, Report, 16 pp.

Foster, I.D.L. and Dearing, J.A. (1987) Lake sediment chemistry; a comparative study of historical catchment processes in Midland England. *GeoJournal* **14**, 285–297.

Foster, I.D.L., Dearing, J.A., Simpson, A.D., Carter, A.D. and Appleby, P.G. (1985) Lake catchment based studies of erosion and denudation in the Merevale catchment, N Warwickshire, UK. *Earth Surf. Processes Landforms* **10**, 45–68.

Foster, I.D.L., Dearing, J.A. and Appleby, P.G. (1986) Historical trends in catchment sediment yields; a case study in reconstruction from lake sediment records in Warwickshire, UK. *Hydrol. Sci. J.* **31**, 427–443.

Foster, I.D.L., Dearing, J.A., Charlesworth, S.M. and Kelly, L.A. (1987) Paired lake catchment experiments; a framework for investigating chemical fluxes in small drainage basins. *Appl. Geogr.* **7**, 115–133.

Foster, I.D.L., Dearing, J.A. and Grew, R.G. (1988) Lake-catchments: an evaluation of their contribution to studies of sediment yield and delivery processes. In: *Sediment Budgets (Proceedings of the Porto Alegre Symposium)*, IAHS Publication no. 174, pp. 413–424.

Foster, I.D.L., Dearing, J.A., Grew, R. and Orend, K. (1990a) The sedimentary data base: an appraisal of lake and reservoir sediment based studies of sediment yield. In: *Erosion, Transport and Deposition Processes (Proceedings of the Jerusalem Workshop)*, IAHS Publication no. 189, pp. 19–43.

Foster, I.D.L. and Walling, D.E. (1994) Using reservoir deposits to reconstruct changing sediment yields and sources in the catchment of the Old Mill reservoir, South Devon, UK over the past 50 years. *Hydrol. Sci. J.* **39**, 347–368.

Foster, I.D.L., Grew, R. and Dearing, J.A. (1990b) Magnitude and frequency of sediment transport in agricultural catchments. In: Boardman, J., Foster, I.D.L. and Dearing, J.A. (eds), *Soil Erosion on Agricultural Land*, John Wiley & Sons, Chichester, pp. 153–171.

Foster, I.D.L., Dalgleish, H., Dearing, J.A and Jones, E.D. (1994) Quantifying soil erosion and sediment transport in drainage basins; some observations on the use of Cs-137. In: *Variability in Stream Erosion and Sediment Transport (Proceedings of Canberra Symposium)*, IAHS Publication no. 224, pp. 55–64.

Gaskell, R. (1992) Lake sediment properties from an urban catchment: south Telford, Shropshire. Unpublished BSc dissertation, Coventry University.

Grew, R. (1990) Sediment yields and sources over short and medium timescales in a small agricultural catchment in N Warwickshire, UK. Unpublished PhD Thesis, Coventry University.

Grönlund, E. (1991) Sediment characteristics in relation to cultivation history in two varved lake sediments from East Finland. *Hydrobiologia* **214**, 137–142.

Heathwaite, A.L. (1993) Catchment controls on recent sediment history of Slapton Ley, south-west England. In: Thomas, D.S.G. and Allison, R.J. (eds), *Landscape Sensitivity*, John Wiley & Sons, Chichester, pp. 241–259.

Heathwaite, A.L., Burt, T.P. and Trudgill, S.T. (1990) The effect of land use on nitrogen, phosphorus and suspended sediment delivery to streams in a small catchment in southwest England. In: Thornes, J.B. (ed.), *Vegetation and Erosion, Processes and Environments*, John Wiley & Sons, Chichester, pp. 161–178.

Holt, C. (1994) Tracing suspended sediment sources in the Old Mill catchment, South Devon; a magnetic and chemical investigation. Unpublished BSc dissertation, Coventry University.

Knox J.C. (1972) Valley alluviation in SW Wisconsin. *Ann. Assoc. Am. Geogr.* **62**, 401–410.

Laronne, J. (1987) Rhythmic couplets: sedimentology and prediction of reservoir design periods in semiarid areas. In: Berkofsky, L. and Wurtele, M.G. (eds), *Progress in Desert Research*, Rowman and Littlefield, Totowa, pp. 229–244.

Laronne, J. (1990) Probability distribution of event sediment yields in the northern Negev, Israel. In: Boardman, J., Foster, I.D.L. and Dearing, J.A. (eds), *Soil Erosion on Agricultural Land*, John Wiley & Sons, Chichester, pp. 481–492.

Livesey, R.H. (1972) Corps of engineers methods for predicting sediment yield. *Proceedings of Sediment Yield Workshop*, US Dept of Agriculture, Mississippi, pp. 16–32.

Livingstone, D.L. and Cambray, R.S. (1978) Confirmation of Cs-137 dating by algal stratigraphy in Rostherne Mere. *Nature* **276**, 259–261.

Lockwood, J.G. (1983) Modelling climate change. In: Gregory, K.J. (ed.), *Background to Palaeohydrology*, John Wiley & Sons, Chichester, pp. 25–50.

Maner, S.B. (1958) Factors affecting the sediment delivery rates in the Red Hills physiographic area. *Trans. Am. Geophys. Union* **39**, 669–675.

Meade, R.H. (1982) Sources, sinks and storage of river sediment in the Atlantic drainage of the United States. *J. Geol.* **90**, 235–252.

Meade, R.H., Dunne, T., Richey, J.E., Santos, U. de M. and Salati, E. (1985) Storage and remobilisation of suspended sediment in the Lower Amazon River of Brazil. *Science* **228**, 488–490.

Neil, D.T. and Mazari, R.K. (1993) Sediment yield mapping using small dam sedimentation surveys, southern Tablelands, New South Wales. *Catena* **20**, 13–25.

O'Sullivan, P.E. (1990) Studies of varved sediments in the Shropshire–Cheshire Meres, UK. In: Saarnisto, M. and Kahra, A. (eds), *Laminated Sediments (Proceedings of Workshop at Lammi Biological Station, 4–6 June, 1990)*, Geological Society of Finland, Special Paper no. 14, pp. 33–46.

O'Sullivan, P.E., Heathwaite, A.L., Appleby, P.G., Brookfield, D., Crick, M.W., Moscrop, C., Mulder, T.B., Vernon, N.J. and Wilmhurst J.M. (1991) Palaeolimnology of Slapton Ley, Devon. *Hydrobiologia* **214**, 115–124.

Owen, C. (1985) Lake sediment based erosion studies at Seeswood Pool: an appraisal of historical data sources. Unpublished BSc Dissertation, Coventry University.

Owens, P.N. (1994) Towards improved interpretation of caesium-137 measurements in soil erosion studies. Unpublished PhD thesis, Exeter University.

Owens, P.N. and Slaymaker, O. (1993) Lacustrine sediment budgets in the Coast Mountains of British Columbia. In: McManus, J. and Duck, R.W. (eds), *Geomorphology and Sedimentology of Lakes and Reservoirs*, John Wiley & Sons, Chichester, pp. 105–124.

Pennington, W., Cambray, R.S. and Fisher, E.M. (1973) Observations on lake sediments using fallout Cs-137 as a tracer. *Nature* **242**, 324–326.

Rausch, D.L. and Heinemann, H.G. (1984) Measurement of reservoir sedimentation. In: Hadley, R.F. and Walling, D.E. (eds), *Erosion and Sediment Yields: Some Methods of Measurement and Modelling*, Geo Books, Norwich, pp. 179–200.

Rennie, I. (1988) Agricultural and vegetational changes to Seeswood Pool catchment inferred from pollen and charcoal analysis of a dated sediment core. Unpublished BSc dissertation, Coventry University.

Ritchie, J.C., McHenry, J.R. and Gill, A.C. (1973) Dating recent reservoir sediments. *Limnol. Oceanogr.* **18**, 254–263.

Rowan, J.S., Higgitt, D.L. and Walling, D.E. (1993) Incorporation of Chernobyl derived radiocaesium into reservoir sedimentary sequences. In: McManus, J. and Duck, R.W. (eds), *Geomorphology and Sedimentology of Lakes and Reservoirs*, John Wiley & Sons, Chichester, pp. 55–72.

Saarnisto, M. (1986) Annually laminated lake sediments. In: Berglund, B. (ed.), *Handbook of Holocene Palaeoecology and Palaeohydrology*, John Wiley & Sons, Chichester, pp. 343–370.

Schumm, S. (1979) Geomorphic thresholds: the concept and its application. *Inst. Br. Geogr. Trans.* **4**, 485–515.

Sheard, N. (1994) Rainfall in Coventry 1867–1992: a statistical analysis. Unpublished BSc dissertation, Coventry University.

Smith, J.P., Bradley, S.B., Macklin, M.G. and Cox, J.J. (1983) The influence of catastrophic floods on water quality as recorded in the the sediments of Blagdon Lake, England. In: *Dissolved Loads of Rivers and Surface Water Quality/Quantity Relationships (Proceedings of Hamburg Symposium)*, IAHS Publication no. 141, pp. 421–430.

Starkel, L. (1983) The reflection of hydrologic changes in the fluvial environment of the temperate zone during the last 15 000 years. In: Gregory, K.J. (ed.) *Background to Palaeohydrology*, John Wiley & Sons, Chichester, pp. 213–236.

Talbot, M. (1992) Sediment yield in a paired lake catchment, Warwickshire: climate or land use control? Unpublished BSc dissertation, Coventry University.

Thornes, J.B. (ed.) (1990) *Vegetation and Erosion, Processes and Environments*, John Wiley & Sons, Chichester.

Trimble, S. (1981) Changes in sediment storage in the Coon Creek Basin, Driftless area, Wisconsin. In: *Proceedings of 3rd Interagency Sedimentation Conference*, pp. 5-100–112.

Walling, D.E. (1978) Reliability considerations in the evaluation and analysis of river loads. *Z. Geomorphol.* **29**, 29–42.

Walling, D.E. (1983) The sediment delivery problem. *J. Hydrol.* **65**, 209–237.

Walling, D.E. (1988) Measuring sediment yield from river basins. In: Lal, R. (ed.), *Soil Erosion Research Methods*, Soil and Water Conservation Society, Ankeny, Iowa, pp. 39–73.

Walling, D.E. (1990) Linking the field to the river: sediment delivery from agricultural land. In: Boardman, J., Foster, I.D.L. and Dearing, J.A. (eds), *Soil Erosion on Agricultural Land*, John Wiley & Sons, Chichester, pp. 129–152.

Walling, D.E. and Bradley, S.B. (1990) Some applications of caesium-137 measurements in the study of fluvial erosion, transport and deposition. In: *Erosion, Transport and Deposition Processes (Proceedings of Jerusalem Workshop)*, IAHS Publication no. 189, pp. 179–203.

Walling, D.E. and He, Q. (1993) Towards improved interpretation of Cs-137 profiles in lake sediments. In: McManus, J. and Duck, R.W. (eds), *Geomorphology and Sedimentology of Lakes and Reservoirs*, John Wiley & Sons, Chichester, pp. 31–53.

Walling, D.E. and Quine, T.A. (1990) Use of Cs-137 to investigate patterns and rates of soil erosion on arable fields. In: Boardman, J., Foster, I.D.L. and Dearing, J.A. (eds), *Soil Erosion on Agricultural Land*, John Wiley & Sons, Chichester, pp. 33–53.

Walling, D.E. and Quine, T.A. (1992) The use of Cs-137 measurements in soil erosion surveys. In: *Erosion and Sediment Transport Monitoring Programmes in River Basins (Proceedings of Oslo Symposium)*, IAHS Publication no. 210, pp. 143–152.

Walling, D.E. and Webb, B.W. (1981) The reliability of suspended sediment load data. In: *Erosion and Sediment Transport Measurement (Proceedings of Florence Symposium)*, IAHS Publication no. 133, pp. 177–194.

Walling, D.E. and Webb, B.W. (1983) Patterns of sediment yield. In: Gregory, K.J. (ed.), *Background to Palaeohydrology, A Perspective*, John Wiley & Sons, Chichester, pp. 69–100.

Walling, D.E. and Webb, B.W. (1987) Material transport by the world's rivers: evolving perspectives. In: *Water For the Future: Hydrology in Perspective (Proceedings of Rome Symposium)*, IAHS Publication no. 164, pp. 313–329.

Walling, D.E. and Woodward, J.C. (1992) Use of radiometric fingerprints to derive information on suspended sediment sources. In: *Erosion and Sediment Transport Monitoring Programmes in River Basins (Proceedings of Oslo Symposium)*, IAHS Publication no. 210, pp. 143–152.

Walling, D.E., Bradley, S.B. and Wilkinson, C.J. (1986) A caesium-137 budget approach to the investigation of sediment delivery from a small agricultural drainage basin in Devon, UK. In: *Drainage Basin Sediment Delivery (Proceedings of Albuquerque Symposium)*, IAHS Publication no. 159, pp. 423–435.

Wigley, T., Lough, J.M. and Jones, P.D. (1982) Spatial patterns of precipitation in England and Wales and a revised homogenous England and Wales precipitation series. *J. Climatol.* **4**, 1–25.

Wise, S.M. (1980) Caesium-137 and Pb-210: a review of the techniques and some applications in geomorphology. In Cullingford, R.A., Davidson, D.A. and Lewin, J. (eds), *Timescales in Geomorphology*, John Wiley & Sons, Chichester, pp. 109–127.

Section V

RADIONUCLIDE STUDIES

16 The Development and Application of Caesium-137 Measurements in Erosion Investigations

DAVID L. HIGGITT

Department of Geography, University of Durham, UK

INTRODUCTION

During the last two decades, the application of caesium-137 (^{137}Cs) measurements as a tracer in soil erosion investigations has become increasingly popular. Over the same period, concern about the problems of soil erosion and sedimentation has also intensified. It is likely that the growing perception of an erosion problem in many parts of the world and in many different environments is at least partly due to the increased attention afforded by research scientists and agricultural advisers, as well as to the detrimental effects of changing land management. Despite the increasing concern, the accurate assessment and quantification of rates of soil loss have remained difficult, not least because of the spatial and temporal variability of the erosion processes.

Strategies for environmental management require robust techniques that allow rapid and reliable estimation of the erosion status of a particular area. Reviews of research requirements and technical developments throughout the period (e.g. Wolman, 1977; Walling, 1983, 1988; Loughran, 1989; Higgitt, 1993) have continued to emphasise the importance of identifying and understanding the pathways by which sediments are moved from source areas to channels via periods of intermittent storage. The nature of this 'sediment delivery problem' is given clear expression in Walling's (1983) widely cited paper of that name. Disaggregation of the spatial lumping inherent in erosion estimates derived from catchment sediment yields or by the extrapolation of plot data is an important consideration. To this end, a number of developments have occurred that have improved the potential to derive information about the spatial patterns of soil loss and degradation. These include the improvement of aerial photograph interpretation, remote sensing, airborne laser profiles of surface topography and various GIS (geographical information systems) and modelling applications (e.g. de Roo *et al.*, 1989; Garg and Harrison, 1992; Graham, 1992; Ritchie *et al.*, 1992). However, it can be argued that one of the most significant recent developments in the field-based estimation of erosion rates is the use of ^{137}Cs measurements.

Comprehensive reviews of the use of ^{137}Cs measurements in erosion investigations have been produced by a number of authors at various times (e.g. Wise, 1980; McHenry, 1985; Ritchie and McHenry, 1990; Walling and Quine, 1992). The intention of this chapter is not to produce another review of the applications, but rather to

Sediment and Water Quality in River Catchments. Edited by I.D.L. Foster, A.M. Gurnell and B.W. Webb.
© 1995 John Wiley & Sons Ltd.

consider the way in which such applications have evolved in relation to assumptions underlying the technique and to the research agenda of the erosion–sedimentation field. Following a brief section outlining the basis of the technique and its utility, the historical development and application of ^{137}Cs measurements are discussed. It is apparent that the technique has progressed through a number of overlapping phases of development, which might be considered typical of the evolution of an innovative idea in the sciences. Early papers sought to validate the potential of the technique, and were followed by a number of enquiries seeking to apply the ^{137}Cs technique to particular issues of sediment delivery. More recently, emphasis has been paid to questions concerning the representativeness of the assumptions of the technique. This includes consideration of the effect of particle-size sorting, methods to calculate soil loss from ^{137}Cs residuals, questions of statistical uncertainty and sample size, and the impact of Chernobyl fallout. The implications of these questions are discussed in the second half of the chapter.

CAESIUM-137 AS A SEDIMENT TRACER: THE THEORY

Caesium-137 originated as fallout from the atmospheric testing of nuclear weapons. As a result of the tests, large amounts of ^{137}Cs were distributed globally from 1954 onwards. Most of the fallout occurred in the decade between 1956 and 1965, with maximum fallout in 1963, the year of the Nuclear Test Ban Treaty. Global fallout has decreased considerably since 1972, but the Chernobyl reactor accident in 1986 resulted in significant amounts of ^{137}Cs, and other radionuclides, being dispersed across the Northern Hemisphere, and Europe in particular. The temporal variability in ^{137}Cs fallout, combined with its tendency to be strongly adsorbed on fine-grained particulate matter and its relatively long half-life of 30.2 years, has promoted the use of ^{137}Cs in two main areas of geomorphological investigation. First, the utilisation of the vertical profile of ^{137}Cs in lake sediments (or other depositional sequences) has been used as a means of dating recent sediments and estimating accretion rates. Secondly, the strong fixation to sediment particles enables ^{137}Cs to act as a tracer of surficial sediment movement.

Soil science research in the 1950s and 1960s established that caesium, which occurs as a simple monovalent cation Cs^+, does not undergo changes through redox reactions or complexation. It is strongly adsorbed in soil by ion exchange on clay and organic colloidal matter, and is essentially non-exchangeable (Amphlett and McDonald, 1956; Schultz et al., 1960; Tamura, 1964; Rogowski and Tamura, 1970a). Because ^{137}Cs is adsorbed near the soil surface and generally resists leaching through the profile, its redistribution, subsequent to its deposition as fallout, is likely to be controlled by the erosion, transport and deposition of sediment particles. Empirical evidence for the retention of ^{137}Cs in surface soil is plentiful (e.g. Ritchie et al., 1970; Frissel and Pennders, 1983; Cawse and Horrill, 1986), suggesting its potential as a medium-term tracer for soil erosion investigations.

The potential of using ^{137}Cs as a tracer is further reinforced by the relative ease with which it can be measured by gamma spectrometry. The disintegration of ^{137}Cs emits a gamma-ray of sufficient energy to permit non-destructive quantitative analysis after limited sample preparation. By comparing the ^{137}Cs content of soil cores from different locations with a reference value representing the baseline fallout to that area, it is possible to assemble information on the spatial variability of ^{137}Cs redistribution and hence of soil loss or accumulation. The cost of obtaining this information is relatively low, in terms of both financial and labour time inputs, when compared with other methods for estimating soil loss, at least once the cost of the gamma detection system has been offset. Furthermore, since the ^{137}Cs inventory of a soil core represents the net loss or gain over the period since the onset of global fallout (i.e. 1954), a 40-year average rate of soil redistribution can be estimated on the basis of a single site visit.

These aspects of the environmental mobility of ^{137}Cs suggest that it is an ideal tracer for establishing the spatial pattern of soil movement and identifying sediment sources and storage areas, providing that a set of assumptions can be justified:

- Baseline fallout is relatively uniform across a study site and can be determined from soil cores collected at level, undisturbed sites.
- Adsorption near the soil surface occurs rapidly, restricting any downslope movement of ^{137}Cs before adsorption.
- Preferential adsorption of ^{137}Cs on fine-grained particulate matter does not unduly bias its representativeness as a tracer of total soil movement.
- Reliable methods can be used to translate measurements of ^{137}Cs redistribution into soil loss estimates.

Early applications of the ^{137}Cs technique tended to take such assumptions for granted, but more recent investigations have attempted to consider the validity, and hence the implications, of these assumptions more fully.

APPLICATIONS OF CAESIUM-137 MEASUREMENTS IN SOIL EROSION STUDIES

Investigation of the environmental mobility of ^{137}Cs undertaken by soil scientists and radioecologists suggested its potential use in geomorphological investigations as a means of dating recent sediment accumulation or as a tracer of surficial sediment movement. To a large extent, these two objectives have been represented by separate research developments, with the application of ^{137}Cs for dating sediment preceding its use as a tracer, although these applications have been combined in the study of catchment sediment budgets.

The pioneering studies using ^{137}Cs measurements in erosion investigations were undertaken by J.R. McHenry and J.C. Ritchie and associates, based at the USDA Sedimentation Laboratory in Mississippi. Having initially been concerned with using ^{137}Cs to determine sedimentation rates in reservoirs, they became interested in the possibility of using ^{137}Cs as a tag for surficial soil movement. Much of their early work consisted of empirical investigations of the vertical distribution of ^{137}Cs in the soils of numerous small catchments in southeastern USA (e.g. Ritchie *et al.*, 1970, 1972; McHenry *et al.*, 1973), and was followed by a paper (Ritchie *et al.*, 1974) demonstrating that ^{137}Cs redistribution was logarithmically related to soil loss estimated by the Universal Soil Loss Equation. The regression relationship was further improved by incorporating data from earlier radioecological studies of the removal of ^{137}Cs and ^{90}Sr from soil plots by runoff (Menzel, 1960; Graham, 1963; Rogowski and Tamura, 1970b).

The development of ^{137}Cs applications in erosion studies can be divided into a series of overlapping phases, which are depicted in Figure 16.1. These phases are described as:

1 Demonstration and validation (mainly on agricultural land).
2 Extensification (to new geomorphological problems or different process environments).
3 Quantification (of erosion rates from ^{137}Cs residuals).
4 Methodological examination.

Once the potential for using ^{137}Cs redistribution as a surrogate for soil erosion and deposition had been identified, a number of studies sought to verify the principle. The first phase of application concerned attempts to study patterns of erosion along transects or grids on agricultural land (McHenry and Ritchie, 1977; McHenry et al., 1978; Mitchell et al., 1980; McCallan et al., 1980; de Jong et al., 1982; Longmore et al., 1983). The technique was soon extended to consider the overall behaviour of catchments and to construct sediment budgets (Brown et al., 1981; Campbell et al., 1982; Ritchie et al., 1982; Wilkin and Hebel, 1982).

The ability to determine spatial variation in sediment transfer has numerous applications in geomorphology. An early review was provided by Wise (1977), while, in a general review of hydrological research, Walling (1982) predicted that ^{137}Cs had considerable potential. Accordingly, the 1980s saw the ^{137}Cs technique extended to a variety both of geomorphological applications and environmental settings. During this period the majority of ^{137}Cs erosion research was concentrated in three research teams led by E. de Jong (Canada), B.L. Campbell and R.J. Loughran (Australia) and D.E. Walling (UK). Geomorphological applications included the ability to map sediment sources and elucidate on catchment sediment dynamics (Campbell et al., 1986; Loughran et al., 1986, 1987; Walling et al., 1986b), to determine the origin of suspended sediment by its ^{137}Cs concentration (Peart and Walling,1986), to investigate patterns of floodplain deposition (Walling et al., 1986a) and the role of wetlands as nutrient traps (Johnston et al., 1984; Cooper et al., 1987), and to examine the relationship between topography and erosion rates (Martz and de Jong, 1987; Pennock and de Jong, 1987).

While most of the previous work had concentrated on the erosion of agricultural land in humid temperate environments, the late 1980s onwards has seen a number of papers concerned with non-agricultural land uses and other climatic zones. Attempts have been made to extend the ^{137}Cs technique to examine erosion rates on rangeland (Lance et al., 1986; Loughran et al., 1990b), clear-cut forestry slopes (Mcintyre et al., 1987), upland environments (Bonnett et al., 1989) and playa environments (Longmore et al., 1986). The prospective value of ^{137}Cs measurements for assembling information on soil loss in developing countries, where problems are often acute and data scarce, was suggested by Campbell (1983), although research investigations have lagged. Early attempts to examine the potential of the technique include work in South Korea (Menzel et al., 1987), Lesotho (Kulander and Strömquist, 1989) and China (Zhang et al., 1989, 1990). In recent years, coverage has considerably expanded.

A dominant feature of the papers cited above is their aim to provide empirical support to evaluate the potential of a new technique. Few discuss the limitations of the inherent assumptions or the statistical significance of results obtained in any great

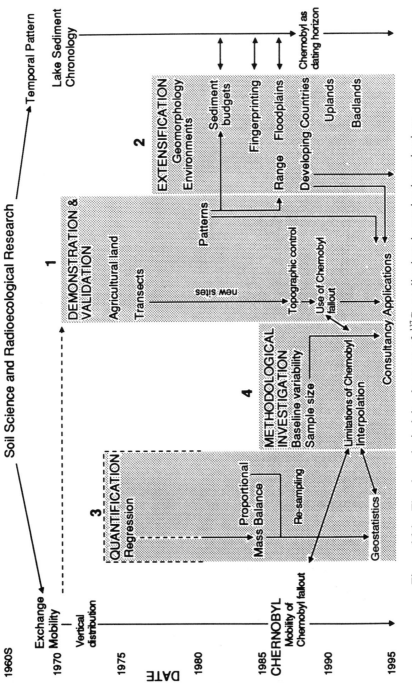

Figure 16.1 The chronological development of ^{137}Cs applications in erosion investigations

detail. It is in the agricultural environments, where the earliest applications of the technique were tested, that consolidation of the methodology has been most fully developed. In particular, questions concerning the reliability of methods to translate ^{137}Cs measurements into soil loss estimates and of the sampling strategies required to reduce uncertainty, have begun to be investigated.

METHODOLOGICAL CONSIDERATIONS

The versatility of using ^{137}Cs measurements in a variety of environmental settings and over an array of scales is an advantage that has attracted many commentators (Charman, 1988; Loughran, 1989; Allison, 1991). Figure 16.2 depicts the continuum of scales over which ^{137}Cs measurements might be employed to elucidate some aspect of erosion or sedimentation, ranging from intra-field patterns of soil redistribution to the construction of regional sediment budgets. However, the range of applications is conditioned by the validity of the assumptions about the environmental mobility of ^{137}Cs.

Questions concerning the methodological rigour of ^{137}Cs applications are discussed here with particular reference to two contrasting sites in the UK where the technique has been applied. The first site is a reclaimed opencast mine spoil at Waunafon, Gwent. The mine operated between 1942 and 1947 and was subsequently infilled. An attempt was made to resculpture the infill to relocate the pre-existing watercourse, but the reclamation was unsuccessful and a large part of the site has remained devoid of vegetation. A series of long-term erosion pin measurements (Haigh, 1979) were available to examine the performance of ^{137}Cs measurements on mine spoil environments. The second site is agricultural land on the brown sand of the Bridgnorth soil association, at Dalicot, Shropshire. A grid network of 83 soil cores were collected from a field (field 1) as part of a countrywide survey of agricultural soil erosion (Quine and Walling, 1991). Additional samples were also collected from a second field (field 2) that had been more seriously affected by rill erosion. Detailed descriptions of the field sites are given in Higgitt *et al.* (1994) and Walling and Quine (1991) respectively. Site locations are shown in Figure 16.3.

IMPLICATIONS OF PARTICLE-SIZE SORTING

Because ^{137}Cs is preferentially adsorbed on fine-grained particulate matter, there is a possibility that ^{137}Cs redistribution tags the movement of fines rather than of the net soil. Some reported estimates of erosion rates on UK agricultural soils (Quine and Walling, 1991) suggest an overestimation of ^{137}Cs-derived estimates on clay-rich soils compared with survey results. In the majority of studies, soil samples are prepared by disaggregation and screening through a 2 mm sieve, but McFarlane *et al.* (1992) found that the gravel fraction of soil in Western Australia contained up to a third of the total ^{137}Cs activity. Detailed examination of the distribution of ^{137}Cs content within different size fractions has been limited.

Bulk samples of surface material were collected at each of the sites described in the previous section. At Waunafon two samples were collected from the crest of the slope and from sediment deposited in the bed of an ephemeral channel at the slope base. At

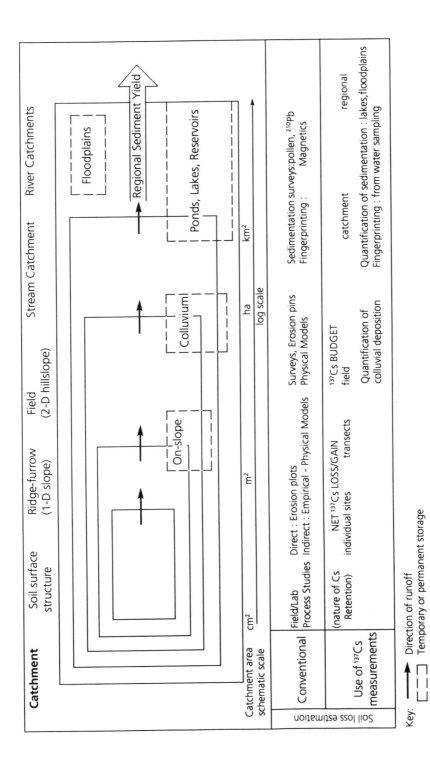

Figure 16.2 Schematic representation of potential ^{137}Cs applications over a continuum of scales

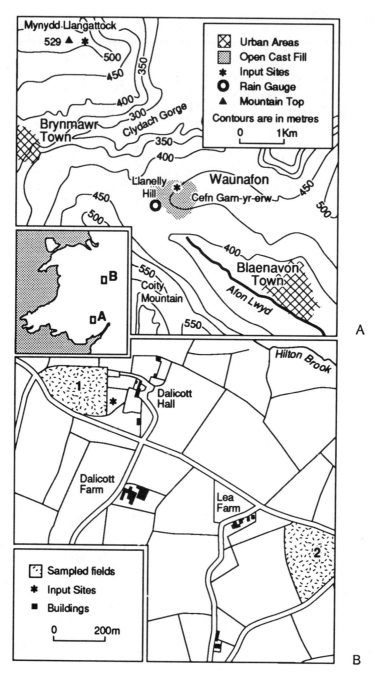

Figure 16.3 Location of study sites: (A) Waunafon opencast fill, Gwent; (B) Dalicott Farm, Shropshire

Dalicott a single sample was collected from the spur ridge in field 2. Composite samples of air-dried material were passed through a bank of sieves of 2.0, 1.7, 1.0, 0.5 and 0.25 mm (plus 0.125 mm for the Dalicott sample), which were shaken mechanically for 15 min. The fractions retained on each sieve were weighed and packed into beakers for ^{137}Cs analysis. After analysis these fractions, excluding the finest, were remixed and placed on a 2 mm sieve with the mesh just immersed in water for 24 hours to allow capillary wetting. The material was then transferred back to the bank of sieves and wet sieved for 5 min. The material left on each sieve was washed off into a weighed beaker, freeze-dried and once again analysed for ^{137}Cs content.

The cumulative percentage of material retained above each size fraction after the dry and wet sieving procedures, together with the percentage of ^{137}Cs activity in those fractions, is plotted in Figure 16.4. If ^{137}Cs content is concentrated in the finest fractions, then its redistribution will not be representative of overall sediment

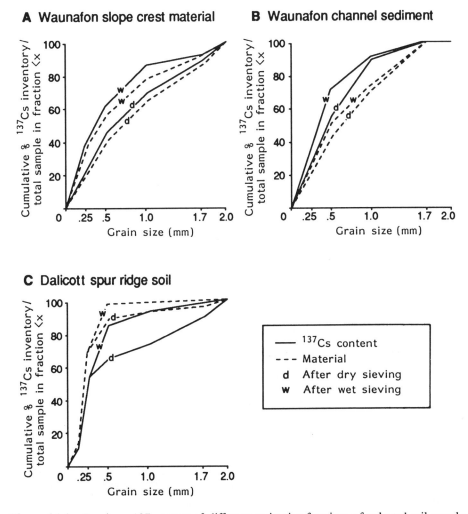

Figure 16.4 Caesium–137 content of different grain-size fractions of selected soil samples

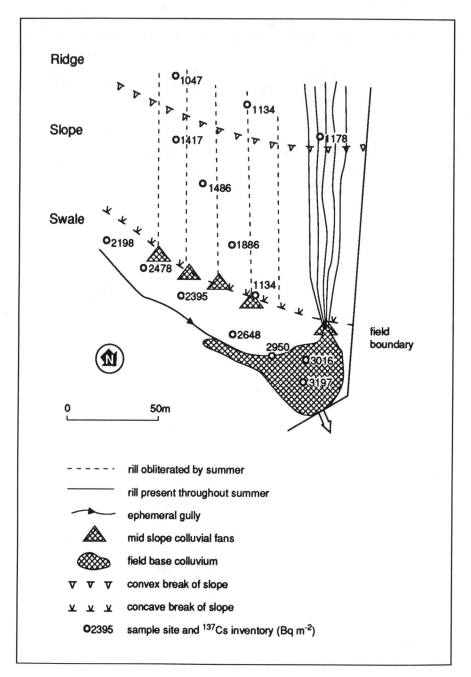

Figure 16.5 Morphological map of part of field 2, Dalicott, Shropshire

movement. Although there is a slight concentration of ^{137}Cs within the finest fraction of the Waunafon slope crest sample, 30% of the total content is contained within the >1.0 mm fraction after dry sieving, presumably adsorbed on constituents of aggregates or as particle coatings. After wet sieving, which simulates the breakdown of unstable aggregates, the mass of the >1.0 mm fraction has been reduced by a third and its ^{137}Cs content by half. In the Shropshire soil, a substantial amount of the ^{137}Cs content (35%) is contained within the >0.5 mm fraction, which only contains 11% by mass. When these fractions were wet sieved, approximately two-thirds of ^{137}Cs present was lost to finer fractions. The cumulative distribution curves for the Waunafon channel sediment indicate a more consistent concentration of ^{137}Cs in finer fractions.

It is suggested from these limited data that the preferential transport of fine particles by erosion processes will not have an especially significant bias on the redistribution of ^{137}Cs, since it appears that much of the ^{137}Cs inventory is adsorbed onto particles that are constituents of aggregates. However, if the use of ^{137}Cs measurements were to be extended from estimating on-slope erosion in coarse materials to interpretation of the rates of colluvial deposition and storage, the effects of particle-size sorting would require further consideration. This can be illustrated by the relationship between the ^{137}Cs content and particle-size characteristics of 14 sample sites in field 2 at Dalicott. A morphological map of the field is presented in Figure 16.5, indicating the ^{137}Cs inventories of the samples collected. A Mann–Whitney U test demonstrates that the ^{137}Cs inventories of soil cores in the slope and ridge units are significantly different from those in the swale unit ($\alpha = 0.05$). As the accumulation of ^{137}Cs at the base of the swale is associated with the deposition of predominantly coarse material, ^{137}Cs inventory was found to increase with percentage sand content and decline with percentage clay content. Erosional sites appear to lose particles of all sizes, thus maintaining a relatively high clay content but low ^{137}Cs inventory. This implies that, although high levels of ^{137}Cs do occur at the base of the swale, indicative of aggregation, they may underestimate the actual rate of deposition, as ^{137}Cs-enriched fines are transported beyond field boundaries. It should also be noted that the processes of detachment and entrainment of soil particles on slopes are transport-limited, such that slope sites experiencing net soil loss will also experience some redeposition of material from upslope. Any particle-size sorting between material arriving from upslope and material being transported downslope may lead to relative ^{137}Cs enrichment, which will have some impact on the quantification of erosion rates. Clearly further data are required so that procedures to account for ^{137}Cs concentration distribution can be incorporated into quantification methods.

REDUCTION OF UNCERTAINTY

The variation in the ^{137}Cs inventories of a series of samples may be due to the non-uniform deposition of baseline fallout to the area, to uneven uptake of ^{137}Cs at the soil surface, to microscale redistribution of topsoil, to redistribution of ^{137}Cs in solution prior to adsorption, and to analytical errors associated with the collection, preparation and counting of samples, in addition to the variability attributed to ^{137}Cs redistribution in association with erosion processes. A certain degree of uncertainty therefore surrounds the extent to which the ^{137}Cs measurement represents actual soil redistribution. Many research papers have simply assumed the other sources of variability to be negligible, but more recently attention has been paid to the implications of inherent variability and strategies to overcome it.

Very few papers have attempted to determine the relative importance of different causes of variability at specific sites, in order to examine the suitability of the ^{137}Cs technique on particular soil or ground types, climatic conditions or management

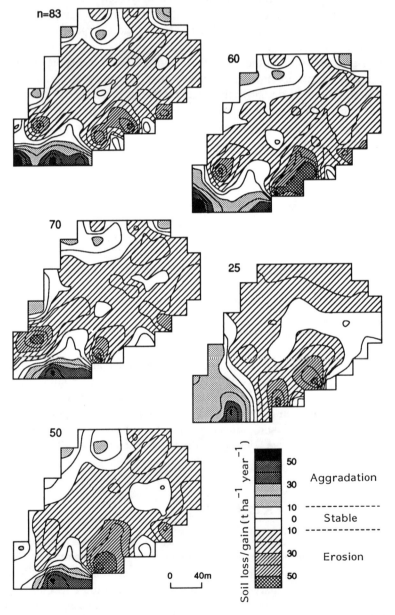

Figure 16.6 Interpolated maps of intra-field erosion, field 1, Dalicott, Shropshire, for different sample sizes

practices. Sutherland (1991) undertook a specific study of ^{137}Cs variability in control sites in order to estimate the number of samples required to estimate the baseline inventory. Adopting an allowable error of 10% at 95% confidence, estimation of the population mean requires four samples when the coefficient of variation (*CV*) is 10%, 16 samples where *CV* is 20%, and 35 samples where *CV* is 30%. In Britain, coefficients of variation are typically <15% (Higgitt, 1990). Specific studies of ^{137}Cs variability in non-eroding sites report *CV* values of 20% in Germany (Bachhuber et al., 1987), 20% on grassland in Oklahoma (Lance et al., 1986), 34% and 54% for forest and pasture sites, respectively, in New South Wales (Fredericks et al., 1988), and from 21% to 64% for pasture sites in Alberta (R.W. Howitt, personal communication). Sutherland (1991) goes on to identify a number of studies where too few reference samples were undertaken to estimate the baseline fallout for the criteria above.

In many cases the number of samples required to satisfy these constraints would far exceed the availability of detector time. If the ^{137}Cs technique is to be used in a consultancy capacity to supply data on erosion rates to an end-user, the uncertainty surrounding the estimates is an important consideration. Here it is important to consider the specific requirements of that end-user. If detector time is restricted, it is possible to composite a large number of soil cores to provide an accurate average ^{137}Cs assessment for those sites. In many cases, interest will be focused on assessing point erosion rates and identifying the relative importance of potential source areas. Strategies available include stratifying the sampling framework according to morphological or land-use units. An example of the influence of sample size is provided by a series of maps (Figure 16.6) derived from a network of soil cores collected from field 1 at Dalicott. The complete 20 × 20 m grid comprises 83 samples and the pattern of soil redistribution has been interpolated using the package UNIRAS. A random-number generator was used to identify samples, which were successively withdrawn from the

Table 16.1 Estimated erosion rates for field 1, Dalicott, Shropshire, based on different sample sizes

	Sample size				
Data source	83	70	60	50	25
^{137}Cs data					
Mean	11.50	11.20	12.07	10.46	10.83
Standard error	2.38	2.45	2.81	2.98	4.94
Median	13.56	12.88	13.54	12.66	13.20
Upper quartile	19.96	20.31	20.68	18.39	21.15
Lower quartile	3.75	2.04	2.71	3.40	3.26
Interpolated map					
Aggradation (%)	18.34	21.13	19.76	18.10	18.06
Erosion (%)	81.66	78.87	80.24	81.90	81.94
Gross erosion (t ha^{-1} yr^{-1})	13.86	13.49	14.09	12.89	13.83
Deposition (t ha^{-1} yr^{-1})	3.26	2.92	3.03	2.56	2.86
Net erosion (t ha^{-1} yr^{-1})	10.59	10.57	11.06	10.33	10.98
SDR (%)	76.4	78.4	78.5	80.1	79.4
95% CI for error of erosion estimates	7.35	8.00	8.64	9.47	13.39

grid. The pattern of soil distribution was mapped for sample sizes of 70, 60, 50 and 25. The interpolated maps maintain a broadly similar pattern until the sample size is reduced to 25, where a large portion of the left side is assumed to be depositional. However, the summary statistics for the field remain quite similar at all sample sizes (Table 16.1). In many applications, ^{137}Cs measurements may be required to estimate average gross or net erosion rates for particular sites or soil types, and hence consideration of the minimum sampling effort required to achieve that goal within given confidence limits is an important consideration. In this situation, a reliable and robust method of quantifying soil loss from ^{137}Cs inventory is necessary.

QUANTIFICATION OF EROSION RATES

A network of ^{137}Cs measurements permits rapid appraisal of the relative pattern of soil redistribution, but their translation into estimates of actual soil loss (or gain) is desirable. A detailed discussion of the issues involved is given by Quine (this volume). Suffice it to say that, after 20 years of attempts to derive erosion estimates from ^{137}Cs inventories, no single method has gained universal approval.

The evolutionary sequence of quantification methodology has led from empirical regression equations to proportional models to mass-balance models. Regression models are generally unsatisfactory because the relationship between annual soil loss and cumulative ^{137}Cs loss is time-dependent and site-specific (Kachanoski and de Jong, 1984). The mass-balance approach avoids this problem by calculating the temporal relationship between the change in ^{137}Cs inventory at a given site and the average annual erosion rate, taking into account atmospheric fallout deposition, radioactive decay, the initial vertical profile of ^{137}Cs within the soil profile and any subsequent mixing by tillage operations. Such a procedure can also be used to simulate the ambient ^{137}Cs concentration of eroded material over time, which can then be employed to interpret depositional profiles (Higgitt and Walling, 1993). Empirical evidence supports the exponential distribution of ^{137}Cs in soil profiles, but no sampling methods have been devised to examine the ^{137}Cs content of the uppermost 0.1 mm. Furthermore, mass-balance models are usually constrained to assume a long-term average rate of erosion. In many investigations, changes in land use have affected erosion rates and hence the nature of ^{137}Cs depletion. If suitable empirical evidence exists, models can be adapted to simulate management scenarios, but this is rarely available.

The difficulties of parametrisation in mass-balance methods compound the problems of uncertainty discussed in the previous section. Confidence in ^{137}Cs-derived estimates is not improved by the diversity of models employed, a diversity that is best summarised by Walling and Quine (1990). Two examples from New South Wales indicate the problem. Loughran et al. (1990a) compared a regression method and proportional method for estimating soil loss at Merriwa, noting that the latter method yielded erosion estimates up to 20 times larger. A later paper examining erosion in potato fields at Dorrigo (Elliott and Cole-Clark, 1993) discriminated between the two approaches for different land uses, in order to make a statement about the severity of erosion on arable land. The danger is that comparison of ^{137}Cs-derived estimates of erosion rates might indicate as much a variation in the suitability of quantification methods employed as the actual dynamics of the erosion processes. A novel approach to evaluating the performance of ^{137}Cs as an indicator of cumulative erosion is provided by resampling locations after fixed intervals (Kachanoski, 1987, 1993).

CHERNOBYL: SEQUEL OR SPOILER?

In 1986, the Chernobyl nuclear power plant accident resulted in an additional 70 PBq of ^{137}Cs being released to the atmosphere (Cambray et al., 1989). The fallout was confined to the Northern Hemisphere and affected parts of Europe in particular. The pattern of Chernobyl fallout across Europe is complex, reflecting the passage of the plume and its interaction with rainfall. As much of the transfer was accomplished by individual rain events, the baseline deposition is highly variable.

In many ways the reaction of the research community to Chernobyl was similar to research following 'atom bomb' test fallout, except that the volume and rapidity of output was much enhanced. There were numerous reports of inventories and variability of Chernobyl fallout soon after the accident (e.g. Hohenemser et al., 1986; Frissel et al., 1987), which have led to a renewed interest in the environmental behaviour of ^{137}Cs in soils and vegetation (e.g. Livens and Rimmer 1988; Kirk and Saunton, 1989; Bunzl et al., 1992). Those working in the erosion–sedimentation field initially considered the additional opportunities afforded by Chernobyl fallout. A number of papers have attempted to demonstrate its potential (Vanden Berghe and Gulinck, 1987; Bonnett et al., 1989; Branca and Voltaggio, 1993). However, the high variability of fallout deposition has restricted the ability of ^{137}Cs measurements to provide intricate patterns of soil loss (de Roo, 1991; Higgitt et al., 1992, 1993), and Walling and Quine (1991) have suggested that the technique should be used with caution in areas where the Chernobyl-derived ^{137}Cs input exceeds 25% of the total inventory.

In general, it appears that the addition of Chernobyl fallout has not been conducive to the application of ^{137}Cs measurements in erosion investigations. By contrast, studies have demonstrated the opportunity to exploit the Chernobyl input as a second dating marker in depositional chronologies (Rowan et al., 1993), and the increased interest in the environmental behaviour a caesium provides a basis for re-evaluating the assumptions of the ^{137}Cs technique.

THE FUTURE

It is clear that the logistic advantages of the ^{137}Cs technique, which permit the rapid appraisal of rates and patterns of erosion, are such that it will continue to elucidate aspects of erosion and sedimentation problems. To date, most of the researchers employing ^{137}Cs measurements have reported on their interpretations of erosion and sediment delivery at a particular site. Few have embarked on fieldwork with the specific objective of rigorously testing one or more of the assumptions of the technique. In many cases, such as the examples presented here, preliminary data to evaluate assumptions have only been collected as an aside to the geomorphological objectives of a particular project and perhaps fall short of critical evaluation.

This output of ^{137}Cs research from the University of Exeter, pioneered by Des Walling, has helped to explore the complexities of sediment delivery systems. While future applications will continue to test the potential of the technique in new situations and environments, further contemplation of the methodological questions outlined in this chapter would seem appropriate.

ACKNOWLEDGEMENTS

Much of the work reported in this chapter was undertaken as part of a PhD study at Exeter University, funded by NERC and supervised by Des Walling. The diagrams were prepared by Matthew Ball and Nicola Shadbolt, Department of Geography, Lancaster University.

REFERENCES

Allison, R.J. (1991) Slopes and slope processes. *Prog. Phys. Geog.* **15**, 423–437.
Amphlett, C.B. and McDonald, L.A. (1956) Equilibrium studies on natural ion exchange materials: 1 Caesium and strontium. *J. Inorg. Nucl. Chem.* **2**, 403–414.
Bachhuber, H., Bunzl, K. and Schimmack, W. (1987) Spatial variability of fallout ^{137}Cs in the soil of a cultivated field. *Environ. Monit. Assess.* **8**, 93–107.
Bonnett, P.J.P., Leeks, G.J.L. and Cambray, R.S. (1989) Transport processes for Chernobyl labelled sediments: preliminary evidence from upland mid-Wales. *Land Degrad. Rehab.* **1**, 39–50.
Branca, M. and Voltaggio, M. (1993) Erosion rates in badlands of central Italy—estimated by radiocaesium isotope ratio from Chernobyl nuclear accident. *Appl. Geochem.* **8**, 437–445.
Brown, R.B., Kling, G.F. and Cutshall, N.H. (1981) Agricultural erosion indicated by ^{137}Cs redistribution: II. Estimates of erosion rates. *J. Soil Sci. Soc. Am.* **45**, 1191–1197.
Bunzl, K., Kracke, W. and Schimmack, W. (1992) Vertical movement of plutonium-239 + 240, americium-241 and caesium-137 fallout in a forest soil under spruce. *The Analyst* **117**, 469–474.
Cambray, R.S., Playford, K., Lewis, G.N.J. and Carpenter, R.C. (1989) *Radioactive Fallout in Air and Rain: Results to the End of 1987*, AERE Report R13226, UK Atomic Energy Authority, Harwell.
Campbell, B.L. (1983) *Applications of Environmental Caesium—137 for the Determination of Sedimentation Rates in Reservoirs and Related Catchment Studies in Developing Countries*, International Atomic Energy Authority, Vienna, Tech-Doc 291.
Campbell, B.L., Loughran, R.J. and Elliott, G.L. (1982) Caesium-137 as an indicator of geomorphic processes in a drainage basin. *Aust. Geogr. Stud.* **20**, 49–64.
Campbell, B.L., Loughran, R.J., Elliott, G.L and Shelly, D.J. (1986) Mapping drainage basin sediment sources using caesium-137. In: *Drainage Basin Sediment Delivery (Proceedings of Alberqueque Symp.)*, IAHS Publication no. 159, pp. 437–446.
Cawse, P.A. and Horrill, A.D. (1986) *A survey of Cs-137 and Plutonium in British Soils in 1977*, AERE Report R10155, UK Atomic Energy Authority, Harwell.
Charman, P.E.V. (1988) Soil conservation research—past, present and future. *J. Soil Conserv. NSW* **44**, 28–41.
Cooper, J.R., Gilliam, J.W., Daniels, R.B. and Robarge, W.P. (1987) Riparian areas as filters for agricultural sediment. *J. Soil Sci. Soc. Am.* **51**, 416–420.
de Jong, E., Villar, H. and Bettany, J.R. (1982) Preliminary investigations on the use of ^{137}Cs to estimate erosion in Saskatchewan. *Can. J. Soil Sci.* **62**, 673–683.
de Roo, A.P.J. (1991) The use of ^{137}Cs in an erosion study in south Limburg (The Netherlands) and the influence of Chernobyl fallout. *Hydrol. Processes* **5**, 215–227.
de Roo, A.P.J., Hazelhoff, L. and Burrough, P.A. (1989) Soil erosion modelling using 'ANSWERS' and geographical information systems. *Earth Surf. Processes Landforms* **14**, 517–532.
Elliott, G.L. and Cole-Clark, B.E. (1993) Estimates of erosion on potato lands on krasnozems at

Dorrigo, N.S.W., using the caesium-137 technique. *Australian Journal of Soil Research* **31**, 209–233.
Fredericks, D.J., Norris, V. and Perrens, S.J. (1988) Estimating erosion using caesium-137: I. Measuring caesium-137 activity in a soil. In: *Sediment Budgets (Proceedings of Porto Alegre Symp.)*, IAHS Publication no. 174, pp. 225–231.
Frissel, M.J. and Pennders, R. (1983) Models for the accumulation and migration of ^{90}Sr, ^{137}Cs, ^{239,240}Pu, and ^{241}Am in the upper layers of soils. In: Coughtrey, P.J., Bell, J.N.B. and Roberts, T.M. (eds), *Ecological Aspects of Radionuclide Release*, Blackwell, Oxford, pp. 63–72.
Frissel, M.J., Stoutjesdijk, J.F., Koolwijk, A.C. and Koster, H.W. (1987) The Cs-137 contamination of soils in the Netherlands and its consequences for the contamination of crop products. *Neth. J. Agric. Sci.* **35**, 339–346.
Garg, P.K. and Harrison, A.R. (1992) Land degradation and erosion risk analysis in south-east Spain: a geographical information system approach. *Catena* **19**, 411–425.
Graham, E.R. (1963) Factors affecting Sr-85 and I-131 removed by runoff water. *Water Sewage Works* **11**, 407–410.
Graham, O.P. (1992) Survey of land degradation in New South Wales, Australia. *Environ. Manage.* **16**, 205–233.
Haigh, M.J. (1979) Ground retreat and slope evolution on regraded surface mine dumps, Waunafon, Gwent. *Earth Surf. Processes Landforms* **4**, 183–189.
Higgitt, D.L. (1990) The use of caesium-137 measurements in erosion investigations. Unpublished PhD thesis, University of Exeter.
Higgitt, D.L. (1993) Soil erosion and soil problems. *Prog. Phys. Geogr.* **17**, 461–472.
Higgitt, D.L. and Walling, D.E. (1993) The value of caesium-137 measurements for estimating soil erosion and sediment delivery in an agricultural catchment, Avon, UK. In: Wicherek, S. (ed.), *Farm Land Erosion: In Temperate Plains Environment and Hills*, Elsevier, Amsterdam, pp. 301–315.
Higgitt, D.L., Froehlich, W. and Walling, D.E. (1992) Applications and limitations of Chernobyl radiocaesium measurements in a Carpathian erosion investigation, Poland. *Land Degrad. Rehab.* **3**, 15–26.
Higgitt, D.L., Rowan, J.S. and Walling, D.E. (1993) Catchment scale deposition and redistribution of Chernobyl radiocaesium in upland Britain. *Environ. Int.* **19**, 155–166.
Higgitt, D.L., Walling, D.E. and Haigh, M.J. (1994) Estimating rates of ground retreat on mining spoils using caesium-137. *Appl. Geogr.* **14**, 294–307.
Hohenemser, C., Deicher, M., Hofsass, H. and Lindner, G. (1986) Agricultural impact of Chernobyl: a warning. *Nature* **321**, 817.
Johnston, C.A., Bubenzer, G.D., Lee, G.B., Madison, F.W. and McHenry, J.R. (1984) Nutrient trapping by sediment deposition in a seasonally flooded lakeside wetland. *J. Environ. Qual.* **13**, 283–290.
Kachanoski, R.G. (1987) Comparison of measured soil 137-cesium losses and erosion rates. *Can. J. Soil Sci.* **67**, 199–203.
Kachanoski, R.G. (1993) Estimating soil loss from changes in soil cesium-137. *Can. J. Soil Sci.* **73**, 629–632.
Kachanoski, R.G. and de Jong, E. (1984) Predicting the temporal relationship between soil cesium-137 and erosion rate. *J. Environ. Qual.* **13**, 301–304.
Kirk, G.J.D. and Staunton. S. (1989) On predicting the fate of radioactive caesium in soil beneath grassland. *J. Soil Sci.* **40**, 71–84.
Kulander, L. and Strömquist, L. (1989) Exploring the use of top-soil ^{137}Cs content as an indicator of sediment transfer rates in a small Lesotho catchment. *Z. Geomorphol.* **33**, 455–462.
Lance, J.C., Mcintyre, S.C., Naney, J.W. and Rousseva, S.S. (1986) Measuring sediment movement at low erosion rates using cesium-137. *J. Soil Sci. Soc. Am.* **50**, 1303–1309.
Livens, F.R. and Rimmer, D.L. (1988) Physico-chemical controls on artificial radionuclides in soil. *Soil Use Manage.* **4**, 63–69.
Longmore, M.E., O'Leary, B.W., Rose, C.W. and Chindica, A.L. (1983) Mapping soil erosion and accumulation with the fallout isotope caesium-137. *Aust. J. Soil Res.* **21**, 373–385.
Longmore, M.E., Luly, J.G. and O'Leary, B.W. (1986) Caesium-137 redistribution in the

sediments of the playa, Lake Tyrell, northwestern Victoria. II. Patterns of caesium-137 and pollen redistribution. *Palaeogeog. Palaeoclimatol. Palaeoecol.* **54**, 197–218.

Loughran, R.J. (1989) The measurement of soil erosion. *Prog. Phys. Geogr.* **13**, 216–233.

Loughran, R.J., Campbell, B.L. and Elliott, G.L. (1986) Sediment dynamics in a partially cultivated catchment in New South Wales, Australia. *J. Hydrol.* **83**, 285–297.

Loughran, R.J., Campbell, B.L. and Walling, D.E. (1987) Soil erosion and sedimentation indicated by caesium-137: Jackmoor Brook catchment, south Devon, England. *Catena* **14**, 201–212.

Loughran, R.J., Campbell, B.L. and Elliott, G.L. (1990a) The calculation of net soil loss using caesium-137. In: Boardman, J., Foster, l.D.L. and Dearing, J.A. (eds), *Soil Erosion on Agricultural Land*, John Wiley & Sons, Chichester, pp. 119–126.

Loughran, R.J., Campbell, B.L., Elliott, G.L. and Shelly, D.J. (1990b) Determination of the rate of sheet erosion on grazing land using caesium-137. *Appl. Geogr.* **10**, 125–133.

McCallan, M.E., O'Leary, B.M. and Rose, C.W. (1980) Redistribution of caesium-137 by erosion and deposition in an Australian soil. *Aust. J. Soil Res.* **18**, 119–128.

McFarlane, D.J., Loughran, R.J. and Campbell, B.L. (1992) Soil erosion of agricultural land in Western Australia estimated by caesium-137. *Aust. J. Soil Res.* **30**, 533–546.

McHenry, J.R. (1985) Quantification of soil erosion and deposition—the future. USA experience and its relevance to world needs. In: *Drainage Basin Erosion and Sedimentation (Conference and Review Papers)*, University of Newcastle, NSW, vol. 2, pp. 33–41.

McHenry, J.R. and Ritchie, J.C. (1977) Estimating field erosion losses from fallout cesium-137 measurements. In: *Erosion and Solid Matter Transport in Inland Waters (Proc. Paris Symp.)*, IAHS Publication no. 122, pp. 26–33.

McHenry, J.R., Ritchie, J.C. and Gill, A.C. (1973) Accumulation of fallout cesium-137 in soils and sediments in selected catchments. *Water Resour. Res.* **9**, 676–686.

McHenry, J.R., Ritchie, J.C. and Bubenzer, G.D. (1978) Redistribution of cesium-137 due to erosional processes in a Wisconsin watershed. In: Adriano, D.C. and Brisbin, I.L., Jr, (eds), *Environmental Chemistry and Cycling Processes*, DOE CONF-760429, Washington DC, pp. 495–503.

Mcintyre, S.C., Lance, J.C., Campbell, B.L. and Miller, R.L. (1987) Using cesium-137 to estimate soil erosion on a clearcut hillside. *J. Soil Water Conserv.* **42**, 117–120.

Martz, L.W. and de Jong, E. (1987) Using cesium-137 to assess the variability of net soil erosion and its association with topography in a Canadian Prairie landscape. *Catena* **14**, 439–451.

Menzel, R.G. (1960) Transport of strontium-90 in runoff. *Science* **131**, 499–500.

Menzel, R.G., Jung, P-K., Ryu, K,S. and Um, K,T. (1987) Estimating soil erosion losses in Korea with fallout cesium-137. *J. Appl. Radiat. Isotopes* **38**, 451–454.

Mitchell, J.K., Bubenzer, G.D., McHenry, J.R. and Ritchie, J.C. (1980) Soil loss estimation from cesium-137 measurements. In: de Boodt, M. and Gabriels, D. (eds), *Assessment of Erosion*, John Wiley & Sons, Chichester, pp. 393–401.

Peart, M.R. and Walling, D.E. (1986) Fingerprinting sediment sources: the example of a small drainage basin in Devon, UK. In: *Drainage Basin Sediment Delivery (Proc. Alberqueque Symp.)*, IAHS Publication no. 159, pp. 41–55.

Pennock, D.J. and de Jong, E. (1987) The influence of slope curvature on soil erosion and deposition in hummock terrain. *Soil Sci.* **144**, 209–217.

Quine, T.A. and Walling, D.E. (1991) Rates of soil erosion on arable fields in Britain: quantitative data from caesium-137 measurements. *Soil Use Manage.* **7**, 169–176.

Ritchie, J.C. and McHenry, J.R. (1990) Application of radionuclide fallout cesium-137 for measuring soil erosion and sediment accumulation rates and patterns. *J. Environ. Qual.* **19**, 215–233.

Ritchie, J.C., Clebsch, E.E.C. and Rudolph, W.K. (1970) Distribution of fallout and natural gamma radionuclides in litter, humus and surface soil layers under natural vegetation in the Great Smoky Mountains, North Carolina–Tennessee. *Health Phys.* **18**, 479–489.

Ritchie, J.C., McHenry, J.R. and Gill, A.C. (1972) The distribution of ^{137}Cs in the litter and upper 10 cm of soil under different cover types in northern Mississippi. *Health Phys.* **22**, 197–198.

Ritchie, J.C., Spraberry, J.C. and McHenry, J.R. (1974) Estimating soil erosion from the redistribution of fallout ^{137}Cs. *Proc. Soil Sci. Am.* **38**, 137–139.

Ritchie, J.C., McHenry, J.R. and Bubenzer, G.D. (1982) Redistribution of fallout ^{137}Cs in Brunner Creek watershed in Wisconsin. *Wisc. Acad. Sci. Arts Lett.* **70**, 161–166.

Ritchie, J.C., Jackson, T.J., Everitt, J.H., Escobar, D.E., Murphey, J.B. and Grissinger, E.H. (1992) Airborne laser: a tool to study landscape surface features. *J. Soil Water Conserv.* **47**, 104–107.

Rogowski, A.S. and Tamum, T. (1970a) Environmental mobility of cesium-137. *Radiat. Bot.* **10**, 34–45.

Rogowski, A.S. and Tamura, T. (1970b) Erosional behaviour of Cesium-137. *Health Phys.* **18**, 467–477.

Rowan, J.S., Higgitt, D.L. and Walling, D.E. (1993) The incorporation of Chernobyl-derived radiocaesium into reservoir sedimentary profiles. In: Duck, R.W. and McManus, J. (eds), *Geomorphology and Sedimentology of Lakes and Reservoirs*, John Wiley & Sons, Chichester, pp. 55–71.

Schultz, R.K., Overstreet R. and Barshad, I. (1960) On the soil chemistry of Cs-137. *Soil Sci.* **89**, 16–27.

Sutherland, R.A. (1991) Examination of caesium-137 areal activities in control (uneroded) locations. *Soil Technol.* **4**, 33–50.

Tamura, T. (1964) Selective sorption reactions of cesium with soil minerals. *Nucl. Safety* **5**, 262–268.

Vanden Berghe, I. and Gulinck, H. (1987) Fallout ^{137}Cs as a tracer for soil mobility in the landscape framework of the Belgian loamy region. *Pédologie* **37**, 5–20.

Walling, D.E. (1982) Physical hydrology. *Prog. Phys. Geog.* **6**, 122–133.

Walling, D.E. (1983) The sediment delivery problem. *J. Hydrol.* **65**, 209–237.

Walling, D.E. (1988) Erosion and sediment yield research—some recent perspectives. *J. Hydrol.* **199**, 113–141.

Walling, D.E. and Quine, T.A. (1990) Calibration of caesium-137 measurements to provide quantitative erosion rate data. *Land Degrad. Rehab.* **2**. 161–175.

Walling, D.E. and Quine, T.A. (1991) The use of ^{137}Cs measurements to investigate soil erosion on arable fields in the UK: potential applications and limitations. *J. Soil Sci.* **42**, 147–165.

Walling, D.E. and Quine, T.A. (1992) The use of caesium-137 measurements in soil erosion surveys. In: *Erosion and Sediment Transport Monitoring Programmes in River Basins (Proc. Oslo Symp.)*, IAHS Publication no. 210, pp. 143–152.

Walling, D.E., Bradley, S.B. and Lambert, C.P. (1986a) Conveyance losses of suspended sediment within a floodplain system. In: *Drainage Basin Sediment Delivery (Proc. Alberqueque Symp.)*, IAHS Publication no. 159, pp. 119–131.

Walling, D.E., Bradley, S.B. and Williams, C.J. (1986b) A caesium-137 budget approach to the investigation of sediment delivery from a small agricultural drainage basin in Devon, UK. In: *Drainage Basin Sediment Delivery (Proc. Alberqueque Symp.)*, IAHS Publication no. 159, pp. 423–435.

Wilkin, D.C. and Hebel, S.J. (1982) Erosion, deposition and delivery of sediment to midwestern streams. *Water Resour. Res.* **18**, 1278–1282.

Wise, S.M. (1977) The Use of Fallout Radionuclides Pb-210 and Cs-137 in Estimating Denudation Rates and in Soil Erosion Measurements, Department of Geography, King's College London, Occasional Paper, no. 7.

Wise, S.M. (1980) Caesium-137 and lead-210: a review of the techniques and some applications in geomorphology. In: Cullingford, R.A., Davidson, D.A. and Lewin, J. (eds), *Timescales in Geomorphology*, John Wiley & Sons, Chichester, pp. 109–127.

Wolman, M.G. (1977) Changing needs and opportunities in the sediment field. *Water Resour. Res.* **13**, 50–54.

Zhang, X., Li, S., Wang, C., Tan, W., Zhao, Q., Zhang, Y., Yan, M., Liu, Y., Jiang, J., Xiao, J. and Zhou, J. (1989) Use of caesium-137 measurements to investigate erosion and sediment sources within a small drainage basin in the Loess Plateau of China. *Hydrol. Processes* **3**, 317–323.

Zhang, X., Higgitt, D.L. and Walling, D.E. (1990) A preliminary assessment of the potential for using caesium-137 to estimate rates of soil erosion in the Loess Plateau of China. *Hydrol. Sci. J.* **35**, 243–252.

17 Estimation of Erosion Rates from Caesium-137 Data: The Calibration Question

TIMOTHY A. QUINE
Geography Department, University of Exeter, UK

INTRODUCTION

Caesium-137 (^{137}Cs) has been employed in erosion investigations in environments ranging from temperate (cf. Martz and de Jong, 1987, 1991; Sutherland, 1991a; Walling and Quine, 1990a, 1991; Froehlich and Walling, 1992) through semi-arid (cf. Loughran *et al.*, 1981, 1988; Campbell *et al.*, 1982, 1986; Navas and Walling, 1992; Quine *et al.*, 1993a, 1994a; Zhang *et al.*, 1994) to humid subtropical (cf. Menzel *et al.*, 1987; Quine *et al.*, 1992) during more than two decades since the pioneering work of Rogowski and Tamura (1965, 1970) and the early studies by Ritchie and McHenry (1973, 1975). The wide range of studies has revealed the very considerable potential of the technique and the advantages over more conventional approaches to erosion investigation (Loughran, 1989; Walling and Quine, 1990a). The basis of the technique is now well documented (cf. Walling and Quine, 1991) and recent studies have examined some of the fundamental assumptions in greater detail (Sutherland, 1991b; Walling and Quine, 1992). Despite this evidence of the growing maturity of the approach (Higgitt, this volume), there remains uncertainty concerning one key area, namely the estimation of quantitative soil erosion rates from ^{137}Cs data, often referred to as calibration, and Walling and Quine (1990b) have recently examined the limitations and inconsistencies of some of the existing approaches. This chapter will examine the factors that must be taken into account if reliable estimates of erosion rates are to be derived from ^{137}Cs data from cultivated land, and it indicates that an adaptable approach to calibration is needed that incorporates the identified factors.

THE CALIBRATION PROBLEM

The context of calibration

Detailed description of the ^{137}Cs technique is beyond the scope of this chapter but an outline of the application of the technique provides a valuable context for the discussion of calibration. Typically, seven stages are involved in the application of the

Sediment and Water Quality in River Catchments. Edited by I.D.L. Foster, A.M. Gurnell and B.W. Webb.
© 1995 John Wiley & Sons Ltd.

^{137}Cs technique:

1. Collection of soil core samples from the study field and from undisturbed 'reference' sites in the surrounding area.
2. Measurement of the total remaining ^{137}Cs fallout input (per unit area) to the location by analysis of soil samples from the 'reference' sites.
3. Measurement of the ^{137}Cs inventory (activity per unit area) for each location sampled in the study field.
4. Identification of ^{137}Cs loss and gain for each sampled location in the study field by comparison of the measurements from the reference sites with those from the study field.
5. Development of site-specific 'calibration' relationships between ^{137}Cs loss and gain, and soil erosion and deposition rates.
6. Use of the calibration relationships from stages with the ^{137}Cs loss and gain data from stage 4 to estimate the rate of soil erosion or deposition for each location in the study field.
7. Integration of the point erosion and deposition data to obtain field-based estimates of soil erosion and deposition.

Clearly, the development of reliable calibration relationships is fundamental if two of the main advantages of the approach are to be realised, viz. the retrospective assessment of medium-term average erosion rates for individual fields, and the documentation of patterns of erosion and deposition rates across the study fields. Furthermore, the full potential of the technique will only be realised when there is general acceptance that the derived rates are reliable. This has not always been the case, despite a growing body of evidence in support of ^{137}Cs derived erosion rate estimates (Quine and Walling, 1993; Quine et al., 1994b). It is, therefore, important both to develop optimum procedures for calibration and to be explicit about the approach employed.

Approaches to calibration

Although the simplest approach to calibration would be the application of an empirically derived relationship between soil loss and ^{137}Cs loss (Ritchie and McHenry, 1975; Campbell et al., 1986; Loughran et al., 1988), this is rarely, if ever, practicable, for two reasons. First, this approach demands collection of samples for ^{137}Cs analysis from sites with detailed quantitative erosion records extending back to 1953, and such sites are few and far between. Secondly, and more fundamentally, any relationship established for the control site would only be applicable at sites with identical land-management strategies and for sites sampled at the same time. Calibration is, therefore, more often undertaken using a theoretically derived relationship, and two approaches have been used, namely, the directly proportional method and the mass-balance model.

The directly proportional relationship may be summarised as follows:

$$E = M\left(\frac{C_i - C_r}{C_r}\right) \qquad (17.1)$$

where E = total soil loss per unit area since initial ^{137}Cs fallout, M = mass of plough layer per unit area, C_i = ^{137}Cs inventory at sample point and C_r = ^{137}Cs input (inventory at reference site). This relationship, or slight variations on it, has been widely applied in erosion studies using ^{137}Cs (cf. Mitchell et al., 1980; de Jong et al., 1983; Martz and de Jong, 1987; Kachanoski, 1987; Fredericks and Perrens, 1988). However, despite the attraction of its simplicity, the method is seriously flawed because it is dependent on the assumption that all ^{137}Cs received by the soil profile is mixed evenly through the plough layer before being subject to erosion. It therefore fails to take account of potential loss of soil-associated ^{137}Cs from the soil surface, prior to mixing through the plough layer, during the period of atmospheric fallout. Where surface processes (water and wind erosion) play an important part in soil redistribution, the 'directly-proportional method' will provide overestimates of erosion rates.

A more adaptable approach is the use of a mass-balance model to simulate ^{137}Cs loss and gain with soil erosion and deposition. In its simplest form the general model may be expressed as follows:

$$C_t = (C_{t-1} + C_f - E_t C_e)D \qquad (17.2)$$

where C_t = ^{137}Cs inventory at time t (Bq m^{-2}), C_f = input of fallout ^{137}Cs between time $t-1$ and time t (Bq m^{-2}), C_e = ^{137}Cs content of soil eroded between time $t-1$ and time t (Bq kg^{-1}), E_t = erosion rate between time $t-1$ and time t (kg m^{-2}) and D = radioactive decay constant. This approach was pioneered by Kachanoski and de Jong (1984), and a number of versions of varying complexity have been developed (cf. Fredericks and Perrens, 1988; Quine, 1989; Quine et al., 1993b). The greatest uncertainty exists regarding the simulation of the ^{137}Cs content of the eroded soil (C_e), with some authors stressing particle-size enrichment and others the nature of the depth distribution and no general agreement has been achieved. If this approach to calibration is to be pursued, there is a need for a comprehensive and adaptable approach to the simulation of ^{137}Cs accumulation and depletion in the soil profile that accounts for the controls on the ^{137}Cs content of eroded and aggraded soil, and which can incorporate new evidence as it becomes available. The following section, therefore, examines the accumulation and depletion of ^{137}Cs in the soil profile and the controls on the content of eroded and aggraded sediment.

CAESIUM-137 ACCUMULATION AND DEPLETION IN THE SOIL PROFILE

Figure 17.1a illustrates the major processes of accumulation, depletion and distribution of ^{137}Cs in a cultivated soil profile, and Figure 17.1b summarises the controls on the ^{137}Cs content of eroded sediment. The following discussion will focus on the former processes and indicate their significance for the determination of the ^{137}Cs content of eroded sediment. Reference will also be made to the inclusion of the processes and factors in existing calibration procedures.

Atmospheric fallout and radioactive decay

Caesium–137 is a radioactive isotope with a half-life of 30.17 years and is present in the environment only as a result of human activity. Atmospheric fallout and

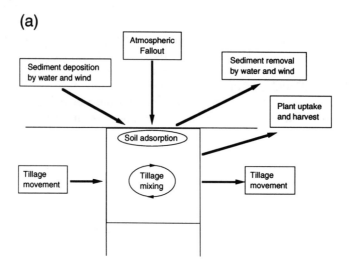

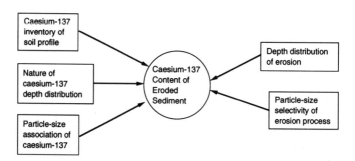

Figure 17.1 (a) Summary diagram of the major processes of accumulation, depletion and distribution of ^{137}Cs in a cultivated soil profile; (b) the controls on the ^{137}Cs content of eroded sediment

radioactive decay are, therefore, vital processes that must be included in all calibration simulations.

There have been two main causes of release of ^{137}Cs into the environment, namely the atmospheric testing of nuclear weapons and the accident at the Chernobyl nuclear power station. Although both sources could potentially be valuable in erosion studies (Vanden Berghe and Gulinck, 1987), uncertainties surround the local spatial uniformity of the deposition of 'Chernobyl-derived' ^{137}Cs (Walling and Quine, 1991; de Roo, 1991; Higgitt, this volume). Most erosion studies have, therefore, used 'bomb-derived' ^{137}Cs, and this is the focus of the present discussion.

As Figure 17.2a illustrates, fallout deposition of bomb-derived ^{137}Cs occurred over a period in excess of 20 years and is characterised by marked temporal variability. Consequently, fallout inputs represent a major control on the temporal variation in ^{137}Cs inventories, and Figure 17.2a illustrates the annual variation in ^{137}Cs inventory that would be expected in a stable (no erosion or deposition) soil profile as a result of atmospheric fallout and radioactive decay, based on data from Milford Haven, UK.

Fallout deposition also exhibited marked intra-annual variation, which is equally significant regarding temporal variability in ^{137}Cs inventories (Figure 17.2b) and surface concentrations. Fallout amounts were influenced by the timing of nuclear tests and, therefore, although most fallout occurred in association with precipitation, the intra-annual variation cannot simply be derived from precipitation amounts (Figure 17.3). This makes prediction of the intra-annual pattern problematic, and more information is required from other sites to establish the representativeness of the Milford Haven data used to derive Figure 17.3. Nevertheless, it is important to take account of the variation in fallout inputs because it will have played a major role in determining the ^{137}Cs content of sediment eroded from the soil surface, and this is considered at greater length in the discussion of the seasonal timing of erosion. Although most existing mass-balance models account for annual variation in fallout input, the significance of intra-annual variation is rarely recognised.

Soil adsorption and tillage mixing—implications for depth distributions

The nature of the depth distribution of ^{137}Cs in a cultivated soil profile may be expected to have varied cyclically during the period of fallout deposition. At each tillage event, the radiocaesium deposited at the surface since the preceding tillage event would be mixed through the plough layer (although the degree of homogenisation is uncertain), but in the period between tillage events gradual deposition of ^{137}Cs would have led to surface accumulation of ^{137}Cs as a result of fallout adsorption onto soil particles. However, most mass-balance models (cf. Kachanoski and de Jong, 1984) and the 'directly-proportional' method ignore this cyclical pattern and assume that fallout inputs were instantaneously mixed through the plough layer. Indeed, only Fredericks and Perrens (1988), Quine (1989) and Walling and Quine (1990b) explicitly recognise any surface accumulation between tillage events.

Clearly, the nature of the depth distribution of the initial surface accumulation is important in the prediction of ^{137}Cs content of sediment eroded from the surface. Some indication of the form of the depth distribution may be obtained from laboratory studies of ^{137}Cs behaviour, which indicate that it is rapidly and strongly adsorbed by soil particles, especially the fine fraction (Bachhuber et al., 1982; Frissel and Pennders,

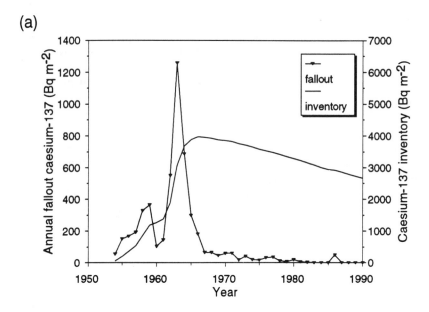

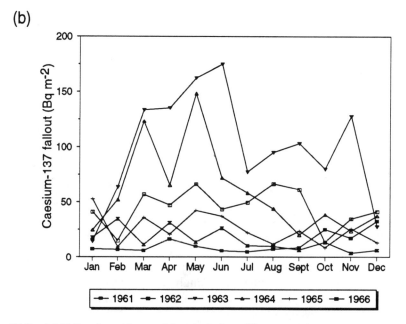

Figure 17.2 (a) Fallut deposition oof bomb-derived ^{137}Cs at Milford Haven (Playford *et al.*, 1993) and the expected resultant annual variation in ^{137}Cs inventory in a stable soil profile; (b) intra-annual variation in fallout inputs at Milford Haven (K. Playford, personal communication)

EROSION RATES FROM ^{137}Cs DATA: CALIBRATION

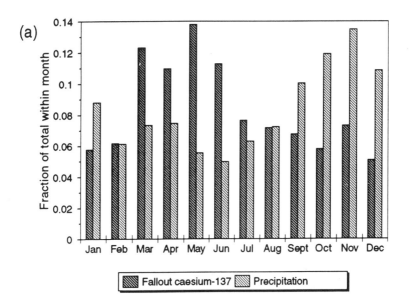

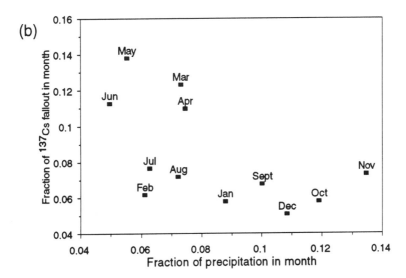

Figure 17.3 The fraction of ^{137}Cs fallout and total precipitation occurring in each month at Milford Haven over the period from 1959 to 1967 (K. Playford, personal communication): (a) the monthly pattern; (b) ^{137}Cs fallout against precipitation

1983; Squire and Middleton, 1966). This strong adsorption is reflected in the current depth distributions from undisturbed sites which show a sharp decline in concentration with increasing depth below the soil surface (cf. Walling and Quine, 1992). Nevertheless, the nature of the depth distribution of ^{137}Cs immediately after fallout remains uncertain because replication of fallout conditions is problematic and sampling of agricultural fields at millimetre precision poses serious problems. However, on the basis of the strong affinity of ^{137}Cs for mineral sediment and the low concentrations in the precipitation (3.4 Bq l^{-1} maximum at Milford Haven), it may be expected that the radiocaesium would concentrate at the surface and that the activity would decrease exponentially with depth. This distribution, which was employed by Walling and Quine (1990b), may be described as follows:

$$C_d = C_f k \, e^{-kd} \quad (17.3)$$

with

$$k = \frac{\ln(10)}{NL} \quad (17.4)$$

where $C_d = {}^{137}$Cs activity at depth d (Bq m^{-2} cm^{-1}), $C_f = {}^{137}$Cs fallout input (Bq m^{-2}), d = depth (cm), k = constant and NL = depth above which 90% of fallout input is found (cm). If this equation is used in conjunction with the monthly data from Milford Haven (K. Playford, personal communication), it is possible to obtain an indication of the likely variation in surface concentrations of ^{137}Cs as a result of the fallout–adsorption–tillage cycle at that location (Figure 17.4). It is clear from Figure 17.4 that simulation of the ^{137}Cs content of sediment eroded from the surface must take into account this probable intra-annual variation in the concentration of ^{137}Cs at the soil surface during the period of fallout deposition. Although it is evident that empirical support and parameterization of Eqn (17.3) are difficult to obtain, the results of field experiments undertaken by members of Walling's research team at Exeter University are consistent

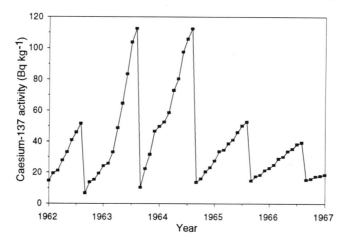

Figure 17.4 The likely variation in ^{137}Cs activity in the upper 0.5 cm of the soil as a result of the fallout–adsorption–tillage cycle from 1961 to 1966 based on the monthly data from Milord Haven (K. Playford, personal communication). Assumptions: plough depth, 0.2 m; ploughing on 1 September; NL = 0.8 cm in Eqn (17.4); bulk density, 1300 kg m^{-3}

with rapid adsorption of ^{137}Cs fallout at the surface and the formation of an initial exponential depth distribution, with a value of NL of the order of 0.8 cm (He, 1993; Owens, 1994).

Work undertaken by Walling and colleagues at Exeter has made an important contribution to the understanding of another component of soil adsorption of ^{137}Cs, namely the role of particle size. Walling and Woodward (1992) separated a surface soil sample into five size fractions, <8, 8–16, 16–32, 32–63, and >63 μm, and measured ^{137}Cs activity of each fraction. Their study confirmed preferential adsorption of ^{137}Cs by the finer fraction but also demonstrated that ^{137}Cs was adsorbed by all of the fractions examined. These data are particularly valuable when considering the impact of size-selective processes on the removal and deposition of ^{137}Cs in association with sediment.

Soil erosion and deposition by water

In order to determine the impact of water erosion on the ^{137}Cs inventory of the soil profile and to calculate the ^{137}Cs content of water-eroded sediment, it is necessary to take into account the following factors, which have been identified in the previous subsections:

- The temporal variation in ^{137}Cs inventories and surface concentrations.
- The non-uniform depth distribution of ^{137}Cs during the period of atmospheric fallout.
- The preferential adsorption of ^{137}Cs by the fine fraction.

On the basis of these factors, it can be suggested that there are four properties of the erosion processes that must be taken into account, as follows:

- Annual variation in erosion.
- The seasonal timing of erosion.
- The depth and spatial extent of erosion.
- The size selectivity of erosion and deposition processes.

The significance of these properties and previous attempts at inclusion in calibration procedures will now be examined in turn.

Annual variation in erosion

Long-term field surveys have provided ample evidence for annual variability in soil erosion rates (Colborne and Staines, 1985; Evans, 1988; Boardman, 1990) and, in view of the marked temporal variability in fallout deposition (Figure 17.2a), these should be recognised in the development of an optimum simulation procedure. However, taking account of annual variation in erosion is problematic because precise records of erosion will rarely be available and few approaches have included this variable. Nevertheless, Fredericks and Perrens (1988) have provided a useful basis for addressing this problem by using an index of erosion based on rainfall erosivity and the crop cover factor (C) from the USLE (Universal Soil Loss Equation). In a different context, Boardman and Favis Mortlock (1993) have found a rainfall index to be useful in predicting inter-annual variation in rates of soil erosion on the South Downs in the

UK. These studies demonstrate that it is feasible to account for annual variation in erosion by using fairly simple indices, although it will be important to identify appropriate local thresholds for the indices selected.

Seasonal timing of erosion

Field surveys have provided valuable evidence concerning seasonal variability of erosion (cf. Boardman, 1990; Vandaele, 1993; Wicherek, 1993), but again this has rarely been accounted for in calibration procedures (Walling and Quine, 1990b). However, the significance of this property of erosion is evident when the marked intra-annual variation in fallout deposition (Figures 17.2b and 17.3) and probable resultant variation in surface concentrations (Figure 17.4) are considered. If no account of the seasonal timing of erosion is made, and it is assumed that the entire annual fallout is subject to erosion prior to mixing through the plough layer, ^{137}Cs loss per unit of soil will be overestimated and erosion rates underestimated. It is, therefore, necessary to identify the period of the year in which most erosion takes place (an 'erosion season') in the locality under investigation, and to combine this with information concerning monthly fallout inputs, to predict the percentage of the annual fallout input which is subject to erosion prior to mixing through the plough layer (referred to in the following text as 'surface-eroded percentage'). Only the fallout that is deposited between tillage mixing and the end of the 'erosion season' will be subject to erosion prior to mixing and, in the case of the South Downs, Boardman and Favis Mortlock (1993) suggest that this period extends from 1 September to 1 March. If it were assumed that fallout deposition showed no intra-annual variation, then the 'surface-eroded percentage' would be 50%; using rainfall as a surrogate for fallout records would give a figure closer to 60%; whereas the monthly fallout data from Milford Haven suggest that the true figure would be closer to 40%. If the 'erosion season' occurred later in the year, then the predictions would be quite different. For example, the 'surface-eroded percentage' for the months February to July would be as follows: 50% with no intra-annual variation; 37% if proportional to rainfall; and 62% based on measured fallout at Milford Haven. The precision of any estimate of the 'surface-eroded percentage' will, clearly, depend on the availability of relevant monthly fallout records and the predictability of erosion in the study area. Nevertheless, the cited examples emphasise the importance of the inclusion of a seasonality parameter.

When considering seasonality of erosion it is also important to recognise that erosion and fallout would have occurred synchronously. As a result, not all of the 'surface-eroded percentage' fallout would be subject to all of the annual erosion (unless all of the erosion occurred in a single event at the end of the 'erosion season'). Therefore, the simulation will be more realistic if it is assumed that a specified number of erosion events took place in each year, interspersed by periods of fallout deposition. While any such measure of the number of erosion events can only be a generalisation, it provides a better approximation of reality than the single-episode approach.

The depth and spatial extent of erosion

Most mass-balance models have taken no account of the variation in spatial extent and depth of erosion by water, which is seen on agricultural fields. In many cases this will not have influenced the results because of the assumption of uniform mixing of ^{137}Cs through the plough layer. However, if the simulation is to be made more realistic, by inclusion of non-uniform depth distributions of ^{137}Cs during fallout deposition, then the depth and spatial extent of erosion must be taken into account. In such circumstances, if the simulation assumes that erosion occurs to a uniform depth across the entire area, ignoring rill incision and potential protected areas, then ^{137}Cs loss per unit of soil will be overestimated and erosion rates underestimated.

In order to account for variability in erosion extent and depth, it is necessary to distinguish between surface (splash and inter-rill) and incised (rill and gully) erosion and to establish a relationship between width and depth of incision. In distinguishing between surface and incised erosion, it is reasonable to assume that, for any one profile, surface processes will dominate at low rates of erosion and incised erosion will increase in importance with increasing rates. It may, therefore, be possible to distinguish between the processes for the purposes of simulation by specifying a maximum rate of surface erosion above which increased rates are attributed to rill and gully processes. Identification of a relationship between width and depth of incision may be more problematic because of the scarcity of relevant data. However, a detailed investigation of rill dimensions on a field near Ansai in Shaanxi Province, which was undertaken with Zhang Xinbao, produced the data illustrated in Figure 17.5. There is a clear need for more information concerning the width–depth relationship, although the data presented in Figure 17.5 provide a useful initial basis for simulation of variation in depth and extent of erosion.

The size selectivity of erosion and deposition processes

The preferential removal of fine sediment by some water erosion processes and deposition of course sediment may be significant because of the well-known affinity of ^{137}Cs for the finer fractions, which was referred to above. This was recognised in the earliest calibration simulation by Kachanoski and de Jong (1984) and subsequently by

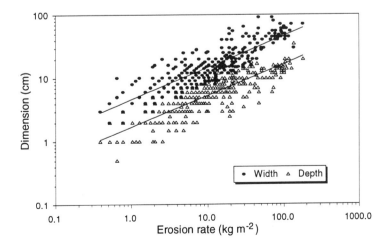

Figure 17.5 The relationship between rill width and depth and erosion rate for a field on the Loess Plateau near Ansai

Fredericks and Perrens (1988). However, their use of a fixed enrichment factor of 1.1 to differentiate between the ^{137}Cs content of eroded sediment and the source material does not take into account the variation in size selectivity of different erosion processes. Although there is evidence for some preferential removal of fine sediment by inter-rill erosion (cf. Gilley and Gee, 1976), rill and gully erosion are generally considered to be aselective with regard to particle size. Any enrichment factor should, therefore, only be used to correct for size selectivity of inter-rill processes, and should be applied in conjunction with the process differentiation referred to in the previous section. The use of a size selectivity enrichment factor for eroded sediment is further limited by the problems associated with the definition of an appropriate value, because particle-size selectivity will vary according to both the nature of the source material and the local slope (which will influence the transport capacity of the inter-rill wash). This is evidently an area where further research is required, and the simulation procedure must be able to accommodate relevant data as they become available.

Particle-size selectivity is potentially more important with regard to deposition because selectivity in deposition may be expected to influence sediment derived from all erosion processes. It is, therefore, necessary to introduce a size selectivity factor (DF) in calculating the ^{137}Cs content of deposited sediment:

$$D_c = E_{cs} DF \tag{17.5}$$

where D_c = ^{137}Cs content of deposited sediment (Bq kg^{-1}), E_{cs} = ^{137}Cs content of eroded sediment (Bq kg^{-1}) and DF = particle-size selectivity factor for deposition. Ideally the value of DF should be based on: (1) comparison of the particle-size distribution of deposited sediment with the mean particle-size distribution of the sediment eroded from the source area; and (2) the relationship between ^{137}Cs content and particle-size characteristics. The former may be addressed by appropriate field sampling of deposited sediment and the source area, but there are few data on which to base the latter. However, the data obtained by Walling and Woodward (1992), in their study of ^{137}Cs content of the particle-size fractions of source sediment from the Jackmoor Brook basin in Devon (UK), have been used to derive example DF values (Figure 17.6). These indicate the range of values of DF that may be expected, but there is clearly a need for further studies to examine the representativeness of the data obtained by Walling and Woodward.

Soil erosion and deposition by wind

The impact of soil erosion and deposition by wind is rarely considered explicitly in calibration procedures. However, where such erosion takes place, it may be an important factor in soil and ^{137}Cs redistribution. For the purposes of simulation, many of the factors that are relevant for water erosion also apply, although the timing of the 'erosion seasons' will differ significantly. Where relevant, the impact of wind erosion as a size-selective, spatially discontinuous surface process should be considered.

Tillage redistribution

The process of tillage leads to differential displacement of the plough soil and there is increasing evidence that this displacement may be as important as, or more important

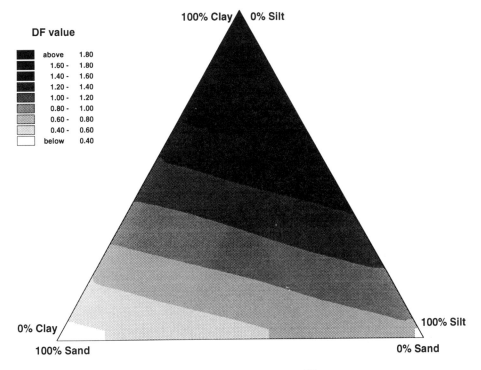

Figure 17.6 The influence of particle-size composition on ^{137}Cs content of sediment based on data from Walling and Woodward (1992): DF values calculated for an eroding source area characterised by sediment with equal contents of clay, silt and sand

than, water erosion processes in determining long-term patterns of soil redistribution on agricultural land (Govers *et al.*, 1993, 1994; Lindstrom *et al.*, 1990, 1992; Quine *et al.*, 1993b, 1994b). Tillage redistribution at each position on a cultivated field is the result of an influx of soil from upslope and an outflux downslope, with the magnitude of the fluxes defined by the slope angle (Lindstrom *et al.*, 1992; Govers *et al.*, 1993, 1994). The relative magnitude of the fluxes will determine the local mass balance: net soil loss will occur where the outflux exceeds the influx and vice versa. For a one-dimensional simulation:

$$T_r = \frac{(T_i - T_o)}{L} \tag{17.6}$$

where T_r = net tillage redistribution rate (kg m^{-2}), T_i = tillage influx (kg m^{-1}), T_o = Tillage outflux (kg m^{-1}) and L = length of slope segment (m).

The change in inventory may, therefore, be assumed to reflect the ^{137}Cs content of the local plough layer and the adjacent upslope plough layer, as follows (assuming slope segments of equal length and width):

$$C_{tc} = \frac{T_i U_c - T_o C_p}{\rho P L} \tag{17.7}$$

where C_{tc} = gain in ^{137}Cs inventory due to tillage redistribution (Bq m^{-2}), U_c = ^{137}Cs inventory of adjacent upslope plough layer (Bq m^{-2}), C_p = ^{137}Cs inventory of profile plough layer (Bq m^{-2}), P = plough depth (m) and ρ = bulk density (kg m^{-3}).

Therefore, optimum simulation of the impact of tillage displacement on ^{137}Cs inventories should take account of the ^{137}Cs content of the plough layer of the upslope adjacent profile (the source of the influx). This is not possible using a point simulation method. However, where change in ^{137}Cs inventories occurs gradually in space (i.e. in large fields with moderate topography), an acceptable approximation may be made using the point method by assuming that the ^{137}Cs inventory of the plough layer of the adjacent profile is the same as that of the simulated profile. The change in inventory may then be calculated using a defined net tillage redistribution rate T_r (kg m^{-2}), where a positive value of T_r indicates aggradation, and a negative value indicates erosion:

$$C_{tc} = \frac{T_r C_p}{\rho P} \qquad (17.8)$$

Although all existing approaches to calibration have taken into account the mixing effect of tillage in redistributing ^{137}Cs through the plough layer, only Quine et al. (1993b) address this redistribution of soil by tillage. However, the importance of including tillage redistribution is evident for two reasons in particular. First, because tillage erosion and deposition result from the differential displacement of the entire plough layer, the relationship between ^{137}Cs and soil loss or gain as a result of tillage will differ significantly from the relationship as a result of surface processes of water and wind erosion. Secondly, the spatial distributions of erosion and deposition as a result of water erosion and tillage differ significantly. Indeed, slope concavities, which are often zones of maximum water erosion, tend to be characterised by tillage deposition (cf. Quine et al., 1994b). An optimum point simulation procedure must, therefore, be able to account for water erosion and tillage redistribution independently and be able to accommodate sites that are subject to erosion by water and aggradation by tillage and vice versa, rather than simply recognising eroding and aggrading sites.

Harvest and leaching losses

The removal of the crop from a cultivated area can lead to ^{137}Cs export by two routes. First, uptake of ^{137}Cs by the plant during the growing season may account for loss of ^{137}Cs of the order of 0.2% per year (Coughtrey and Thorne, 1983). Secondly, ^{137}Cs may be removed from the field in association with soil adhering to the surface of root crops. There is little information available concerning the magnitude of such soil losses, but Vanden Berghe and Gulinck (1987) have suggested that the rate of removal could be as high as 5 t ha^{-1} per crop. For a plough depth of 20 cm this would represent export of 0.2% of the plough layer with each crop, and this is, therefore, of the same order of magnitude as plant uptake. These factors should be taken into account because, although they may be considered to be spatially uniform, they will influence the ^{137}Cs budget and, if not accounted for, may lead to erroneous interpretation of net ^{137}Cs and soil loss when inventories are compared to the reference location. However, when making this comparison, it will also be important to consider any potential ^{137}Cs loss due to plant uptake and removal from the reference location (e.g. by grazing animals or vegetation removal). Both field and laboratory experiments have demonstrated that ^{137}Cs exhibits a very low rate of vertical migration (cf. Bachhuber *et al.*, 1982; Frissel and Pennders, 1983). Leaching losses may, therefore, be considered to be negligible over the period since the initiation of fallout, especially because sampling of the soil profiles is undertaken to depths well beyond the measurable limit of ^{137}Cs (this can be verified by analysis of a subsample from the base of each core).

AN ADAPTABLE APPROACH TO CALIBRATION

On the basis of the preceding discussion, it is evident that there is no existing, generally accepted, satisfactory calibration procedure. The most common limitations occur as a result of failure to account for the following factors:

- The concentration of ^{137}Cs at the soil surface during the period of atmospheric fallout.
- The seasonal variation in erosion and fallout.
- The variation in depth and extent of erosion.
- The redistribution of soil by tillage.
- The potential for both erosion and aggradation to take place at one location.

There is, therefore, a need for an adaptable approach to calibration, which is based on simulation of the identified processes of ^{137}Cs accumulation and depletion in the soil. In addition to comprehensive coverage of the identified factors and processes, the approach must be able to accommodate new information as it becomes available.

An adaptable profile simulation model

In order to address the evident need, a profile simulation model has been developed which, rather than being based a single relationship of the type shown in Eqn. (17.2), consists of a series of modules that are linked in a computer programme. At its simplest the model simulates two profiles in parallel, one of which is subject to erosion by water and one to deposition and each is subject to either erosion or aggradation by tillage (Figure 17.7a). Each module accounts for one aspect of the ^{137}Cs accumulation/depletion cycle, with parameters defined on the basis of existing knowledge (Figure 17.7b). Each simulation is conducted iteratively from the time of the initiation of atmospheric fallout to the time of sampling. The impact of erosion and deposition on the ^{137}Cs inventory of a soil profile can be investigated by running the simulation for a range of erosion and deposition rates. The resultant predicted inventories can then be combined with the specified erosion and deposition rates to produce calibration curves, representing different combinations of tillage redistribution and water erosion (Figure 17.8). Detailed description of the model lies beyond the scope of this chapter, but the associated computer software may be obtained from the author. It is suggested that the model offers significant advantages over existing approaches in addressing the limitations outlined above and accommodating local factors and new information as they become available. The importance of some of these parameters is illustrated in Figure 17.9, which shows the change in the predicted relationship between ^{137}Cs loss and soil loss that occurs when the parameters are included. Furthermore, each of the key parameters relate to properties and processes of

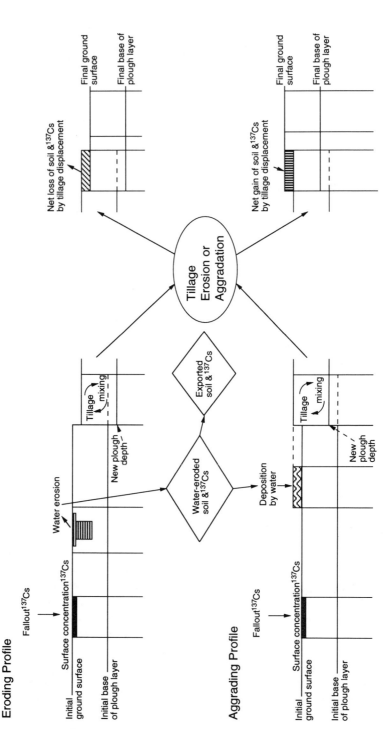

Figure 17.7 (a) Outline of the profile simulation model

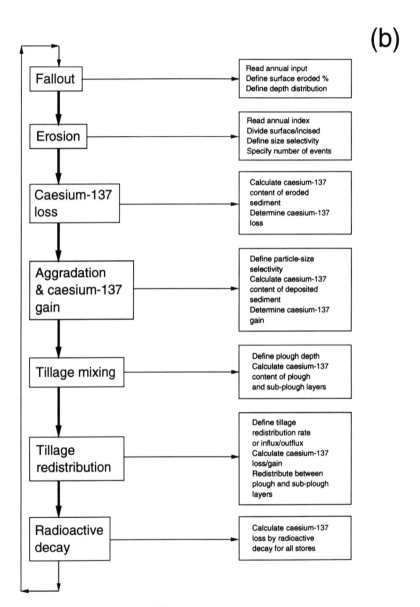

Figure 17.7 (b) the individual modules

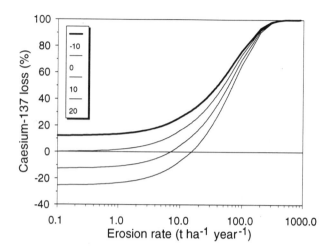

Figure 17.8 The relationship between ^{137}Cs loss and soil loss predicted by the profile model for a range of water erosion (x-axis) and tillage redistribution (Key: t ha^{-1} year^{-1}) rates. Assumptions: $NL = 0.8$ cm in Eqn (17.4); plough depth, 0.2 m; percentage area subject to erosion, 80%; percentage fallout subject to erosion prior to mixing, 40%; number of erosion events, three; maximum rate of inter-rill erosion, 10 t ha^{-1} yr^{-1}

erosion and fallout that potentially can be measured, and sources of relevant information for each of these parameters have been discussed in this chapter (Table 17.1). The basis of simulation may, therefore, be made explicit and subject to scrutiny and the quality of simulation may be improved as further information becomes available.

Addressing the limitations of the profile approach

The major limitations of the individual profile approach relate to the problems of simulation of a single point in isolation from it surroundings. This is of particular

Table 17.1 Key parameters in profile simulation and potential sources of relevant data

Parameter	Source
Annual pattern of fallout input	Published data
Annual variation in erosion	Erosion surveys/rainfall indices
Amount of annual fallout input in erosion season	Monthly fallout data and erosion survey data
Form of initial depth distribution	Experimental work
Maximum rate of inter-rill erosion	Field and laboratory study
Relationship between rill width and depth	Field and laboratory study
Fraction of the area protected from erosion	Field and laboratory study
Particle-size selectivity of erosion	Field and laboratory study
Particle-size selectivity of deposition	Field and laboratory study
Number of erosion events during the year	Erosion surveys
Rate of tillage erosion or aggradation	Field experimental data and simulation

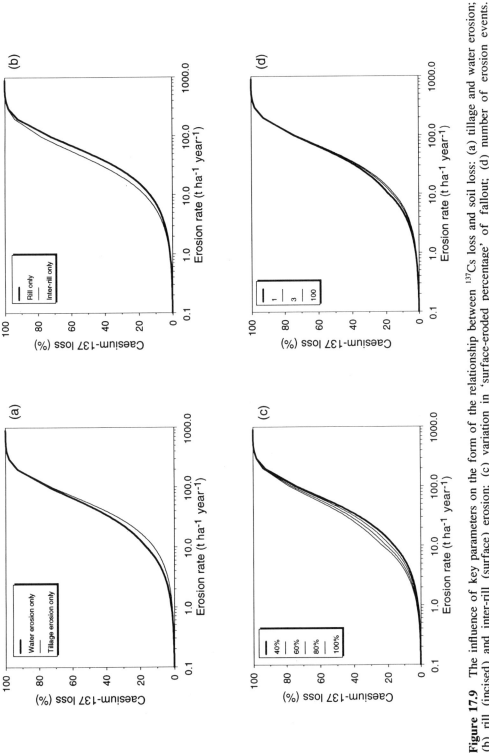

Figure 17.9 The influence of key parameters on the form of the relationship between ^{137}Cs loss and soil loss: (a) tillage and water erosion; (b) rill (incised) and inter-rill (surface) erosion; (c) variation in 'surface-eroded percentage' of fallout; (d) number of erosion events. Assumptions: no tillage redistribution except in (a); other parameters as for Figure 17.8

significance when simulating tillage redistribution and deposition of sediment by water. Prediction of the ^{137}Cs content of sediment deposited by water should be based on the predicted ^{137}Cs content of the sediment derived from the entire source area and, as has been indicated, simulation of the impact of tillage redistribution requires information concerning adjacent profiles.

Both these limitations have been addressed for simple slopes by the adaptation of the profile model to take account of the topographic context of each profile and to simulate adjacent profiles along a slope transect (Quine et al., 1993b). This provides a one-dimensional simulation of ^{137}Cs deposition and redistribution along a slope transect in association with water-eroded sediment and tillage movement. The main advantages of this approach are the improved simulation of the ^{137}Cs content of sediment deposited by water (because the whole contributing area is simulated) and of ^{137}Cs redistribution by tillage (because the ^{137}Cs content of both influx and outflux is known). However, application in areas of complex topography must await the development of a two-dimensional model, which requires prediction of flow paths to delineate sediment-contributing areas for deposition sites and two-dimensional simulation of tillage redistribution. This represents an important goal, which is still being pursued.

CONCLUSIONS

This chapter has attempted to highlight the central importance of calibration in the further development and wider acceptance of the ^{137}Cs technique. The major factors that must be taken into account in the development of effective calibration approaches have been identified and these have revealed the inadequacies of existing methods, in particular the widely used 'directly proportional' approach. An adaptable approach to calibration has, therefore, been proposed, which is based on simulation of ^{137}Cs accumulation and depletion in a soil profile through a series of modules linked by an iterative computer programme. One of the most important inclusions in the proposed approach is the facility to account independently for both water erosion and tillage redistribution. However, the growing awareness of the role of tillage in soil redistribution highlights the importance of a 'field-based' approach to calibration in which ^{137}Cs may be routed through the field rather than being examined for each location in isolation. The profile simulation model has, therefore, been adapted for one-dimensional field simulation and two-dimensional simulation is currently being pursued. Although discussion here has focused on calibration of ^{137}Cs data, the profile accumulation/depletion simulation approach has wider potential application in the study of other naturally and artificially occurring radionuclides, and potentially in the investigation of other soil chemical and physical constituents.

ACKNOWLEDGEMENTS

The author has enjoyed fruitful collaboration with Professor Des Walling, and his contribution to developing ideas regarding calibration is gratefully acknowledged. Thanks are also due to Professor Gerard Govers and Professor Zhang Xinbao, who have provided further valuable insights, and to Keith Playford for providing fallout data.

REFERENCES

Bachhuber, H., Bunzl, K., Schimmack, W. and Gans, I. (1982) The migration of ^{137}Cs and ^{90}Sr in multilayered soils: results from batch, column, and fallout investigations. *Nucl. Technol.* **59**, 291–301.
Boardman, J. (1990) Soil erosion on the South Downs: a review. In: *Soil Erosion on Agricultural Land*, Boardman, J., Foster, I.D.L. and Dearing, J.A. (eds), John Wiley & Sons, Chichester, pp. 87–105.
Boardman, J. and Favis Mortlock, D.T. (1993) Simple methods of characterizing erosive rainfall with reference to the South Downs, southern England. In: *Farm Land Erosion in Temperate Plains and Hills*, Wicherek, S. (ed.), Elsevier, Amsterdam, pp. 17–29.
Campbell, B.L., Loughran, R.J. and Elliott, G.L. (1982) Caesium-137 as an indicator of geomorphic processes in a drainage basin system. *Aust. Geogr. Stud.* **20**, 49–64.
Campbell, B.L., Loughran, R.J., Elliott, G.L. and Shelly, D.J. (1986) Mapping drainage basin sources using caesium-137. In: *Drainage Basin Sediment Delivery (Proc. Albuquerque Symposium, August 1986)*, Hadley, R.F. (ed.), IAHS Publication no. 159, pp. 437–446.
Colborne, G.J.N. and Staines, S.J. (1985) Soil erosion in south Somerset. *J. Agric. Sci. Cambs.* **104**, 107–112.
Coughtrey, P.J. and Thorne, M.C. (1983) *Radionuclide Distribution and Transport in Terrestrial and Aquatic Ecosystems: a Critical Review of Data*, vol. 1, Balkema, Rotterdam.
de Jong, E., Begg, C.B.M. and Kachanoski, R.G. (1983) Estimates of soil erosion and deposition for some Saskawatchewan soils. *Can. J. Soil Sci.* **63**, 617–607.
de Roo, A.P.J. (1991) The use of ^{137}Cs in an erosion study in south Limburg (The Netherlands) and the influence of Chernobyl fallout. *Hydrol. Processes* **5**, 215–227.
Evans, R. (1988) *Water Erosion in England and Wales, 1982–1984*, Report for Soil Survey and Land Research Centre, Silsoe.
Fredericks, D.J. and Perrens, S.J. (1988) Estimating erosion using caesium-137: II. Estimating rates of soil loss. In: *Sediment Budgets (Proc. Porto Alegre Symposium)*, Bordas, M.P. and Walling, D.E. (eds), IAHS Publication no. 3, **174**, pp. 233–240.
Frissel, M.J. and Pennders, R. (1983) Models for the accumulation and migration of ^{90}Sr, ^{137}Cs, ^{239,249}Pu and ^{241}Am in the upper layer of soils. In: *Ecological Aspects of Radionuclide Release*, Coughtrey, P.J. (ed.), Special Publication no. 3, British Ecological Society, Oxford, pp. 63–72.
Froehlich, W. and Walling, D.E. (1992) The use of fallout radionuclides in investigations in the Polish Flysch Carpathians. In: *Erosion, Debris Flows and Environment in Mountain Regions (Proc. Chengdu Symposium, July 1992)*, Walling, D.E., Davies, T.R. and Hasholt, B. (eds), IAHS Publication no. 209, pp. 61–76.
Gilley, J.E. and Gee, G.W. (1976) Particle size distribution of eroded spoil materials. *N. Dakota Farm Res.* **34**, 35–36.
Govers, G., Quine, T.A., and Walling, D.E. (1993) The effect of water erosion and tillage movement on hillslope profile development: a comparison of field observations and model results. In: *Farmland Erosion in Temperate Plains Environment and Hills*, Wicherek, S. (ed.), Elsevier, Amsterdam, pp. 285–300.
Govers, G., Vandaele, K., Desmet, P.J.J., Poesen, J., and Bunte, K. (1994) The role of tillage in soil redistribution on hillslopes. *Eur. J. Soil Sci.* **45**, 469–478.
He, Q. (1993) Interpretation of fallout radionuclide profiles in sediments from lake and floodplain environments. Unpublished PhD thesis, University of Exeter.
Kachanoski, R.G. (1987) Comparison of measured soil cesium-137 losses and erosion rates. *Can. J. Soil Sci.* **67**, 199–203.
Kachanoski, R.G. and de Jong, E. (1984) Predicting the temporal relationship between soil cesium-137 and erosion rate. *J. Environ. Qual.* **13**, 301–304.
Lindstrom, M.J., Nelson, W.W., Schumacher, T.E. and Lemme, G.D. (1990) Soil movement by tillage as affected by slope. *Soil Tillage Res.* **17**, 255–264.
Lindstrom, M.J., Nelson, W.W. and Schumacher, T.E. (1992) Quantifying tillage erosion rates due to moldboard plowing. *Soil Tillage Res.* **24**, 243–255.
Loughran, R.J. (1989) The measurement of soil erosion. *Prog. Phys. Geogr.* **13**, 216–233.

Loughran, R.J., Campbell, B.L. and Elliott, G.L. (1981) Sediment erosion, storage and transport in a small steep drainage basin at Pokolbin, NSW, Australia. In: *Erosion and Sediment Transport in Pacific Rim Steeplands (Proc. Christchurch Symposium)*, IAHS Publication no. 132, pp. 252–268.

Loughran, R.J., Elliott, G.L., Campbell, B.L., and Shelly,, D.J. (1988) Estimation of soil erosion from caesium-137 measurements in a small cultivated catchment in Australia. *J. Appl. Radiat. Isotopes* **39**, 1153–1157.

Martz, L.W. and de Jong, E. (1987) Using cesium-137 to assess the variability of net soil erosion and its association with topography in a Canadian Prairie landscape, *Catena* **14**, 439–451.

Martz, L.W. and de Jong, E. (1991) Using cesium-137 and landform classification to develop a net soil erosion budget for a small Canadian Prairie watershed. *Catena* **18**, 289–308.

Menzel, R.G., Jung, P., Ryu, K. and Um, K. (1987) Estimating soil erosion losses in Korea with fallout cesium-137. *J. Appl. Radiat. Isotopes* **38**, 451–454.

Mitchell, J.K., Bubenzer, G.D., McHenry, J.R. and Ritchie, J.C. (1980) Soil loss estimation from fallout cesium-137 measurements. In: *Assessment of Erosion*, DeBoodt, M. and Gabriels, D. (eds), John Wiley & Sons, Chichester, pp. 393–401.

Navas, A. and Walling, D.E. (1992) Using caesium-137 to assess sediment movement on slopes in a semi-arid upland environment in Spain. In: *Erosion, Debris Flows and Environment in Mountain Regions (Proc. Chengdu Symposium, July 1992)*, Walling, D.E., Davies, T.R. and Hasholt, B. (eds), IAHS Publication no. 209, pp. 129–138.

Owens, P. (1994) Toward improved interpretation of caesium-137 measurements in soil erosion studies. Unpublished PhD thesis, University of Exeter.

Playford, K., Toole, J. and Adsley, I. (1993) *Radioactive Fallout in Air and Rain: Results to the End of 1991*, Report AEA-EE-0498, DOE/RAS/93.003, HMSO, London.

Quine, T.A. (1989) Use of a simple model to estimate rates of soil erosion from caesium-137 data. *J. Water Resour.* **8**, 54–81.

Quine, T.A. and Walling, D.E. (1993) Assessing recent rates of soil loss from areas of arable cultivation in the UK. In: *Farm Land Erosion in Temperate Plains and Hills*, Wicherek, S. (ed.), Elsevier, Amsterdam, pp. 357–371.

Quine, T.A., Walling, D.E., Zhang, X. and Wang, Y. (1992) Investigation of soil erosion on terraced fields near Yanting, Sichuan Province, China, using caesium-137. In: *Erosion, Debris Flows and Environment in Mountain Region (Proc. Chengdu Symposium, July 1992)*, Walling, D.E., Davies, T.E. and Hasholt, B. (eds), IAHS Publication no. 209, pp. 155–168.

Quine, T.A., Walling, D.E. and Mandiringana, O.T. (1993a) An investigation of the influence of edaphic, topographic and land-use controls on soil erosion on agricultural land in the Borrowdale and Chinamora areas, Zimbabwe, based on caesium-137 measurements. In: *Sediment Problems: Strategies for Monitoring, Prediction and Control (Proc. Yokohama Symposium, July 1993)*, Hadley, R.F. and Mizuyama, T. (eds), IAHS Publication no. 217, pp. 185–196.

Quine, T.A., Walling, D.E. and Zhang, X. (1993b) The role of tillage in soil redistribution within terraced fields on the Loess Plateau, China: an investigation using caesium-137. In: *Runoff and Sediment Yield Modelling*, Banasik, K. and Zbikowski, A. (eds), Warsaw Agricultural University Press, pp. 149–155.

Quine, T.A., Navas, A., Walling, D.E. and Machin, J. (1994a) Soil erosion and redistribution on cultivated and uncultivated land near Las Bardenas in the central Ebro River Basin, Spain, *Land Degrad. Rehab.* **5**, 41–55.

Quine, T.A., Desmet, P.J.J., Vandaele, K., Govers, G. and Walling, D.E. (1994b) A comparison of the roles of tillage and water erosion in landform development and sediment export on agricultural land near Leuven, Belgium. In: *Variability in Stream Erosion and Sediment Transport (Proc. Canberra Symposium, December 1994)*, Olive, L. (ed.), IAHS Publication no. 224, 77–86.

Ritchie, J.C. and McHenry, J.R. (1973) Vertical distribution of fallout cesium-137 in cultivated soils. *Radiat. Data Rep.* **14**, 727–728.

Ritchie, J.C. and McHenry, J.R. (1975) Fallout Cs-137: a tool in conservation research, *J. Soil Water Conserv.* **30**, 283–286.

Rogowski, A.S. and Tamura, T. (1965) Movement of ^{137}Cs by runoff, erosion and infiltration on the alluvial Captina silt loam. *Health Phys.* **11**, 1333–1340.

Rogowski, A.S. and Tamura, T. (1970) Erosional behavior of cesium-137. *Health Phys.* **18**, 467–477.

Squire, H.M. and Middleton, L.J. (1966) Behaviour of ^{137}Cs in soils and pastures—a long term experiment. *Radiat. Bot.* **6**, 413–423.

Sutherland, R.A. (1991a) Caesium-137 and sediment budgeting within a partially closed drainage basin. *Z. Geomorphol.* **35**, 47–63.

Sutherland, R.A. (1991b) Examination of caesium-137 areal activities in control (uneroded) locations. *Soil Technol.* **4**, 33–50.

Vandaele, K. (1993) Assessment of factors affecting gully erosion in cultivated catchments of the Belgian Loam Belt. In: *Farm Land Erosion in Temperate Plains and Hills*, Wicherek, S. (ed.), Elsevier, Amsterdam, pp. 125–136.

Vanden Berghe, I. and Gulinck, H. (1987) Fallout ^{137}Cs as a tracer for soil mobility in the landscape framework of the Belgian Loamy region. *Pedologie* **37**, 5–20.

Walling, D.E. and Quine, T.A. (1990a) Use of caesium-137 to investigate patterns and rates of soil erosion on arable fields. In: *Soil Erosion on Agricultural Land*, Boardman, J., Foster, I.D.L. and Dearing, J.A. (eds), John Wiley & Sons, Chichester, pp. 33–53.

Walling, D.E. and Quine, T.A. (1990b) Calibration of caesium-137 measurements to provide quantitative erosion rate data. *Land Degrad. Rehab.* **2**, 161–175.

Walling, D.E. and Quine, T.A. (1991) The use of caesium-137 measurements to investigate soil erosion on arable fields in the UK: potential applications and limitations. *J. Soil Sci.* **42**, 147–165.

Walling, D.E. and Quine, T.A. (1992) The use of caesium-137 measurements in soil erosion surveys. In: *Erosion and Sediment Transport Monitoring Programmes in River Basins (Proc. Oslo Symposium, August 1992)*, Bogen, J., Walling, D.E. and Day, T. (eds), IAHS Publication no. 210, pp. 143–152.

Walling, D.E. and Woodward, J.C. (1992) Use of radiometric fingerprints to derive information on suspended sediment sources. In: *Erosion and Sediment Transport Monitoring Programmes in River Basins (Proc. Oslo Symposium)*, Bogen, J., Walling, D.E. and Day, T. (eds), IAHS Publication no. 210, pp. 153–164.

Wicherek, S. (1993) Impact of agriculture on soil degradation: modelisation at the watershed scale for a spatial management and development. In: *Farm Land Erosion in Temperate Plains and Hills*, Wicherek, S. (ed.), Elsevier, Amsterdam, pp. 137–153.

Zhang, X., Quine, T.A., Walling, D.E. and Li, Z. (1994) Application of the caesium-137 technique in a study of soil erosion on gully slopes in a yuan area of the Loess Plateau near Xifeng, Gansu Province, China. *Geogr. Ann.* **76A**, 103–120.

18 The Erosional Transport of Radiocaesium in Catchment Systems: A Case Study of the Exe Basin, Devon

JOHN S. ROWAN

Environmental Science Division, Lancaster University, UK

INTRODUCTION

The central role of suspended particulate matter in the dispersal of non-point-source pollution is now well established (Shear and Watson, 1977). Accordingly, attention has increasingly focused on the pathways, sinks and time-scales for sediment delivery from its sources to the basin outlet (Walling, 1983). Caesium-137 (^{137}Cs) is a relatively long-lived radionuclide (half-life of 30.2 years) that has been globally dispersed in the environment since 1954 as a result of nuclear weapons tests. Owing to its importance in dosimetry, ^{137}Cs has long been the focus of considerable research efforts (Peirson and Salmon, 1959). However, renewed interest in the environmental fate of radiocaesium was stimulated by the 1986 accident at the Chernobyl nuclear power plant, which was responsible for a 40% increase in the global inventory of ^{137}Cs (Cambray et al., 1987). Moreover, the accident highlighted the paucity of quality field data upon which to model long-term behaviour (Santchi et al., 1990).

The importance of fluvial pathways to redistributed fallout ^{137}Cs was quickly appreciated (Rogowski and Tamura, 1970). It is now widely accepted that, upon reaching the ground surface, ^{137}Cs fallout will be rapidly and irreversibly adsorbed by surface soils (Frissel and Pennders, 1983). If ^{137}Cs uptake by plants and losses via the grazing chain are discounted, then any subsequent mobility is associated with the erosion, transport and deposition of fine-grained particles. In this way the tracing properties of ^{137}Cs were quickly realised (Ritchie et al., 1970).

The range of studies employing the ^{137}Cs technique is now extensive, with the most frequent application relating to soil erosion within slope systems and small experimental catchments (Brown et al., 1981; Walling and Quine, 1992). The incorporation of ^{137}Cs-labelled sediment to deposition sites has also provided a basis for estimating sedimentation rates in lakes and reservoirs (Pennington et al., 1973; Rowan et al., 1992) and elucidating floodplain accretion rates (Ritchie et al., 1975; Walling and Bradley, 1989).

The fundamental assumption underlying all qualitative and quantitative applications of the ^{137}Cs technique is that any net loss or gain in the known fallout inventory (Bq m^{-2}) is the result of sediment transfer (de Jong et al., 1983). However, few studies report the ^{137}Cs content of suspended sediment and fewer still have considered

Sediment and Water Quality in River Catchments. Edited by I.D.L. Foster, A.M. Gurnell and B.W. Webb.
© 1995 John Wiley & Sons Ltd.

temporal variability during 'storm periods' when episodic transport processes are most dynamic.

This chapter reports the findings of an intensive event-based suspended sediment sampling programme in the Exe basin, Devon, UK. An account is given of the observed variability in sediment-associated ^{137}Cs levels over a range of time-scales. These measurements were supported by an atmospheric fallout survey, which provided the basis for establishing the annual ^{137}Cs flux and the erosional residence time of fallout in the basin (cf. Dominik *et al.*, 1987). A simple model was used to reconstruct the mean annual ^{137}Cs content of suspended sediment since 1954 based on the cumulative catchment fallout inventory. These results are interpreted with reference to dated overbank sediment sequences in the Exe floodplain. The discussion reviews these findings in relation to comparable studies and makes suggestions for future research.

BACKGROUND AND SETTING

With a catchment area of 601 km^2, the Exe basin is a medium-sized basin by British standards (Figure 18.1). The presence of several major tributary systems and a marked heterogeneity in geology, topography, soils, land use and hydrology indicated that the

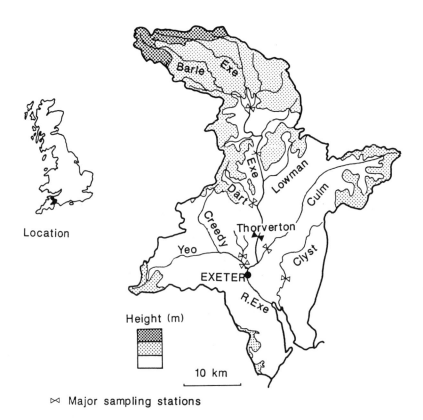

Figure 18.1 Location of study site and main monitoring stations in the Exe basin

Exe basin would be highly suited to the study of ^{137}Cs redistribution. Moreover, the main channel and sub-basins have been extensively studied by D.E. Walling and co-workers over the past 20 years, who have provided a wealth of information on the sediment delivery characteristics of the basin (e.g. Walling and Webb, 1985; Lambert and Walling, 1988; Walling and Moorehead, 1989).

The lithology of the northern headwaters of the catchment is dominated by resilient Devonian sandstones, siltstones and slates forming the Exmoor plateau. The plateau contains extensive areas above 400 m dissected by the River Barle and the Upper Exe. Mean annual precipitation totals reach a maximum of *ca.* 2000 mm. The major soil types are ferric stagnopodsols (Hafren association) and cambic humic gleys (Wilcocks 2) in the valley bottoms. The northern uplands are used primarily for stock rearing on rough grazing and improved pasture. South of the Exmoor uplands, the lithology conformably succeeds to Carboniferous and Permian sandstones and shales. The southern half of the basin is characterised by a rolling topography with elevations typically below 200 m. Mean annual precipitation declines to values of *ca.* 900 mm, and brown earths of the Neath and Crediton associations become prevalent. Arable cultivation is widespread in the southern and eastern areas of the basin.

EXE BASIN FALLOUT SURVEY

The quantity of atmospheric fallout that accumulates at any point on the ground surface is the product of several discrete, though interrelated, environmental controls. Global-scale latitudinal variations occur owing to stratospheric circulation systems and synoptic-scale weather conditions (Peirson and Cambray, 1967). At progressively smaller levels of resolution, regional controls influenced by such factors as topography and mean annual precipitation totals become more important (Kiss *et al.*, 1988). At the plot or field scale, the penetration of fallout into the soil column is further influenced by variations in vegetation type, microtopography and infiltration capacity (Loughran, 1989).

The 'reference sites' of McHenry and Ritchie (1975), the 'input sites' of Campbell *et al.* (1982) and the 'control sites' of de Jong *et al.* (1983), are all similar in that they are used to establish base-line estimates of atmospheric fallout. According to Walling *et al.* (1986), these sites are typically on hilltop locations, have minimal slopes and are generally in pastures undisturbed by cultivation. As such, they experience no net gain or loss of their ^{137}Cs inventory other than by losses through decay.

Previous work on the distribution of weapons-test fallout has suggested that mean annual precipitation is the principal factor controlling the deposition of ^{137}Cs (Cawse and Horrill, 1986). Consequently it is commonly assumed that local variability (i.e. at the field scale) will be low and that broad homogeneous zones with similar fallout totals will be found within the environment. The validity of this assumption was questioned in the case of a climatically and morphologically diverse catchment such as the Exe. An input sampling programme based on a 5×5 km^2 grid and involving 42 sampling sites within the catchment was undertaken to provide information on the variability in ^{137}Cs input. The input survey was completed in early 1989.

Caesium-137 inventories were determined from cores (i.d. 6.9 cm) collected to 30 cm depth at sites conforming with the criteria previously described. Prior to the commencement of the catchment survey, multiple cores ($n = 12$) were collected from a single undisturbed pasture site, yielding a 14% coefficient of variation in ^{137}Cs inventory (D.L. Higgitt, pers com.). By taking a minimum of two cores at each site, the sampling method therefore carries a precision level of 20% at the 95% confidence interval (cf. Bachhubber et al., 1987). After oven drying, the soil samples were sieved and the stone fraction (>2 mm) removed. The homogenised <2 mm residue was then subsampled into 1 kg Marinelli beakers and allowed to settle prior to gamma assay.

Radiocaesium determinations were made using Ortec and Canberra coaxial HPGe detector systems. Counting efficiencies were calibrated appropriate to sample size and geometry, with count times typically between 30 000 and 50 000 s. Analytical precision levels were generally below 5% at the 95% confidence interval. Additional information on the physicochemical characteristics of the soils and suspended sediment samples included determination of organic carbon (C) and nitrogen (N) by pyrolysis/thermal conductivity using a Carlo Erba analyser, and grain-size analysis using a Micro-Meritics SediGraph 5000D apparatus. Particle-size distributions refer to the chemically dispersed mineral fraction.

Atmospheric fallout: distribution and environmental controls

A summary of data collected for the catchment fallout survey is presented in Table 18.1. The wide ranges of elevation and mean annual precipitation reflect the heterogeneous physiography and climate of the catchment. Similarly the high coefficients of variation for organic C and stone content indicate substantial spatial variation in these parameters. The areal deposition of ^{137}Cs also demonstrated considerable variability, with inventories found over a three-fold range from 1900 to 6900 Bq m^{-2}, and a mean deposition (± 1 SE) of 3414 ± 70 Bq m^{-2}. The mean concentration of ^{137}Cs in soils (0–30 cm) was 15.1 ± 1.3 mBq g^{-1}, with the range spanning close to an order of magnitude.

Table 18.1 Statistical summary of measured variables in Exe basin ^{137}Cs atmospheric fallout survey

Variable ($n = 42$)	Min.	Max.	Mean	SD	CV (%)
Easting	2716	3040	–	–	–
Northing	1034	1430	–	–	–
Height (m)	80	510	278	109.4	39.4
Mean annual precipitation (mm)	900	2000	1273	300.6	23.6
Carbon (%)	1.70	16.2	4.4	3.5	79.9
Nitrogen (%)	0.12	0.74	0.28	0.13	46.4
C/N ratio	120	2.2	15.6	1.8	57.1
Bulk density (g cm^{-3})	1.23	1.65	1.24	0.24	19.4
Stone, >2 m (%)	0.30	45.5	17.8	11.3	63.2
Sand (%)	9.1	50.7	28.9	9.4	32.6
Silt (%)	20.7	61.6	36.0	9.1	25.3
Clay (%)	17.2	51.3	35.3	7.9	22.5
^{137}CS activity (Bq m^{-2})	1892	6879	3414	1099	32.2
^{137}Cs activity (mBq g^{-1})	6.3	51.5	15.1	8.4	56.0

A geostatistical analysis of the fallout data was undertaken using kriging (Oliver et al., 1989). The semi-variograms for both fallout inventory and concentration conformed to unbounded Gaussian distributions indicating drift in the data (McBratney and Webster, 1986). The interpolated patterns are presented in Figure 18.2 and show significant north-west to south-east declines in both ^{137}Cs inventory and soil concentration. The highest totals (>6000 Bq m^{-2}) occurred in the northwestern extremes of the Exmoor plateau, whereas the lowest totals (<3000 Bq m^{-2}) were found in the southern and eastern margins of the catchment. No soil samples yielded any detectable quantity of ^{134}Cs, which would indicate fallout from the 1986 Chernobyl accident. Mass-specific soil ^{137}Cs concentrations also demonstrated a strong north to south gradient. The northern areas yielded values in the order of 15–35 mBq g^{-1}, whereas southern and eastern sub-basins typically contained less than 10 mBq g^{-1}.

Environmental controls on fallout distribution

Correlation and stepwise regression were used to explore the factors controlling the ^{137}Cs distribution. A high degree of collinearity was evident in the data, reflecting the interaction between climate, topography and soils. Stepwise regression, by forward selection, was used to isolate the main explanatory variables, resulting in only two of the 11 dependent variables being selected for ^{137}Cs inventory values ($p > 0.05$). Mean annual precipitation, taken from the International Standard Period of 1941–70, explained 75% of the variance in the fallout inventory ($p > 0.001$). Selecting clay percentage improved the R^2 value to 0.79, but was discounted as having no functional

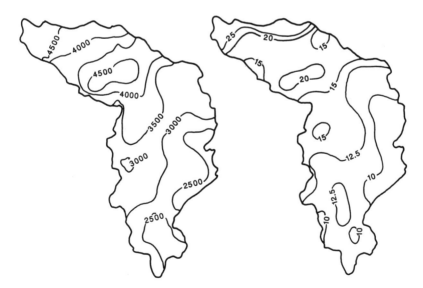

(a) Caesium–137 inventory (Bq m^{-2}) (b) Caesium–137 concentration (mBq g^{-1})

Figure 18.2 Spatial patterns of ^{137}Cs inventory and soil concentrations (0–30 cm) in the Exe basin, 1989

contribution to the input distribution, thus giving:

$$^{137}\text{Cs inventory (Bq m}^{-2}) = -657 + 3.21 MAP \tag{18.1}$$

where MAP = mean annual precipitation (mm yr^{-1}).

The factors controlling mass-specific activities were also investigated. Here the stepwise regression accounted for 82% ($R^2 = 0.82$) of the variance ($p > 0.001$) using the multivariate expression:

$$^{137}\text{Cs (mBq g}^{-1}) = 23.6 - 25Bd + 0.02MAP + 0.22SC \tag{18.2}$$

where Bd = bulk density (g cm^{-3}), MAP = mean annual precipitation (mm yr^{-1}) and SC = stone content (%). The high level of explanation afforded by this relationship can be understood in terms of the highest fallout inputs occurring in the organic-rich soils in the northern headwaters. Many of the upland soils are relatively thin and a high stone content also naturally enhances fallout onto matrix fines.

It must be noted that the mass-specific ^{137}Cs data refer to the average (0–30 cm) soil core. In undisturbed sites the concentration profile is observed to decline exponentially with depth from a maximum value near the surface layer (cf. Frissel and Pennders, 1983). Using the relationship:

$$I_H = I(1 - e^{-\lambda H}) \tag{18.3}$$

where $I_H = {}^{137}$Cs inventory (Bq m^{-2}) above a given depth H (cm), I is the total fallout inventory (Bq m^{-2}) and λ is the exponential coefficient (Zhang et al., 1990), it is possible to estimate activity level of the surface layers. From depth-incremented profiles in the Exe basin (e.g. Walling et al., 1986) the λ coefficient is estimated to lie in the range 0.25–0.35, indicating that mean surficial concentrations (0–2 cm) are 4–5 times higher than 0–30 cm depth-integrated values presently reported.

The range of values found in the survey confirms an uneven pattern of fallout over the basin area. Clearly the assumption of uniform fallout breaks down in a heterogeneous catchment system such as the Exe basin, where diverse topographic and climatic controls combine to vary the input function. Based on weighted mean values, the total ^{137}Cs inventory of the Exe basin in 1989 was estimated to lie within $(1.85–2.26) \times 10^{12}$ Bq (95% CI). This inventory total and the illustrated fallout patterns form the basis for assessing the importance of erosion as a mechanism for ^{137}Cs redistribution.

Routeing ^{137}Cs to the catchment outlet

Two main modes of ^{137}Cs dispersal are recognised at the catchment scale. In the first, a fraction of the annual fallout falls directly in the channel network, lake/reservoir surfaces, or is conveyed rapidly from riparian areas either in solution or sorbed to readily mobilised sediment sources such as road dust. This mode of transport has been variously termed 'direct runoff' (Jacobi, 1972) or the 'fast hydrological component' (Smith et al., 1987). Carlsson (1978) suggests that, on an annual basis, as much as 2% of the freshly deposited fallout will be lost from the basin by this mode of transport.

The importance of this 'flushing' mechanism is most relevant to the years 1958–64, when the bulk of weapons-test ^{137}Cs was deposited (Cambray et al., 1989). It was also observed in the aftermath of the Chernobyl accident, when several studies identified peaks in suspended sediment-associated radiocaesium levels, which rapidly declined due to the single-pulse nature of the input (e.g. Walling *et al.*, 1989). Santchi *et al.* (1988) suggest that as much as 1% of the atmospheric loading from Chernobyl may have been lost in this way.

The second mode of dispersal relates to erosional losses in the accumulated inventory of catchment soils and sediment sinks, which is released slowly, but over more extended time-scales. Particle-reactive radionuclides such as ^{137}Cs are rapidly sorbed to surface soils, and thus the fate of this component is intimately linked to the pathways and residence times of eroded sediment as it is routed through the catchment. This slower but dominant form of release is known as 'delayed removal' (Jacobi, 1972) or the 'slow erosional component' (Smith *et al.*, 1987). It is this component of the ^{137}Cs flux that offers the greatest potential for linking on-site rates of erosion to the sediment ultimately delivered to the basin outlet (Walling, 1990).

EXE BASIN SUSPENDED SEDIMENT SAMPLING PROGRAMME

During the period August 1987 to June 1989, suspended sediment samples were collected from 28 discrete flood events monitored at the Thorverton gauging station. A wide range of hydrograph and sedigraph forms were observed and are considered representative of the long-term flood series. Discharge and turbidity were measured continuously, but sampling was restricted to storm events when suspended sediment concentrations were high (*ca.* 200–1000 mg l^{-1}). Bulk river water samples (*ca.* 100 l) were collected using a submersible pump. Larger volumes were also collected during periods of low sediment concentration (i.e. 30 mg l^{-1}) to examine for seasonal effects. Suspended sediment was extracted using an Alfa-Laval continuous flow centrifuge and then freeze-dried. By this method bulk samples of 5–150 g dry mass were obtained. A statistical summary of the main data collected during this period is provided in Table 18.2.

Table 18.2 Statistical summary of hydrological and sediment properties measured during storm runoff events 1987–89

Variable ($n = 89$)	Min.	Max.	Mean	SD	CV (%)
^{137}Cs activity (mBq g^{-1})	9.23	31.68	21.03	5.26	25.0
Discharge (m^3 s^{-1})	7.29	164.56	73.77	29.25	39.7
Turbidity (mg l^{-1})	13.00	961.40	253.80	170.70	67.3
Sediment load (kg s^{-1})	0.90	78.09	19.90	15.69	78.8
Sand content (%)	1.28	14.97	5.50	2.78	50.5
Silt content (%)	28.86	61.37	47.68	7.62	16.0
Clay content (%)	28.86	61.36	47.03	7.66	16.3
Silt/clay ratio	0.38	2.23	1.45	0.23	15.9
Carbon content (%)	3.89	9.34	6.07	0.97	16.0
Nitrogen content (%)	0.29	1.04	0.54	0.11	20.4
C/N ratio	9.11	12.79	11.23	0.81	0.1

Annual and seasonal ^{137}Cs transport patterns

The variability of ^{137}Cs measured during the study period is shown in Figure 18.3. The pattern reveals a well-defined range with upper and lower limits falling between 32 and 9 mBq g^{-1}. The mean ^{137}Cs activity was determined as 21.03 ± 1.11 mBq g^{-1} (95% CI), with a (coefficient of variation of 25%). The general pattern appears to exhibit no seasonal effects. Instead, the variability was essentially stationary around the observed mean. Only a small number of samples were collected during the summer months, but recent evidence presented by Walling *et al.* (1992) confirms the absence of major seasonal variation. Very few studies offer comparable data sets; however, Dominik *et al.* (1987) did identify seasonal modulation in the Alpine Rhône, Switzerland. This difference is most likely attributed to significant differences in catchment characteristics, as discussed in the final section of this chapter.

Inter-storm variability in sediment-associated ^{137}Cs

The previous section showed no evidence of systematic variation in ^{137}Cs levels over the sampling period. Considerable scatter was, however, present within the data, and correlation analysis was used as a means to identify possible controls (Table 18.3). Correlation coefficients were generally low, but a number of important relationships were found. The first was that discharge appeared to have no influence on ^{137}Cs levels, instead, concentrations varied widely over the entire range of observed flows. Secondly, ^{137}Cs content exhibited significant positive correlation with organic carbon, the C/N ratio, silt content and silt/clay ratio, whereas it is inversely correlated with clay content and suspended sediment concentration (all $p > 0.05$). The importance of these findings lies in the conclusions that can be made regarding sediment provenance.

Data on the distribution of C, C/N and silt/clay ratios of catchment soils obtained during the fallout survey indicated strong zonation, with the upper range of all three

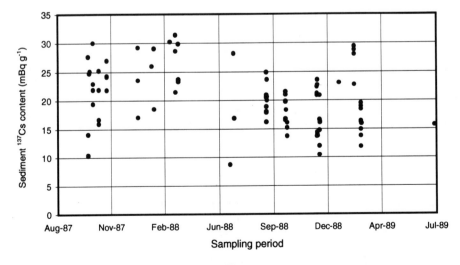

Figure 18.3 Suspended sediment associated ^{137}Cs levels observed in River Exe at Thorverton (October 1987–July 1989)

Table 18.3 Correlation matrix: hydrological and physicochemical properties of suspended sediment: River Exe at Thorverton [a]

	^{137}Cs sediment content	Discharge	Sediment load	Sediment concentration	C/N ratio	Nitrogen	Carbon	Silt/clay	Sand	Silt
Discharge ($m^3 s^{-1}$)	−0.035									
Sediment load ($kg s^{-1}$)	0.206*	0.628**								
Sediment concentration ($mg l^{-1}$)	−0.271**	0.239*	0.892**							
Carbon/nitrogen ratio	0.215*	0.332**	0.178	0.085						
Nitrogen (%)	0.018	−0.573**	−0.244*	−0.111	−0.273**					
Carbon (%)	0.257*	−0.575**	−0.233*	−0.089	0.134	0.753**				
Silt/clay ratio	0.298**	0.251*	0.130	0.039	0.455**	−0.095	0.024			
Sand content (%)	−0.003	0.160	−0.174	−0.311**	0.068	−0.336**	−0.331**	0.089		
Silt content (%)	0.283**	0.341**	0.233*	0.132	0.427**	−0.033	0.035	0.950**	−0.128	
Clay content (%)	−0.272**	−0.368**	−0.150	−0.009	−0.437**	0.163	0.098	−0.946**	−0.244**	−0.930**

[a] Pearson's product moment correlation coefficients. Degrees of freedom $(n − 2) = 87$. P levels (two-tailed): * 0.05, critical $r = 0.211$; ** 0.01, critical $r = 0.275$

parameters consistently lying in the northern headwaters of the basin. Much lower values were found in the southern and eastern areas of the basin. The positive correlation between ^{137}Cs levels and C and C/N ratios thus tentatively links higher ^{137}Cs levels with sediment derived from the northern tributaries.

Support for an underlying source area control is strengthened with reference to the particle-size data. The strong grain-size dependence of surface reaction and sorption processes is well documented; for example, Livens and Baxter (1988) showed that clays can contain 30 times the radiocaesium activity of a comparable mass of sand. In this study ^{137}Cs was inversely correlated with clay percentage, but showed positive correlations with silt and silt/clay ratios. This behaviour contrasts with the expected situation in small catchments where, if fallout is uniform and sediment is derived from a limited range of sources, ^{137}Cs levels are intimately linked with the proportion of clay eroded during a given event (McHenry and Ritchie, 1977). In larger catchments, with uneven fallout patterns, the relationship between ^{137}Cs content and the particle-size distribution is less predictable, and the ^{137}Cs content of sediment may instead be related to the ^{137}Cs content in sediment supply areas, even if these sediment sources have a coarser grain-size distribution.

Lambert (1986) has previously demonstrated the utility of the silt/clay and C/N ratios to distinguish the provenance of suspended sediment in the Exe basin. By dividing the catchment on a simple north–south basis, he showed that high silt/clay and C/N ratios on suspended sediment were diagnostic of sediment derived from predominantly northern runoff events, whereas the opposite was true of sediments of a predominantly southern provenance. The negative relationship with turbidity adds further credence to the argument because southern storms typically yield higher suspended sediment concentrations relative to northern storms of comparable discharge magnitude (Walling and Moorehead, 1989).

These observations indicate that sediment provenance, rather than hydraulic factors, is the principal determinant of the ^{137}Cs content measured at the catchment outlet. This working hypothesis was tested by classifying the 28 observed flood events into three groups, i.e. predominantly northern runoff events and predominantly southern runoff events based on knowledge of precipitation patterns, and the C/N and silt/clay ratios previously discussed. The third group represented data from floods resulting from a catchment-wide runoff event and represented the control class. Difference of means tests revealed significant differences in mean ^{137}Cs, C/N and silt/clay ratios ($p > 0.01$) between the northern and southern storms thus defined. The mean ^{137}Cs content for northern storms was 25.6 ± 0.78 mBq g^{-1}, while southern storms had values of 17.5 ± 1 mBq g^{-1}. The average ^{137}Cs content of samples from catchment-wide events was indistinguishable from the overall mean.

Intra-storm variability in ^{137}Cs

The observed behaviour of ^{137}Cs at the intra-storm period confirms the role of sediment provenance as the major control on measured activity. In Figures 18.4a and b ^{137}Cs levels are plotted in relation to the storm sedigraph. In both of these events, the first sample collected contained a substantially higher ^{137}Cs content (21–25 mBq g^{-1}) than those immediately following. This early turbidity peak is a common phenomenon on the Exe, and has been attributed to the rapid conveyance of sediment from the Dart

EROSIONAL TRANSPORT OF ^{137}Cs IN CATCHMENT SYSTEMS

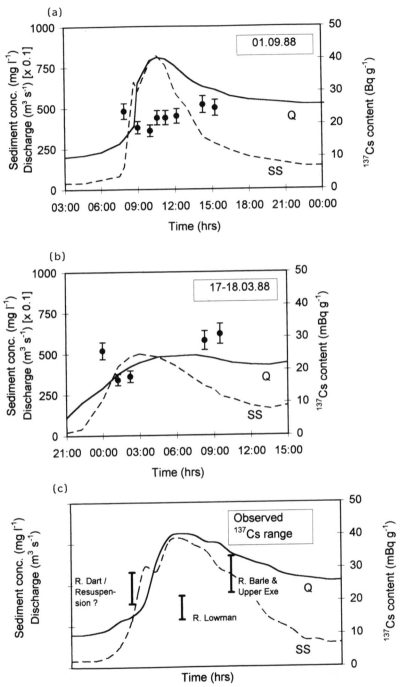

Figure 18.4 The ^{137}Cs variability of suspended sediment at the intra-storm period

basin, which passes the gauging station ahead of the main sediment pulse (Lambert, 1986). Considering that the ^{137}Cs content of this sediment was twice the mean value for Dart sediments reported by Walling and Woodward (1992), an alternative explanation may be that this pulse represents a flush of fine-grained sediment deposited during the recession period of the preceding flood.

The arrival time of the main sediment pulse at Thorverton generally lags the initial rise in discharge by between 2 and 3 h, whereas the turbidity maximum generally precedes the peak discharge by about 1 h. During this period ^{137}Cs levels typically decline to levels below 20 mBq g^{-1}. These mid-storm samples are thought to derive from the main subcatchments in the southern part of the basin, particularly the River Lowman. The River Lowman was expected to yield the lowest ^{137}Cs levels within the catchments because arable land is the main sediment source. It is well known that tillage mixes ^{137}Cs throughout the plough depth: moreover, the Lowman is also the basin with the lowest fallout inventories. During events relating exclusively to southern runoff, ^{137}Cs levels actually declined to values ca. 10 mBq g^{-1}, probably as a result of exhaustion of surface sediment sources (cf. Novotny, 1980) and the consequently greater contribution from channel bank erosion sources only weakly labelled with ^{137}Cs.

Sediment derived from the northern headwaters generally takes the longest time to arrive at the catchment outlet. The turbidity peak occurs after the discharge maximum owing to kinematic lag effects. The highest ^{137}Cs concentrations >25 mBq g^{-1} were typically found during this period associated with the arrival of sediment pulses derived from the Upper Exe and River Barle. During this period the rise in ^{137}Cs level was paralleled by increases in both the silt/clay and C/N ratios, which is consistent with the observations made in the previous section.

Lambert and Walling (1988) presented evidence indicating that the main channel of the River Exe acts as an efficient conveyance system through which sediment is rapidly flushed. Under such conditions the ^{137}Cs content of suspended sediment will vary through the storm sedigraph in response to the routeing of sediment from different parts of the catchment. Figure 18.4c is an idealised representation of ^{137}Cs behaviour involving catchment-wide runoff events. Further work is required to examine ^{137}Cs levels within individual sub-basins and to consider the role of multivariate 'fingerprinting' techniques to gain further insight into the relative importance of different source areas and sediment production mechanisms.

A CAESIUM-137 BUDGET FOR THE EXE BASIN

The evidence presented above indicates that unit-area losses of ^{137}Cs are unlikely to be uniform across the catchment surface, and that instead relative depletion is a function of the erosion and delivery of sediment from local sources. The limited variability associated with the mean value does, however, permit a ^{137}Cs flux rate to be established if annual suspended sediment load is known (cf. Chapman, 1992). Sediment yield information is available from the River Exe since the introduction of continuous turbidity and discharge monitoring. The sediment yield during the study period was determined as 28 t km^{-2} yr^{-1}, or 16 828 t yr^{-1} (cf. Walling and Webb, 1987). On the basis of the 95% confidence intervals of ^{137}Cs content (19.9–22.2 mBq g^{-1}), the mean annual export rate was estimated at $(3.35–3.73) \times 10^8$ Bq yr^{-1}.

Dominik et al. (1987) presented a simple lumped input–output model, from which, assuming steady state, the catchment erosion flux, termed the annual erosional removal factor, (k) of a radionuclide can be determined:

$$F_A - I(\lambda + k) = 0 \tag{18.4}$$

where F_A is the atmospheric flux (Bq m^{-2} yr^{-1}), λ is the decay constant (yr^{-1}), and I is the catchment inventory (Bq m^{-2}). If the input and annual fluvial output ($F_E = kI$) are measured, the mean erosional residence time Γ (yr) can be obtained from:

$$\Gamma = \frac{1}{k} = \frac{F_A - F_E}{\lambda F_E} \tag{18.5}$$

In the case of ^{137}Cs the atmospheric flux is not constant, but has instead varied according to the history of nuclear-weapons testing. Thus the change in catchment inventory (where $\lambda = 0.997$ yr^{-1}) can be expressed as:

$$\frac{dI}{dt} = F_A - I(\lambda + k) \tag{18.6}$$

It follows that, if the loss by erosion is much smaller than that by decay, i.e. $\lambda \gg k$, and knowing the fallout history $F_A(t)$, a solution for I can be found for any time t. Thus the erosional removal constant can be calculated as:

$$k = \frac{F_E}{I} \tag{18.7}$$

In this study the catchment inventory was established from the input sampling programme and the 1989 value established as $(1.85–2.26) \times 10^{12}$ Bq (95% CI). The annual export of ^{137}Cs by sediment-associated transport was determined as $(3.35–3.73) \times 10^8$ Bq yr^{-1} (95% CI). Thus ^{137}Cs is estimated to have a mean erosional residence time of 6000 years, $k = 0.00015–0.0002$ yr^{-1}. This means that between 0.015 and 0.02% of the catchment inventory is lost each year by the delivery of sediment to the basin outlet.

If a steady-state system can be assumed, the erosional removal factor becomes a constant. It is then possible to reconstruct the mean annual ^{137}Cs activity of suspended sediment $C_s(t)$ (mBq g^{-1}), for any year (t) as a function of the decay-corrected cumulative fallout catchment inventory and the annual sediment yield F_Y (t km^{-2} yr^{-1}):

$$C_s(t) = \frac{1000 kI}{F_Y} \tag{18.8}$$

where 1000 is a scaling factor.

The assumption of a steady-state system is probably conservative (cf. Helton et al., 1985), as the main sediment sources will be depleted in ^{137}Cs by erosion (Foster and Hakonson, 1984; Walling and Quine, 1992). However, the simple relationship is considered valid because of the previously demonstrated stationarity of the mean annual suspended sediment ^{137}Cs content. The time-dependent fallout history of the catchment was derived by reference to the long-term fallout monitoring programme at Milford Haven (Cambray et al., 1989; Playford et al., 1993). The mean annual accumulated fallout inventory for the Exe basin was normalised to the Milford Haven input using a correction factor of 1.2. Annual suspended sediment yield data were not available throughout the post-1954 period, which is particularly relevant to the period 1958–66 when the bulk of the atmospheric fallout was deposited. The long-term annual sediment yield of 28 t km^{-2} yr^{-1} was used if annual data were unavailable (cf. Walling and Webb, 1987).

The results of the model reconstruction are shown in Figure 18.5. Of particular interest is the fact that the maximum annual ^{137}Cs activity of sediments was predicted to occur in 1966, i.e. three years after the 1963 fallout peak. The retention of the fallout in catchment soils has an important buffering effect, evidenced by the fact that the 1966 peak concentration of 30.4 mBq g^{-1} was only 30% greater than contemporary values. After 1966 there is an effective decoupling between the atmospheric input and the ^{137}Cs content of suspended sediment. Erosion processes sustain the supply of ^{137}Cs-labelled sediment, but radioactive decay of the inventory becomes the dominant control on activity levels, hence the relatively steady decline to values of ca. 21 mBq g^{-1} by 1990.

To test the validity of the approach, ^{137}Cs profiles were collected from a rapidly accreting floodplain site 3 km down-river from the gauging station. This site was inundated by overbank flows several times per year. A number of cores were collected and sliced to 2 cm increments prior to gamma assay. The base of the ^{137}Cs profile was established at 36 cm. A distinct peak of 133 mBq g^{-1} occurred at 26 cm and thereafter activity levels declined to values of ca. 20 mBq g^{-1} at the sediment surface. The conventional interpretation of such a profile is that the maximum depth corresponds to the initial deposition of atmospheric fallout in 1954 and the peak value to the 1963 fallout maximum (Ritchie et al., 1975). This time control allows a mean sedimentation rate of ca. 1.0 cm yr^{-1} to be inferred, and hence time (years) has been substituted for depth in Figure 18.6.

An accreting floodplain surface incorporates atmospheric input along with contribution of ^{137}Cs associated with deposited overbank sediments. By filtering the yearly atmospheric contribution and correcting for decay since time of deposition, it

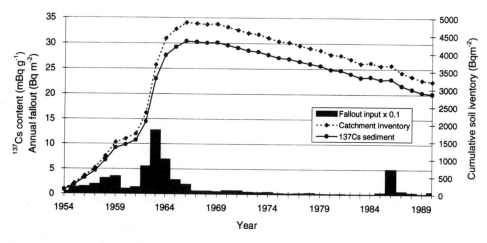

Figure 18.5 Reconstructed sediment-associated ^{137}Cs levels based on annual fallout totals and cumulative catchment inventory

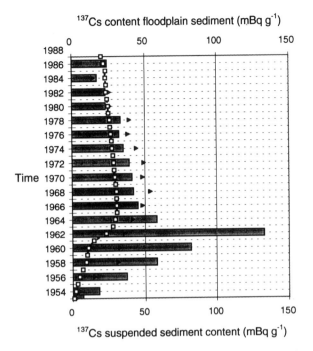

Figure 18.6 Depth profile of the ^{137}Cs content of an Exe basin floodplain core (□) reconstruction of ^{137}Cs values based on erosional removal constant method; (▲) ^{137}Cs activity of sediments after removal of the annual atmospheric fallout component

was thus possible to derive a curve of the ^{137}Cs activity of suspended sediment deposited during overbank flow events.

Least-squares analysis between the two reconstruction methods yielded an $r^2 = 0.81$ ($p > 0.05$), indicating that both curves exhibited the same basic pattern. The floodplain-based ^{137}Cs values typically exceeded the erosional removal predictions by a factor of 1.6, particularly in the lower (oldest) sections of the profile. Several mechanisms could explain these differences. The first relates to enrichment processes involving the preferential deposition of fines in floodplain depressions (cf. Walling and Bradley, 1989). Alternatively, the higher levels may be ascribed to a variable erosional removal factor, but it is more likely that the higher ^{137}Cs concentrations found in the floodplain sediments contain ^{137}Cs resulting from a 'direct runoff' contribution during the peak fallout years (cf. Jacobi, 1972).

RADIONUCLIDE TRANSPORT AT THE CATCHMENT SCALE: AN OVERVIEW

Although few direct measurements of sediment-associated ^{137}Cs levels exist, several studies have estimated fallout removal rates on the basis of inventory budgets and reservoir sedimentation surveys. A selection of these are presented in Table 18.4, which also includes pertinent data on other globally dispersed radionuclides. Comparisons between these studies is subject to considerable uncertainty because of

Table 18.4 Selected review of catchment-scale radionuclide studies

Location	Area (km^{-2})	Loss (% yr^{-1})	Time	Nuclide	Source
Ulkesjön, Sweden	1.3	1.9 0.56	1961–75	'direct runoff' 'delayed runoff'	Carlsson (1978)
Powerline, USA	1.2	0.14	1953–57	^{137}Cs	Ritchie et al. (1974)
Murphy, USA	0.5	0.57		^{137}Cs	
Smith, USA	0.8	0.71		^{137}Cs	
Brunner Creek, USA	5.8	0.50	1953–75	^{137}Cs	Ritchie et al. (1982)
Cyprus Creek, USA	0.8	0.8	1965–85	^{137}Cs	Cooper et al. (1987)
Panther Swamp, USA	1.4	0.5		^{137}Cs	
Yendacott, UK	2.4	0.11	1985	^{137}Cs	Walling et al. (1986)
Brotherswater, UK	11.5	0.1–0.2	1976–88	^{137}Cs	Bonnett (1990)
Great Miami, USA	1401	0.05	1974–75	^{239,240}Pu	Sprugel and Bartel (1978)
Wye UK	174	0.056	1986–89	Chernobyl ^{137}Cs	Rowan and Walling (1992)
Wye UK	2083	0.048	1986–89		
Rhône, Switzerland	5220	0.124	1982–83	^{137}Cs	Dominik et al. (1987)
Rhine, Germany	50000	1.0 0.8	1956–65	β-'direct runoff' β-'delayed runoff'	Jacobi (1972)
Tornionjoli, Finland	40100	0.22	1954–80	^{137}Cs, 1.7% ^{90}Sr	Salo et al. (1984)
Kemijoki, Finland	51400	0.27		^{137}Cs, 1.0% ^{90}Sr	
Oulijoli, Finland	27100	0.33		^{137}Cs, 1.5% ^{90}Sr	
Kokemaenjoki, Finland	37235	0.11		^{137}Cs, 1.0% ^{90}Sr	
Kymijoki, Finland	37235	0.3		^{137}Cs, 2.0% ^{90}Sr	
Hudson, USA	35653	0.0037	1978–81	^{137}Cs	Linsalata et al. (1985)
Sanguenay, Canada	78000	0.046	1978–81	^{137}Cs	Smith and Ellis (1982)

differences in sampling technique (Bonnett, 1990). However, much of the variability in ^{137}Cs flux rates can be explained in terms of variable sediment sources and contrasts in hydrology and sediment delivery regimes.

Comparison of the high percentage loss rates of ^{90}Sr compared to ^{137}Cs reflects the higher solubility of strontium compared with ^{137}Cs, though both are deposited as ions. The negative relationship between annual ^{137}Cs loss and basin size shown in Figure 18.7 is analogous to the trends exhibited by the sediment delivery ratio and the unit-area sediment yield, which both similarly decline with catchment area (cf. Walling, 1983). For example, the relatively high loss rate of ^{137}Cs of 0.5% yr^{-1} in the headwater Brunner Creek catchment (Ritchie et al., 1982) is two orders of magnitude higher than the comparable estimate reported for the Hudson River (Linsalata et al., 1985).

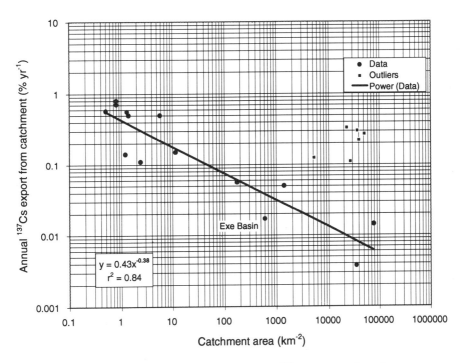

Figure 18.7 Relationship between mean annual ^{137}Cs export and catchment area

Ritchie et al. (1974) report sediment yields within the headwaters of the Upper Mississippi of 1500 t km^{-2} yr^{-1}, whereas the equivalent yields within the Great Miami River were only 50 t km^{-2} yr^{-1} (Sprugel and Bartelt, 1978). Other factors cited as having a role in controlling ^{137}Cs export rates include topography, land use, soil erodibility, and the magnitude–frequency characteristics of runoff events during the study period (de Roo, 1991; Bonnett, 1990). It is therefore suggested that the ^{137}Cs export rate of any catchment will be a non-linear function of basin area and unit-area sediment yield.

A general relationship in terms of the annual percentage loss is expressed as:

$$^{137}\text{Cs loss } (\%) = 0.43 (\text{area})^{-0.38} \qquad (r^2 = 0.84, p > 0.05) \qquad (18.9)$$

The data derived for the present study lie close to the regression line, though plotting on the lower boundary owing to the relatively low mean annual sediment yield. Some data relating to the erosional flux of Chernobyl radiocaesium are also presented (Rowan and Walling, 1992), suggesting that this general relationship also holds for Chernobyl deposition. The regression relationship did, however, exclude the data of Salo et al. (1984) and Dominik et al. (1987), because their studies represent major outliers. This exclusion is justified because these high-latitude and alpine systems have runoff characteristics dominated by snowmelt, and large areas of the basins are lake surfaces, glaciers or exposed rock. The retention of atmospheric fallout on the soil surface is therefore dramatically reduced, with a greater role for the direct runoff component. This effect may be responsible for the seasonal variability of sediment-associated ^{137}Cs content observed by Dominik et al. (1987), but not found in the Exe basin, with its much higher retention capacity.

The development of a series of curves appropriate to a common set of environmental factors (i.e. $f(x)$ = runoff, soil types, erosion rates, source areas, ..., n) is one possible way forward to improve predictions of the fate of environmental contaminants such as ^{137}Cs. However, the utility of this approach is limited by the conceptual lumping of spatially and temporally dynamic erosion and storage processes to an average catchment response. A more effective way forward is to integrate the results of carefully designed sampling programmes as a means to calibrate and test physically based distributed models (Santschi *et al.*, 1990).

CONCLUSIONS AND PERSPECTIVE

This study offers insight into the factors controlling sediment-associated ^{137}Cs levels within a medium-sized UK drainage basin. The focus of the research was on the temporal variability in ^{137}Cs transport over a range of time-scales. The analysis clearly points to an important role for runoff provenance. These observations are in good agreement with the findings from the atmospheric fallout survey, which revealed significant variations in the ^{137}Cs inventory within the catchment. A reconstruction of long-term ^{137}Cs activity trends was attempted using a simple input–output budget and an interpretation of suspended sediment sequences stored in floodplain profiles. These essentially independent methods provided similar results, and suggest that erosional removal of ^{137}Cs is insignificant when compared to natural decay losses. However, the potential of fluvial processes to redistributed considerable quantities of radiocaesium to downstream reaches, with the attendant storage of contaminants in downstream sediment sinks such as channel, floodplain and reservoir sites, indicates the need for further research in this important field.

ACKNOWLEDGEMENTS

Field data were collected while the author was in receipt of an NERC Research Studentship based in the Department of Geography, University of Exeter. The considerable input and support of Professor Des Walling is warmly acknowledged.

REFERENCES

Bachhuber, H., Bunzl, K. and Schimmack, W. (1987) Spatial variability of fallout ^{137}Cs in the soil of a cultivated field. *Environ. Monit. Assess.* **8**, 93–101.

Bonnett, P.J.P. (1990) A review of the erosional behaviour of radionuclides in selected drainage basins. *J. Environ. Radioact.* **11**, 251–266.

Brown, R.B., Cutshall, N.H. and Kling, G.F. (1981) Agricultural erosion indicated by ^{137}Cs redistribution: I. Levels and distribution of ^{137}Cs in soils. *Soil Sci. Soc. Am. J.* **45**, 1184–1190.

Cambray, R.S., Cawse, P.A., Garland, J.A., Gibon, J.A.B., Johnson, P., Lewis, G.N.J., Newton, D., Salmon, L. and Wade, B.O. (1987) Observations on radioactivity from the Chernobyl accident. *Nucl. Energy*, **26**, 77–101.

Cambray, R.S., Playford, K., Lewis, G.N.J. and Carpenter R.C. (1989) *Radioactive Fallout in Air and Rain: Results to the End of 1987*, AERE Report R13226, HMSO, London.

Campbell, B.L., Loughran, R.J. and Elliot, G.L. (1982) Caesium-137 as an indicator of geomorphic processes in a drainage basin system. *Aust. Geogr. Stud.* **20**(1), 49–64.

Carlsson, S. (1978) A model for the movement and loss of ^{137}Cs in a small watershed. *Health Phys.* **34**, 33–37.

Cawse, P.A. and Horrill, A.D. (1986) *A Survey of Caesium-137 and Plutonium in British Soils in 1977*, AERE Report R10155, HMSO, London.

Chapman, D. (ed.) (1992) *Water Quality Assessments*, Chapman and Hall, London.

Cooper, J.R., Gilliam, J.W., Daniels, R.B. and Robarge, W.P. (1987) Riparian areas as filters for agricultural sediment. *Soil Sci. Soc. Am. J.* **51**, 416–420.

de Jong, E., Begg, C.B.M. and Kachanoski, R.G. (1983) Estimates of soil erosion and deposition for some Saskatchewan soils. *Can. J. Soil Sci.* **63**, 603–617.

de Roo, A.P.J. (1991) The use of ^{137}Cs in an erosion study in south Limburg (The Netherlands) and the influence of Chernobyl fallout. *Hydrol. Processes* **5**, 215–227.

Dominik, J., Burrus, D. and Vernet, J.-P. (1987) Transport of the environmental radionuclides in an alpine watershed. *Earth Planet. Sci. Lett.* **84**, 165–180.

Foster, G.R. and Hakonson, T.E., (1984) Predicted erosion and sediment delivery of fallout plutonium. *J. Environ. Qual.* **13**(4), 595–602.

Frissel, M.J. and Pennders, R. (1983) Models for the accumulation and migration of ^{90}Sr, ^{137}Cs, ^{239,240}Pu and ^{241}Am in the upper layers of soils. In: *Ecological Aspects of Radionuclide Release*, Special Publication no. 3, British Ecological Society, Oxford, pp. 63–72.

Helton, J.C., Muller, A.B. and Bayer, A.(1985) Contamination of surface-water bodies after reactor accidents by the erosion of atmospherically deposited radionuclides. *Health Phys.* **48**(6), 757–771.

Jacobi, W. (1972) Transfer of fission products from atmospheric fallout into river water. In: *Radioecology Applied to the Protection of Man and His Environment (Symposium)*, CEC, pp. 1153–1163.

Kiss, J.J., de Jong, E. and Martz, L.W. (1988) The distribution of fallout cesium-137 in southern Saskatchewan, Canada. *J. Environ. Qual.* **17**(3), 445–452.

Lambert, C.P. (1986) The suspended sediment delivery dynamics of river channels in the Exe basin. Unpublished PhD thesis, University of Exeter.

Lambert, C.P and Walling, D.E. (1988) Measurement of channel storage of suspended sediment in a gravel bed river. *Catena* **15**, 68–80.

Linsalata, P., Simpson, H.J., Olsen, C.R., Cohen, H. and Trier, R.M. (1985) Plutonium and radiocaesium in the water column of the Hudson River estuary. *Environ. Geol. Water Sci.* **7**(4), 193–204.

Livens, F.R. and Baxter, M.S. (1988) Particle size and radionuclide levels in some west Cumbrian soils. *Sci. Total Environ.* **70**, 1–17.

Loughran, R.J. (1989) The measurement of soil erosion. *Prog. Phys. Geogr.* **13**, 216–233.

McBratney, A.B. and Webster, R. (1986) Choosing functions for semi-variograms of soil properties and fitting them to sample estmates. *Journal of Soil Science* **37**, 617–639.

McHenry, J.R. and Ritchie, J.C. (1975) Redistribution of caesium-137 in southeastern watersheds. In: Howells, F.G., Gentry, J.B. and Smith, M.H. (eds), *Mineral Cycling in Southeastern Biosystems*, ERDA Symposium Series, Conf. 74 0513, 452–461.

McHenry, J.R. and Ritchie, J.C. (1977) Physical and chemical parameters affecting transport of Cs-137 in arid watersheds. *Water Resour. Res.* **13**, 923–927.

Novotny, V. (1980) Delivery of suspended sediment and pollutants from nonpoint sources during overland flow. *Water Resour. Bull.* **16**(6), 1057–1065.

Oliver, M.A., Webster, R. and Gerrard. A.J.W. (1989) Geostatistics in physical geography: Part 1: theory. *T.I.B.G* **14**, 259–269.

Peirson, D.H. and Cambray, R.S. (1967) Interhemispheric transfer of debris from nuclear explosions using a simple atmospheric model. *Nature* **205**, 433–440.

Peirson, D.H. and Salmon, L. (1959) Gamma radiation from deposited fallout. *Nature* **184**, 1678–1679.

Pennington, W., Cambray, R.S. and Fisher, E.M. (1973) Observations on lake sediment using fallout ^{137}Cs as a tracer. *Nature* **242**, 342–346.

Playford, K., Toole, S. and Adsley, I. (1993) *Radioactive Fallout in Air and Rain: Results to the End of 1991*, Report AEA-EE-0498, AEA Technology, Oxford.

Ritchie, J.C., McHenry, J.R., Gill, A.C. and Hawkes, P.H. (1970) The use of fallout cesium-137 as a tracer of sediment movement and deposition. *Proceedings of the Mississippi Water Resources Conference*, pp. 149–162.

Ritchie, J.C., McHenry, J.R., Gill, A.C. and Hawkes, P.H. (1974) Fallout ^{137}Cs in the soils and sediments of three small watersheds. *Ecology* **55**, 887–890.

Ritchie, J.C., Hawks, P.H. and McHenry, J.R. (1975) Deposition rates in valleys determined using fallout cesium-137. *Geol. Soc. Am. Bull.* **86**, 1128–1130.

Ritchie, J.C., McHenry, J.R. and Babenzer, G.D. (1982) Redistribution of fallout ^{137}Cs in Brunner Creek watershed in Wisconsin. *Wisc. Acad. Sci. Arts Lett.* **70**, 161–166.

Rogowski, A.S. and Tamura, T. (1970) Erosional behaviour of cesium-137. *Health Phys.* **18**, 467–477.

Rowan, J.S. and Walling, D.E. (1992) The transport and fluvial redistribution of Chernobyl-derived radiocaesium within the Wye basin, UK. *Sci. Total Environ.* **121**, 109–131.

Rowan, J.S., Bradley, S.B. and Walling, D.E. (1992) Fluvial redistribution of Chernobyl fallout: reservoir evidence in the Severn basin. *J. Inst. Water Environ. Manage.* **6**, 659–666.

Salo, A., Saxen, R. and Pukahainen, M. (1984) Transport of airborne Sr-90 and Cs-137 deposited in the basins of the five largest rivers in Finland, *Aqua Fenn.* **1**, 21–31.

Santschi, P., Bollhalder, S., Farrenkothen, K., Lueck, A., Zingg, S. and Sturm, M. (1988) Chernobyl radionuclides in the environment—tracers for the tight coupling of atmospheric, terrestrial, and aquatic geochemical processes. *Environ. Sci. Technol.* **22**(5), 510–516.

Santschi, P.H., Bollhalder, S., Zingg, S., Luck, A. and Farrenkothen, K (1990) The self-cleansing capacity of surface waters after radioactive fallout—evidence from European waters after Chernobyl, 1986–1988. *Environ. Sci. Technol.* **24**(4), 519–527.

Shear, H. and Watson, A.E.P. (eds) (1977) *The Fluvial Transport of Sediment-Associated Nutrients and Contaminants*, International Joint Commission on the Great Lakes, Windsor, Ontario, Canada.

Smith, J.N. and Ellis, K.M. (1982) Transport mechanism for ^{210}Pb, ^{137}Cs and Pu fallout radionuclides through fluvial–marine systems. *Geochi. Cosmochimi. Acta* **46**, 941–954.

Smith, J.N., Ellis, K.M. and Nelson, D.M. (1987) Time dependent modelling of fallout radionuclide transport in a drainage basin: significance of 'slow' erosional and 'fast' hydrological components. *Chem. Geol.* **63**, 157–180.

Sprugel, D.G and Bartelt, G.E. (1978) Erosional removal of fallout plutonium from a large Mid-Western watershed. *J. Environ. Qual.* **7**(2), 175–177.

Walling, D.E. (1983) The sediment delivery problem. *J. Hydrol.* **65**, 209–237.

Walling, D.E. (1990) Linking fields to the river: sediment delivery from agricultural land. In: Boardman, J., Foster, I.D.L. and Dearing, J.A. (eds), *Soil Erosion on Agricultural Land*, John Wiley & Sons, Chichester, pp. 129–153.

Walling, D.E. and Bradley, S.B. (1989) Rates and patterns of contemporary floodplain sedimentation: a case study of the River Culm, Devon, UK. *GeoJournal*, **19**(1), 53–62.

Walling, D.E. and Moorehead, P.M. (1989) The particle size characteristics of fluvial suspended sediment. *Hydrobiologia* **176**/177, 125–149.

Walling, D.E. and Quine, T.A. (1992) The use of caesium-137 measurements in soil erosion surveys. In: Bogen, J., Walling, D.E. and Day, T. (eds), *Erosion and Sediment Transport Monitoring Programmes in River Basins*, IAHS Publication no. 210, pp. 143–152.

Walling, D.E. and Webb, B.W. (1985) Estimating the discharge of contaminants to coastal waters: some cautionary comments. *Marine Pollut. Bull.* **16**(12), 488–492.

Walling, D.E. and Webb, B.W. (1987) Suspended load in gravel bed rivers: UK experience. In: Thorne, C.R., Bathurst, J.C. and Hey, R.D. (eds), *Sediment Transport in Gravel-Bed Rivers*, John Wiley & Sons, Chichester, pp. 767–782.

Walling, D.E. and Woodward, J.C. (1992) Use of radiometric fingerprints to derive information on suspended sediment sources. In: Bogen, J., Walling, D.E. and Day, T. (eds), *Erosion and Sediment Transport Monitoring Programmes in River Basins*, IAHS Publication no. 210, pp. 153–164.

Walling, D.E., Bradley, S.B. and Wilkinson, C.P. (1986) A caesium-137 budget approach to the investigation of sediment delivery from a small agricultural drainage basin in Devon, UK. In: Hadley, R.F. (ed.), *Drainage Basin Sediment Delivery*, IAHS Publication no. 159, pp. 423–435.

Walling, D.E., Rowan, J.S. and Bradley, S.B. (1989) Sediment-associated transport and redistribution of Chernobyl fallout radionuclides. In: Hadley, R.F. and Ongley, E.D. (eds), *Sediment in the Environment*, IAHS Publication no. 184, pp. 37–46.

Walling, D.E., Webb, B.W. and Woodward, J.C. (1992) Some sampling consideration in the design of effective strategies for monitoring sediment-associated transport. In: Bogen, J., Walling, D.E. and Day, T. (eds), *Erosion and Sediment Transport Monitoring Programmes in River Basins*, IAHS Publication no. 210, pp. 279–288.

Zhang, X., Higgitt, D.L. and Walling, D.E. (1990) A preliminary assessment of the potential for using caesium-137 to estimate rates of soil erosion in the Loess Plateau, China. *Hydrol. Sci.* **35**(3), 243–252.

19 Use of Caesium-137 to Investigate Sediment Sources in the Hekouzhen–Longmen Basin of the Middle Yellow River, China

X. ZHANG[1] AND Y. ZHANG[2]

[1]*Chengdu Institute of Mountain Hazards and Environment, China;* and [2]*Department of Physics, Sichuan University, Chengdu, China*

INTRODUCTION

The Huang He or Yellow River (Figure 19.1a) is known worldwide for the enormity of its sediment load, 1.6×10^9 t yr^{-1}, and the Hekouzhen–Longmen basin (Figure 19.1b) of the middle reaches is one of the main sediment source areas. Although the Hekouzhen–Longmen basin is relatively small (1.3×10^5 km^2), and represents only 17.2% of the area of the Yellow River, it contributes 61.2% (9.8×10^8 t yr^{-1}) of the annual sediment load (Gong and Xiong, 1980) and is, therefore, of great interest to those involved in the development of land-use policies and soil conservation measures. Furthermore, there is a clear need for information concerning the relative importance of the major sediment sources in the basin if optimum land-use strategies are to be devised. This chapter examines the potential for using caesium-137 (^{137}Cs) measurements to meet this need in a study of the sediment contribution of hill and gully areas, in the Hekouzhen–Longmen basin, to the sediment load of the Yellow and Qinjian Rivers.

Fallout ^{137}Cs, associated with the atmospheric testing of nuclear weapons, has proved to be a valuable, globally distributed, tracer for studies of soil erosion and sediment delivery (McHenry and Ritchie, 1977; Loughran *et al.*, 1988; de Jong *et al.*, 1983). Most ^{137}Cs fallout occurred during the period from 1954 to 1970, with maximum deposition in 1963–64 and relatively little fallout since 1970. On reaching the land surface, ^{137}Cs is strongly adsorbed within the upper horizons of the soil and is subject to minimal downward leaching (Tamura, 1964; Frissel and Pennders, 1983). Subsequent redistribution of ^{137}Cs is, therefore, associated with the erosion, transport and deposition of soil and sediment particles (Walling and Quine, 1992). Because of these properties and its relatively long half-life (*ca.* 30 years), ^{137}Cs has proved to be a valuable tracer for studies of contemporary soil-particle movement (Walling and Bradley, 1988).

Sediment and Water Quality in River Catchments. Edited by I.D.L. Foster, A.M. Gurnell and B.W. Webb.
© 1995 John Wiley & Sons Ltd.

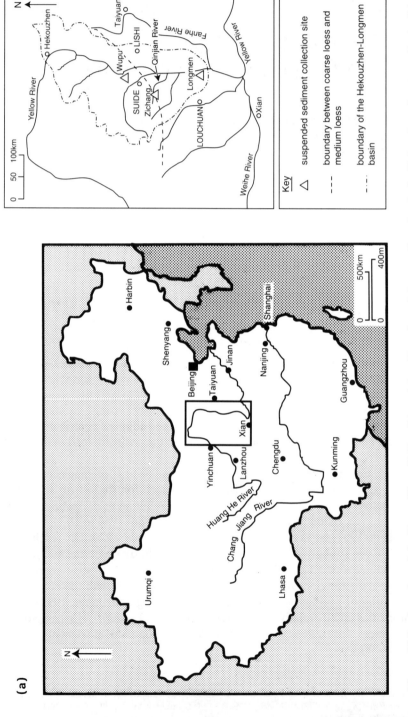

Figure 19.1 Location of the Huang He or Yellow River (a) and location of the sampling sites in the Hekouzhen–Longmen Basin (b)

Table 19.1 Runoff and sediment characteristics of the Yellow River and the Qinjian River

River and station	Catchment area (km^2)	Runoff (×10^8 m^3)		Mean discharge (m^3 s^{-1})		Sediment load (×10^8 t)		Mean sediment concentration (kg m^{-3})	Observed maximum (m^3 s^{-1})
		Annual	Jun–Sept	Annual	Jun–Sept	Annual	Jun–Sept		
Yellow River									
Hekouzhen	367 898	249.2	129.5	790.2	1228.5	1.67	1.14	5.37	–
Wupu	433 514	312.3	162.5	949.0	1541.6	6.73	5.48	21.5	24 000 (1976)
Longmen	497 552	344.1	177.4	1050.0	1682.9	11.46	9.84	33.3	21 000 (1967)
Hekouzhen–Longmen basin	129 654	94.9	47.9	300.9	454.4	9.79	8.70	103.2	–
Qinjian River									
Zichang	913	0.423	0.284	1.35	2.69	0.12	0.118	293.0	3 150 (1969)

THE STUDY BASIN

The Hekouzhen–Longmen basin is situated in the eastern part of the Loess Plateau (Figure 19.1b) and has a semi-arid temperate climate. Mean annual precipitation lies in the range 400–500 mm, but there is marked seasonal variation, with 65–70% falling in the period from June to October. The basin is a deeply dissected hilly plateau characterised by two geomorphological units of essentially equal area, namely the hill area and the gully area. The hill area has relatively gentle slopes (0°–25°), which have been cultivated for several centuries, and erosion is dominated by sheet and rill processes, with depth of incision rarely exceeding the plough depth. The gully area is characterised by valley and gully slopes with angles in excess of 25°. Where the slope angle is lower than 35° the surface is covered by shrubs and grasses and there is very little erosion, but on the steeper slopes there is very little or no vegetation and gully and gravitational (slumps, slides, avalanches and earth falls) erosion processes predominate. In the absence of artificial sediment traps, the majority of the eroded sediment is transported beyond the basin to the Yellow River, and it has been suggested that the sediment delivery ratio is close to 1 (Jing and Cheng, 1993).

The basin has the highest erosion rates in the Loess Plateau, with a mean of 7548 $t\,km^2\,yr^{-1}$. Within the basin the severity of erosion varies with the nature of the loess. In the north, which is characterised by coarse loess, rates of erosion reach 25 000 $t\,km^2\,yr^{-1}$; while in the south basin, where medium loess is dominant, erosion rates of 4000 $t\,km^2\,yr^{-1}$ are more typical (Jing and Cheng, 1993). The importance of the Hekouzhen–Longmen basin for sediment supply to the Yellow River is further emphasised by the data summarised in Table 19.1 (pre-1980 data). These show that the upper Yellow River basin, upstream of Hekouzhen, contributes more runoff but less sediment. The recorded runoff at Hekouzhen is 72% of the value at Longmen, but the sediment load at Hekouzhen is only 15% of the load at Longmen. Similarly, the mean sediment concentration at Hekouzhen of 5.37 $kg\,m^{-3}$ is only 16% of the mean concentration at Longmen (33.3 $kg\,m^{-3}$). The data in Table 19.1 also demonstrate marked seasonal variation, and this reflects the role of a few large flood events during the flood season, from June to September, in contributing a large percentage of the sediment load of the Yellow River. Finally, it should be noted that in the period since 1970 there has been a considerable reduction in sediment load from the Hekouzhen–Longmen basin (although detailed data are not yet available), and this probably reflects both the low precipitation during the period and the retention of large quantities of sediment by dams constructed in the gullies since 1970 (Cheng and Jing, 1989).

CAESIUM-137 CONTENT OF SOURCE MATERIALS AND SUSPENDED SEDIMENTS

A study of undisturbed 'reference' sites across the Hekouzhen–Longmen basin, undertaken in 1988, indicated that measurable levels of ^{137}Cs had been deposited across the entire area. The 'reference' inventories lay in the range 1669–2692 $Bq\,m^{-2}$, with an increase from north to south mirroring the variation in annual precipitation. These results confirmed the potential applicability of the ^{137}Cs approach in the basin.

Soils of the hill area

The hill area is dominated by cultivated land and the ^{137}Cs present in the soil profiles is found to be evenly distributed to the base of the plough layer (15–20 cm). Because erosion in this area is dominated by sheet and rill processes and incision rarely exceeds the base of the plough layer, the ^{137}Cs content of representative samples of ploughed soil may be used to provide an indication of the ^{137}Cs content of sediment originating from the hill area.

Samples were, therefore, collected from the plough layer of cultivated fields at two representative sites in the central area of the Hekouzhen–Longmen basin: Lishi, Shanxi Province, and Suide, Shaanxi Province (Figure 19.1b). At Lishi, in 1988, the ^{137}Cs content of the ploughed soil was found to be in the range 4.81–6.84 Bq kg^{-1} ($n = 8$) with a mean of 5.78 Bq kg^{-1} (Zhang et al., 1989). At Suide, in 1989, the range was found to be 3.16–6.84 Bq kg^{-1} ($n = 30$) with a mean of 4.81 Bq kg^{-1} (Wang and Zhang, 1991). When the mean values are corrected for radioactive decay to 1993 levels, the contents are as follows: 4.23 Bq kg^{-1} at Lishi and 3.60 Bq kg^{-1} at Suide. Although only limited numbers of samples have been employed to represent a very large area, the selected sites are typical of the region and the similarity in the mean values for the two sites (ca. 80 km apart) is encouraging and suggests that a value of 3.9 Bq kg^{-1} may be a reasonable indicator of the ^{137}Cs content of sediment derived from the hill area of the basin.

Sources in the gully area

When attempting to characterise the ^{137}Cs content of sediment derived from the gully area, the vegetated gentler slopes were excluded because the full vegetation cover is indicative of minimal erosion. Samples were, therefore, collected from the surface of bare gully slopes, and in each case no ^{137}Cs was found. Furthermore, it can be assumed that sediment derived from subsurface piping and gravitational erosion would contain negligible levels of ^{137}Cs. The ^{137}Cs content of sediment derived from the gully areas may, therefore, be assumed to be zero.

Suspended sediment samples

Suspended sediment samples were collected at three hydrological stations during the period from July to September 1993. At each station, three floods were sampled and five samples were collected for each flood. Two of the stations were located on the main channel of the Yellow River, at Wupu and Longmen (Figure 19.1b). The three floods sampled at these sites represent the range in magnitude for the year from small to large; however, even the largest had a return period of ca. 1 yr. The third station was located at Zichang on the Qinjian River, a tributary of the Yellow River (Figure 19.1b). The three floods sampled at this site were the largest in the year, but none exceeded the mean annual flood.

The ^{137}Cs contents of the suspended sediment samples are summarised in Table 19.2 with relevant hydrological data. The overall range in the ^{137}Cs content of the suspended sediment was 0–2.696 Bq kg^{-1}, and the arithmetic mean content of all the samples was 0.76 Bq kg^{-1}. However, the ^{137}Cs content of the suspended sediment varied both within

Table 19.2 Caesium-137 contents of flood sediments collected from the Yellow River and the Qinjian River in 1993

Sampling location and river	Flood period	Maximum discharge ($m^3 s^{-1}$)	Maximum sediment concentration ($kg\ m^3$)	^{137}Cs concentration of suspended sediment ($Bq\ kg^{-1}$)	
				Range	Mean value
Wupu Yellow River	30 Jul 0:00 to 2 Aug 12:00	2300	36.0	0–0.275	0.148
	10 Aug 8:00 to 13 Aug 8:00	1640	14.2	0–0.653	0.027
	22 Sep 8:00 to 23 Sep 12:00	1220	11.8	0.335–1.394	0.9194
Longmen Yellow River	1 Aug 9:00 to 2 Aug 2:00	2100	65.3	0.364–2.006	1.324
	4 Aug 10:30 to Aug 16:00	4430	321.6	0–1.387	0.423
	22 Sep 8:00 to 24 Sep 8:00	1290	14.6	0.658–2.696	1.677
Zichang, Qinjian	26 Jul 7:12 to 26 Jul 17:42	370	884	0.437–1.115	0.733
	31 Jul 16:18 to 1 Aug 8:00	536	681	0.144–1.930	1.029
	21 Aug 0:12 to 21 Aug 12:00	690	757	0–0.919	0.445

and between floods. This may in part be explained by variation in the particle-size characteristics of the suspended sediments. In general, the larger floods transported coarser sediment with a lower ^{137}Cs content. Furthermore, variation in particle size within events accounts for much of the variation in ^{137}Cs content. This is illustrated in Figure 19.2 and in the regression of ^{137}Cs content (Y, Bq kg^{-1}) against the fine-particle (<0.01 mm) content (X, per cent) for each of the stations:

- For the Qinjian River at Zichang ($r = 0.754$)

$$Y = 0.227 + 0.020X \tag{19.1}$$

- For the Yellow River at Wupu ($r = 0.851$)

$$Y = 0.049 + 0.028X \tag{19.2}$$

- For the Yellow River at Longmen ($r = 0.908$)

$$Y = 0.008 + 0.040X \tag{19.3}$$

The good correlation between the ^{137}Cs content and fine-particle content reflects the well-known affinity of ^{137}Cs for the finer fraction, and indicates the need for particle-size correction when undertaking comparison with the source samples. The high correlation coefficients for the sites on the Yellow River are consistent with a relatively

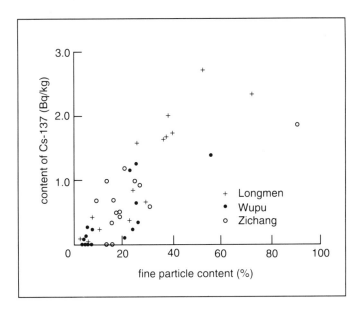

Figure 19.2 The relationship between the ^{137}Cs and fine-sediment content (%<0.01 mm) of suspended sediment samples from Longmen and Wupu on the Yellow River, and Zichang on the Qinjian River

constant sediment source because particle-size variation appears to be the major control on variation in ^{137}Cs content. Conversely, the relatively low correlation coefficient between ^{137}Cs and fine-particle content at Zichang may indicate that intra-flood variation in sediment sources plays a more important role in determining the variability of ^{137}Cs content in suspended sediment in the smaller basin of the Qinjian River.

RELATIVE CONTRIBUTIONS OF SEDIMENT FROM THE HILL AND GULLY AREAS

The clear difference in ^{137}Cs content of potential sediment sources in the two geomorphological units of the Hekouzhen–Longmen basin, the hill areas and the gully areas, allows the use of a simple mixing model to determine the relative contributions of sediment from the two units in the suspended load of the rivers studied:

$$C_{ss} = f_h C_h + f_g C_g \qquad (19.4)$$

where C_{ss} = ^{137}Cs content of suspended sediment, corrected for particle-size enrichment (Bq kg^{-1}), C_h = ^{137}Cs content of sediment derived from the hill area (Bq kg^{-1}), f_h = fraction of suspended sediment derived from the hill area, C_g = ^{137}Cs content of sediment derived from the gully area (Bq kg^{-1}) and f_g = fraction of suspended sediment derived from the gully area.

In this case, because the ^{137}Cs content of sediment derived from the gully area may be considered to be zero, the equation may be solved and the fractional contributions of the gully area and hill area calculated as follows:

$$f_h = C_{ss}/C_h \qquad (19.5)$$

$$f_g = 1 - f_h \qquad (19.6)$$

In order to determine the fractional contributions of sediment from the hill and gully areas of the whole Hekouzhen–Longmen basin, the 'particle-size-corrected' ^{137}Cs content of suspended sediment at Longmen is used in Eqn (19.5). This is considered to be valid because 85.4% of the sediment load of the Yellow River at Longmen is derived from the basin. Correction for particle size was undertaken by using a fine-particle content of 18.7% (Jing and Cheng, 1993) in Eqn (19.3). This provides a 'particle-size-corrected' ^{137}Cs content for the suspended sediment of 0.756 Bq kg^{-1}. As indicated above, on the basis of investigations at Lishi and Suide, a value of 3.9 Bq kg^{-1} may be used to represent the ^{137}Cs content of sediment derived from the hill area. Using these figures, the calculated percentage contributions of sediment derived from the hill (Eqn 19.5) and gully (Eqn 19.6) areas are 19% and 81%, respectively. The ^{137}Cs content of suspended sediment collected at Wupu (mean: 0.394 Bq kg^{-1}; range 0–1.394 Bq kg^{-1}) was lower than at Longmen (mean 1.14 Bq kg^{-1}; range 0–2.696 Bq kg^{-1}). This may indicate that gully and gravitational erosion are more important in the north basin than in the south (use of Eqn 19.5 would indicate an 85% contribution from the gully areas, again using a ^{137}Cs content of 3.9 Bq kg^{-1} for hill-area sediment). However, it is possible that the lower ^{137}Cs content of the suspended sediment at Wupu may reflect the lower ^{137}Cs inventories and contents in cultivated soils in the north basin, which result from the lower ^{137}Cs fallout inputs in that area. The data available do not allow differentiation between these hypotheses, and there is clearly a need for sampling of cultivated soils from representative sites across the entire basin to establish the degree of variation in ^{137}Cs content of the ploughed soils resulting from the variation in fallout inputs.

The possible greater variability in source areas indicated by the suspended sediment samples from Zichang on the Qinjian River has been referred to above. However, the overall pattern is similar to that seen at the other sites. Using a 'particle-size-corrected' ^{137}Cs content of 0.542 Bq kg^{-1} for the suspended sediment at Zichang, the calculated contributions from the hill and gully areas are 14% and 86%, respectively.

CONCLUSION

This investigation has demonstrated the potential value of ^{137}Cs studies in elucidating the relative importance of suspended sediment sources in an area of very serious erosion. By providing rapid access to quantitative information concerning sediment sources, the ^{137}Cs approach can make a significant contribution to the development of more efficient soil erosion and sediment control strategies. This information is of particular importance in the middle reaches of the Yellow River.

The results of this investigation, which identify the gully areas as the source of 81–86% of the suspended sediment, are in close accord with other published work. For example, observations made from 1963 to 1968 in the Yangdao gully, a small deeply dissected catchment near Lishi, and more recent investigations using ^{137}Cs (Zhang et al., 1989) at the same site, indicated that 80% of the sediment was derived from the gully area and 20% from the hill area of the catchment. This agreement is encouraging. However, there is clearly scope for further development of the approach in this environment. First, there is a need to establish the degree of variation exhibited in the ^{137}Cs content of the potential source areas through extensive sampling across the basin. Secondly, more detailed sampling of suspended sediment during individual events may provide valuable information concerning changing sediment sources, and collection of samples from several floods may indicate seasonal variation in sources. However, it must be recognised that such detailed interpretation may only be possible if smaller tributaries, such as the Qinjian River, are investigated, because local variations may be concealed in the composite signature represented by the suspended sediment of the Yellow River.

ACKNOWLEDGEMENTS

The authors are very grateful to the Yuchi headquarters of the Yellow River Conservancy Committee for arranging the collection of sediment samples at the hydrological stations, and to Dr Tim Quine for assistance with the manuscript.

REFERENCES

Cheng, Y. and Jing, K. (1989) *Coarse Sediment Sources of the Yellow River and their Recent Changes*, Meteorology Press, Beijing, China, 144 pp. (in Chinese).

de Jong, E., Begg, C.B.M. and Kachanoski, R.G. (1983) Estimates of soil erosion and deposition for some Saskatchewan soils. *Can. J. Soil Sci.* **63**, 617–607.

Frissel, M.J. and Pennders, R. (1983) Models for the accumulation and migration of ^{90}Sr, ^{137}Cs, ^{239}Pu, ^{240}Pu and ^{24}Am in the upper layer of soils. In: *Ecological Aspects of Radionuclide Release*, (ed.), Coughtrey, P.J. Special Publication no. 3, British Ecological Society, Oxford, pp. 63–72.

Gong, S. and Xiong, G. (1980) The origin and transport of sediment in the Yellow River, In: *Proceedings of the International Symposium on River Sedimentation*, Guaghua Press, Beijing, pp. 45–52.

Jing, K. and Cheng, Y. (1993) *Sediment and Environment of the Yellow River*, Science Press, Beijing, China, 248 pp. (in Chinese).

Loughran, R.J., Elliott, G.L., Campbell, B.L. and Shelly, D.J. (1988) Estimation of soil erosion from caesium-137 measurements in a small cultivated catchment in Australia. *Appl. Radiat. Isotopes* **39**, 1153–1157.

McHenry, J.R. and Ritchie, J.C. (1977) Estimating field erosion losses from fallout cesium-137 measurements. In: *Erosion and Solid Matter Transport in Inland Waters (Proceedings of Paris Symposium, 1977)*, IAHS Publication no. 122, pp. 26–33.

Tamura, T. (1964) Selective sorption reactions of caesium with mineral soils. *Nucl. Safety* **5**, 263–268.

Walling, D.E. and Bradley, S.B. (1988) The use of caesium-137 measurements to investigate sediment delivery from cultivated areas in Devon, UK. In: *Sediment Budgets (Proceedings of Porto Alegre Symposium, December 1988)*, Bordas, M.P. and Walling, D.E. (eds), IAHS Publication no. 174, pp. 325–335.

Walling, D.E. and Quine, T.A. (1992) The use of caesium-137 measurements in soil erosion surveys. In: *Erosion and Sediment Transport Monitoring Programmes in River Basins (Proceedings of Oslo Symposium, August 1992)*, Bogen, J. Walling, D.E. and Day, T. (eds), IAHS Publication no. 210, pp. 143–152.

Wang, Y. and Zhang, X. (1991) A study of soil erosion in the hilly area of the Loess Plateau. *Bull. Soil Water Conserv.* **11**, 34–37 (in Chinese).

Zhang, X., Li, S., Wang, C., Tan, W., Zhao, Q., Zhang, Y., Yan, M., Liu, Y., Jiang, J., Xiao, J. and Zhou, J. (1989) Use of caesium-137 measurements to investigate erosion and sediment sources within a small drainage basin in the Loess Plateau of China. *Hydrol. Processes* **3**, 317–323.

Section VI

NATIONAL AND GLOBAL PERSPECTIVES

20 Patterns of Erosion and Suspended Sediment Yield in Mediterranean River Basins

JAMIE C. WOODWARD

School of Geography, University of Leeds, UK

INTRODUCTION

> The greater the effects of human intervention and the more extreme the conditions of the climate, the more serious the consequences of erosion have been. Thus the countries around the Mediterranean are the most affected by erosion... (Zachar, 1982, p. 389)

This extract is taken from a discussion of erosion and sediment transfer in Dusan Zachar's classic book *Soil Erosion* and succinctly highlights two important elements that have combined with other factors to control rates of erosion and sediment transfer within the river basins of the Mediterranean region. However, in addition to the frequently high erosivity of the precipitation regimes and the intensity of human catchment disturbance, it can be argued that much of the Mediterranean landscape is naturally vulnerable to processes of erosion. The widespread occurrence of steep slopes, high relative relief, fissile sedimentary rocks and thin erodible soil mantles—all within an active tectonic setting—has created headwater river basins where erosive forces and sediment availability are both high. While rates of erosion can be considerable even in areas remote from any disturbance by human action—especially where steep terrain and dispersable rocks occur (cf. Douglas, 1990)—it has been suggested that average stream sediment loads may have increased by a factor of 2 to 10 since humans began widespread deforestation and farming (cf. Milliman and Syvitski, 1992).

Over the last four decades or so, information on water erosion and river sediment loads has been gathered from many parts of the Mediterranean basin and, although many aspects of the database could be improved, it is comprehensive enough to allow some very general statements about basin-wide patterns of erosion and to identify some areas where soil conservation measures are urgently needed. Many of the consequences of erosion that Zachar alludes to are well known, and in the Mediterranean region researchers have described locations where severe land degradation has taken place (including numerous examples of badland formation) and reported problems associated with declining soil fertility and crop productivity following accelerated water erosion (cf. Bonvallot and Hamza, 1977; Imeson, 1986; Rendell, 1986; Katsoulis and Tsangaris, 1994). Badland terrains are present to varying extents in all of the countries that border the Mediterranean Sea and represent an acute form of land deterioration by

---

Sediment and Water Quality in River Catchments. Edited by I.D.L. Foster, A.M. Gurnell and B.W. Webb.
© 1995 John Wiley & Sons Ltd.

water erosion, which may be triggered by human action and/or natural (endogenic and exogenic) processes (cf. Bryan and Yair, 1982; Selby, 1993). The off-site impacts of accelerated soil and bedrock erosion have also been well documented. Increases in flood peaks, the pollution of watercourses by elevated suspended sediment concentrations, and many instances of rapid reservoir sedimentation have been observed throughout the Mediterranean (cf. Colombani, 1977; Ghorbel and Claude, 1977; Lahlou, 1988; Walsh et al., 1992). The last problem is of particular concern in many parts of the region because the pronounced seasonality of the rainfall regime places extra reliance on water storage schemes for both irrigation and potable water supplies. Millington (1990) has described three of the environments in Mediterranean southern Europe where soil erosion is commonplace:

- On the sandy loam and stony soils of the Iberian plateau, where inter-rill erosion is common and there are extensive badlands.
- Soils in the steeply sloping vineyards of many European wine-growing regions are poorly protected from the effects of rainfall and runoff.
- Soils developed on flysch and shale in the Alps and the Mediterranean countries are highly erodible and show extensive erosion by water.

Soil erosion has been observed throughout the Mediterranean region—across a range of physical and cultural settings—yet it can be exceedingly difficult to establish the relative importance of 'anthropogenic' and 'natural' (baseline) contributions to river sediment loads. Dedkov and Mozzherin (1992) have assembled and compared sediment yield data from 1872 mountain streams from various regions of the world and classified the basins into three categories on the basis of the degree of economic development (human disturbance). Development has been indexed using the relative proportions of forest cover and cultivated land in mountain catchments, and has been used as a basis for distinguishing between the 'natural' and 'anthropogenic' components of the specific suspended sediment yield for different regions (Figure 20.1). It is of particular interest to note that the results of this analysis suggest that the Mediterranean mountain streams evidence the highest anthropogenic contribution of any climate vegetation zone.

The Mediterranean basin is currently facing a large number of environmental problems—many of which stem directly from the extra pressures exerted on the region's natural resources in the face of rising populations and tourism (cf. Grove, 1986; Grenon and Batisse, 1989; Katsoulis and Tsangaris, 1994). Perhaps the most pressing challenges for the region are the conservation and effective management of existing soil resources and the provision of adequate water supplies for agricultural, industrial and domestic use. Meeting this last demand is a growing concern, as the rising urban populations of the Mediterranean rim place increasing demands on the already depleted, and often polluted, groundwater sources. In recent decades some of the consequences of water erosion such as land degradation and desertification have intensified across many of the semi-arid parts of Mediterranean North Africa and the Near East as the expansion of local populations has exerted greater stresses on the dwindling soil and water resources (cf. Bonvallot and Hamza, 1977). In the arid and semi-arid parts of Tunisia, for example, the population has risen seven-fold since the late 19th century, placing great pressure on an already fragile ecosystem. Overgrazing around settlements has caused land degradation, and, during dry years, livestock are

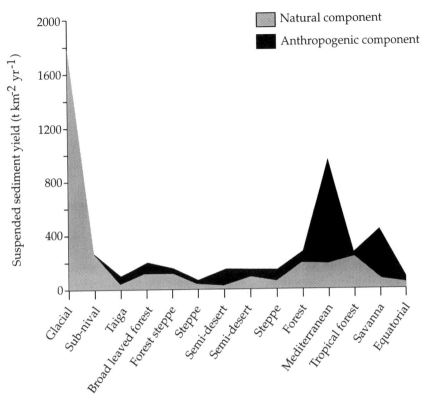

Figure 20.1 The relative importance of natural and human contributions to the suspended sediment yield of mountain river basins in different climatic regions (after Dedkov and Mozzherin, 1992). These data indicate that around 75% of the sediment yield of Mediterranean headwater river basins may be attributed to human activity

forced to rely on the consumption of the protective leaf litter—thus further decreasing soil stability (Wellens and Millington, 1990). The impact of progressive woodland clearance followed by overgrazing in seasonally dry climates has resulted in high rates of soil erosion across the Mediterranean drainage basin.

River sediment yields are controlled by interacting relief, land-use, soil resistance and climatic factors, whose individual effects are often obscured by covariation (Richards, 1982). In addition, the amount of sediment delivered from a source of erosion will be affected by the calibre of the sediment and the distance between the point of erosion on the hillslope and the river channel. A large body of evidence exists to suggest that river sediment yields have increased in the Mediterranean following widespread human disturbance of catchment slopes (Figure 20.2; see Butzer, 1982; Brückner and Hoffman, 1992). With the advent of Neolithic cultures in the Mediterranean and the introduction of various agricultural practices following woodland clearance, catchment water balances were transformed and runoff volumes, rates of soil removal and fluvial suspended sediment loads all increased (Woodward, 1995). Indeed, many researchers have identified a positive relationship between clearance phases and episodes of stream aggradation during the Holocene (cf. Lewin *et al.*, 1995). On the other hand, in later

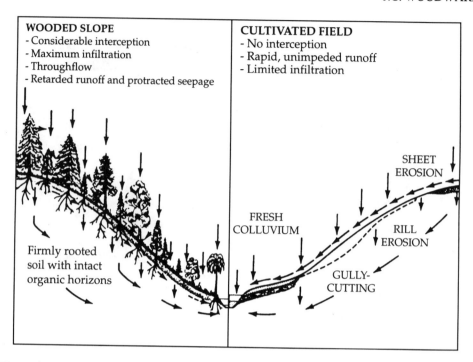

Figure 20.2 Hydrological pathways and erosion processes on steep slopes under a woodland canopy and following cultivation (after Butzer, 1976). Many river basins in the Mediterranean region have been cultivated for thousands of years and thick stores of colluvial sediment are a characteristic landscape feature. Erodible clastic sedimentary lithologies such as flysch and marl are particularly susceptible to rill erosion and gully cutting

periods, the construction of terraces 'uncoupled' the slope/channel system and checked downslope sediment transfer and, where these structures were adequately maintained, served to reduce soil losses (cf. van Andel et al., 1986). Thus it is doubtful whether the existing database provides the basis for an accurate assessment of 'natural' background sediment yield for the region. The only part of the Mediterranean basin that has been relatively little affected by human activity over the last few thousand years is the sparsely populated desert region of Saharan North Africa where the Saharan plateau meets the Mediterranean littoral and provides the only major stretch of coastline that is not flanked by mountainous terrain. As a general rule, few parts of the region have escaped the influence of human activity. In most areas the 'natural' vegetation has been replaced by fire- and/or goat-tolerant scrub, and entire basins have been cultivated for many millennia. The Mediterranean is arguably a unique region because of the extended timescale involved in the direct and indirect human modification of hillslopes and river channels (Figure 20.2). In view of the range of controlling factors outlined above and their potential degree of spatial variation, it is exceedingly difficult to distinguish accurately between 'natural' or background erosion rates and the 'human' contribution across the region. In the absence of, for example, detailed basin-wide lake sediment-based reconstructions (cf. Foster et al., 1986), the model proposed by Dedkov and Mozzherin (1992) for mountain catchments constitutes a most useful attempt (Figure 20.1).

This chapter is concerned with some of the emerging patterns and some of the consequences of water erosion and suspended sediment transport in the Mediterranean region. It is not intended to provide an exhaustive review of the controls, mechanisms and impacts of erosion, and the following account is necessarily selective, briefly examining four main themes. The first two parts describe, respectively, the physiography of the Mediterranean drainage area and then examine the magnitude and pattern of water erosion and fluvial suspended sediment yield. The third and fourth parts discuss two of the consequences of water erosion and suspended sediment transport in the region, namely aspects of badland development and reservoir sedimentation.

THE MEDITERRANEAN DRAINAGE BASIN

The Mediterranean is not so much the sea between the lands, as its name asserts, but the sea between the mountains—except for the long stretch of low-lying desert between Tunisia and Sinai, one is almost never out of sight of mountains if within sight of the Mediterranean coast (McNeill, 1992). Figure 20.3 provides a good illustration of this point, showing the 500 m contour and the major drainage lines and basin watershed of streams that drain to the Mediterranean Sea. This 500 m contour is also used by Milliman and Syvitski (1992) as the dividing line between uplands and lowlands in their elevation-based classification of world rivers and sediment yields. Apart from the exotic River Nile, the headwaters of many rivers that drain into the Mediterranean Sea lie in the young Cenozoic mountains that fringe the Mediterranean littoral. In common with the Pacific-margin steeplands (cf. Beschta et al., 1987), the Mediterranean rim comprises stretches of narrow coastal plain with a backdrop of extensive tracts of mountainous terrain where high-angled bedrock slopes, erodible Tertiary lithologies, rock gorges and cobble- and boulder-bed streams form important components of the geomorphology of the region. In most parts of the region the Mediterranean watershed is less than 200 km from the present coastline, and in many places, such as in parts of southern Spain, Israel and the north-west Balkan peninsula, the drainage divide lies less than 100 km from the sea (Figure 20.3). This relief configuration means that the drainage of the basin is dominated by relatively short, steep-gradient river systems exiting catchments formed in hard limestones and erodible Tertiary sedimentary rocks (cf. Macklin et al., 1995). Many parts of the region are also tectonically active, producing steeply sloping, high-relief terrain in relatively recent rocks (cf. Ergenzinger, 1992). In common with much of the Pacific-rim region, various mass-movement phenomena, including mudslides and debris flows, play an important role in sediment delivery to steepland rivers and alluvial fans—and soil mantles are extremely vulnerable to water erosion following human disturbance.

RATES OF EROSION AND THE SUSPENDED SEDIMENT YIELDS OF MEDITERRANEAN RIVERS

While salinisation, waterlogging and deflation are growing problems in many parts of the region, water erosion currently poses the most serious threat to the soil in the

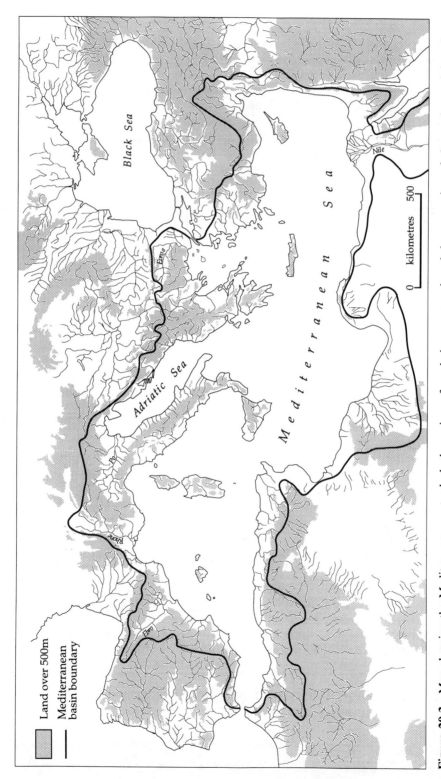

Figure 20.3 Map showing the Mediterranean watershed, the main surface drainage network and the proportion of the catchment lying above 500 m (after Macklin *et al.*, 1995). Much of the basin is characterised by mountainous terrain drained by short, steep-gradient river systems (cf. Milliman and Syvitski, 1992)

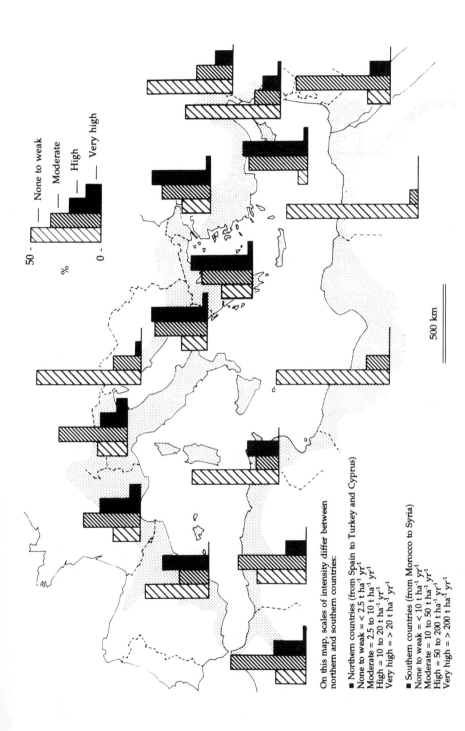

Figure 20.4 Rates of water erosion in the Mediterranean drainage basin (after Grenon and Batisse, 1989 and FAO/Blue Plan). Note that scales of intensity differ between northern and southern countries. The 16 histograms show four erosion intensity classes expressed as a percentage of the total land area of each country within the Mediterranean drainage basin (see Figure 20.3). Note the low erosion rates in the low relief, dryland terrain of Libya and Egypt

countries bordering the Mediterranean Sea (Zachar, 1982; Grenon and Batisse, 1989). In river basins underlain by unconsolidated sedimentary rocks such as marls, siltstones and flysch, rates of soil removal may exceed 250 t ha^{-1} yr^{-1} (Figure 20.4). In Turkey, for example, where 70% of the land is vulnerable to erosion, it has been estimated that around 1000 million tonnes of topsoil is lost from erosion each year (cf. Grenon and Batisse, 1989). It has recently been claimed that the Rif Mountains of Morocco presently lose soil at a greater rate than any other part of the Mediterranean basin (cf. McNeill, 1992). This is supported by recently compiled data (Figure 20.4) that show the intensity of water erosion in various countries within the Mediterranean region. The histograms show the intensity of water erosion in four classes as a percentage of total catchment area. These data, which were compiled in 1980 for the Blue Plan (see Grenon and Batisse, 1989), indicate that a significant proportion of Mediterranean north-west Africa is presently undergoing serious erosion (50–200 t ha^{-1} yr^{-1}). It has been estimated that soil formation proceeds at a rate of approximately 1 cm every 12 to 40 years in the Rif. Even this comparatively rapid rate of pedogenesis is easily outstripped by present-day erosion, which takes place at around 10 times this figure (see McNeill, 1992). Around 60% of the soil lost through water erosion in Morocco is derived from the Rif Mountains and, according to recent figures reported in McNeill (1992), the rivers of this region evidence not only the highest specific yields in the Mediterranean drainage basin, but some of the highest recorded anywhere in the world outside south-east Asia and New Zealand (Table 20.1). The Sebour River, which drains the south flank of the Rif and part of the Middle Atlas, transports a mean annual sediment load of 30 million tonnes, which is equal to the suspended sediment load of the Mississippi at Baton Rouge (McNeill, 1992, p. 323). The database used by Milliman and Syvitski (1992) in their recent global review of suspended sediment flux to the oceans does not provide an accurate representation of the possible upper limit of sediment yields across the Mediterranean region (Table 20.2). For example, Table 20.2 lists six river basins in Mediterranean southern Europe with sediment yields in excess of the highest figure reported for North Africa (1700 t km^{-2} yr^{-1}) by Milliman and Syvitski (1992). A more representative sample of suspended sediment yields for river basins in this region is provided in Table 20.3, which shows that the sediment yields of river systems in the Maghreb are generally much higher than the northshore (European) catchments. In a review of the sediment yields of African rivers, Walling (1984) has mapped the narrow belt of mountainous terrain across the Mediterranean rim of north-west Africa as the highest-yield region on the African continent. Here, sediment yields

Table 20.1 Suspended sediment yields of three rivers in the Rif Mountains of Morocco with drainage basins <2000 km^2 (after McNeill, 1992, p 322)

River	Drainage basin area (km^2)	Suspended sediment yield (t km^{-2} yr^{-1})
Loukas	1820	4 300
Tlata	178	22 000
Nekor	780	7 900–28 500

Table 20.2 Drainage basin areas, annual sediment loads and specific yields of the six highest-yield northshore (European) and southshore (African) Mediterranean rivers from the database of Milliman and Syvitski (1992). Loads and yields have been rounded to the second digit

River	Country	Area (10^6 km^2)	Load (10^6 t yr^{-1})	Yield (t km^{-2} yr^{-1})
Northshore Mediterranean (Europe)				
Semani	Albania	0.0052	22.0	4200
Shkumbini	Albania	0.0019	6.8	3600
Osumi	Albania	0.0020	5.7	2800
Lamone	Italy	0.0052	12.0	2400
Simento	Italy	0.0018	4.0	2000
Savio	Italy	0.0060	11.0	1900
Southshore Mediterranean (Africa)				
Djer	Algeria	0.00039	0.68	1700
El Harrach	Algeria	0.00039	0.63	1600
Isser	Algeria	0.0036	6.1	1700
Sebou	Morocco	0.040	26.0	930
Meddjerdah	Algeria	0.021	13.0	620
Sous	Morocco	0.016	1.6	260

Table 20.3 Mean annual suspended sediment yields for selected southshore river systems draining catchments in Mediterranean north-west Africa (Morocco, Algeria and Tunisia) not listed in the Milliman and Syvitski (1992) database (see Table 20.2). Based on data from Tixeront (1960), Heusch and Milliès-Lacroix (1971) and various other sources cited by Walling (1986)

River	Country	Area (km^2)	Annual runoff (mm)	Suspended sediment yield (t km^{-2} yr^{-1})
Ouerrha	Morocco	1765	326	3590
Aoudour	Morocco	1039	490	3850
Sra	Morocco	493	683	3500
Allalah	Algeria	295	120	4654
Ebda	Algeria	270	338	2493
Leham	Algeria	470	47	2028
Agrioun	Algeria	635	360	5300
Fodda	Algeria	767	110	4700
Bou Namoussa	Algeria	575	234	270
Bou Hamdane	Algeria	1165	89	88
Kebir Ouest	Algeria	1120	238	265
Nebanna	Tunisia	855	43	1330
Kebir	Tunisia	225	77	1313
Kasseb	Tunisia	101	588	5070
Rhezala	Tunisia	138	360	850

typically range from 1000 to 5000 t km^{-2} yr^{-1} (Table 20.3), although the sediment yields of many basins are less than 1000 t km^{-2} yr^{-1} (cf. Tixeront, 1960). Probst and Amiotte Suchet (1992) have estimated that the total annual load of the river systems of the Maghreb which drain to the Mediterranean Sea is about 100 million tonnes.

Following the classic work of Langbein and Schumm (1958), many workers have attempted to explain the marked spatial variations in suspended sediment yield observed between different climatic zones. There has recently been a renewal of interest in the relative importance of the major factors that control the magnitude of fluvial suspended sediment yields, and maximum catchment elevation is now regarded by some workers as the principal control of catchment sediment yield on a global scale (cf. Milliman and Syvitski, 1992). Existing suspended sediment yield data for Mediterranean rivers indicate that the region encompasses a wide range of fluvial denudational rates, covering a broad yet rather narrower range than that recorded at the global scale in recent compilations (Figure 20.5; Walling and Webb, 1983). Suspended sediment load data are not available for many Mediterranean rivers—especially those draining the arid and semi-arid southshore catchments between Algeria and the Nile—not least because of the many practical problems associated with the accurate measurement of sediment yields in desert streams (cf. Inbar, 1982). Elsewhere, however, some nations have long-established sediment monitoring networks which provide some useful insights into regional contrasts in fluvial sediment transfer rates within the Mediterranean environment (Figure 20.5 and Table 20.2). In an early regional study in the former Yugoslavia, Jovanovic and Vukcevic (1958) mapped sediment yield data and reported values ranging from <100 to over 600 t km^{-2} yr^{-1}, with the latter figure associated with severe sheet and gully erosion in the humid montane zone to the east of Albania. More recently, Djorovic (1992) has reported sediment yield values of 31.52 t km^{-2} yr^{-1} for the Jasenica drainage basin of central Serbia. In this drainage basin, hardwood forest, orchards and pasture account for around 65% of the basin area and the remainder is cultivated. These two studies, which were conducted under similar climatic conditions, emphasise the importance of underlying geology and vegetation cover in controlling erosion rates and basin sediment yields (cf. Sala and Calvo, 1990; Andreu et al., 1994). Recent work within the *Eucalyptus globulus* and *Pinus pinaster* forests in the Agueda River basin, north-central Portugal, has highlighted the erosional impacts of fire and land management practices in the Mediterranean environment. Peak streamflows and suspended sediment yields in natural regrowth catchments were similar to those observed in unburned catchments of the Agueda basin—the greatest rates of soil loss followed rip-ploughing of post-burn terrain (Walsh et al., 1992).

It has been estimated that, out of a total land area of 225 804 km^2, there are 5.661×10^6 ha of almost completely denuded karst within the land area of the former Yugoslavia. Of the total area of karst, 1.075×10^6 ha (19%) are used for the cultivation of agricultural crops, whereas another 2.105×10^6 ha (38%) are totally barren (Zachar, 1982). In common with much of the rest of the Balkan peninsula, these figures indicate that a substantial part of the territory of the former Yugoslavia is highly prone to erosion, and sediment loads can be very high by central and northern European standards (cf. Milliman and Syvitski, 1992). In the case of Albania it has been estimated that around 80% of the land area is currently threatened by water erosion (Figure 20.4), and Zachar (1982) has suggested that a combination of unfavourable natural conditions and poor land management have resulted in the highest

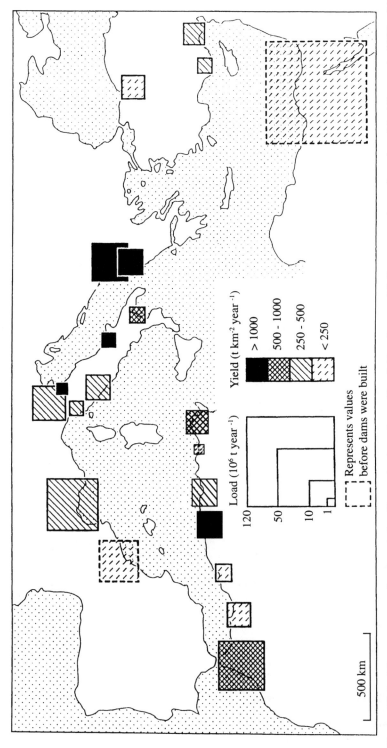

Figure 20.5 The mean annual suspended sediment loads and specific yields of some Mediterranean rivers (based on Macklin et al., 1995 with additions). Load and yield data from Milliman and Syvitski (1992) and Probst and Aniotte Suchet (1992)

specific suspended sediment yields recorded in southern Europe (Figure 20.5). The Semani River near Urage Kucit and the Shkumbini River near Papere have registered figures of 4150 and 3590 t km^{-2} yr^{-1} (Fournier, 1972; Table 20.2). Deep rill and gully erosion on cultivated land is also a common occurrence at lower elevations further south on the Balkan peninsula, especially when fields are laid bare to heavy winter rains. Figure 20.6 shows an example of serious water erosion on soils developed in friable Pliocene sands in the northwestern Peloponnese. Inappropriate land management can rapidly encourage the initial stages of deep gully and badland development—especially when erodible soils on sloping terrain are left unprotected (cf. Katsoulis and Tsangaris, 1994).

While broad climatic and relief variations explain some of the spatial contrasts in suspended sediment yield across the region, differences in rock type can also effect marked variations within a single climatic regime. Figure 20.7 shows the contrasting relationship between mean annual runoff and suspended sediment yield for catchments formed in different lithologies in the Maghreb region of Mediterranean North Africa. There is a strong positive relationship between specific suspended sediment yield and mean annual runoff on all the rock types shown. In this environment values of mean annual runoff span three orders of magnitude from <5 to >500 mm yr^{-1}. These data indicate that suspended sediment yields from river basins draining marl and marl–schist terrains are typically an order of magnitude higher than those draining more resistant

Figure 20.6 Deep rill and gully erosion on a sloping agricultural field in the northwestern Peloponnese, Greece. The substrate is friable marine sand of Pliocene age. The middle gully reaches a maximum depth of around 60 cm. Hand trowel in right gully for scale – view looking downslope (Photograph taken by the author in September 1994)

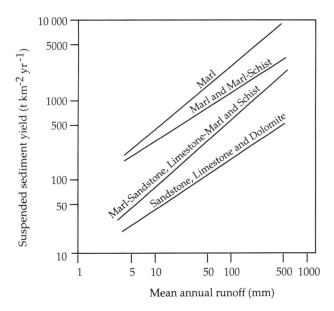

Figure 20.7 The relationship between mean annual runoff and suspended sediment yield for various river basins developed in different rock types in the Maghreb region of Mediterranean north-west Africa (after Heusch and Milliès-Lacroix, 1971)

sandstone and limestone lithologies for a given value of mean annual runoff (Heusch and Milliès-Lacroix, 1971). Figure 20.8 shows the relationship between mean annual precipitation and suspended sediment yield for several streams in Morocco, and demonstrates a strong positive relationship between these two parameters across a precipitation range of *ca.* 350 to 1400 mm (Heusch and Milliès-Lacroix, 1971). The likelihood of covariation must also be considered in such relationships, as precipitation and runoff per unit area tend to increase with altitude and drainage basin relief (cf. Richards, 1982).

Variations in rock type are also a major control on sediment yield elsewhere in the Mediterranean region. The hard limestone rocks that constitute a major proportion of the Mediterranean bedrock generally yield comparatively low amounts of silt- and clay-sized material as solutional weathering processes predominate. Mineralogical analysis of fine (<63 μm) bed sediments from the active floodplain of the Voidomatis River of north-west Greece demonstrate that the suspended sediment load of this system is dominated by flysch-derived sediment even though most of the basin is formed in Cretaceous and Tertiary limestones (Woodward *et al.*, 1992). A recent study by Collier *et al.* (1995), which outlines a new methodology for estimating rates of sediment transfer in regions characterised by extensional tectonics and normal faulting, also suggests that the lithological influence on erosional flux rates (sediment yields) can exceed the influence of drainage basin relief. Richards (1982) has rightly suggested that the strong positive relationship between erosion rates and relative relief identified by Schumm (1954) occurs partly because for his data the steepest terrain occurs in weaker rocks. This relationship does not hold true for present rates of suspended sediment discharge in many parts of the Mediterranean because the steepest relief is

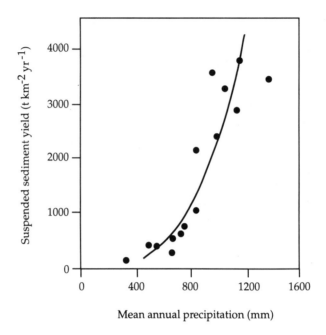

Figure 20.8 The relationship between mean annual precipitation and suspended sediment yield for several Moroccan river systems (after Heusch and Milliès-Lacroix, 1971)

commonly formed in resistant limestones, which make only a very minor contribution to suspended sediment loads. Often the highest suspended sediment yields are derived from intermontane flysch basins of comparatively low relief where the presence of semi-vegetated surfaces on fissile, silt-rich sediments (frequently tightly folded and fractured strata) combine to create high sediment availability (Figure 20.9). In general terms, the catchments with the highest suspended sediment yields in the Mediterranean region are associated with those upland terrains such as the Rif Mountains of Morocco and alpine Albania which are characterised by high-erosivity precipitation events and steep slopes formed in erodible lithologies that have been subjected to human disturbance (Figures 20.4 and 20.5 and Tables 20.2 and 20.3).

Despite the erodible nature of many Mediterranean soils, the marked regional variations in suspended sediment yield described above indicate that many areas are undergoing little fluvial erosion at the present time. In some situations, however, these figures may be accounted for by low sediment delivery ratios and high rates of colluvial and floodplain storage (cf. Trimble, 1990). Sparsely vegetated bedrock (especially limestone) slopes and thick sequences of stratified slope deposits containing large volumes of frequently rubified fine sediment are a characteristic feature of the Mediterranean landscape. In some areas a longer-term sediment exhaustion effect is evident, where the pace of soil removal has outstripped soil formation during the historical period. It should always be appreciated that sediment yield data from gauged river channels may not provide an accurate indication of the magnitude of soil removal (gross erosion) upstream. For example, in an area of the Rif Mountains of Morocco receiving 300 mm of rainfall per year, although severe erosion was occurring on hillslopes, much of the eroded sediment was being deposited and stored on pediments

Figure 20.9 Water erosion on steeply sloping flysch sediments in the Pindus Mountains of north-west Greece, creating a semi-badland terrain (see Figure 20.2). Mature pines on ridge crests and road (top left) for scale (Photograph taken by the author in September 1994)

or in wadis where conveyance losses are high—thus the acute soil loss and sedimentation problems were not reflected in the sediment yields of the river basins (Imeson, 1986). Further east in Mediterranean North Africa, mean annual suspended sediment yield values of only 25 t km^{-2} yr^{-1} have been recorded for both the Mefrouch and Chouly Rivers of Algeria in association with mean annual runoff figures of 218 and 115 mm respectively (Tixeront, 1960; Heusch and Milliès-Lacroix, 1971). More recently, however, Probst and Amiotte Suchet (1992) have suggested that the sediment delivery ratios of river systems in the Maghreb may be relatively high by world standards because of the particular climatological and morphological characteristics of the region.

BADLAND DEVELOPMENT IN THE MEDITERRANEAN REGION

The extended time-scale and intensity of anthropogenic catchment disturbance in the Mediterranean has led to widespread and often irreversible land degradation. The Mediterranean region contains many examples of badland terrain (Table 20.4) and, although badland terrain is actually quite limited in extent (cf. Bryan and Yair, 1982), its existence throughout the region provides a graphic illustration of the potentially disastrous consequences of unchecked erosion of poorly vegetated erodible sediments. Badlands may be described as intensely dissected landscapes that are unsuitable for

Table 20.4 Examples of badland and 'semi-badland' (*) development on a range of lithologies and under various rainfall regimes in the Mediterranean region (based on Campbell, 1989, with additions)

Source	Mean annual rainfall (mm)	Location	Materials
Harris and Vita-Finzi (1968)	1500	Kokkinopilos, Louros valley Epirus, NW Greece	Red silts and clays of uncertain age and provenance
De Ploey (1974)	350	Kasserine area, central Tunisia	Clays, loams and sandy lithosols developed on Cretaceous marls
Alexander (1982)	450	Agri basin, Basilicata, Italy	Plio-Pleistocene marine clays, silt clays, interbedded sands, soft shales and mudstones
Harvey (1982)	170–350	Almeria–Alicante region, Spain	Cenozoic (mostly Tertiary) and Triassic marls, silts, shales and sandstones
Imeson et al. (1982)	300	Rif Mountains, Morocco	Villafranchian, Pliocene and sub-Recent marine sediments, alluvial and colluvial deposits
Yair et al. (1982)	90	Northern Negev, Israel	Palaeocene marls and soft shales
Woodward et al. (1992)*	2000	Pindus Mountains, Epirus, NW Greece	Late Eocene to Miocene flysch sediments: alternations of sandstones and fissile siltstones
Alexander et al. (1994)	170	Tabernas basin, Betic Cordillera, SE Spain	Tortonian marls, shales and turbidites with occasional sandstone bands
Woodward (1995) (this paper)*	1500–2000	East of Konitsa, Pindus Mountains, NW Greece	Tertiary flysch lithologies (shales and sandstones)

agricultural use and they are commonly formed in loosely cemented or unconsolidated materials such as marls, shales and friable sandstones. It is important to appreciate that badland-forming lithologies are widespread throughout the Mediterranean region, and a much larger area can be described as being in a 'semi-badland' condition (Woodward et al., 1992). Figure 20.9 shows an example of semi-badland terrain where steep and straight erosional slopes are present to the east of Konitsa in the Epirus region of northwest Greece. Here, dissection of erodible flysch sediments is encouraged by tectonic uplift and an annual rainfall of *ca.* 1500–2000 mm, which falls between October and May. Grazing by goats and fuelwood gathering have placed additional pressures on this erosion-prone landscape. This fragile terrain lies close to a major geomorphic threshold whereby further vegetation loss or network extension resulting from extreme runoff events could initiate full badland conditions. The state of such marginal areas is a cause for great concern, and sensitive land management schemes are clearly required to ensure confinement and perhaps even rehabilitation of the most severely degraded slopes (see Zachar, 1982; Andreu *et al.*, 1994).

It is far from straightforward to determine the relative importance of natural processes and human action in Mediterranean badlands formation, but many workers believe that these terrains are a direct consequence of accelerated erosion triggered by human action. Attempts to resolve this issue are confounded by the fact that the origin and age of many of the badlands has not been established, although some workers believe that human interference is a prime cause of gully initiation and network extension. Table 20.4 lists nine published examples of gully and badland development from across the region, which effectively span the full range of Mediterranean precipitation regimes (*ca.* 90 to 2000 mm). In most cases the substrates are unconsolidated silt-rich materials of Tertiary age (see Bryan and Yair, 1982). Changes in catchment vegetation cover are one of the most commonly cited causes of increases in the rate and volume of stream channel runoff and decreases in channel stability. Network extension through headward erosion is a frequent outcome. Typically, such gully development takes place relatively rapidly and is not usually regarded as a feature of 'normal' erosion, rather as the result of changes in the environment induced by vegetation clearance (e.g. excessive burning, overgrazing and increases in the size of cultivated areas), climatic shifts affecting vegetation, or extreme storm runoff events following changes in rainfall intensity and periodicity (Selby, 1993). The Mediterranean region contains many spectacular examples of gully formation (e.g. Thomas and Middleton, 1994, p. 138), and gully surveys and lake sediment records have been able to demonstrate their rapid and recent development and their importance as a major suspended sediment source (cf. Foster *et al.*, 1986).

Further complexities arise when rates of tectonic uplift and tilting are spatially variable *within* catchments, creating discrete zones of instability with high sediment availability. Equivalent lithologies under the same climatic regime may be well vegetated and seemingly stable (non-eroding) within the same river basin. This situation can be observed in many of the river basins of northern Greece that drain the western flank of the Pindus Range. In this area, flysch rocks of Late Eocene to Miocene age form an important component of the landscape and are especially susceptible to gullying and sheet erosion (Figure 20.9). Dedkov and Mozzherin (1992) have suggested that the selectivity of erosion in mountain areas is much greater than in lowland terrains, and this feature is particularly prominent in semi-arid mountainous zones and in the Mediterranean where tectonic activity is important.

Wise *et al.* (1982) have used archaeological evidence to demonstrate that the presence of badland landscapes does not always equate with elevated sediment yields. In most cases, however, it is probably true that Mediterranean badlands yield large quantities of suspended sediment to river systems of the region. Yair *et al.* (1982) have identified three types of badlands in the Rif Mountains of north-east Morocco largely on the basis of contrasts in gullying style and the importance of piping. In each case,

Figure 20.10 The badlands of Kokkinopilos in central Epirus, north-west Greece, which were described by Harris and Vita-Finzi (1968). This view looks eastwards towards the limestone ridge that forms the drainage divide with the Louros Valley (Photograph taken by the author in September 1994)

the badland-forming materials displayed high erodibilities and low rates of water acceptance—both related to the decline in organic matter following cultivation. Wise *et al.* (1982) also discuss some of the broader issues and inherent difficulties involved in establishing present-day rates of erosion in areas dominated by infrequent runoff events of high magnitude. Further complications arise when such events are spatially discontinuous and greatly influenced by human activities. Bryan and Yair (1982) have pointed out that some of the Mediterranean badlands appear to be eroding at much lower rates than comparable terrains in other parts of the world. Areas of the badland terrain of Kokkinopolis described by Harris and Vita-Finzi (1968) fall into this category, as many of the ridges have smooth rounded forms and the slopes appear to have reached a comparatively stable state (Figure 20.10).

RESERVOIR SILTATION WITHIN THE MEDITERRANEAN BASIN

The pronounced relief of much of the Mediterranean landscape has created favourable conditions for river impoundments at a range of scales. Water storage schemes have been important throughout the region since ancient times and afford a particularly attractive means of eliminating the marked seasonal differences in water availability produced by the Mediterranean climate. In Morocco, for example, 60% of water demand is met by storage in artificial reservoirs (Lahlou, 1988). The construction of dams and artificial reservoirs has had a marked effect on suspended sediment delivery to the Mediterranean Sea. Perhaps the best-known example of the effects of impoundment on suspended sediment loadings is provided by the River Nile, where typical sediment concentrations fell from *ca.* 600 ppm to around 0.05 ppm following the completion of the Aswan High Dam in 1971. It has been estimated that only about 8% of the pre-impoundment load now reaches the Mediterranean Sea (Abul-Atta, 1978) and even lower figures are often quoted in the literature. A large proportion of the sediment load of the River Nile derives from the Ethiopian Highlands, and the Kashm el Girba Reservoir in eastern Sudan, whose inflow headwaters lie in this terrain, operates at only 40% capacity two decades after completion (Schwartz et al., 1990). The Aswan High Dam is located about 1000 km south of the Mediterranean coast but has effected major alterations to the hydrographic regime over the continental shelf in the south-east Mediterranean Sea (Din, 1977) and has caused significant changes to the morphology of the Rosetta and Damietta promontories of the Nile Delta.

In common with other parts of the world (cf. Beaumont, 1978), dam construction in Mediterranean river catchments, in terms of both the number and size of the structures and their storage capacities, has increased almost exponentially during the last 100 years, and particularly since 1945 (Schwartz et al., 1990). Many of the major rivers draining to the Mediterranean Sea such as the Ebro, Rhône and Nile are now fed by impounded headwater streams and their flow regimes are regulated. Numerous studies have documented wholesale declines in the transport of suspended sediment to the coastal and shelf zone following impoundment. A useful example of such development is provided by the Ebro River of northern Spain. The Ebro is the largest river on the Iberian peninsula draining to the Mediterranean Sea, and over 24 major dams have been constructed on the trunk stream and its major tributaries such as the Segre, Cinca and Guadalope Rivers (Palanques et al., 1990). Table 20.5 lists monitored water and suspended sediment discharge for the Ebro River and various tributaries at

Table 20.5 Monitored water and suspended sediment discharges in the River Ebro catchment from 1983 to 1986 from gauging stations upstream and the Asco station downstream of the major impoundment scheme at Mequinenza–Ribarroja (after Palanques et al., 1990). Sediment load figures in parentheses are mean annual values for the period of record

Gauging station	Water discharge ($m^3 s^{-1}$)	Suspended sediment concentration ($mg\, l^{-1}$)	Suspended sediment load ($\times 10^5$ t)
Inputs to the Mequinenza–Ribarroja dam complex (1983–86)			
Sastago	216.0	51.30	10.49 (3.49)
Segre	30.0	27.76	0.79 (0.26)
Cinca	49.0	40.76	1.92 (0.64)
Guadalope	1.6	9.30	0.01 (0.003)
Total (three years)	296.6		13.21 (4.40)
Output from the Mequinenza–Ribarroja dam complex (1983–86)			
Asco	396.0	9.71	3.63 (1.21)

five gauging stations between 1983 and 1986. The Sastago, Segre, Cinca and Guadalope gauging stations are located upstream of the major dam complexes at Mequinenza and Ribarroja in the lower Ebro basin. The Asco gauging station is sited immediately downstream of these dams, and more than 97% (84 239 km^2) of the Ebro catchment lies upstream of the Ribarroja dam. Sediment monitoring indicates an annual inflow of around 4.4×10^5 t of suspended sediment to the Mequinenza–Ribarroja dam complex and an outflow of 1.2×10^5 t. Thus, approximately 72% of the suspended sediment input is deposited in the reservoirs behind these dams (Palanques et al., 1990). Although these observations are derived from only three years of monitoring, they provide a valuable example of the profound impact of dam construction on sediment delivery in a major Mediterranean river basin. Furthermore, the Ebro catchment straddles an important ecotone in the transitional zone between the humid Pyrenees and the drier regions of north-east Spain. Thus, marked hydrological and vegetation gradients result in significant spatial variations in runoff and erosion. This is clearly reflected in the sediment load data, as the magnitudes of the annual sediment loads within the major tributaries of the Ebro system span three orders of magnitude, ranging from 0.01×10^5 t (Rio Guadalope) to 10.49×10^5 t (Rio Sastago). This example highlights an important feature of the Mediterranean region—the presence of steep environmental gradients, which create marked spatial variations in both water and sediment discharge.

While there are obvious repercussions for water resource projects, as the useful life-spans of many reservoirs become shortened by sediment accumulation, it is interesting to note that sediment accumulation within impounded reaches may actually have some beneficial effects downstream. The trapping of suspended solids within artificial reservoirs has implications for downstream sediment quality as well as for rates of gross sediment transfer and coastal erosion. For example, it has been shown that impoundments can serve to check the downstream transfer of polluted fine sediments and restrict their delivery to ecologically sensitive coastal areas (Palanques et al., 1990). In general, however, storage losses to accumulated sediments and the shortened life-cycles of expensive hydraulic installations are the major concern of water resource planners in the Mediterranean region, and this represents a serious consequence of erosion in many of the region's river basins (Ghorbel and Claude, 1977).

Table 20.6 lists five river systems in the Mediterranean region with catchment areas greater than 30 000 km^2 whose headwaters have been subjected to impoundment by dam construction for major water storage programmes. In four of the five cases listed, the trap efficiencies of the schemes are very high, with overall reductions in annual sediment load ranging from 91.7 to 100%. The impact of impoundment on the transfer of suspended solids through the delivery system will clearly vary with the nature, purpose and operation of the structures involved (Petts, 1984).

In Mediterranean north-west Africa, where soil erosion rates are a cause for great concern (Figure 20.4), reservoir siltation problems have been encountered in many river basins following dam construction for local water storage and torrent control (cf. Ghorbel and Claude, 1977; Lahlou, 1988). In Algeria, for example, Zachar (1982) reports the construction of 10 water storage reservoirs with a capacity of 700×10^6 m^3, which were intended to store water to irrigate 100 000 ha of land. Some of the reservoirs were choked with sediment after only 20 years, while others had a serviceable life of just 50 years. In the Sig reservoir alone, more than 800 000 m^3 of

Table 20.6 The suspended sediment load of five rivers in the Mediterranean region before and after the construction of major dams. Data from Milliman and Syvitski (1992) and Palanques *et al.* (1990)

River	Country	Basin area (10^6 km^2)	Sediment load prior to dam construction (10^6 t yr^{-1})	Sediment load after dam construction (10^6 t yr^{-1})	Reduction in load (%)
Yesil-Irmak	Turkey	0.034	19.0	0.36	98.0
Sakarya	Turkey	0.046	8.8	6.20	29.5
Kizil-Irmak	Turkey	0.074	23.0	0.46	98.0
Ebro	Spain	0.085	18.0	1.50	91.7
Nile	Egypt	3.0	120.0	0.0	100.0

suspended sediment were deposited each year. All the reservoirs in Algeria are estimated to be losing 2–3% of their storage capacity annually, which equates to around 90×10^6 m^3 of water (Grenon and Batisse, 1989). Elsewhere in North Africa, the Chiba Dam in Tunisia was opened in 1963 and had lost some 70% of its capacity to sedimentation by 1975. The average annual storage loss in the 34 largest reservoirs in Morocco amounts to about 0.5% with considerable variability between catchments draining different lithologies (Lahlou, 1988). Owing to the highly irregular precipitation, the steep terrain and the high sediment yield of the streams, the hill ponds in Algeria require impoundment walls over 15 m high, whereas 7 m is sufficient for the same type of structure in southern Italy (Grenon and Batisse, 1989). While it is difficult to make such direct comparisons between regions, it is clear that the magnitude of river sediment yields can impinge directly on many aspects of human activity, and rapid sediment accumulation prevents dams from performing their economically vital dual role of flood control and water storage in semi-arid steeplands.

PERSPECTIVE

The extract from Zachar (1982) quoted at the beginning of this chapter outlines only two of the major factors influencing rates of water erosion in the Mediterranean basin. Despite the nature of the climate and the role of human action, it can be argued that much of the region is naturally susceptible to erosion, especially where steep slopes and erodible substrates are found in association with aggressive and typically highly seasonal precipitation regimes. Many parts of the region, such as southern Italy and northern Greece, are currently experiencing steady or episodic tectonic uplift, which provides additional energy for gully and stream incision and a wide range of hillslope denudational processes, including landslides and debris flows (Ergenzinger, 1992). Even in the absence of human activity, high relief and steep slopes, large areas of friable soils and unconsolidated sedimentary rock, and high-intensity rainfall events are all factors that make Mediterranean river basins naturally prone to erosion. From the examples discussed above it has been shown that, in addition to the degree of human disturbance, drainage basin area and maximum basin elevation (cf. Milliman and

Syvitski, 1992), lithological variations, runoff volumes and tectonic activity also constitute important variables in controlling rates of erosion and suspended sediment yield in the Mediterranean basin.

Quantitative assessments of suspended sediment yield allow temporal and spatial comparisons of the magnitude of fluvial sediment transfer and provide useful insights into the severity of upstream erosion. In the Mediterranean region, where large expanses of steep terrain with thin soil covers create significant areas of marginal agricultural potential, the development of appropriate land management strategies that promote the sustainable use of such environments is of great importance (cf. Imeson, 1986; Andreu et al., 1994). While badland terrain accounts for only a very small proportion of the Mediterranean drainage basin, its presence throughout the region—across the full range of precipitation regimes—underscores the importance of maintaining a protective vegetation cover on the region's widespread erodible Tertiary lithologies.

Many of the original suspended sediment monitoring networks in the Mediterranean were set up to measure soil loss for agricultural planning (cf. Zachar, 1982). These data can be used to target the most erosion-prone areas and constitute a vital parameter for the evaluation of soil conservation measures and can also provide valuable inputs to environmental impact assessments (cf. Guy, 1977). It is important to stress that the natural disposition of the landscapes bordering the Mediterranean Sea and the associated climatic regime have rendered the region extremely sensitive to human intervention. Present rates of soil loss in environmentally fragile areas such as the Rif Mountains of Morocco are a matter for great concern and improved land management strategies are urgently required. Accurate long-term records of river sediment loads are also vital for the effective design and siting of water storage and power generation schemes. The huge capital investments required to finance the increasingly sophisticated and expensive multipurpose impoundments demand that project proposals are aware of siltation rates and storage loss projections.

ACKNOWLEDGEMENTS

The author is grateful to John Lewin for providing helpful comments on an earlier draft of this chapter and to the referees for their comments and suggestions. Mark Macklin and John Lewin generously provided the base maps for figures 3 and 5 respectively and Ian Reid, Alison Manson and Lois Wright kindly produced the diagrams. Financial support for fieldwork in Greece in 1994 was provided by the Department of Environmental and Geographical Sciences at Manchester Metropolitan University and is gratefully acknowledged.

REFERENCES

Abul-Atta, A.A. (1978) *Egypt and the Nile After the Construction of the High Aswan Dam* Ministry of Irrigation and Land Reclamation, Cairo.

Alexander, D. (1982) Difference between 'calanchi' and 'biancane' badlands in Italy. In: Bryan, R.B. and Yair, A. (eds), *Badland Geomorphology and Piping*, GeoBooks, Norwich, pp. 71–88.

Alexander, R.W., Harvey, A.M., Calvo, A., James, P.A. and Cerda, A. (1994) Natural stabilisation mechanisms on badland slopes: Tabernas, Almeria, Spain. In: Millington, A.C.

and Pye, K. (eds), *Environmental Change in Drylands: Biogeographical and Geomorphological Perspectives*, John Wiley & Sons, Chichester, pp. 85–111.

Andreu, V., Rubio, J.L. and Cerni, R. (1994) Use of a shrub (*Medicago arborea*) to control water erosion on steep slopes. *Soil Use Manage.* **10**, 95–99.

Beaumont, P. (1978) Man's impact on river systems: a world-wide view. *Area* **10**, 38–41.

Beschta, R.L., Blinn, T., Grant, G.E., Ice, G.G. and Swanson, F.J. (1987) *Erosion and Sedimentation in the Pacific Rim (Proceedings of the Corvallis Symposium, August 1987)*, IAHS Publication no. 165.

Bonvallot, J. and Hamza, A. (1977) Causes et modalités de l'erosion dans le bassin versant inférieur de l'Oued El-Hadjel (Tunisie Centrale). In: *Erosion and Solid Matter Transport in Inland Waters (Proceedings of the Paris Symposium, July 1977)*, IAHS Publication no. 122, pp. 260–268.

Brückner, H. and Hoffman, G. (1992) Human-induced erosion processes in Mediterranean countries—evidence from archaeology, pedology and geology. *Geoökoplus* **3**, 97–110.

Bryan, R.B. and Yair, A. (eds) (1982) *Badland Geomorphology and Piping*, GeoBooks, Norwich.

Butzer, K. (1976) *Geomorphology from the Earth*, Harper and Row, New York.

Butzer, K. (1982) *Archaeology as Human Ecology*, Cambridge University Press, Cambridge.

Campbell, I.A. (1989) Badlands and badland gullies. In: Thomas, D.S.G. (ed.), *Arid Zone Geomorphology*, Belhaven Press, London, pp. 158–183.

Collier, R.E.Ll., Leeder, M.R. and Jackson, J.A. (1995) Quaternary drainage development, sediment fluxes and extensional tectonics in Greece. In: Lewin, J., Macklin, M.G. and Woodward, J.C. (eds), *Mediterranean Quaternary River Environments*, Balkema, Rotterdam, pp. 31–44.

Colombani, J. (1977) Effets sur les transports solides des ouvrages hydrauliques en Afrique du Nord. In: *Erosion and Solid Matter Transport in Inland Waters.(Proceedings of the Paris Symposium, July 1977)*, IAHS Publication no. 122, pp. 295–300.

Dedkov, A.P. and Mozzherin, V.I. (1992) Erosion and sediment yield in mountain regions of the world. In: Walling, D.E., Davies, T.R. and Hasholt, B. (eds), *Erosion, Debris Flows and Environment in Mountain Regions (Proceedings of the Chengdu Symposium, July 1992)*, IAHS Publication no. 209, pp. 29–36.

De Ploey, J. (1974) Mechanical properties of hillslopes and their relation to gullying in central semi-arid Tunisia. *Z. Geomorphol.* **21**, 177–190.

Din, S.H. Sharaf (1977) Effects of the Aswan High Dam on the Nile Flood and the estuarine and coastal circulation pattern along the Mediterranean Egyptian coast. *Limnol. Oceanogr.* **22**,(2), 194–207.

Djorovic, M. (1992) Ten years of sediment discharge measurement in the Jasenica research drainage basin, Yugoslavia. In: Walling, D.E., Davies, T.R. and Hasholt, B. (eds), *Erosion, Debris Flows and Environment in Mountain Regions (Proceedings of the Chengdu Symposium, July 1992)*, IAHS Publication no. 209, pp. 37–40.

Douglas, I. (1990) Sediment transfer and siltation. In: Turner, B.L. II, Clark, W.C., Kates R.W., Richards, J.F., Mathews, J.T. and Meyer, W.B. (eds), *The Earth as Transformed by Human Action: Global and Regional Changes in the Biosphere Over the Past 300 Years*, Cambridge University Press, Cambridge, pp. 215–234.

Ergenzinger, P. (1992) A conceptual geomorphological model for the development of a Mediterranean river basin under neotectonic stress (Buonamico basin, Calabria, Italy). In: Walling, D.E., Davies, T.R. and Hasholt, B. (eds), *Erosion, Debris Flows and Environment in Mountain Regions (Proceedings of the Chengdu Symposium, July 1992)*, IAHS Publication no. 209, pp. 51–60.

Foster, I.D.L., Dearing, J.A., Airey, A., Flower, R.J. and Rippey, B. (1986) Sediment sources in a Moroccan lake-catchment: a case study using magnetic measurements. *J. Water Resour.* **5**(1), 220–234.

Fournier, F. (1972) *Les Aspects de la Conservation des Sols dans les Différentes Régions Climatiques et Pedologiques de l'Europe*, Conseil de l'Europe.

Ghorbel, A. and Claude, J. (1977) Mesure de l'envasement dans les retenues de sept barrages en Tunisie: estimation des transports solides. In: *Erosion and Solid Matter Transport in Inland*

Waters *(Proceedings of the Paris Symposium, July 1977)*, IAHS Publication no. 122, pp. 219–232.

Grenon, M. and Batisse, M. (1989) *Futures for the Mediterranean Basin: The Blue Plan*, Oxford University Press, Oxford.

Grove, A.T. (1986) The scale factor in relation to the processes involved in 'desertification' in Europe. In: Fantechi, R. and Margaris, N.S. (eds), *Desertification in Europe*, Reidel, Dordrecht, pp. 9–14.

Guy, H.P. (1977) Sediment information for an environmental impact statement regarding a surface coal mine, western United States. In: *Erosion and Solid Matter Transport in Inland Waters (Proceedings of the Paris Symposium, July 1977)*, IAHS Publication no. 122, pp. 98–108.

Harris, D. and Vita-Finzi, C. (1968) Kokkinopilos—a Greek badland. *Geogr. J.* **134**, 537–546.

Harvey, A.M., (1982) The role of piping in the development of badland and gully systems in southeast Spain. In: Bryan, R.B. and Yair, A. (eds), *Badland Geomorphology and Piping*, GeoBooks, Norwich, pp. 317–336.

Heusch, B. and Milliès-Lacroix, A. (1971) Une Méthode pour Estimer l'Ecoulement et l'Erosion dans un Bassin. Application au Maghreb, Mines et Geologie (Rabat), no. 33, 21–39.

Imeson, A.C. (1986) An eco-geomorphological approach to the soil degradation and erosion problem. In: Fantechi, R. and Margaris, N.S. (eds), *Desertification in Europe*, Reidel, Dordrecht, pp. 110–125.

Imeson, A.C., Kwaad, F.J.P.M. and Verstraten, J.M. (1982) The relationship of soil physical and chemical properties to the development of badlands in Morocco. In: Bryan, R.B. and Yair, A. (eds), *Badland Geomorphology and Piping*, GeoBooks, Norwich, pp. 47–70.

Inbar, M. (1982) Measurement of fluvial sediment transport compared with lacustrine sediment rates: the flow of the River Jordan into Lake Kinnerat. *Hydrol. Sci. J.* 439–449.

Jovanovic, S. and Vukcevic, M. (1958) Suspended sediment regimen on some watercourses in Yugoslavia and analysis of erosion processes. In: *Land Erosion, Instruments, Precipitation* (Proceedings of the General Assembly of Toronto, September 1957), IAHS Publication no. 43, pp. 337–359.

Katsoulis, B.D. and Tsangaris, J.M. (1994) The state of the Greek environment in recent years. *Ambio* **23**, 274–279.

Lahlou, A. (1988) The silting of Moroccan dams. In: Bordas, M.P. and Walling, D.E. (eds), *Sediment Budgets (Proceedings of the Porto Alegre Symposium, December 1988)*, IAHS Publication no. 174, pp. 71–77.

Langbein, W.B. and Schumm, S.A. (1958) Yield of sediment in relation to mean annual precipitation. *Trans. Am. Geophys. Union* **39**, 1076–1084.

Lewin, J., Macklin, M.G. and Woodward, J.C. (eds) (1995) *Mediterranean Quaternary River Environments*, Balkema, Rotterdam.

Macklin, M.G., Lewin, J. and Woodward, J.C. (1995) Quaternary fluvial systems in the Mediterranean basin. In: Lewin, J., Macklin, M.G. and Woodward, J.C. (eds), *Mediterranean Quaternary River Environments*, Balkema, Rotterdam, pp.1–25.

McNeill, J.R. (1992) *The Mountains of the Mediterranean World: An Environmental History*, Cambridge University Press, Cambridge.

Milliman, J.D. and Syvitski, J.P.M. (1992) Geomorphic/tectonic control of sediment discharge to the ocean: the importance of small mountainous rivers. *J. Geol.* **100**, 525–544.

Millington, A.C. (1990) Soil erosion. In: Mannion, A.M. and Bowlby, S.R. (eds), *Environmental Issues in the 1990s*, John Wiley & Sons, Chichester, pp. 227–244.

Palanques, A., Plana, F. and Maldonado, A. (1990) Recent influence of man on the Ebro margin sedimentation system, northwestern Mediterranean Sea. *Marine Geol.* **95**, 247–263.

Petts, G.E. (1984) *Impounded Rivers: Perspectives for Ecological Management*, John Wiley & Sons, Chichester.

Rendell, H.M. (1986) Soil erosion and land degradation in southern Italy. In: Fantechi, R. and Margaris, N.S. (eds), *Desertification in Europe*, Reidel, Dordrecht, pp. 184–193.

Probst, J.L. and Amiotte Suchet, P. (1992) Fluvial suspended sediment transport and mechanical erosion in the Maghreb (North Africa). *Hyrol. Sci. J.* **37**, 621–637.

Richards, K.S. (1982) *Rivers: Form and Process in Alluvial Channels*, Methuen, London.
Sala, M. and Calvo, A. (1990) Response of four different Mediterranean vegetation types to runoff and erosion. In: Thornes, J.B. (ed.), *Vegetation and Erosion: Processes and Environments*, John Wiley & Sons, Chichester, pp. 347–362.
Schumm, S.A. (1954) The relation of drainage basin relief to sediment loss. *Int. Assoc. Sci. Hydrol. Publ.* **36**, 216–219.
Schwartz, H.E., Emel, J., Dickens, W.J., Rogers, P. and Thompson, J. (1990) Water quality and flows. In: Turner, B.L., et al. (eds), *The Earth as Transformed by Human Action*, Cambridge University Press, Cambridge, pp. 253–270.
Selby, M.J. (1993) *Hillslope Materials and Processes*, 2nd edn, Oxford University Press, Oxford.
Thomas, D.S.G. and Middleton, N.J. (1994) *Desertification: Exploding the Myth*, John Wiley & Sons, Chichester.
Tixeront, J. (1960) Débit solide des cours d'eau en Algérie et en Tunisie. *Proceedings of the General Assembly of Helsinki (July–August 1960)*, IAHS Publication no. 53, pp. 26–41.
Trimble, S.W. (1990) Geomorphic effects of vegetation cover and management: some time and space considerations in the prediction of erosion and sediment yield. In: Thornes, J.B. (ed.), *Vegetation and Erosion: Processes and Environments*, John Wiley & Sons, Chichester, pp. 55–65.
van Andel, T.H., Runnels, C.N. and Pope, K.O. (1986) Five thousand years of land use and abuse in the southern Argolid, Greece. *Hesperia* **55**, 103–128.
Walling, D.E. (1984) The sediment yields of African rivers. In: Walling, D.E., Foster, S.S.D. and Wurzel, P. (eds), *Challenges in African Hydrology and Water Resources (Proceedings of the Harare Symposium, July 1984)*, IAHS Publication no. 144, pp. 265–283.
Walling, D.E. (1986) Sediment yields and sediment delivery dynamics in Arab countries: some problems and research needs. *J. of Water Resour.* **5**(1), pp. 775–799.
Walling, D.E. and Webb, B.W. (1983) Patterns of sediment yield. In: Gregory, K.J. (ed), *Background to Palaeohydrology*, John Wiley & Sons, Chichester, pp. 61–100.
Walsh, R.P.D., Coelho, C. de O.A., Shakesby, R.A. and Terry, J.P. (1992) Effects of land use management practices and fire on soil erosion and water quality in the Agueda River basin, Portugal. *Geoökoplus* **3**, 15–36.
Wellens, J. and Millington, A.C. (1990) Desertification. In: Mannion, A.M. and Bowlby, S.R. (eds), *Environmental Issues in the 1990s*, John Wiley & Sons, Chichester, pp. 245–261.
Wise, S.M., Thornes, J.B. and Gilman, A. (1982) How old are the badlands? A case study from southwest Spain. In: Bryan, R.B. and Yair, A. (eds), *Badland Geomorphology and Piping*, GeoBooks, Norwich, pp. 259–277.
Woodward, J.C. (1995) Archaeology and human-river environment interactions. In: Lewin, J., Macklin, M.G. and Woodward, J.C. (eds), *Mediterranean Quaternary River Environments*, Balkema, Rotterdam, pp. 99–102.
Woodward, J.C., Lewin, J. and Macklin, M.G. (1992) Alluvial sediment sources in a glaciated catchment: the Voidomatis Basin, northwest Greece. *Earth Surf. Processes Landforms* **17**, 205–216.
Yair, A., Goldberg, P. and Brimer, B. (1982) Long term denudation rates in the Zin-Hazerim badlands, northern Negev, Israel. In: Bryan, R.B. and Yair, A. (eds), *Badland Geomorphology and Piping*, GeoBooks, Norwich, pp. 279–291.
Zachar, D. (1982) *Soil Erosion* (Developments in Soil Science 10), Elsevier, Amsterdam.

21 Soil Erosion and Sediment Yield in the Philippines

SUE WHITE
Instituto Pirenaico de Ecología, Zaragoza, Spain

INTRODUCTION

Soil erosion and its on-site and downstream consequences are a major problem in the Philippines. The topography, geology, soils, climate and land-use practices all contribute to a very fragile environmental situation. In this chapter the aim is to describe this environment and, by way of data from a field measurement programme, to try to explain the particular features that must be understood for successful management and mitigation of the erosion and sedimentation problems. In the first part of the chapter the physical, social and economic factors affecting soil erosion and sediment transport in the Philippines are described. Then, in the second half of the chapter, a major field measurement programme in northern Luzon is described, with a summary of the main results. Using these data, some basic analysis allows various potential erosion and sediment delivery mechanisms to be discussed. Finally, conclusions about the important factors relating to erosion and sedimentation in the Philippines are made.

BACKGROUND

The Republic of the Philippines is made up of 7107 islands with a total land area of 299 404 km^2. Of these islands only 2000 are inhabited and 500 are less than 1 km^2 in size (Figure 21.1).

Social and economic factors

About two-thirds of the Philippine people live by fishing, agriculture and forestry. Some 70% of the population are dependent on local subsistence agriculture for their daily food, while 52% are thus employed (Davis, 1987). It is estimated that 70% of rural children are malnourished. Many Filipinos work on large plantations producing sugar cane, coconuts and other export crops. The agricultural wages are poor, most workers receiving less than 20 pesos a day (about $0.8 at May 1995 exchange rates).

Recent figures on population growth (World Bank, 1988) show the annual average increase from 1980 to 1987 was 2.4%, by no means the highest in the world, but still of concern.

Sediment and Water Quality in River Catchments. Edited by I.D.L. Foster, A.M. Gurnell and B.W. Webb.
© 1995 John Wiley & Sons Ltd.

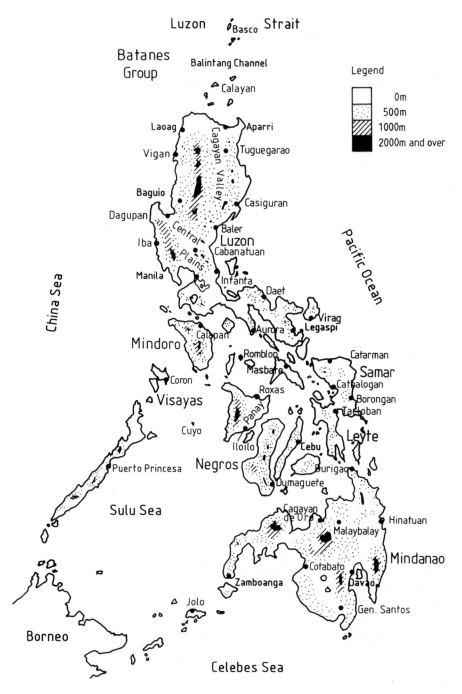

Figure 21.1 Location of the Philippines

Topography and soils

The entire Pacific basin owes much of its structural form to a vast and exceedingly complex series of stress lines, along which faulting, folding and volcanic and earthquake activity have taken place, and still frequently occur. Evidence of geologically recent and pronounced movement in these zones can be seen in the coralline limestone at considerable elevations in Palawan, Mindanao, Cebu and northern Luzon. Central and northern Luzon (the areas within which case studies on soil erosion and sediment yield are presented below) are dominated by several large mountain ranges, with shallow, highly erodible soils and steep slopes.

The main soil types found in the Philippines are a nitosol–acrisol association in northern and southern Luzon, Panay, Cebu and eastern and western Mindanao; a cambisol–luvisol–nitosol association in central Luzon; and an acrisol–nitosol association in the remaining islands and central Mindanao (FAO–Unesco, 1979; Mariano and Valmidiano, 1972). Both nitosols and acrisols, which dominate the mountain soils of northern Luzon, degrade rapidly and are highly susceptible to erosion once the dominant natural forest cover is removed. Such degradation of deforested land results in low success rates for reforestation schemes and abandonment of agricultural land.

Climate—cyclones and thunderstorms

After the physical setting of the Philippines, the climate regime provides the key to understanding the sediment supply and delivery processes. The Philippine islands are completely surrounded by large oceans, which have a fundamental effect on the climate. The most important climatic controls affecting the area are semi-permanent cyclones and anticyclones, air streams, ocean currents, linear systems, tropical cyclones and thunderstorms. In terms of erosion, sediment transport and sediment delivery, the most important features are the cyclones and thunderstorms.

Mean annual rainfall for the country as a whole is 2533.0 mm, with Luzon, Visayas and Mindanao having average annual rainfalls of 2723.4, 2391.7 and 2349.8 mm, respectively. Rainfall in northern Luzon shows a linear dependence on elevation, which varies from year to year (Blyth and White, 1990). The distribution of rainfall throughout the year is generally described as a rainy season from June to December, with a dry season from January to May. Although this is true for the country as a whole, variations of this pattern occur.

The Philippines are located in the region recognised as having the greatest frequency of tropical cyclones in the world (Flores and Balagot, 1969). Various studies (Chin, 1958; Coronas, 1920; Algue, 1904) have reported average annual frequencies of tropical cyclones in the Philippines of 22 (1884–1953), 17.3 (1908–18) and 21.3 (1880–1901), respectively. For the period 1948–88 (Figure 21.2) the annual occurrence of tropical cyclones in the Philippines varied from a minimum of 12 in 1973 to a maximum of 30 in 1964, with a mean annual frequency of 19.8 (e.g. PAGASA, 1975–85); of these, an average of 8.8 per year directly cross the islands. The months June to December are most likely to be subject to a cyclone, with July having the greatest frequency of cyclone occurrence. Models of future climate scenarios suggest that the frequency of cyclones in this region may increase (Broccoli and Manabe, 1990).

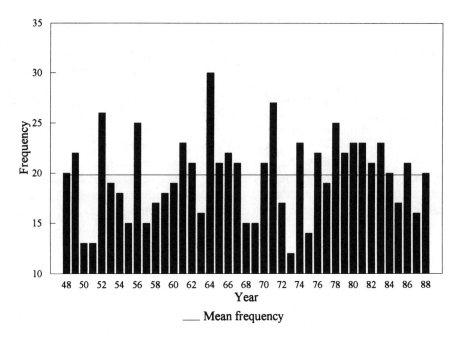

Figure 21.2 Annual cyclone frequency in the Philippines, 1948–88 (data derived from annual reports of the Philippines Atmospheric Geographical and Astronomical Services Administration (PAGASA))

Cyclones are a mixed blessing to the islands, causing great loss of life and destruction of property, but supplying a large proportion of the rainfall needed to sustain life. Individual cyclones affect the Philippines for 1–7 days, depending on their tracks and speed of movement, and they can stall in one location for up to four days. It is also not unknown for a cyclone to double back on itself and hit the islands twice or more. In the months with more frequent cyclones, it is possible for two or three cyclones to follow one another with hardly a break between them.

Tropical cyclones do not affect the whole of the Philippines for the whole year, tending to hit the more southerly islands in January, then moving northwards until August, then travelling southwards again, in association with movements of the 'intertropical convergence zone' (ICZ). From November to April, the ICZ is located to the south of the Philippines, reaching its most southerly position in January or February. During early May it starts to appear in the south-west of the country, and by July or August it has reached its most northerly position, well to the north of the Philippines. By November it is once again well to the south of the islands. However, the variations in position of the ICZ are quite large and disorganised. The presence of this zone is usually characterised by disturbed weather conditions, with widespread clouds, precipitation and moderate to strong winds. Precipitation is convective in nature, and seldom continuous, except where the zone is very well defined. As a result of these ICZ movements and the distribution of cyclones through the year, northeastern Luzon is the area of the country most affected by cyclones, with around 40% of the country total affecting this region.

Another frequent feature of the Philippines climate are thunderstorms. Compared with tropical cyclones, these are short-term small disturbances, but they occur much more frequently throughout the islands and during most of the year, and can be responsible for large rainfall amounts as a result of the moist, unstable conditions of most of the air streams affecting the country, in combination with orographic lifting. Figure 21.3 shows the distribution of mean annual number of days with thunderstorms in the Philippines.

Rivers and discharge patterns

As a result of these rainfall patterns, river discharges are extremely flashy. Continuous discharge monitoring is therefore desirable for an accurate assessment of water resources, and particularly for estimation of sediment transport. Some basic statistics, calculated from published data for discharge sites in the Cagayan valley in northeastern Luzon (National Water Resources Council, 1980), illustrate the problems of measurement and the nature of discharge events. There are 49 monitoring sites in the Cagayan valley, with basin areas ranging from 57 to 27 580 km^2 at the mouth of the Cagayan River. Data are very patchy, with the number of consecutive years of data at a site varying between 0 and 11, with a mean of 4.4 years. As the data gaps are often for cyclone-affected months, it is likely that some of the higher flows at many sites have not been recorded. This means that care must be taken in drawing any firm conclusions from the published data.

The ratio of instantaneous peak discharge to maximum daily discharge ranges from 1.0 to 5.9. These figures indicate the flashiness of the discharge regime and emphasise the effect of averaging discharge, valuable information about variability being lost with even a daily average discharge figure. The ratio of maximum daily discharge to minimum daily discharge indicates the range of flows in these rivers, with typical values of 200 to 1000, and values over 10 000 in three cases. The ratio of instantaneous peak discharge to mean discharge indicates the likely impact of flood events. Values of between 20 and 100 are common, with two sites having a peak/mean flow ratio of over 200. Thus, river channels that develop in response to normal mean annual flood discharges are likely to be completely reworked by major flood events.

Land use

A survey of land use was carried out by the Swedish Space Corporation and the Department of Natural Resources (DENR) of the Philippines Government (Swedish Space Corporation, 1988), on behalf of the World Bank, who produced a natural resources review of the Philippines (World Bank, 1989). Twenty-two different land uses were identified from SPOT satellite images and were grouped into three classes for mapping, namely forest, extensive and intensive land use. For the country as a whole there was estimated to be 25% forested land, 41% extensively cultivated land and 34% intensively cultivated. Owing to the extreme fragility of the environment, much of the land under cultivation will be at threat from erosion. During the Swedish Space Corporation study, 284 km^2 of siltation along the coast or in lakes were identified. If this situation is allowed to continue, many of the rich fishing grounds and coral reefs

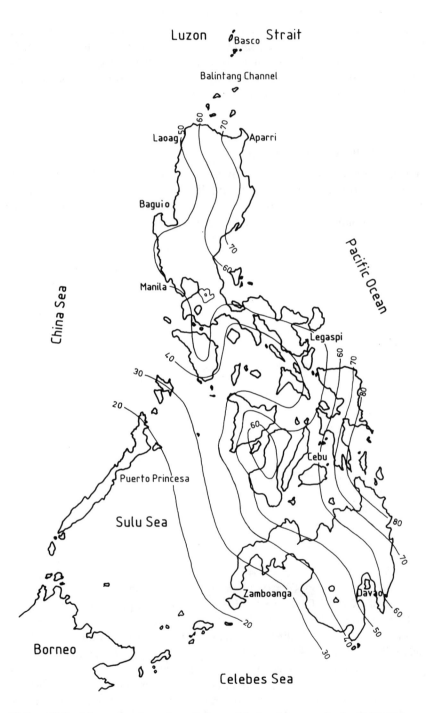

Figure 21.3 Mean annual number of days with thunderstorms in the Philippines

around the coast will be in danger from deposited sediment, which kills the corals and destroys fish spawning grounds.

Much of the land of the Philippines is strictly suited only to forestation, and any disturbance of the forest must be carefully managed. While population numbers remained fairly low, the traditional practice of shifting, slash-and-burn agriculture ('kaingin' in Filipino) could be sustained. But, since 1900, population growth has accelerated, the demands on land have consequently increased, and land has not been left fallow for long enough to regain fertility. This has resulted in increased erosion, clearance of new areas for farming and abandonment of highly degraded land where forest cannot regenerate. Logging of hardwoods for export has also been a considerable problem over recent years. In theory, much of the country is protected from logging, but in the 1980s this was openly flouted. The argument for sustainable simple agriculture for the majority of farmers in the Philippines seems to be very strong. This probably cannot be implemented until problems of security of land tenure are resolved. A recent positive move is the development of the 'SALT' agroforestry system (Tacio, 1991) to allow sustainable cultivation on sloping lands.

Erosion and reservoir sedimentation

The Philippines, like many other Asian countries, suffers because of its oil dependence for energy, although there is the possibility of developing hydro-electric and geothermal power resources to help combat this situation. At present the Philippines is second only to the USA in harnessing geothermal energy resources. The scope for hydro-electric power development is enormous, but potential sites are concentrated in the mountainous and more environmentally fragile regions of the country. There are considerable problems with sedimentation, owing to high erosion rates in the river basins, in many of the reservoirs currently in operation. This situation was not foreseen at the design stage, and the responsible authorities are now having to try to recover from a seemingly impossible situation.

River basin degradation has been such that vast amounts of sediment are now stored in river channel floodplains, and are gradually transported into the reservoirs. Soils in many river basins are severely degraded, reducing crop yields and making the re-establishment of forest extremely difficult. Even if conservation methods could be successfully introduced to river basins, their effect on the reservoirs would not be felt for many years. A combined programme of conservation and engineering structures, such as check dams, is probably the best way forward for the currently afflicted areas. It may also be necessary, in the most severe cases, to carry out some de-silting or sediment movement in the reservoirs, but for planned reservoirs the emphasis must be on finding out why the extent of the sedimentation problem was not foreseen, and how it can be avoided in future developments.

An awareness of the erosion and sedimentation problems in the Philippines appears to be fairly recent. A paper presented by Jayme (1984) during a workshop in the Philippines (Hydraulics Research and National Irrigation Administration, 1984) lists some measured and predicted erosion and sedimentation rates, and Miner and Gulcur (1971) list sediment yields for 26 reservoirs around the Philippines. These data show sedimentation rates from 68 to 22 740 $t\,km^{-2}\,yr^{-1}$, with a mean value of 3427 $t\,km^{-2}\,yr^{-1}$. Many of these sediment yield figures are of concern. Converting these figures to basin erosion rates using conventional sediment delivery theory (e.g. Roehl, 1962) would suggest rates up to 585 t ha^{-1} yr^{-1}. Bearing in mind the fact that much of the sediment supply is likely to be from mass movement, results from plot studies reported by Jayme (1984) and David (1989) (Table 21.1) would confirm this estimate. Of further concern is the likely increase in erosion and sedimentation in the future. It appears that sediment yields in Asia generally are increasing at a rate of approximately 4% per year (Abernethy, 1987). Data from Ambuklao (Table 21.2) suggest that the rate of increase in the Philippines may be even higher than this. That there are serious problems of soil erosion and sedimentation is clear. However, in order to tackle such problems effectively, one needs to understand the processes of soil erosion and sediment transport and delivery which control these phenomena (Walling, 1983; Walling and Webb, 1983).

Table 21.1 Data from erosion plot studies in the Philippines (Jayme, 1984; David, 1989)

Treatment	Slope (%)	Size of plot (m)	Mean annual erosion rate (tonne km^2 yr^{-1})
Logging site (log landing)	40–60	2 × 5	884
Logging site (cableways)	40–60	2 × 5	2630
Load fill slope	40–60	2 × 5	1200
Forested site	40–60	2 × 5	380
Coconut/lanzones	50–60	8 × 8	6.76
Coconut/pineapple	50–60	8 × 8	1020
Coconut	50–60	8 × 8	2.63
Grassland	50–60	8 × 8	8.52
Rice/peanut (contour strips)	na[a]	2 × 4	67.5
Corn/peanut/sitao (contour strips)	na	2 × 4	25.2
Mung bean/soybean (contour strips)	na	2 × 4	85.4
Corn/gabi/banana/citrus	na	2 × 4	50.1
Bare soil	na	35 m^2	860
Primary forest	20	na	0.5
Softwood fallow	20	na	0.72
Imperata grassland	20	na	1.00
New rice kaingin	20	na	2.11
Twelve-year-old rice kaingin	20	na	150

[a] Data not available

Table 21.2 Sediment yield rates for Ambuklao Reservoir, Luzon (Abernethy, 1987)

Measurement dates	Mean annual runoff (mm yr^{-1})	Mean annual sediment yield (t km^{-2} yr^{-1})
1950–52		3990
1954		5230
1963		9800
1964		8740
1956–67	2622	5008

THE MAGAT STUDY

During the period 1984–88 a measurement programme was carried out in the Magat River basin in northern Luzon (Figure 21.4), by Hydraulics Research, Wallingford, UK, and the National Irrigation Administration of the Philippines, to monitor runoff and sediment yield from a number of basins of different sizes. Measurements of erosion and sediment yield rates were carried out at a number of scales, to try and gain an understanding of the complex processes involved and their interactions. The region studied is probably the area of the country at greatest risk from both natural and human causes. There are already several reservoirs in the area whose sedimentation rates are considerably higher than those predicted at the design stage. There are also a number of potential dam sites in the area, and lessons need to be learnt from the existing problem reservoirs before further dam construction is undertaken.

A reservoir was impounded at the bottom of the Magat basin in 1982. Concerns over sedimentation rates in neighbouring reservoirs and the general degradation of the Magat watershed led to concern over potential sedimentation rates. Therefore, in 1984 a reservoir survey, using echo-sounders, was carried out to estimate the amount of sediment deposition that had occurred since impoundment (Wooldridge, 1986). The results suggested an average sediment yield of $38\, t\, ha^{-1}\, yr^{-1}$, and reservoir life was estimated at 25 years, just 25% of that estimated before impoundment. This survey led to an enhanced monitoring programme, to ascertain both the temporal and spatial variability of sediment supply in the basin.

Two small (15.3 and 26.5 ha) reforested basins close to the Magat Reservoir provided baseline data on the reductions in runoff and sediment yield which are achievable with good land management (Amphlett and Dickinson, 1989). Ground cover in the basins was very dense, and tree growth was rapid. Results showed sediment yields of less than $1\, t\, ha^{-1}\, yr^{-1}$, with only few large events contributing to runoff and sediment supply and transport. In contrast, limited data from a series of grazed and burnt basins near Aritao (2.35, 2.69 and 6.31 ha) showed sediment yields of up to 6.2, 1.9 and $5.4\, t\, ha^{-1}$ respectively, in individual events. Although worrying, these results may be expected to be in the middle range of sediment yields expected from the basin as a whole. Agricultural areas, where the soil is cleared by burning and then ploughed before the heavy rains, may produce even higher sediment yields, and large mass movements of soil are common in the largest cyclone events and as a result of earthquakes.

A series of measurements were also made on a set of nested river basins, whose areas increased by a factor of 10 from one to the next (Dickinson *et al.*, 1990; White, 1992). Thus, the site at Santa Fe (18.9 km^2) is included within the Aritao basin (159.2 km^2), which in turn is included within the Baretbet basin (2041 km^2). The basin to Baretbet is about 50% of the total basin area draining to the Magat Reservoir (4143 km^2). For each of the river monitoring sites, stage was measured using gauge boards, and converted to discharge by means of discharge rating curves. Owing to safety and security problems, no data were collected at night or during extreme flood events. Suspended sediment samples were collected from various points across the river section and through the water depth, and sediment rating curves were derived for the three sites. Unusually, the rating curves for wash-load (material less than 63 μm in size) had higher regression coefficients than those for suspended bed material load

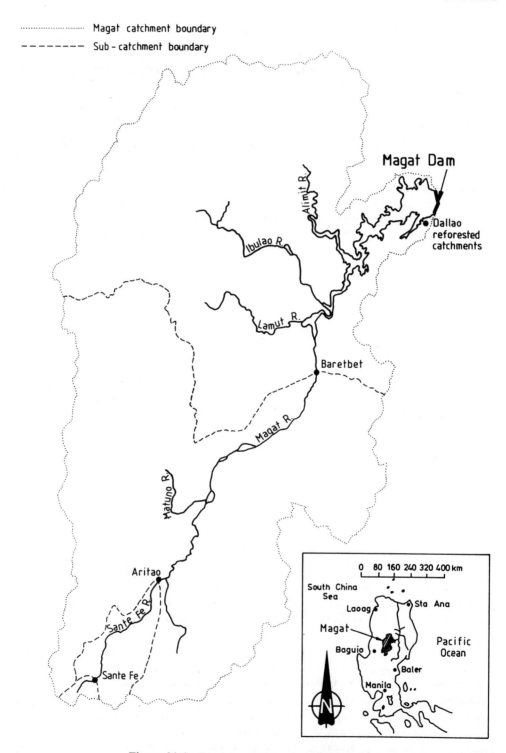

Figure 21.4 The Magat basin measurement sites

Table 21.3 Annual water discharge at the river monitoring sites

Site	Discharge ($m^3 \times 10^6$)			Discharge (m over the basin)		
	1986	1987	1988	1986	1987	1988
Santa Fe	68	19	40	3.6	1.01	2.17
Aritao	266	71	149	1.67	0.45	0.94
Baretbet	4049	2204	4599	1.98	1.08	2.25

Table 21.4 Annual sediment yield at the river monitoring sites

Site	Sediment transport ($t \times 10^3$)			Sediment yield ($t\ ha^{-1}$)		
	1986	1987	1988	1986	1987	1988
Santa Fe	75	2.5	14.8	39.7	1.3	7.8
Aritao	350.3	10	103	22.0	0.6	6.5
Baretbet	6827	821	6180	33.5	4.0	30.3

(>63 μm). Sediment measurements during the rising and falling limbs of flood events showed hysteresis, which may in part account for the poor relationship between suspended bed material concentrations and water discharge.

Using the derived rating curves, annual water and sediment discharge at the three river sites were estimated. Results are summarised in Tables 21.3 and 21.4. Sediment loads should be compared with the estimate of annual sediment delivery to the reservoir of 38 $t\ ha^{-1}\ yr^{-1}$ (1982–84).

DISCUSSION

There were many difficulties in the measurement programme in the Magat basin, owing to both the extreme environment and guerilla activity in the area. This resulted in a less-than-ideal data set. The nature of the collected data, and the lack of continuous discharge and sediment records, preclude sophisticated analysis. However, in spite of the shortcomings, the trend in results is clear and allows various conclusions to be drawn.

The discharge and sediment yield results raise several interesting points. First, 1987 was a much drier year than 1986 or 1988. In fact, in 1987 no cyclone events affected the southern part of the Magat River basin, which contributes discharge to the three river measurement sites; 1988 received the most frequent cyclone events, whilst 1986 showed the highest discharge peaks. This is reflected by the split of sediment transported between wash-load and suspended bed material load. In 1986 sediment transport at all sites was dominated by suspended bed material (91%, 62% and 64% at Santa Fe, Aritao and Baretbet respectively). The only other time such dominance is observed is at Aritao in 1988, when suspended bed material load accounts for 59% of total sediment transport; in the same year wash-load accounts for 55% of the total at Santa Fe and 58% at Baretbet. In the dry year of 1987, wash-load and suspended bed material account for 50% each of sediment transport at Aritao, while at Santa Fe and Baretbet wash-load dominates (92% and 73% respectively).

The river at Aritao is acting as a sediment 'sink', with sediment yields (t h^{-1} yr^{-1}) lowest at this site in all three years. Indeed, Aritao is located at the beginning of the main Magat River floodplain, with rivers from mountain ranges to the south-east and south-west contributing to flow. The sites at Aritao and Baretbet both have ample amounts of sediment stored in the river bed and floodplain. Thus, as discharge increases at these sites, sediment can be picked up until the flow reaches transport capacity. The river bed at Santa Fe, however, is fairly rocky, and any fine sediment must be transported from outside the river.

Sediment yields at Baretbet are of the same order as those estimated from the reservoir survey in 1986 and 1988. Discharge records for the Baretbet site are not available for the period 1982–84, following impoundment of the dam at Magat. However, it is known that in the years 1982, 1983 and 1984, 21, 23 and 20 cyclones, respectively, affected the Philippines. These frequencies are just over the average number observed each year, so it is likely that discharges and sediment transport rates were at the high end of the range.

The cyclones, therefore, appear to be the main sediment-transporting events, but are they the main sediment supply events? In the Philippines the tropical cyclones are seen as the main problem. If a cyclone stalls over an area, then it can indeed cause massive damage to the environment. However, in terms of energy of rainfall in cyclone and non-cyclone events, it can be seen that it is often the non-cyclone events, typically thunderstorms, that provide the most erosive power on an event basis (White 1990) (Figure 21.5). Thunderstorms also occur much more frequently than cyclones; for northern Luzon there are between 60 and 70 thunderstorms each year, and on average the area is affected by between two and four cyclones of varying strength.

The thunderstorm events, while supplying the majority of the rainfall energy that detaches soil particles, are short-lived. Thus, the areas of the river basin contributing to overland flow and the total runoff resulting from such events are both low. In the mountain areas where thunderstorms are most frequent, increased flow in the headwaters of rivers can be seen during and immediately after a storm. A theoretical consideration of such rainfall–runoff events (Baldwin and White, 1991) shows that the flood peak is dissipated within around 10 km downstream from the storm. This means that although sediment transport capacity will be increased as a result of the storm, this is rapidly diminished as the flood peak moves downstream. Therefore, the role of thunderstorms is in detachment of sediment particles, their overland transport from limited areas around the river, and then their transport downstream for distances of around 10 km. The major role of thunderstorms in soil erosion at the field scale is important to remember.

Tropical cyclones are much less frequent than thunderstorms. They can contribute highly intense rainfall with its consequent eroding capability. However, the main difference from thunderstorm events is the (typically) much higher amount of rain that falls and the larger area affected. Thus, cyclone events provide overland flow from a

SOIL EROSION AND SEDIMENT YIELD IN THE PHILIPPINES

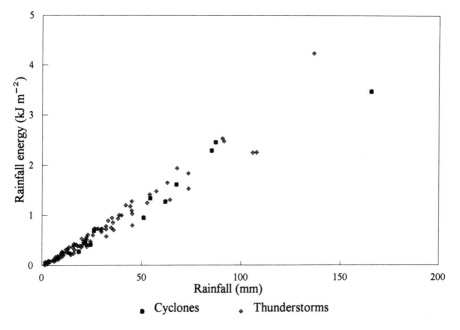

Figure 21.5 Cyclone and non-cyclone rainfall energy

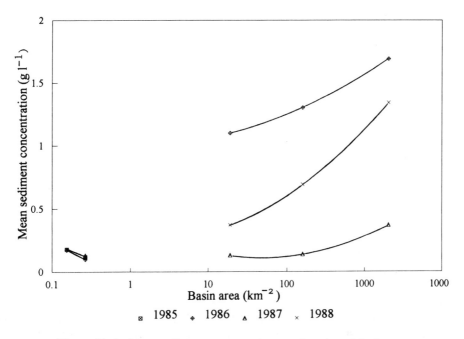

Figure 21.6 Mean sediment concentration as a function of basin area

much greater proportion of the river basin and result in much larger and persistent flood discharges in the rivers. Cyclone flood peaks, although difficult and dangerous to measure accurately, have been estimated in excess of 3000 $m^3 s^{-1}$ at Baretbet, and can erode river beds to a depth of 3 m or more (White, 1992). Therefore, material that may have been detached and moved some way overland or through the river system by thunderstorm events can be moved completely away downstream by large cyclone events. It is the combination of these two types of event that appears to cause such high sediment yield rates in the Philippines.

Of course, much of the damage caused by thunderstorms and cyclones could be avoided if the land was under suitable vegetative cover that was well managed. A summary of the results from the small reforested basins and the river measurement sites is shown in Figure 21.6. Here, mean sediment concentration (annual sediment load/annual discharge) is plotted as a function of basin area. The decrease in sediment concentration with increasing drainage area for the reforested basins confirms the concept of sediment delivery ratios, where sediment yield is expected to decrease with increasing basin area. The river sites, however, show an increase overall, although concentrations (and thus yield) at the intermediate Aritao site are generally low. Results from other areas of the world have shown a linear decrease in sediment yield with area, for river basins of comparable size (e.g. Glymph, 1951; Fleming, 1969).

CONCLUSIONS

The combination of a fragile mountainous environment with cyclones and frequent thunderstorms appears to explain the very high sediment yield rates in the Philippines. Erosive power is supplied by both the short localised thunderstorms and the longer-duration, more widespread cyclone events. High sediment-transporting capacity is provided by the cyclone-related floods. In years without cyclones, fine sediments dominate sediment transport, with suspended bed material load assuming more importance during and after cyclone events.

The results from the Philippines suggest that, under good vegetative cover where land degradation does not occur, sediment yields may well decrease with increasing basin area; indeed, such a pattern could be distinguished in the basin neighbouring Magat (White, 1987). However, once sustainable land cover is disturbed by farming or logging, erosion is rapid and sedimentation of river valleys and reservoirs will quickly follow. In these cases sediment yield may well increase downstream. This has major implications, both for changes in land use and management and for siting of dams. The Filipino environment is so fragile that even the slightest disturbance of the natural forest vegetation must be handled with care. Much damaging forest clearance has already occurred, with consequent soil erosion and sedimentation. Because of the nature of the soils, such changes are not readily reversible, and the sedimentation problems must be dealt with by use of sediment traps, sediment removal structures and dredging. This means that in much of the country sedimentation of reservoirs will continue to be a problem that must be tackled *in situ*. Any new, undeveloped areas considered for dam sites must be properly managed before dam construction, as erosion and sedimentation are sure to follow the opening up of new areas.

ACKNOWLEDGEMENTS

The Magat field study was carried out with funding from the Overseas Development Administration of the British Government. The author gratefully acknowledges the work on this project carried out by colleagues at Hydraulics Research, Wallingford, UK, and the National Irrigation Administration of the Philippines. The PhD study which contributed the background information and some analysis of the Magat results was carried out under the supervision of Professor Des Walling at Exeter University and Dr Kathiravelu Sanmuganathan at Hydraulics Research.

REFERENCES

Abernethy, C.L. (1987) *Soil Erosion and Sediment Yield*, Report for FAO, Rome.
Algue, J. (1904) *The Cyclones of the Far East*, Bureau of Printing, Manila.
Amphlett, M.B. and Dickinson, A. (1989) *Dallao Sub-Catchment Study, Magat Catchment, the Philippines: Summary Report (1984–1987)*, Report no. OD 111, Hydraulics Research, Wallingford.
Baldwin, A.P. and White, S.M. (1991) *Sediment Delivery: Literature Review and Simulation of the Delivery Processes in a Stream Channel*, Report no. OD/TN 57, Hydraulics Research, Wallingford.
Blyth, E.M. and White, S.M. (1990) *Statistical Analysis of Rainfall in the Magat Catchment, Luzon, the Philippines*, Report no. OD/TN 54, Hydraulics Research, Wallingford.
Broccoli, A.J. and Manabe, S. (1990) Can existing climate models be used to study anthropogenic changes in tropical cyclone climate? *Geophys. Res. Lett.* **17**, 1917–1920.
Chin, P.C. (1958) *Tropical Cyclones in the Western Pacific and China Sea Area from 1884 to 1953*, Royal Observatory, Hong Kong.
Coronas, J. (1920) *The Climate and Weather of the Philippines, 1903 to 1918*, Bureau of Printing, Manila.
David, W.P. (1989) *Erosion and Sediment Transport*, University of the Philippines at Los Banos, Laguna, Philippines.
Davis, L. (1987) *The Philippines: People, Poverty and Politics*, Macmillan, New York.
Dickinson, A., Amphlett, M.B. and Bolton, P. (1990) *Sediment Discharge Measurements, Magat Catchment: Summary Report 1986–1988*, Report no. OD 122, Hydraulics Research, Wallingford.
FAO–UNESCO (1979) *Soil Map of the World*, vol. IX, Southeast Asia, FAO, Rome.
Fleming, G. (1969) Design curves for suspended load estimation. *Proc. ICE* **43**, 1–9.
Flores, J.F. and Balagot, V.F. (1969) Climate of the Philippines. *World Survey of Climatology*, vol. 8, *Climates of Northern and Eastern Asia*, Elsevier, Amsterdam, Ch. 3.
Glymph, L.M. (1951) *Relation of Sedimentation to Accelerated Erosion in the Missouri River Basin*, Technical Paper 102, United States Department of Agriculture, Soil Conservation Service, Washington, DC.
Hydraulics Research and National Irrigation Administration (1984) *Workshop on Soil Erosion Measurement and Data Analysis. (Proceedings of workshop, Echague, Philippines, June 1984)*, Hydraulics Research, Wallingford, and National Irrigation Administration, EDSA, Manila.
Jayme, R.S. (1984) Review on erosion and sedimentation data collection and studies in Central Visayas. In: *Workshop on Soil Erosion Measurement and Data Analysis (Proceedings of Workshop, Echague, Philippines, June 1984)*, Hydraulics Research, Wallingford, and National Irrigation Administration, EDSA, Manila.
Mariano, J.A. and Valmidiano, A.T. (1972) *Classification of Philippines Soils in the Higher Categories*, Bureau of Soils, Manila.
Miner, N.Y. and Gulcur, M.Y. (1971) *Demonstration and Training in Forest, Forest Range and Watershed Management in the Philippines*, UNDP/FAO Project SF/PHI/16, Technical Report no. 6.

National Water Resources Council (1980) *Philippine Water Resources Summary Data*, vol. 1, *Streamflow and Lake or River Stage*, Report no. 9, National Water Resources Council, Manila.

PAGASA (Philippine Atmospheric, Geophysical and Astronomical Services Administration) (1975-1985) *Tropical Cyclone Reports*, PAGASA, Manila.

Roehl, J.W. (1962) Sediment source areas, delivery ratios and influencing morphological factors. Proceedings of the Commission of Land Erosion Symposium of Bari, 1-8 October. IAHS Publication no. 59, IAHS Press, Wallingford, pp. 202-213.

Swedish Space Corporation (1988) *Mapping of the Natural Conditions of the Philippines*, Final Report (FSF101), Swedish Space Corporation.

Tacio, H.D. (1991) The SALT system—agroforestry for sloping lands. *Agroforestry Today* January-March, pp. 12-13.

Walling, D.E. (1983) The sediment delivery problem. *J. Hydrol.* **65**, 209-237.

Walling, D.E. and Webb, B.W. (1983) Patterns of sediment yield. In: *Background to Paleohydrology*, Gregory, K.J. (ed.), John Wiley & Sons, Chichester, Ch. 4.

White, S.M. (1987) *Casecnan Watershed Sedimentation Study*, vols 1 and 2, Report no. EX1596, Hydraulics Research, Wallingford.

White, S.M. (1990) The influence of tropical cyclones as soil eroding and sediment transporting events. An example from the Philippines. In: *Research Needs and Applications to Reduce Erosion and Sedimentation in Tropical Steeplands*, IAHS Publication no. 192, IAHS Press, Wallingford.

White, S.M. (1992) Sediment yield estimation from limited data sets: a Philippines case study. PhD Thesis, University of Exeter.

Wooldridge, R. (1986) *Sedimentation in Reservoirs: Magat Reservoir, Cagayan Valley, Luzon, Philippines. 1984 Reservoir Survey and Data Analysis*, Report no. OD 69, Hydraulics Research, Wallingford.

World Bank (1988) *The World Bank Atlas, 1988*, The World Bank, Washington, DC.

World Bank (1989) *Philippines Forestry, Fisheries and Agricultural Management Study (ffARM)*, The World Bank, Washington, DC, (confidential).

22 Sediment Yield from Alpine Glacier Basins

ANGELA M. GURNELL
School of Geography, University of Birmingham, UK

INTRODUCTION

In many regions of the world, meltwater from areas of permanent snow and ice provides the primary water resource. Meltwater draining from glacier basins is characterised by sediment loads that are well above the global average (Gurnell, 1987; Lawler *et al.*, 1992), which presents major problems for the management and use of the meltwater (Bezinge *et al.*, 1989; Bogen, 1989). The estimation and prediction of both sediment transport and meltwater discharge from glacier basins is, therefore, of major management importance. There is also a geomorphological rationale for such studies in terms of understanding the denudation systems of alpine areas and in drawing comparisons with other morphoclimatic zones. Geomorphological interest in the sediment loads of proglacial streams centres on the very dynamic properties they confer on the sediment response and morphology of proglacial rivers when coupled with a variable discharge regime (e.g. Gurnell and Fenn. 1987, Maizels, 1983), and on the degree to which this variability combined with other factors can be used to make inferences about subglacial processes (e.g. Gurnell, 1987, Gurnell *et al.*, 1992a; Gurnell and Warburton, 1990).

Glacier hydrology and sediment transfer

The discharge in high alpine proglacial rivers is mainly generated by the melting of snow and ice. As a result of the high altitude, precipitation rarely falls as rain, and the primary energy source for snow and icemelt is incident radiation rather than advected energy, although high-intensity rain falling on ice and snow can produce major floods (Röthlisberger and Lang, 1987; Haeberli, 1983). Thus the proglacial discharge regime at the annual and diurnal time-scales reflects the temporal pattern in solar energy receipt, and also incorporates lag and attenuation effects that result from the varying routes and storage areas through which the meltwater passes as it drains to the proglacial river (Gurnell *et al.*, 1992a).

Ice- and snowmelt occur at varying rates in different locations within the glacier basin, and at different times within the ablation season (Figure 22.1). Once meltwater is produced, it can drain to the proglacial river along a variety of routes (extraglacial, supraglacial, englacial and subglacial). It drains at different rates according to the route taken and the size of the pathways through which it drains: intergranular spaces in

---

Sediment and Water Quality in River Catchments. Edited by I.D.L. Foster, A.M. Gurnell and B.W. Webb.
© 1995 John Wiley & Sons Ltd.

snowpack, ice or subglacial moraine; small to very large channels, conduits and cavities; sheets of water at the snow/ice or ice/bedrock/moraine interfaces; open channels. As the meltwater drains along these pathways, it is exposed to a variety of sediment sources, and its ability to tap these sources varies according to the energy of the flow and the availability of sediments in the source areas. In the alpine glacial environment, the major source areas of sediment for fluvial transport are hillslopes

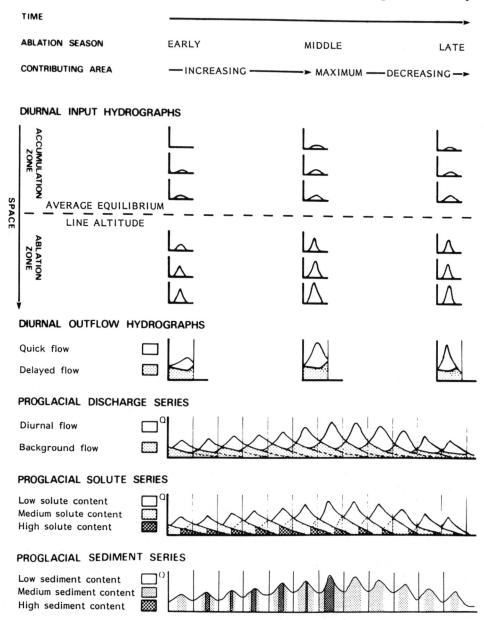

Figure 22.1 Hydrograph-based models of the nature of proglacial discharge, solute and sediment series (reproduced with permission of John Wiley & Sons from Fenn, 1987)

(including areas covered by snow patches and small hanging glaciers) and the subglacial environment. The availability of sediments at the glacier bed is a function of the activity of the glacier in eroding and transporting sediments and the activity of meltwater in removing them.

Once meltwater drains from the glacier to the proglacial stream, a range of proglacial sediment sources can be tapped, and there are also many locations where deposition of sediments can occur. Thus, sediment transport is expected to vary along the proglacial river (Bogen, 1980; Gurnell and Fenn, 1987), and this variability is well illustrated by two recent studies of contrasting spatial scale. Desloges and Church (1992) describe the transformation and subsequent recovery in river channel and floodplain morphology within a 3 km length of the Noeick River as a result of a major meltwater outburst flood. The influence of sediment going into and coming out of storage within the channel and valley train on sediment transport by the river is clearly illustrated by this study. Such influences are also identified in the very detailed study by Lane et al. (1994) of a 50 m length of the proglacial river of the Haut Glacier d'Arolla, Switzerland. Here the interactions between detailed river channel topography, flow hydraulics and sediment transport indicated by cut and fill is illustrated to a maximum spatial resolution (through the use of photogrammetry) of over 20 elevation points per square metre. This study illustrates clear spatial patterns in cut and fill over periods of one or more days.

Sediment transport and discharge in proglacial rivers

The seasonal and diurnal variability that is characteristic of the discharge of alpine proglacial rivers can also be identified in observations of sediment transport. Detailed temporal observations of bedload transport in proglacial rivers are rare (see below). As a result, the character of suspended sediment transport has received the most research attention, and the similarity that has been identified in the temporal patterns of discharge and suspended sediment concentration has influenced approaches to estimating suspended sediment loads in proglacial rivers. For example, suspended sediment rating curves have been estimated by linear regression analysis of the suspended sediment concentration and discharge time series. The problem with this approach is that the relationship between suspended sediment concentration and discharge in proglacial rivers is neither reliably linear nor stable (Fenn et al., 1985). Hysteresis in the relationship between suspended sediment concentration and discharge exists at a range of time-scales from diurnal, through sub-seasonal and seasonal to inter-annual. As a result, researchers have attempted to develop the sediment rating curve methodology to incorporate other factors that may induce hysteresis, including the influence of antecedent conditions (e.g. Richards, 1984) and the potential influence of englacially and subglacially routed meltwater (Gurnell and Fenn, 1984a). Furthermore, the presence of complex hysteresis in the residuals from sediment rating curves has led to the adoption of a time-series approach that estimates Box–Jenkins transfer functions between discharge and suspended sediment concentration (e.g. Gurnell and Fenn, 1984b; Gurnell et al., 1992a). Such manipulations of the proglacial time-series information produce improvements in the accuracy with which suspended sediment concentration can be estimated from discharge. However, the manipulations do not explicitly incorporate information on the many other processes that are believed to influence sediment transfer rates in proglacial streams. A particularly important omission is the influence of the availability of sediment for the rivers to transport.

Sediment availability and sediment yield

The role of sediment availability and storage in controlling sediment transfer rates is a recurring research theme, which is concerned with the importance of changes in the sediment budget of whole glacier basins or of sub-areas of these basins in controlling sediment yield.

Church and Ryder (1972) defined the concept of 'paraglacial sedimentation' to describe the impact of glaciation in increasing sediment availability by the deposition of sediment in storage in glacial drift, from where it can be transported at an enhanced rate by proglacial and postglacial rivers. This concept represents a fundamental change in the sediment budget of drainage basins as a result of glaciation, which can have a significant impact on the magnitude of basin sediment yield over very long periods of time. Such temporal instabilities in sediment availability and transport can be identified within the extraglacial, glacial and proglacial zones of glacier basins at many different time-scales. Thus the previously described impact of flood scour and deposition of sediment within the proglacial zone as a result of a meltwater outburst (Desloges and Church, 1992) is a shorter-term example of a change in sediment budget within a component of a glacier basin. The overriding importance of sediment storage for sediment transfer rates led Harbor and Warburton (1993) to argue that short-term sediment yield information from glacier basins is of limited interpretative value without a knowledge of the amount and location of sediment in storage. Furthermore, Warburton (1990a) constructed a fluvial sediment budget for a section of the proglacial valley train of the Bas Glacier d'Arolla, which illustrates the important role of sediment storage. However, this latter research also illustrates the labour-intensive nature of such an approach. Indeed, the estimation of a full sediment budget for a glacier basin is virtually impossible because of the difficulties in quantifying many of the components of the budget.

The discussion presented below illustrates that accurate monitoring of glacier basin sediment yield alone is a daunting task. However, by adopting a comparative approach between basins and time-scales, this chapter employs detailed observations of glacier basin sediment and discharge output in an attempt to identify the role of sediment availability as well as hydrological processes in governing sediment yield from alpine glacier basins.

To achieve further understanding of the controls on sediment yield from alpine glacier basins, this chapter considers three themes. First, the monitoring techniques and sampling frameworks used to describe sediment transport in proglacial rivers are emphasised, because these can have very significant effects on the estimates of sediment yield that are produced. Secondly, and against the background of the significance of field monitoring techniques, the variability in sediment yield from a sample of glacier basins located within one small area of Switzerland and assessed by similar monitoring approaches is presented. Comparative information from these basins at a variety of time-scales is used to illustrate the enormous variability in sediment yield that can occur within one small area, and to suggest potential controlling factors. Thirdly, more detailed information from one of these basins is used to associate glacier drainage evolution with sediment delivery to the proglacial river. This illustrates the potential to develop more physically based approaches to estimating glacier basin sediment yield.

THE SIGNIFICANCE OF MONITORING TECHNIQUES AND SAMPLING FRAMEWORKS

Suspended sediment transport in proglacial rivers has been monitored by the filtration of water samples taken by hand or by automatic pump sampling, and by the use of optical turbidity meters. An enormous variety of instruments and sampling frameworks have been employed (Gurnell, 1987), but detailed temporal studies have illustrated the importance of careful choice of instruments and sampling locations, and also the high temporal variability of suspended sediment concentration in many proglacial rivers (Bogen, 1988).

As a result of the turbulent nature of the flow in many stretches of proglacial rivers, single-site, fixed-depth sampling of meltwater has the potential to provide water samples that are representative of the average suspended sediment concentration of the river cross-section at the sampling time (Bogen, 1988). The results of multiple-site sampling across sections of some small proglacial rivers suggest that suspended sediment concentration in the turbulent flow does not vary greatly (Gurnell et al., 1992b; Østrem et al., 1971), but there are very few published studies that explore this potential source of error in single-site sampling. In order to ensure sampling that is representative of suspended sediment concentration and particle size in the river cross-section, some researchers have taken the additional precaution of sampling at a site where turbulence and mixing of water and sediment are deliberately induced (e.g. Bogen, 1992). Given the relatively small variability in depth-integrated estimates of suspended sediment concentration across small, turbulent proglacial rivers, it appears that differences in suspended sediment concentration estimates from filtration of meltwater samples taken from the same site at different times are very likely to reflect true temporal differences in suspended sediment concentration in the river (Gurnell et al., 1992b). However, this approach to characterising suspended sediment transport in proglacial rivers is time-consuming and only gives point estimates in time, whereas the suspended sediment concentration in proglacial rivers can vary greatly and rapidly in time. As a result, optical turbidity meters are often used in combination with water sampling to generate a detailed temporal picture of suspended sediment concentration variations. Such an approach can provide high-quality, temporally detailed estimates of suspended sediment concentration if careful calibration of the turbidity record is undertaken (Gurnell et al., 1992b). The turbidity record reflects not only the suspended sediment concentration of the meltwater, but also its colour and particle-size characteristics (Foster et al., 1992). Furthermore, the record can be adversely affected by air bubbles in the flow, ambient light, accumulation of sediment or algal growth on the turbidity sensor, and variability in the instrument's power supply. It is, therefore, essential that frequent meltwater samples are taken to adjust the calibration of the turbidity record continuously to observed suspended sediment concentration. For example, Table 22.1 lists calibration relationships that were estimated between two- or three-hourly determinations of suspended sediment concentration from water samples

Table 22.1 Estimated relationships[a] between suspended sediment concentration and turbidity for subperiods of the 1989 and 1990 ablation seasons

1989
Periods 1 and 2: 01.00 h 1 June–21.00 h 16 June
$S = 3.73T + 2.37\nabla Q_t - 226.62$ $R^2 = 0.640$
Period 3: 22.00 h 16 June–15.00 h 8 July
$S = 10.14T + 0.49\nabla Q_t - 429.63$ $R^2 = 0.959$
Period 4: 16.00 h 8 July–24.00 h 6 August
$S = 9.87T + 0.20Q_t - 957.21$ $R^2 = 0.733$
Period 5: 01.00 h 7 August–24.00 h 31 August
$S = 8.54T + 0.29Q_t - 885.30$ $R^2 = 0.834$

1990
Period 1: 16.00 h 29 May–07.00 h 19 June
$S = 6.61T - 0.20 \text{(count)} - 122.0$ $R^2 = 0.850$
Period 2: 08.00 h 19 June–15.00 h 2 July
$\log_{10} S = 1.646 \log_{10} T - 0.596$ $R^2 = 0.938$
Period 3: 16.00 h 2 July–17.00 h 27 July
$S = 11.64T - 315.6$ $R^2 = 0.843$
Period 4: 18.00 h 27 July–11.00 h 13 August
$S = 11.37T - 203.8$ $R^2 = 0.830$
Period 5. 12.00 h 13 August–10.00 h 26 August
$S = 10.68T - 4.5$ $R^2 = 0.745$

[a] All slope coefficients are significantly different from zero ($P<0.01$). R^2 values are adjusted for the degrees of freedom of the estimated model. S = suspended sediment concentration (mg l^{-1}); T = turbidity as a deflection across an arbitrary scale fixed at the start of the season; Q_t = concurrent discharge (l s^{-1}); ∇Q_t = change in discharge over the preceding hour (l s^{-1}); count = number of hours since start of period

taken by an Isco automatic pump sampler, and turbidity estimates from the probe of a Partech 7000 model 3RP suspended sediment monitor mounted next to the intake nozzle of the pump sampler, over the 1989 and 1990 ablation seasons on the proglacial river of the Haut Glacier d'Arolla. By a careful combination of the use of pump samplers and turbidity meters, it is possible to derive high-quality, continuous records of suspended sediment concentration in proglacial rivers. Such records can then be degraded to whatever temporal resolution is required.

Studies of bedload transport from glacier basins are very rare. There have been a few studies that have used purpose-designed sediment samplers, such as the study by Warburton (1990b) of bedload transport in the proglacial river of the Bas Glacier d'Arolla, which employed a Helley–Smith sampler. However, the high rates of relatively coarse bedload transport, the typically uneven and mobile beds of proglacial rivers, and the frequently highly turbulent flow make the use of such samplers difficult. Thus, other less-conventional approaches have been applied to assess gross bedload transport rates, including the accumulation of sediments behind mesh fences constructed across proglacial rivers (e.g. Hammer and Smith, 1983), observations of accumulation of deltaic deposits in proglacial lakes (e.g. Bogen, 1989; Kjeldsen, 1981; Kjeldsen and Østrem, 1980) and monitoring of the accumulation of coarse sediments in sediment traps (e.g. Kjeldsen, 1981; Kjeldsen and Østrem, 1980; Bezinge et al.,

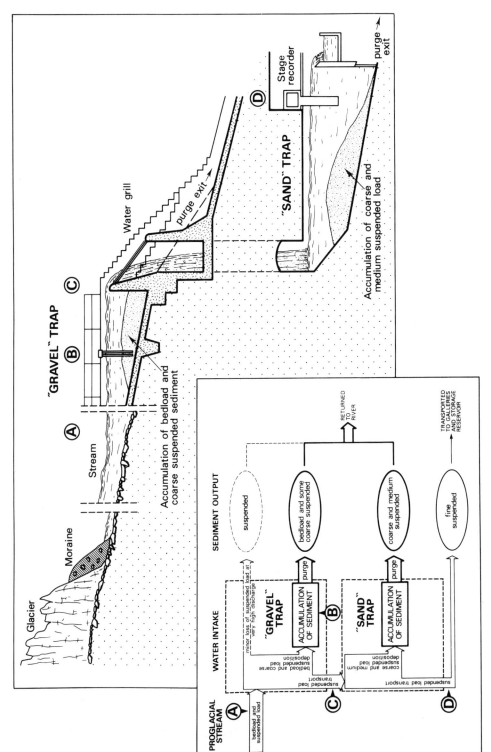

Figure 22.2 Schematic diagram of the Tsidjiore Nouve and Bas Arolla meltwater intake structures (reproduced with permission of the International Glaciological Society from Bezinge *et al.*, 1989)

1989). In the latter case, sediment traps incorporated in the design of hydro-power meltwater intake structures can yield high-quality bedload transport estimates for proglacial rivers, although the temporal resolution of such data is relatively crude. For example, Bezinge et al. (1989) and Gurnell et al. (1988) describe the calibration of 'sand' and 'gravel' traps that are incorporated in some of the meltwater intakes of the Grande Dixence hydro-power scheme, Switzerland, through which entire proglacial rivers are diverted (Figure 22.2). Both of these traps are designed to empty automatically once sediment has accumulated to a certain level. In the case of the 'gravel' trap, repeat surveys of the accumulation of sediment coupled with observations of the packing density of sediment in the trap permits estimation of the weight of sediment released by each automatic purge (emptying) of the trap. Monitoring of suspended sediment input and output to this trap (via a calibrated continuous turbidity record) allows the determination of the proportion of suspended sediment within the purged material, so that the weight of total bedload and trapped suspended sediment load can be estimated. Differences in the input and output of suspended sediment to the sand trap allow the estimation of the total weight of suspended sediment associated with each purge of the trap and also the amount of suspended sediment passing completely through the two traps and into the hydro-power scheme. In this way past records of emptying of the two traps can be used to reconstruct the suspended sediment and bedload transport record for the site. This approach has been used to construct 10 years of estimates of suspended sediment and bedload transport for two glacier basins in Switzerland. Unfortunately, changes in the protocol for operating the traps have been introduced in recent years, so that such an approach cannot reliably be applied after 1986–87.

SEDIMENT YIELD FROM ALPINE GLACIER BASINS AT DIFFERENT TIMESCALES

The problems of accurately monitoring sediment output from glacier basins, and the very wide range of approaches that have been used in different studies (Gurnell, 1987), suggest that inferences concerning the potential controls on alpine glacier basin sediment yield should ideally be based on the interpretation of data sets collected using similar instrumentation and sampling designs or by using studies where cross-calibration between different approaches has rendered the data sets comparable. This section compares discharge and sediment yield records from four Alpine glacier basins (Bas Glacier d'Arolla, Haut Glacier d'Arolla, Glacier de Ferpècle and Glacier de Tsidjiore Nouve) located close to one another in the Val d'Hérens Switzerland (Figure 22.3). Similar approaches have been used to estimate the discharge and sediment yield from these basins. In this way, contrasts between the discharge and sediment outputs of the basins can be more confidently attributed to differences in the character of the basins and in the relative significance of the processes operating within the basins.

The four drainage basins vary in their dimensions and cover of permanent ice and snow (Table 22.2). The natural Bas Arolla basin includes the Haut Arolla basin, but, because of the operation of the Grande Dixence hydro-power scheme, the currently active (i.e. managed) basin, which excludes the Haut Arolla basin and some other small basins upstream of their hydro-power intakes, is used for the estimates presented here.

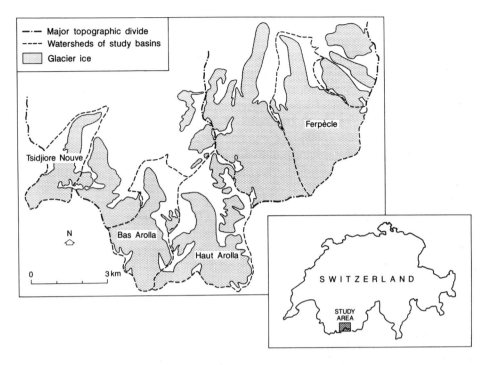

Figure 22.3 The four study basins within the context of glacier basins in the upper Val d'Hérens, Switzerland

Table 22.2 Some catchment characteristics for the four study basins

Characteristics	Glacier basin			
	Tsidjiore Nouve	Bas Arolla	Haut Arolla	Ferpècle
Basin area (nearest km^2)	5	25 (natural[a]) 8 (managed)	12	13
Permanent snow and ice cover (nearest 5%)	70	55 (natural) 70 (managed)	55	80
Altitude of highest point on watershed	3795	3538 (managed)	3838	4098
Altitude of proglacial monitoring site[b]	2140	2100	2560	2040

[a] The natural catchment of the Bas Arolla basin includes the Haut Glacier d'Arolla basin, and the basins of the Glacier de Vuibe and Glacier de Bertol, all of which feed meltwater to the Grande Dixence hydro-power scheme via intakes that are upstream of the Bas Arolla meltwater intake
[b] The monitoring sites at Tsidjiore Nouve and Bas Arolla are within 400 m of the glacier snout and those at Ferpècle and Haut Arolla are within 100 m of the glacier snout throughout the period of records presented in this chapter

The basins are located on a complex mix of schists, gneisses and granites of the Arolla series, giving a wide variability in susceptibility to weathering and erosion, which results in one glacier (Ferpècle) that is believed to have a predominantly hard bed and the remaining three to be underlain by predominantly soft (moraine) beds. Useful comparisons can be made: (i) between the Haut Arolla and Ferpècle basins, which are of similar size, orientation and geometry, but have different bed character; and (ii) between the Tsidjiore Nouve and (managed) Bas Arolla basins, which both have soft beds and similar geometry (both have a major icefall between their accumulation and ablation areas), but have slightly different orientations and sizes (Figure 22.3, Table 22.2).

A series of field studies have produced data sets of different length and temporal resolution for each glacier basin. The data that are presented here are based on field monitoring of suspended sediment concentration by a combination of continuous turbidity monitoring and direct suspended sediment concentration estimation from filtration of meltwater sampled by an automatic pump sampler. A combination of these approaches with calibration of sedimentation rates within sand and gravel traps operated by the Grande Dixence hydro-power company on their proglacial meltwater intakes is used to estimate longer-term suspended sediment and bedload transport.

Inter-annual variations in suspended sediment and bedload transport

From the very different periods of record for the four basins, there appear to be major contrasts in annual suspended sediment yields: 1800 t km^{-2} yr^{-1} (10 years of record), 1100 t km^{-2} yr^{-1} (10 years), 3000 t km^{-2} yr^{-1} (two years) and 280 t km^{-2} yr^{-1} (one year) for the Tsidjiore Nouve, Bas Arolla, Haut Arolla and Ferpècle basins, respectively.

The proglacial rivers of three of the four basins (Tsidjiore Nouve, Bas Arolla and Haut Arolla) drain directly into water intakes for the Grande Dixence hydro-power scheme. The Tsidjiore Nouve and Bas Arolla basins have separate 'sand' and 'gravel' traps, whereas the Haut Arolla basin has a single, combined sediment trap. The automatic initiation of sediment purges at the Tsidjiore Nouve and Bas Arolla traps, and the semi-automatic purging at Haut Arolla, allow estimation of trends in sediment yield over an 11-year period from 1977 to 1987 inclusive (although information is missing for Tsidjiore Nouve in 1978 and for Bas Arolla in 1982). Calibration of the Tsidjiore Nouve and Bas Arolla 'sand' and 'gravel' traps permits estimation of rates of suspended sediment and bedload transport, whereas the single, and thus uncalibrated, Haut Arolla trap can be used to underpin analysis of purge frequency as an indicator of variability in total sediment transport.

Figure 22.4 plots the annual purge frequency of the sediment traps in the three basins over the 11-year period 1977–87. Figure 22.4 also shows the annual discharge volumes from each of the three basins and, for Bas Arolla and Tsidjiore Nouve, the annual movement of the glacier snout is also shown. Haut Arolla shows a general correspondence in the pattern of annual discharge volume and purge frequency; Bas Arolla shows a rising trend in all four variables over the period; and Tsidjiore Nouve shows an increase followed by a decrease in sand trap purges, which shows some correspondence with movement in the position of the glacier snout, and a general increase in discharge volume and bedload trap purging over the 1977–87 period. These

SEDIMENT YIELD FROM ALPINE GLACIER BASINS 417

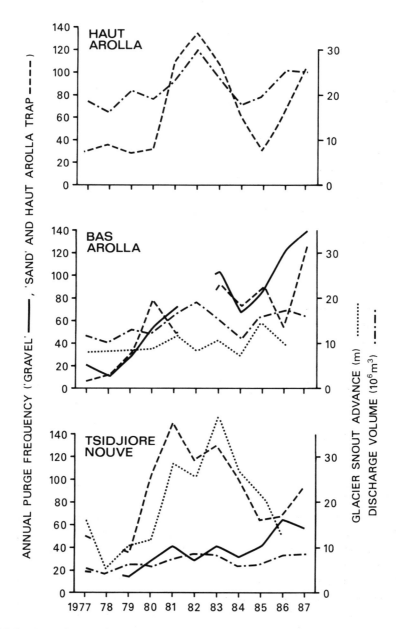

Figure 22.4 Annual purge frequency in relation to annual discharge volume and glacier snout advance—Haut Arolla, Bas Arolla and Tsidjiore Nouve basins (reproduced with permission of the International Glaciological Society from Bezinge *et al.*, 1989)

data illustrate clear contrasts between the basins in terms of both the pattern through the 11-year period in each of the variables, and the degree of correspondence in the patterns between variables for individual basins. If the trap purges are translated into estimates of suspended sediment load and bedload for the Tsidjiore Nouve and Bas Arolla basins, very strong differences between the basins begin to emerge. Bezinge *et al.* (1989) discuss the likely error associated with these estimates, but, for the present discussion, average trap calibrations are used. To the nearest 100 t km^{-2} yr^{-1}, Tsidjiore Nouve has a lower mean annual total sediment yield (2500 in comparison with 2900 t km^{-2} yr^{-1}), but higher mean annual suspended sediment yield (1700 in comparison with 1300 t km^{-2} yr^{-1}), giving the proportion of the total load transported as suspended sediment as approximately 67 and 45% for the two basins, respectively. Tsidjiore Nouve also yields a higher mean total sediment concentration (1.77 in comparison with 1.59 g l^{-1}) than the Bas Arolla basin. The higher concentration for Tsidjiore Nouve is particularly surprising, given that much of the sediment from the Haut Arolla basin is purged into the Bas Arolla basin, but virtually all of the meltwater

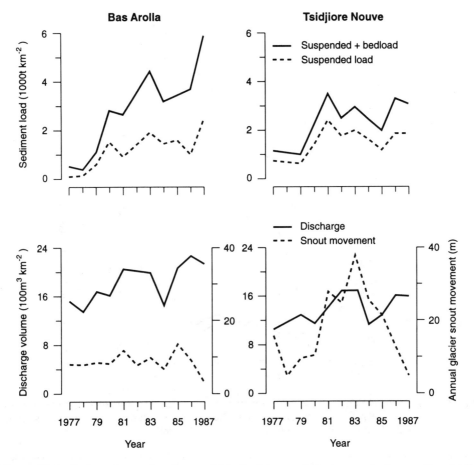

Figure 22.5 Estimated annual sediment yield, discharge and glacier snout advance, Bas Arolla and Tsidjiore Nouve basins, 1977–87

from the Haut Arolla basin is removed through a meltwater intake. Figure 22.5 plots the estimates of sediment yield, discharge volume and annual movement in snout position for the two glacier basins. At this mean annual time-scale, there is some correspondence between the temporal pattern in sediment transport and discharge, and there is a significant simple regression relationship between these variables for both basins (Table 22.3). There also appears to be some correspondence in the pattern of movement of the glacier snout and sediment yield for the Tsidjiore Nouve basin, but snout position movement is not a significant explanatory variable for sediment yield when included on its own or with discharge in regression analyses.

Intra-annual variations in suspended sediment and bedload transport

Trap purge and discharge information also illustrate similar mean ablation season patterns of water and sediment yield for the Tsidjiore Nouve, Bas Arolla and Haut Arolla basins (Figure 22.6), although subtle differences can be identified. In general, there is a relatively steep rise in discharge from mid/late May to reach a maximum

Table 22.3 Simple linear regression relationships[a] between sediment yield (t) and discharge (m^3), Tsidjiore Nouve and Bas Arolla Basins, 1977–87

Information analysed		a	t_a	b	t_b	n	R^2
Annual totals 1977–87							
Tsidjiore Nouve		−3383	0.53	2.27	2.47	10	0.36
Bas Arolla		−31646	1.74	3.88	2.99	10	0.47
Weekly totals							
Tsidjiore Nouve	1977	−7	0.04	1.16	2.28	19	0.23
	1979	−151	1.14	1.32	3.72	20	0.43
	1980	−201	0.87	2.79	4.38	20	0.52
	1981	−612	2.27	4.31	6.31	20	0.68
	1982	−133	0.76	1.83	5.07	20	0.59
	1983	−282	1.48	2.47	6.59	20	0.70
	1984	−42	0.21	2.30	4.06	20	0.46
	1985	−148	1.34	2.08	7.11	20	0.73
	1986	−272	1.29	2.79	6.09	20	0.66
	1987	−324	1.09	2.84	4.46	20	0.51
Bas Arolla	1977	−149	1.51	0.65	4.46	20	0.54
	1978	−69	0.51	0.47	2.22	20	0.23
	1979	−311	1.62	1.12	4.84	20	0.59
	1980	−1153	1.75	3.65	4.20	20	0.49
	1981	−688	1.53	2.27	4.21	20	0.48
	1983	−487	1.84	2.95	9.64	20	0.83
	1984	−582	1.64	2.94	7.00	20	0.79
	1985	−195	0.72	1.95	6.35	20	0.68
	1986	−799	2.42	2.62	7.68	20	0.76
	1987	−484	1.01	3.39	6.70	20	0.70

[a] Symbols are as follows: a = regression intercept; b = regression slope; t_a and t_b = t statistics for intercept and slope; n = number of observations; R^2 = coefficient of determination

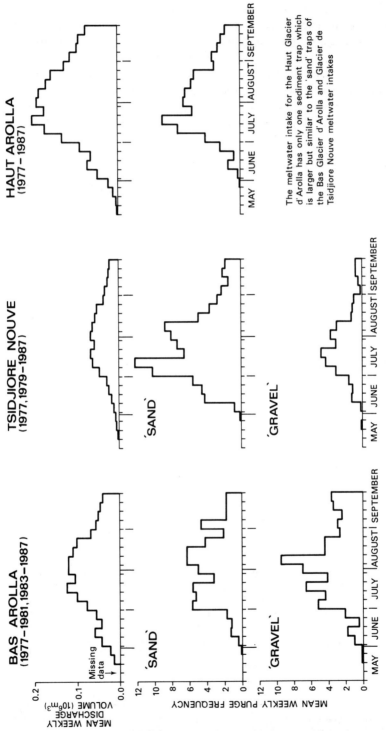

Figure 22.6 Mean weekly meltwater discharge and purge frequency—Haut Arolla, Bas Arolla and Tsidjiore Nouve, 1977–87 (reproduced with permission of the International Glaciological Society from Bezinge *et al.*, 1989)

value in mid/late July. This is followed by a more gradual decline in discharge to the end of the ablation season. Sediment trap purge frequency follows a similar pattern, although purging commences after discharge in all three basins. For Bas Arolla there appears to be enhanced sediment purging in comparison with the other two basins in August and September. For Haut Arolla, the seasonal increase in sediment trap purge frequency is steep, commencing some three weeks after discharge but peaking at the same time. Furthermore, the entire seasonal distribution of discharge and purge frequency is delayed in comparison with Bas Arolla and Tsidjiore Nouve. This presumably reflects the fact that the Haut Glacier d'Arolla is located at a higher altitude than the other two glaciers. The altitudes of the snouts of the three glaciers are approximately 2220, 2140 and 2560 m for Tsidjiore Nouve, Bas Arolla and Haut

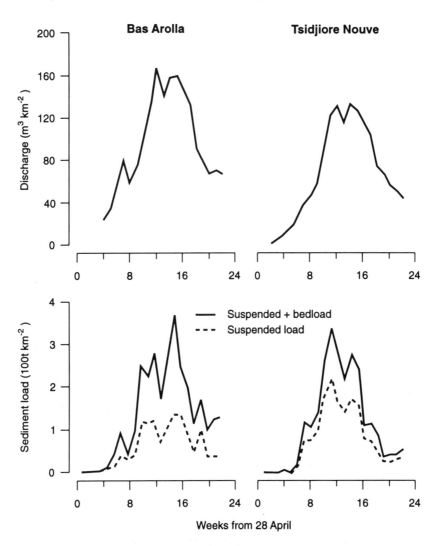

Figure 22.7 Mean weekly discharge and estimated sediment load through the ablation season, Bas Arolla and Tsidjiore Nouve basins, 1977–87

Arolla, respectively. For Tsidjiore Nouve the rise in trap purging is even steeper than Haut Arolla, commencing approximately three weeks after and peaking slightly in advance of discharge. These results suggest different patterns of sediment availability to meltwater under the three glaciers, which are investigated further below.

Trap calibrations for Tsidjiore Nouve and Haut Arolla (Figure 22.7) confirm the delayed peak and the higher late-season levels in average weekly sediment yield for Bas

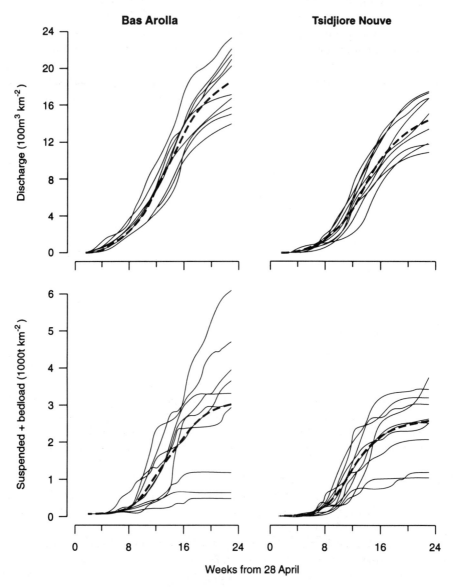

Figure 22.8 Ablation-season cumulative plots of weekly discharge and estimated weekly total sediment load, Bas Arolla and Tsidjiore Nouve basins, 1977–87. Full curves represent individual years; the bold broken curve is the average

Arolla in comparison with Tsidjiore Nouve. Cumulative plots of weekly discharge and total sediment yield from the two basins (Figure 22.8) show a consistent reduction in sediment transport from Tsidjiore Nouve towards the end of the ablation season, whereas this does not always occur in the more variable sediment transport from the Bas Arolla basin. Thus, Tsidjiore Nouve is highly productive of both suspended sediment and bedload early in the ablation season, whereas Bas Arolla appears to have a more persistent availability of sediment for transport throughout the melt season. This difference is also indicated by simple linear regression relationships estimated between weekly sediment yield and discharge for the two basins for each of the years of record (Table 22.3). Although the residuals from all of these regression relationships are autocorrelated, the stronger seasonal hysteresis for the Tsidjiore Nouve basin supports lower coefficients of determination (R^2) in six out of the nine years for which data sets are available for both basins. The enormous variability in the slope coefficient b between years and basins illustrates the contrasts in sediment yield that can occur in relation to discharge. A major cause of the difference in the seasonal pattern in sediment yield between the two basins is probably the operation of the hydro-power scheme. Early in the ablation season, meltwater discharge and sediment are generated entirely within the Bas Arolla basin. As the season advances, purges of sediment from the Haut Arolla glacier begin to make a contribution to the Bas Arolla drainage system. There is probably an initial period when the drainage of the purge water is not well connected to the Bas Arolla drainage system, but once drainage routes beneath the Bas Glacier d'Arolla extend to link with the purge water, sediment can be delivered efficiently to the proglacial river of the Bas Arolla from purges of the Haut Arolla sediment trap. In the absence of a meltwater intake, early-season meltwater from the Haut Arolla would probably become rapidly integrated into the Bas Arolla drainage network providing higher sediment concentrations in late July, shortly after the peak in sediment yield (as inferred from purge activity) from Haut Arolla. Figure 22.9 shows suspended sediment concentration, monitored close to the glacier snout, and discharge from the Haut Glacier d'Arolla in 1989. This confirms the delayed timing and steep rise in suspended sediment transport from the basin. These data on mean seasonal patterns in discharge and sediment yield from three glaciers with soft beds suggest that, although there are typical patterns in both variables between melt seasons, subtle differences relate to the altitudinal distribution (and probably the slope and aspect) of the permanent snow and ice within the basin, and thus the timing of melt and access of the meltwater to sediment at different locations within the basin. Gurnell *et al.* (1988) discuss such contrasts between the Tsidjiore Nouve and Bas Arolla basins in more detail.

The influence of bed character on sediment yield can be illustrated by contrasting suspended sediment yield estimates from the Ferpècle basin with the Haut Arolla basin over the same ablation season. These basins are of similar size and orientation, but Ferpècle is believed to have a predominantly hard bed. Unfortunately, suspended sediment yield from the Ferpècle basin has only been monitored for one year, and the extremely turbulent nature of the proglacial stream makes accurate monitoring of discharge difficult, but very strong contrasts in the sediment output from the two basins are observed. Table 22.4 illustrates that, early in the 1989 ablation season, suspended sediment yields from these two basins are comparable, but whereas the Haut Arolla basin yields increasing amounts of sediment through the melt season, the amount of

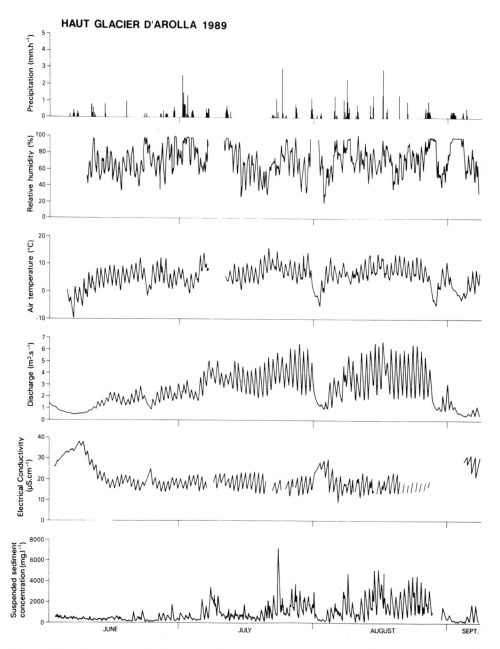

Figure 22.9 Some hydroclimatological time series for the 1989 ablation season, Haut Glacier d'Arolla basin. (reproduced with permission of John Wiley & Sons from Gurnell *et al.*, 1992a)

Table 22.4 Contrasts in the sediment yield between the Ferpècle and Haut Arolla basins, 1989

Variable	June	July	August
Average discharge (m^3 s^{-1})			
Haut Arolla	1.373	3.400	2.928
Ferpècle	1.067	2.239	2.073
Ferpècle^a	1.565	3.876	3.338
Average suspended sediment concentration (mg l^{-1})			
Haut Arolla	390	1061	1628
Ferpècle	320	254	154
Average suspended sediment load (t d^{-1})			
Haut Arolla	55.5	358.6	560.5
Ferpècle	33.7	52.1	30.7
Ferpècle^a	49.6	89.6	49.7

^{a} As a result of the extremely turbulent flow in the Ferpècle proglacial stream, discharge is difficult to measure accurately. Therefore, for comparative purposes, this second estimate assumes that the discharge per unit area from the Ferpecle basin is the same as that from the Haut Arolla basin

sediment generated by the Ferpècle basin remains low and relatively stable. The hard bed clearly limits the amount of sediment available for transport, so that the typical seasonal pattern in sediment yield described above for three soft-bed glaciers appears to be truncated for Ferpècle. The very high suspended sediment yield estimated for the 1989 and 1990 ablation seasons (2500 and 3500 t km^{-2}, respectively) from the Haut Arolla basin also requires some comment. These estimates suggest that the Haut Arolla basin has the highest annual suspended sediment yield of the four study basins, while the Ferpècle basin has the lowest. A possible explanation for the high sediment productivity of the Haut Arolla basin in comparison with the Bas Arolla and Tsidjiore Nouve basins is its more regular long profile. Unlike the Bas Glacier d'Arolla and the Glacier de Tsidjiore Nouve, the Haut Glacier d'Arolla has no icefall between its accumulation and ablation zones. This more regular long profile is likely to have a significant influence on the evolution of the subglacial drainage network. Certainly detailed dye tracer studies within the Haut Arolla basin (Nienow et al., 1995a, b) indicate the evolution of an integrated subglacial drainage system across virtually the entire glacier bed during the ablation season, and also the headward extension of major conduits at the expense of a distributed drainage system. Further implications of these observations are discussed below, but their relevance here is that the evolution and extension of the drainage system provide the potential for subglacial streams to tap sediment widely across the glacier bed during each ablation season.

All of the data presented above represent mean values for quite long time periods, but there are sharp contrasts in short-term variability in sediment yield between the basins. Suspended sediment transport from the Tsidjiore Nouve basin, particularly in the early 1980s, when the glacier was advancing quite rapidly, is characterised by frequent pulses or flushes of sediment that persist over time-scales ranging from minutes to days. Such flushes or pulses are far less common in the other three basins. Hourly suspended sediment concentration information for 1981 (Figure 22.10) illustrates the longer-term pulses well. Two meltwater outbursts (15–18 June and 2–7 August) were monitored in

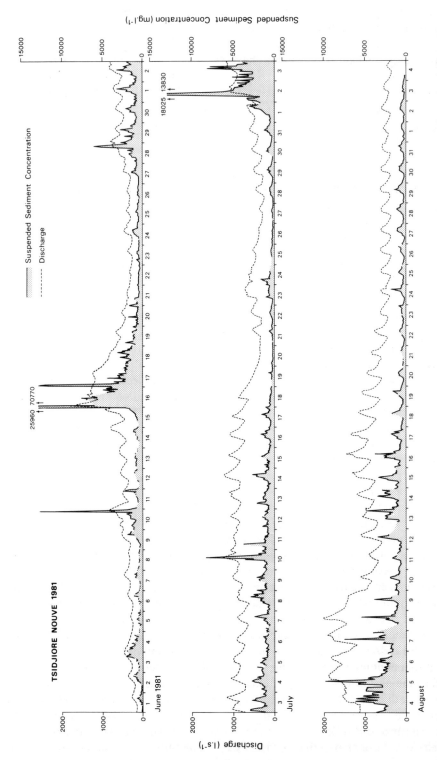

Figure 22.10 Hourly discharge and suspended sediment concentration observations for the 1981 ablation season, Glacier de Tsidjiore Nouve basin (suspended sediment concentration determinations undertaken by I. Beecroft)

1981, during which storage and release of meltwater generated large amounts of sediment. In the earlier outburst, the disrupted routeing of the meltwater was observed (Beecroft, 1983), and in the latter case, snowfall caused a major reduction in ablation rates (20–31 July), which is believed to have led to a reduction in the capacity of the glacial drainage network and thus to have resulted in modifications of the network as the subsequent high flows were transmitted (Figure 20.10). Certainly, observations of suspended sediment concentration in different tributary streams draining from the glacier snout during the 1981 melt season (Gurnell, 1982) suggest that changes in the glacier drainage network were continually occurring and generating shorter-term (1 or 2 h) flushes from single or multiple tributaries. Flushes of sediment occurring over time-scales of minutes are also characteristic of the suspended sediment transport from the Tsidjiore Nouve basin, and are well illustrated by the analysis of continuous turbidity records. Gurnell and Warburton (1990) analyse the magnitude, shape and frequency characteristics of 571 flushes of suspended sediment from the Tsidjiore Nouve basin over a 22 day period in 1986. Although many of these small and short-term flushes seem to be associated with changes in the braiding pattern, bank erosion, and cut and fill in proglacial channels, the larger flushes are believed to be glacial in origin and to reflect changes in the subglacial drainage system (e.g. reorganisation of glacier drainage, including movements in subglacial channels, drainage capture from one channel to another, development of new drainage channels, and inputs of sediment to existing channels as a result of fluvial and glacial processes). These inferences on the causes of frequent flush events from the Tsidjiore Nouve basin, particularly during the period of rapid glacier advance in the early 1980s, suggest the presence of a fairly chaotic drainage system beneath the ablation zone of the glacier, particularly during the first half of the ablation season when five or more tributary streams of varying absolute and relative discharge issued from the glacier snout. Towards the second half of the ablation season, one of the centrally located tributary streams was usually observed to take an increasing proportion of the discharge; and in more recent years, a single stream has dominated throughout the ablation season.

In contrast, pulses of sediment from the Bas Arolla basin (Gurnell and Warburton, 1990) are less frequent (256 pulses between 25 May and 4 September 1986) and many (30% in 1986) can be linked to purges of the Haut Arolla meltwater intake. A fuller analysis of the sediment flushes from the Bas Arolla basin is presented by Warburton (1989). He notes that only 13 purges of the Haut Arolla sediment trap in 1986 did not produce a clear response in the turbidity record for the Bas Arolla basin. Travel times of the linked pulses were short, varying between 11 and 50 min, and, although times tended to be longer in July than in August, there was no very clear systematic pattern through these months, suggesting well-developed drainage beneath the tongue of the Bas Glacier d'Arolla from early in the ablation season, probably through a major conduit or conduits. The lower number of sediment pulses and the consistent linkage of many of these pulses with purging of meltwaters from the Haut Arolla basin suggest that the system draining the tongue of the Bas Glacier d'Arolla is in general more open and stable than that draining the tongue of the Glacier de Tsidjiore Nouve. This, combined with differences in the levels of activity of the two glaciers, as indicated by annual movements in their snout positions (Figure 22.5), and the relatively late seasonal inputs of sediment to Bas Arolla from the Haut Arolla basin, may explain differences in the magnitude and timing of sediment yield from the basins.

Controls on sediment yield

Many of the interpretations presented above are speculative, but it seems that a variety of factors influence the sediment yield from the four glacier basins. First, from the single year of data available, it appears that the presence of a hard bed can result in suspended sediment yields that are an order of magnitude lower than for glaciers of similar size but contrasting bed character. Secondly, for glaciers with a soft bed, the size of the glacier and the volume of discharge are fundamental controls on total annual sediment yield. The level of activity of the glacier also appears to influence the amount and timing of sediment yield. This is well illustrated at Tsidjiore Nouve by the high and variable suspended sediment concentration records, particularly during the period of relatively rapid glacier advance in the early 1980s.

From the limited information available, and the inferences made from that information, it seems that the nature and evolution of the glacial drainage system has a significant impact on the amount and timing of sediment transport. Major conduits are probably more stable in their position than smaller conduits and the components of a linked cavity drainage network, and thus the latter types of drainage system are likely to be associated with a higher and more variable production of suspended sediment in proportion to the discharge than the former. Transitions from one drainage system to another are also likely to result in major increases in sediment availability to the changing drainage network. Thus the low production of suspended sediment early in the ablation season from the Bas Arolla glacier may represent reoccupation of a pre-existing conduit system beneath the glacier tongue. In the case of Tsidjiore Nouve, particularly during the early 1980s, the drainage system appears to have been continually adjusting as a result of the relatively high rates of movement of the glacier, and this adjustment coupled with possible subglacial channel movement may lead to pulses in sediment transport from a drainage system that certainly did not consist of major conduits, although a number of minor ones may have existed below the glacier tongue during the period of significant glacier advance. Finally, direct observations of the evolution of the subglacial drainage system beneath the Haut Glacier d'Arolla suggest that the mode of drainage network evolution and the very extensive spatial dimensions of the network may contribute to high sediment availability and thus the high suspended sediment yield from this basin.

From the inferences presented on the patterns of discharge and sediment yield from four glacier basins, it appears that the character and stability of the subglacial drainage system may be particularly significant in influencing sediment transfer rates. Integrated studies of the storage and transfer of meltwater in the Haut Arolla basin are used in the following section to illustrate the potential to couple information on ablation-season glacier drainage evolution with sediment delivery processes, and so to produce more physically based approaches to estimating sediment yield.

THE POTENTIAL TO COUPLE GLACIER DRAINAGE EVOLUTION WITH SEDIMENT DELIVERY PROCESSES

Subglacial drainage below temperate glaciers may take place in major conduits or tunnel systems (Röthlisberger, 1972; Walder and Fowler, 1989), in smaller conduits that link cavities (Kamb, 1987; Walder, 1986), or, to a lesser extent, as a thin film or sheet between the ice and the glacier bed (Weertman, 1972). The proportion of meltwater carried by each of these components of the subglacial drainage system varies between glaciers, between different areas of the bed of the same glacier, and with time (Seaberg et al., 1988; Hooke, 1989; Willis et al., 1990). Differences in the stability, water pressure characteristics and seasonal dynamics of these components of the subglacial drainage system would be expected to have a major influence on the degree to which the meltwater gains access to fine sediments as it is transmitted through the glacier drainage system. Observations from the Haut Glacier d'Arolla basin illustrate how an understanding of the evolution of the subglacial drainage system may assist in improving estimates of suspended sediment transport in proglacial rivers.

Nienow et al. (1995a, b) describe the results of dye tracer studies that have defined the seasonal evolution of two distinct drainage systems beneath the Haut Glacier d'Arolla—a distributed system and a conduit system. Nienow et al. (1995a) illustrate that the upper limit of the conduit system corresponds approximately with the position of the ambient snowline. The upper limit of the distributed system is limited by the area of firn and snow generating meltwater. Nienow et al. (1995b) show that by mid-August in 1989, 1990 and 1991, the lowest 1 km of the glacier bed was entirely drained by major conduits, upstream of this was a 2 km section that was predominantly drained by major conduits, and the remaining headward area of the glacier was largely drained by a distributed system. Travel times of dye tracer from injection sites below the snowline are less than approximately 100–150 min, and above the snowline are over 200 min in 1990 (similar rates were observed in 1989). Typical dye tracer travel times from the area of the glacier drained by the distributed system are 300–400 min.

Hydrological studies of alpine glacier basins (e.g. Lundquist, 1982; Oerter el al., 1981; Baker et al., 1982) have shown how the discharge from glacier basins can often be represented as the result of routeing snow and ice ablation through a number of linear reservoirs. Such an approach assumes that, when there is no recharge, the outflow (Q_t) at time t from each reservoir can be expressed as a function of a preceding recession flow (Q_0) at time t_0 and a storage constant K:

$$Q_t = Q_0 \exp[-(t-t_0)/K]$$

Thus, the storage constant for a linear reservoir can be estimated from information on the character of recession flows. The degree to which such discrete reservoirs may exist, may be linear or non-linear, and may be connected in parallel or in series, can present complications for such analyses, but an investigation of recession flows from the Haut Glacier d'Arolla (Gurnell, 1993) suggests that the discharge record may be described by the outflow from up to four linear reservoirs. The storage 'constants' for each of these reservoirs (which are indicative of the residence time of water) appear to decrease through the ablation season, reaching a minimum value towards the end of August (Figure 22.11). The average K values for the four reservoirs were estimated as 3, 8, 24 and 203 h. Thus, the two most rapidly responding reservoirs have estimated residence times that appear to correspond with the observed dye tracer travel times for the conduit and distributed components of the subglacial drainage system (Nienow et al., 1995b). The remainder of this section presents an analysis of the discharge and suspended sediment record for 1989, and interprets the results of that analysis by assuming that the two fastest reservoirs identified from analysis of discharge records are the conduit and distributed drainage systems that have been recognised through dye tracer observations.

In 1989, the suspended sediment productivity of the glacier varied through five distinct phases (periods 1 to 5) in the melt season (Table 22.5; Gurnell et al., 1992a, 1995). From determinations of the residence times in the three fastest reservoirs from analysis of discharge recessions (henceforth numbered 1, 2 and 3), it is possible to estimate the amount of water in storage in the most rapidly-responding reservoir (1)

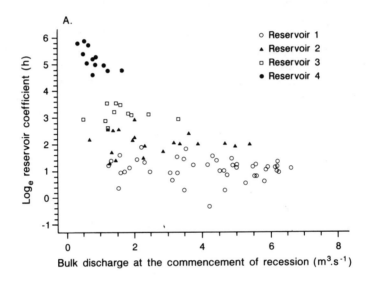

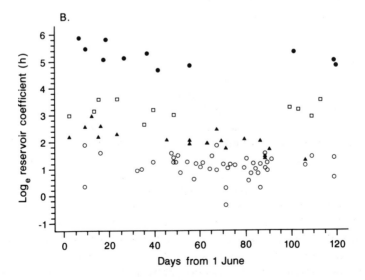

Figure 22.11 Estimates of the storage constant K for four linear reservoirs; A. in relation to discharge; B. in relation to time during the ablation season; Haut Glacier d'Aroka

Table 22.5 Simple regression relationships[a] estimated between $\log_{10}$ (suspended sediment concentration) and $\log_{10}$ (discharge) at the best temporal match position, Haut Glacier d'Arolla, 1989

Dependent variable (lag (h))	Independent variable	Period	a	t_a	b	t_b	n	R^2
$\log_{10} S$ (0)	$\log_{10} Q$	1	1.75	6.6	0.31	3.2	137	0.072
$\log_{10} S$ (0)	$\log_{10} Q$	2	−0.15	0.6	0.86	11.7	116	0.545
$\log_{10} S$ (0)	$\log_{10} Q$	3	−3.95	28.4	1.98	47.2	506	0.815
$\log_{10} S$ (−2)	$\log_{10} Q$	4	−0.87	5.2	1.09	22.9	658	0.445
$\log_{10} S$ (0)	$\log_{10} Q$	5	−1.07	7.3	1.21	29.2	490	0.443

[a] Symbols a, b, t_a, t_b, n, R^2 defined as for Table 22.3

and the total storage in the three reservoirs (Figure 22.12). If it is assumed that reservoirs 1 and 2 represent the conduit and distributed systems, then Figure 22.12 provides information on the development and capacity of these drainage network components. Periods of increasing meltwater storage early in the ablation season (periods when more meltwater is entering the glacier than can readily leave it) are associated with periods during which a distributed drainage system is developing under the glacier, but the conduit system is not developed to any significant extent and so is not affecting drainage (period 3, Figure 22.9), and with periods when the suspended sediment concentration is relatively responsive to variations in the meltwater discharge (period 3, Table 22.5). Periods of relatively stable total storage in reservoirs 1 to 3 (periods 4 and 5, Figure 22.12) appear to relate to phases when the area of the glacier drained by a distributed system migrates up-glacier as the conduit system extends headwards. The increasing extent of the conduit system is reflected in the higher water storage estimates in reservoir 1 during period 5 than during period 4 (Figure 22.12). The phase of low storage in Figure 22.12, at the junction between periods 4 and 5, reflects a period of summer snowfall and flow recession from the glacier. In summary, the evidence presented in Table 22.5 and Figures 22.11 and 22.12 suggests that the nature and activity of the subglacial drainage system is probably the key to understanding the varying location of subglacial sediment source areas and the delivery of sediments from those source areas to the proglacial stream. In particular, it seems that the distributed drainage system is probably the main source area for suspended sediments.

This analysis of information from the Haut Glacier d'Arolla provides one possibility for developing an improved approach to estimating sediment yield from some alpine glacier basins. It suggests that the evolution of the subglacial drainage system is a very significant influence on sediment yield, and that analysis of meltwater discharge records can provide a great deal of information on the types and residence times of drainage system components, and on the evolution of the capacity and ability of those components to transmit meltwater through the ablation season. These types of information provide clues concerning likely water pressure conditions under the glacier and the degree to which sediment may be tapped and transferred to the snout of the glacier.

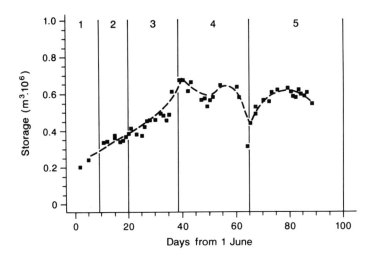
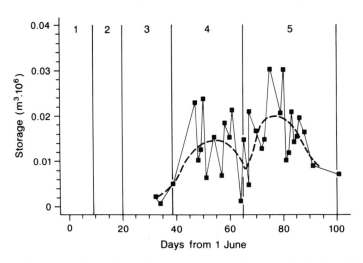

Figure 22.12 Estimates of (A) the total water in storage in reservoirs 1, 2 and 3, and (B) the water storage in reservoir 1, through the 1989 ablation season. Haut Glacier d'Arolla

ACKNOWLEDGEMENTS

Research on the Haut Glacier d'Arolla was undertaken during the tenure of NERC Research Grant GR3/7004. Research in Switzerland over 18 years has been made possible by the logistical support and provision of discharge and other data by Grande Dixence SA. My interest in hydrology was stimulated by Des Walling, when I was an undergraduate at the University of Exeter. His supervision of my PhD research and his encouragement and guidance over the ensuing 21 years have been a continual source of support in my academic career.

REFERENCES

Baker, D.H., Escher-Vetter, H., Moser, H., Oerter, H. and Reinwarth, O. (1982) A glacier discharge model based on results from field studies of energy balance, water storage and flow. In: Glen, J.W. (ed.), *Hydrological Aspects of Alpine and High-Mountain Areas*, IAHS Publication no. 138, pp. 103–112.

Beecroft, I. (1983) Sediment transport during an outburst from Glacier de Tsidjiore Nouve, Switzerland, 16–19 June, 1981. *J. Glaciol.* **29**, 185–190.

Bezinge, A., Clark, M.J., Gurnell, A.M. and Warburton, J. (1989) The management of sediment transported by glacial meltwater streams and its significance for the estimation of sediment yield. *Ann. Glaciol.* **13**, 1–5.

Bogen, J. (1980) The hysteresis effect of sediment transport systems. *Norsk Geogr. Tidsskr.* **34**, 45–54.

Bogen, J. (1988) A monitoring programme of suspended sediment transport in Norway. In: Bordas, M.P. and Walling, D.E. (eds), *Sediment Budgets*, IAHS Publication no. 174, 149–159.

Bogen, J. (1989) Glacial sediment production and development of hydro-electric power in glacierized areas. *Ann. Glaciol.* **13**, 6–11.

Bogen, J. (1992) Monitoring grain size of suspended sediments in rivers. In: Bogen, J., Walling, D.E. and Day, T. (eds), *Erosion and Sediment Transport Monitoring Programmes in River Basins*, IAHS Publication no. 210, 183–190.

Church, M. and Ryder, J.M. (1972) Paraglacial sedimentation: a consideration of fluvial processes conditioned by glaciation. *Geol. Soc. Am. Bull.* **83**, 3059–3072.

Desloges, J.R. and Church, M. (1992) Geomorphic implications of glacier outburst flooding: Noeick River valley, British Columbia. *Can. J. Earth Sci.* **29**, 551–564.

Fenn, C.R. (1987) Sediment transfer processes in alpine glacier basins. In: Gurnell, A.M. and Clark, M.J. (eds), *Glacio-Fluvial Sediment Transfer: An Alpine Perspective*, John Wiley & Sons, Chichester, pp. 59–85.

Fenn, C.R., Gurnell, A.M. and Beecroft, I. (1985) An evaluation of the use of suspended sediment rating curves for the prediction of suspended sediment concentration in a proglacial stream. *Geogr. Ann.* **67A**, 71–82.

Foster, I.D.L., Millington, R. and Grew, R.C. (1992) The impact of particle size controls on stream turbidity measurements: some implications for suspended sediment yield estimation. In: Bogen, J., Walling, D.E. and Day, T. (eds), *Erosion and Sediment Transport Monitoring Programmes in River Basins*, IAHS Publication no. 210, pp. 51–62.

Gurnell, A.M. (1982) The dynamics of suspended sediment concentration in a proglacial stream system. In: Glen, J.W. (ed.), *Hydrological Aspects of Alpine and High-Mountain Areas*. IAHS Publication no. 138, pp. 319–330.

Gurnell, A.M. (1987) Suspended sediment. In: Gurnell, A.M. and Clark, M.J. (eds), *Glacio-Fluvial Sediment Transfer: An Alpine Perspective*, John Wiley & Sons, Chichester, pp. 305–354.

Gurnell, A.M. (1993) How many reservoirs? An analysis of flow recessions from a glacier basin. *J. Glaciol.* **39**, 409–414.

Gurnell, A.M. and Fenn, C.R. (1984a) Flow separation, sediment source areas and suspended sediment transport in a pro-glacial stream. *Catena* **5** Suppl. 109–119.

Gurnell, A.M. and Fenn, C.R. (1984b) Box-Jenkins transfer function models applied to suspended sediment concentration—discharge relationships in a proglacial stream. *Arctic Alpine Res.* **16**, 93–106.

Gurnell, A.M. and Fenn, C.R. (1987) Proglacial channel processes. In: Gurnell, A.M. and Clark, M.J. (eds), *Glacio-Fluvial Sediment Transfer: An alpine Perspective*, John Wiley & Sons, Chichester, pp. 423–472.

Gurnell, A.M. and Warburton, J. (1990) The significance of suspended sediment pulses for estimating suspended sediment load and identifying suspended sediment sources in Alpine glacier basins. In: Lang, H. and Musy, A. (eds), *Hydrology in Mountainous Regions. I: Hydrological Measurements. The Water Cycle*, IAHS Publication no. 193, pp. 463–470.

Gurnell, A.M., Warburton, J. and Clark, M.J. (1988) A comparison of the sediment transport and yield characteristics of two adjacent glacier basins, Val d'Hérens, Switzerland. In: Bordas, M.P. and Walling, D.E. (eds), *Sediment Budgets*. IAHS Publication no. 174, pp. 431–441.

Gurnell, A.M., Clark, M.J. and Hill, C.T. (1992a) Analysis and interpretation of patterns within and between hydroclimatological time series in an Alpine glacier basin. *Earth Surf. Processes Landforms* **17**, 821–839.

Gurnell, A.M., Clark, M.J., Hill, C.T. and Greenhalgh, J. (1992b) Reliability and representativeness of a suspended sediment concentration monitoring programme for a remote alpine proglacial river. In: Bogen, J., Walling, D.E. and Day, T. (eds) *Erosion and Sediment Transport Monitoring Programmes in River Basins*, IAHS Publication no. 210, pp. 191–200.

Gurnell, A.M., Hodson, A., Clark, M.J., Bogen, J., Hagen, J.O. and Tranter, M. (1995) Water and sediment discharge from glacier basins: an arctic and alpine comparison. In: *Variability in Stream Erosion and Sediment Transport (Proc. Canberra Symposium, December 1994)*, IAHS Publication no. 224, 325–334.

Haeberli, W. (1983) Frequency and characteristics of glacier floods in the Swiss Alps. *Ann. Glaciol.* **4**, 85–90.

Hammer, K.M. and Smith, N.D. (1983) Sediment production and transport in a proglacial stream: Hilda Glacier, Alberta, Canada. *Boreas* **12**, 91–106.

Harbor, J. and Warburton, J. (1993) Relative rates of glacial and nonglacial erosion in alpine environments. *Arctic Alpine Res.* **25**, 1–7.

Hooke, R. LeB. (1989) Englacial and subglacial hydrology: a qualitative review. *Arctic Alpine Res.* **21**, 221–223.

Kamb, B. (1987) Glacier surge mechanism based on linked-cavity configuration of the basal water conduit system. *J. Geophys. Res.* **92**(B9), 9083–9100.

Kjeldsen, O. (1981) *Materialtranspørtundersokelser i Norske Bre-elver 1980*, Vassdrags Directoratet Hydrologisk Avdeling Rapport 4–81, 41 pp.

Kjeldsen, O. and Østrem, G. (1980) *Materialtranspørtundersokelser i Norske Bre-elver 1979*, Vassdrags Directoratet Hydrologisk Avdeling Rapport 1–80, 43 pp.

Lane, S.N., Chandler, J.H. and Richards, K.S. (1994) Developments in monitoring and modelling small-scale river bed topography. *Earth Sur. Processes Landforms* **19**, 349–368.

Lawler, D.M., Dolan, M., Tomasson, H. and Zophoniasson, S. (1992) Temporal variability of suspended sediment flux from a subarctic glacial river, southern Iceland. In: Bogen, J., Walling, D.E. and Day, T. (eds), *Erosion and Sediment Transport Monitoring (Programmes in River Basins*, IAHS Publication no. 210, pp. 233–243.

Lundquist, D. (1982) Modelling runoff from a glacierised basin. In: Glen, J.W. (ed.), *Hydrological Aspects of Alpine and High-Mountain Areas*, IAHS Publication no. 138, pp. 131–136.

Maizels, J.K. (1983) Proglacial channel systems: change and thresholds for change over long, intermediate and short timescales. *Special Publication of the International Association of Sedimentologists* no. 6, pp. 251–266.

Nienow, P., Sharp, M.J., Willis, I.C., and Richards, K.S. (1995a) Dye tracer investigations of the Haut Glacier d'Arolla, Switzerland. I. The late-summer configuration of the subglacial drainage system, in submission.

Nienow, P., Sharp, M.J. and Willis, I.C. (1995b) Dye tracer investigations of the Haut Glacier d'Arolla, Switzerland. II. Seasonal changes in the morphology of the drainage system, in submission.

Oerter, H., Baker, D., Moser, H. and Reinwarth, O. (1981) Glacial–hydrological investigations at the Vernagtferner glacier as a basis for a discharge model. *Nordic Hydrol.* **12**, 335–348.

Østrem, G., Ziegler, T., Ekman, S.R., Olsen, H., Andersson, J E. and Lundén, B. (1971) *Slamtransportstudier i Norska Glaciärälvar 1970*, Stockholms Universitet Naturgeografiska Insfitutionen, 133 pp.

Richards, K. (1984) Some observations on suspended sediment dynamics in Storbregrova. Jotunheim. *Earth Surf. Processes Landforms* **9** 101–112.

Röthlisberger, H. (1972) Water pressure in intra- and subglacial channels. *J. Glaciol*, **11**, 177–203.

Röthlisberger, H. and Lang, H. (1987) Glacial hydrology. In: Gurnell, A.M. and Clark, M.J.

(eds), *Glacio-Fluvial Sediment Transfer: An Alpine Perspective*, John Wiley & Sons, Chichester, pp. 207–284.

Seaberg, S.Z., Seaberg, J.Z., Hooke, R. LeB. and Wiberg, D.W. (1988) Character of the englacial and subglacial drainage system in the lower part of the ablation area of Storglaciaren, Sweden, as revealed by dye tracer studies. *J. Glaciol.* **34**, 217–227.

Walder, J.S. (1986) Hydraulics of subglacial cavities. *J. Glaciol.* **32**, 439–445.

Walder, J.S. and Fowler, A. (1989) Channelised subglacial drainage over a deformable bed. *Eos* **70**, 1084.

Warburton, J. (1989) Alpine proglacial fluvial sediment transfer. Unpublished PhD thesis, University of Southampton.

Warburton, J. (1990a) An alpine proglacial fluvial sediment budget. *Geogr. Ann.* **72A**, 261–272.

Warburton, J. (1990b) Comparison of bedload yield estimates for a glacial meltwater stream. In: Lang, H. and Musy, A. (eds), *Hydrology in Mountainous Regions. I: Hydrological Measurements. The Water Cycle*, IAHS Publication no. 193, pp. 315–323.

Weertman, J. (1972) General theory of water flow at the base of a glacier or ice sheet. *Rev. Geophys. Space Phys.* **10**, 287–333.

Willis, I.C., Sharp, M.J. and Richards, K.S. (1990) Configuration of the drainage system of Midtdalsbreen, Norway, as indicated by dye-tracing experiments. *J. Glaciol.* **36**, 89–101.

23 Sediment Transport and Deposition in Mountain Rivers

JIM BOGEN
Norwegian Water Resources and Energy Administration, Oslo, Norway

INTRODUCTION: THE JOSTEDØLA MOUNTAIN RIVER BASIN

In many ways, sediment transport dynamics in mountain environments differ from those of the lowlands. The high relief and the steep slopes are an obvious feature of mountain areas. The cold climate is another factor controlling runoff and erosional processes, and the distribution of vegetation is also controlled by the prevailing climate. Above the tree line, processes of erosion are more intense than those operating below. However, mountain areas are of varying character and the conditions differ greatly from one region to another and from low to high latitudes. One of the main difficulties when attempting to generalise with respect to processes of erosion and sediment transport in mountain areas is the dominating influence of bedrock geology and the thickness and nature of the overburden. The Scandinavian mountains differ in this respect from the European Alpine region: the rocks are less erodible and lakes are more frequent. Sediment sources are also unevenly distributed across drainage basins, so that only parts of the river course may be alluvial and, as a result, water discharge and sediment transport can be extremely variable.

A typical mountain river basin in Scandinavia may thus be composed of a number of local erosion and sedimentation subsystems, and it may not be possible to characterise the entire system from records collected at downstream stations. One way to deal with such problems is to recognise the various types of subsystem and to take the effect of each of them into account when river basins of various regions are compared.

The Jostedøla River basin has been selected as an example of a mountain river in western Norway (Figure 23.1). Unlike the European Alps, the relief is more subdued, with an extensive plateau at 1200–1500 m above sea level (a.s.l.). A glacial valley, Jostedal, is deeply incised into this surface.

Sediment in the Jostedøla basin is essentially supplied from two types of sources: subglacial erosion, and erosion of moraines and glaciofluvial deposits. This is material deposited by Pleistocene glaciers and by the glacial advance during the Little Ice Age. The transport system is the watercourse of the gravel-bed River Jostedøla. In some reaches it is subject to extensive braiding, and several lakes constitute local sedimentation basins. The river basin (with a catchment area to the fjord of 860 km^2; about 27% of which is glacier-covered) is situated on the northern side of the

---

Sediment and Water Quality in River Catchments. Edited by I.D.L. Foster, A.M. Gurnell and B.W. Webb.
© 1995 John Wiley & Sons Ltd.

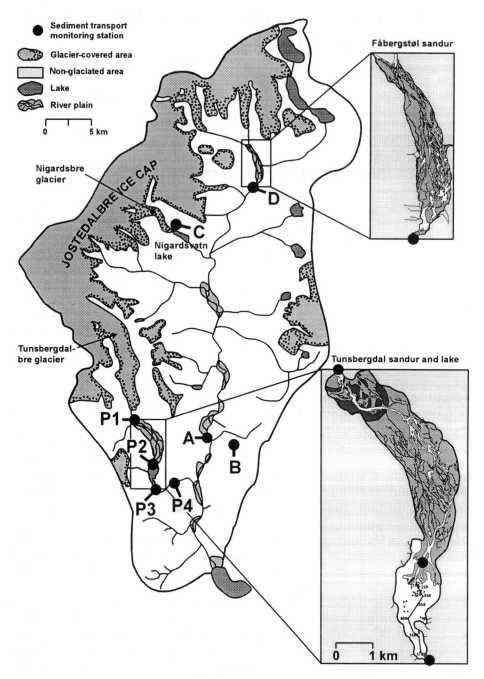

Figure 23.1 The Jostedøla River basin, showing sediment transport monitoring stations active at present: (A) Haukåsgjelet, (B) Jostedal power station (diverted water), (C) Nigardsbreen glacier, (D) Fåbergstølen sandur. Numbering along the main flow axis indicates distance from river mouth. P1–P4 are the sediment transport stations referred to in Figure 23.6

SEDIMENT IN MOUNTAIN RIVERS 439

Sognefjord. Specific runoff within the river basin varies from 25 $1 s^{-1} km^{-2}$ at the fjord to nearly 100 $1 s^{-1} km^{-2}$ at the summit of the ice-cap.

This chapter reviews data from two decades of sediment transport studies. The monitoring stations are shown in Figure 23.1. Sediment transport sampling methods from before 1980 are described by Nilsson (1972), Nordseth (1974) and Østrem (1975). After 1980 automatic samplers were brought into use (Bogen 1986a, 1988, 1992). Some data have also been collected at stations outside the Jostedal area.

GLACIAL SEDIMENT PRODUCTION

Glaciers are the most important sediment source in the Norwegian mountains, and the sediment transport in glacier meltwater rivers has been studied for many years. The sediment concentration in glacier meltwater streams is often subject to large fluctuations within short time intervals. There is some dependence on discharge but no obvious direct correlation. The relationship between water discharge and sediment transport is subject to continuous change through the season and from year to year (Figure 23.2; Østrem, 1975; Bogen, 1980; Raubakken, 1982).

This complex pattern may be explained by the seasonal development of the subglacial drainage system. When the water pressure increases during a melt period, sediments are supplied to the drainage system by expansion of subglacial cavities and

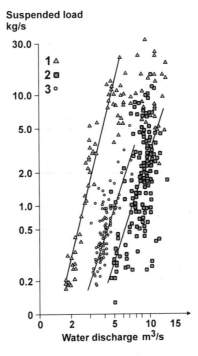

Figure 23.2 Water discharge plotted against suspended sediment transport in the meltwater river from Erdalsbreen glacier in 1969: (1) 12 Jun–23 Jun, (2) 24 Jun–3 Aug, (3) 4 May–28 Aug

tunnels. Sediments may be entrained from the glacier bed or released by the melting of ice that is loaded with debris (Figure 23.3). When glacial melting slows down in response to a drop in air temperature, subglacial tunnels are deformed by glacier movement and the weight of the overlying ice, so that after a time they may actually close. During a subsequent period of higher melt, a regeneration and possibly a new expansion of the drainage system may take place. A flood occurring later in the melt season has to exceed the magnitude of preceding floods if it is to transport a higher sediment load. The reason for this is presumably that the tunnels and cavities have to expand to allow more debris to melt out of the ice or to be washed out from the bed.

During individual flood events, or parts of the melt season, correlations between sediment transport and water discharge may exist. Correlations are sometimes stronger during falling than during rising discharges. Bogen (1980) suggests that, whereas the amount of sediment remaining in transport during the falling stage of floods is determined by hydraulic variables, erosion of spatially variable sediment sources during the rising discharge stage may lead to fluctuations in sediment concentration and, hence, exhibit a lower correlation with discharge.

A sediment transport monitoring programme for the meltwater river from Nigardsbreen has been in continuous operation since 1968. The transport of suspended load and bedload is subject to large variations from year to year (Figure 23.4). The mean annual transport of suspended sediment and bedload is 10.908×10^6 kg and 8.956×10^6 kg respectively. Annual suspended load seems to be dependent on the number of flash floods rather than on the total annual discharge, so that the highest rates

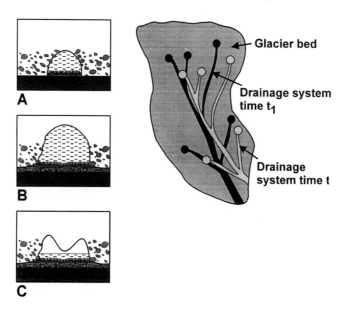

Figure 23.3 Symbolic diagram indicating the way a cavity or an ice tunnel melts out sediments incorporated in the ice. (A, B) Water pressure is rising and the subglacial tunnel or cavity is expanding. (C) Water pressure is falling and the tunnel or cavity is deformed. The continuous opening and closing of channel networks and changes in the position of a channel system may be the reason for the complicated hysteresis effects of sediment transport in glacier meltwater rivers

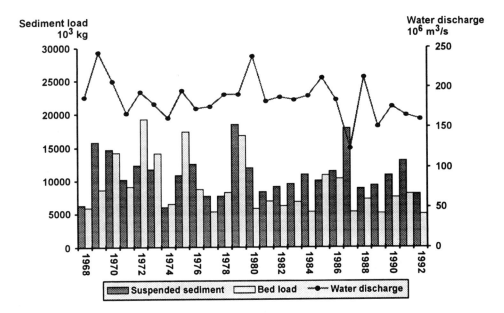

Figure 23.4 Annual totals of bedload and suspended load during the years 1968–92 in the meltwater river from the Nigardsbreen glacier

occur during years with several flash flood events. This pattern corresponds very well to the model of the subglacial drainage system described above. The number of closures is important in controlling the way that new sediments are melted out of the ice. If the discharge is less variable, the same tunnel may be kept open and so, after some time, it may be cleaned of sediments.

Estimates of bedload have been obtained by annual measurements of deltaic growth in the Lake Nigardsvatn downstream from the Nigardsbreen glacier. During years with large volumes of runoff, the magnitude of the bedload transport is high. The largest recorded bedload transport rates occurred during the years 1970, 1972, 1973 and 1975. These years differ from others in the record in that the discharge exceeded $1.8 \times 10^6 \, m^3 \, d^{-1}$ for more than 30 days of the melt season. The year 1969 was exceptional in that the bedload transport rate was relatively low despite a long-lasting high discharge. This indicates that the availability of sediments is an important factor controlling the transport rate.

AREAS WITHOUT GLACIERS

The igneous and metamorphic bedrocks in the Jostedal area are not very susceptible to weathering and erosion. Thus, the sediment supplied from areas that are not covered by present-day glaciers is, to a large extent, derived from the loose material of the overburden. This cover is discontinuous and consists of moraines and glaciofluvial deposits from the meltdown of the Pleistocene ice sheet and glacial deposits from the Little Ice Age. The streams draining these areas often have an armour layer of immobile material preventing further erosion. The streams, therefore, have to change their position to contact erodible material.

Water samples from tributaries with low glacier cover on the western side of Jostedøla rarely have suspended sediment concentrations exceeding 5 mg l^{-1} during normal runoff situations, and so the water remains clear most of the time. However, during large-magnitude floods, conditions may change. Extensive erosion took place during a flood in August 1979. Large amounts of sediment came down the steep mountain sides, and agricultural land on the river fans was strewn with boulders and, in some locations, covered by a thick layer of overbank sediments. An examination of sediment sources revealed that some of the rivers had undergone removal of their armour layer. The abrupt lowering of the river bed of Røykjedalselvi induced mass movement on slopes adjacent to the river channel. A number of other types of sediment source were also active during the flood. The heavy rain triggered a large number of slides and river channel changes, including the formation of new channels. In some areas the bedrock was stripped entirely of overburden and increased areas of bare rock surface were revealed. Inside the neoglacial moraines, the loose material deposited is less consolidated than the older sediments and is more susceptible to erosion. Several erosion scars were observed inside the neoglacial moraines after the flood.

Gullying is rare in this area, probably as a result of the patchy overburden. In other parts of Scandinavia, gullies have been seen to develop during major floods (Rapp and Strømquist, 1976; Bogen, 1986b). The relatively resistant rocks also make the conditions in Jostedøla very different from the torrents of the Alps. Torrent erosion results from rock failure, often caused by inherent instability of geological formations (Heede, 1980).

As systematic measurements are lacking in the non-glacial area of Jostedøla, details of the transport dynamics cannot be given. However, data from a mountain stream, the

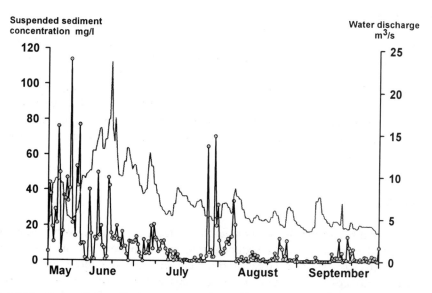

Figure 23.5 Suspended sediment concentrations (dotted curve) in the River Atna in the Rondane Mountains measured at Lia bridge. Discharge is shown as the full curve

River Atna, in the Rondane area, provides some information about how the sediment transport dynamics relate to the process of mass movement in high-mountain areas (Bogen, 1989a, b). This river is situated almost entirely above the treeline (about 900 m a.s.l.) and the catchment area is 132 km^2. A major part of the transported sediment is derived from mass movement and channel erosion downstream from the major sediment sources (Bogen, 1983a b). Suspended sediment concentrations are poorly correlated with water discharge and vary in an apparently unpredictable manner. Figure 23.5 illustrates that the highest suspended sediment concentrations occurred during snowmelt in May and decrease later on even when the water discharge increases. In late July and August, a similar pattern is repeated. At this time the high concentrations most probably occur as a result of slumping, which introduces large amounts of sediment into the channel. As this source is exhausted, the suspended sediment concentration decreases. The second rise may be due to another slide of a smaller volume. The effect of this slide was of shorter duration and suspended sediment concentrations fell to a low level after some days.

THE SANDURS

Sandurs are braided river systems formed at the front of glaciers. Krigstrøm (1962) made a distinction between sandurs, where the channel system expands freely, and valley sandurs, where the development of the channel network is confined by valley walls. Almost all sandurs developed in front of Norwegian glaciers are valley sandurs, whereas the free type is most common on the coastal plains in Iceland. Characteristic features of the braided river systems are broad and shallow channels with a relatively steep mean gradient. The carrying capacity and competence of glacial ice and the subglacial tunnels are larger than the receiving meltwater streams. Some of the boulders that are supplied by the ice may thus be immobile in a proglacial fluvial environment and are deposited in the upper part of the sandurs. In the glacier-fed rivers in the Norwegian mountains, the same zones as found by Church (1972) for sandurs in the Canadian Arctic may be recognised. There is an upper steep part composed of coarse-grained fractions where channel bifurcation takes place. Lateral channel shifting is frequent in this part and there is no vegetation on the river banks. This zone grades into the second zone, where the number of channels is at a maximum. The grain fractions in river banks and bars are smaller and some vegetation is present. In the third zone, vegetation is more abundant and a layer of overbank sediments is deposited on the river plain.

Whereas the transport and deposition of bedload in braided river reaches have been subject to extensive studies, the fate of suspended sediments has not been given similar attention. Obviously, the presence of overbank sediments in the third zone indicates deposition and a downstream decrease in the sediment load of the river when water discharges exceed bankful.

A large sandur system is composed of a network of channels of variable size. During low discharge the current velocity may be low in some channels. Thus, the system may act as a sediment trap at low discharges and as a sediment source when discharge rises. When the water discharge increases, the suspended sediment concentration is dependent not only on the amount of sediment supplied from

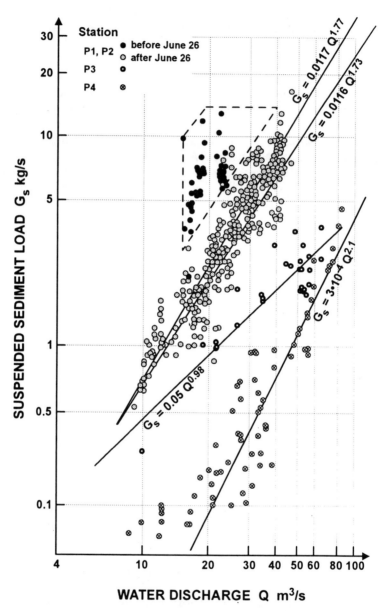

Figure 23.6 Sediment rating curves showing the changes in suspended sediment load when the river passes through a sandur, a lake and a reach downstream from the lake where sediments are eroded. Station P1 and P2: the upper and lower curves are upstream and downstream of the sandur. Station P3 is at the lake outlet and P4 is 2 km downstream from the lake. See Figure 23.1 for location

upstream, but also on the amount of sediment deposited in, or eroded from, the braided river system. The actual concentration reached during the passage of a flood may be expressed as the sum of the input concentration (C_i) and the amount of sediment (M) added to the water volume $(Q + \Delta Q)\Delta t$ passing through a cross-section during the time interval t:

$$C_u = C_i + \Delta M/[(Q + \Delta Q)\Delta t] \qquad (1)$$

where C_u is the output concentration of suspended sediment, Q is the water discharge in the river and ΔQ the discharge increment. It can be seen from this equation that the output concentration (C_u), during a sequence of water discharge increments, is dependent upon the length of time necessary to accomplish the increase. During a rapid increase the sediments are added to a smaller volume of water than during a slow increase. Thus, the highest concentrations are reached during a rapid increase in water discharge.

Apart from the discharge increase, the suspended sediment concentration is also dependent on the amount of sediment available for erosion (M). This amount is dependent on the deposition in the channel during the preceding time interval.

Calculations carried out by Bogen (1980) showed that the added loads are at their highest when the discharge during the preceding period is of intermediate size. During high flow, the current velocity is high, which does not permit sedimentation. During low flow the supply of sediment is too small. It is therefore evident that a sandur system may influence the sediment transport dynamics in a river over a short time and on a seasonal basis. On a year-to-year basis, data from the Tunsbergdal sandur shows that little permanent deposition takes place. The sediment rating curves from an upstream and a downstream station are almost coincident, except for some deposition at high discharges (Figure 23.6).

During major changes in the channel network, large amounts of overbank sediments may be exposed to erosion. In 1987 the channel system of the Fåbergstøen sandur moved from a position on the eastern side to the western side of the valley. This incident was associated with an increase in suspended load from between 30 000 and

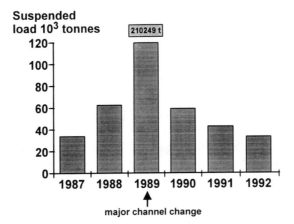

Figure 23.7 Annual suspended sediment loads at the monitoring station downstream from Fåbergstølen sandur. A major change in the channel network caused extremely high sediment transport in 1989

60 000 t yr^{-1} during 1987 and 1988 to 210 000 t in 1989. During the following years the sediment load fell to the same level as that before the channel changes took place (Figure 23.7). The shift in the channel network was caused by a permanent change in the position of the main channel at the upstream end of the sandur. At the same time, tapping of water for the Styggevatn hydro-power reservoir influenced the lower part of the sandur (Bogen, 1993).

The role of water abstraction in the channel change is not clear. However, the channel shift did initiate severe bank erosion and removal of overbank sediments. When the new channel was established and swept free of suspended sediment, this source apparently became exhausted.

THE LAKES

Lakes form sediment traps where a part of the sediment load is deposited. The sediment rating curves in Figure 23.6 illustrate the impact of the passage of water through a lake. When the water masses flow through a lake, suspended sediment concentrations are reduced by the settlement of particles. As grain size normally increases with discharge, the relative rate of deposition is greater at high than at low discharge. Thus, the slope of the sediment rating curve at the lake outlet is less than that at the inlet. As the grain size of the sediment load may be subject to seasonal changes, the rating curve is not well correlated with discharge. Much fine material in the suspended load at the inlet may create relatively high concentrations at the outlet even at moderate discharges. Another sampling station was established about 2 km downstream from the lake. Owing to erosion on the intervening reach, the sediment rating curve increases its slope between the lake outlet and this downstream site.

Varves that reflect annual sedimentation have often been reported in proglacial lakes. In Lake Tunsbergdalsvatn very distinct varves have been found (Bogen, 1976). The varves reflect annual changes in grain size and mineralogy. The layers deposited during late autumn and winter conditions contained more fine grain sizes and more biotite than the summer layers. Inside some of the varves some thinner layers were present, so-called pseudovarves. These layers were found to be related to discharge variations during the summer season. As the biotite layers were much thinner during summer deposition, it was relatively easy to identify the winter layer. Similar conditions have been described by Gilbert (1973), Østrem (1975) and Østrem and Olsen (1987).

Bogen (1976, 1979) presented a method for estimating the sediment load from varve thickness to obtain suspended sediment load estimates in years when no sampling has been carried out. Obviously, the sediment rating curve may be used to obtain a rough estimate of sediment load during the years when water discharge was recorded. However, this curve is only valid for parts of the season. During the first part of the melt season in particular there is no simple correlation with discharge. The varve thickness of the individual years may be used to correct for this deficiency. A sediment core taken 1400 m from the river mouth was used for this purpose. In this core the varve thicknesses in 1965 and 1972 (when direct measurements of sediment transport were made) were identified. Sediment transport measurements from 1965 were obtained from Hatling (1967). The results of the calculations and interpretation of the sediment core are given in Figure 23.8. The corrections to the sediment rating curve

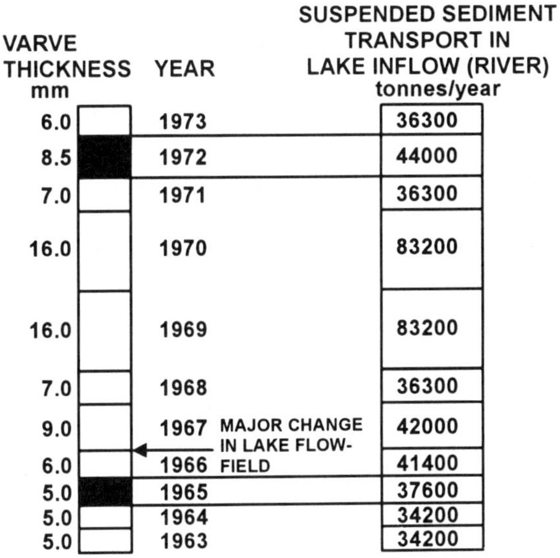

VARVE THICKNESS mm	YEAR		SUSPENDED SEDIMENT TRANSPORT IN LAKE INFLOW (RIVER) tonnes/year
6.0	1973		36300
8.5	1972		44000
7.0	1971		36300
16.0	1970		83200
16.0	1969		83200
7.0	1968		36300
9.0	1967	MAJOR CHANGE IN LAKE FLOW- FIELD	42000
6.0	1966		41400
5.0	1965		37600
5.0	1964		34200
5.0	1963		34200

Figure 23.8 Suspended sediment transport estimated from varves in the bed sediments of Lake Tunsbergdalsvatn

estimates based on bed sediments led to an increase in estimated load in some years and a decrease in others. For the two years 1970 and 1971, large corrections were found to be necessary. In 1970 a large flood event occurred early in the melt season, and it is very likely that large amounts of sediment were flushed out during this flood. In 1971 the sedimentation in the lake was less than that computed from the sediment rating curve. The reason for this is most probably that the large flood event in that year, which occurred in November, did not transport much sediment because at this time of the year the subglacial channel system may be closed so that sediment from this source is not available.

From Hatling's (1967) descriptions it appeared that a change in the flow field of the lake took place in 1967. Prior to that year the main flow followed the western side of the lake and its velocity was retarded more than in the subsequent years. Thus, less sediment was carried in suspension to be deposited at the location of the sediment core, and the varves were correspondingly thinner. The application of two years of calibration were used to compensate for this effect of unequal deposition.

Estimation of sediment load in the years before the discharge record commenced may be possible, but may involve the collection of a large number of sediment cores to compensate for the influence of channel changes on sedimentation rates.

JOSTEDØLA: A COMPOSITE RIVER BASIN

The runoff in the River Jostedøla is subject to considerable variations throughout the melt season. About 90% of the runoff is concentrated in the months May to September. The largest discharge events are most often rain-induced floods or a combination of glacier melting and rain (Harsten, 1979; Gjessing and Wold, 1980). The sediment monitoring station at Haukåsgjelet records the transport in the main stem of River Jostedøla (location A, Figure 23.1). A number of different sources contribute to the sediment transport at this station. The main contributors are the Fåbergstølbreen and the glaciers upstream of the Fåbergstølen sandur. On average less than 20% of the sediments supplied from Nigardsbreen pass Lake Nigardsvatn and are carried into the main river. On the basis of seasonal variations in air temperature (Figure 23.9) three periods may be recognised: the snowmelt period in May, the glacier melt from June to early September, and the period of autumn rain.

The suspended sediment concentrations (Figure 23.9) vary in a complicated manner throughout the season. During snowmelt conditions in May, concentrations are low (less than 40 mg l^{-1}) in comparison to the later period of glacier melting, when the daily mean is most often in the range of 50–100 mg l^{-1}. The highest concentrations exceed 500 mg l^{-1} and occur during episodes associated with floods generated by a combination of both rain and glacier melting. During rain floods in the late season,

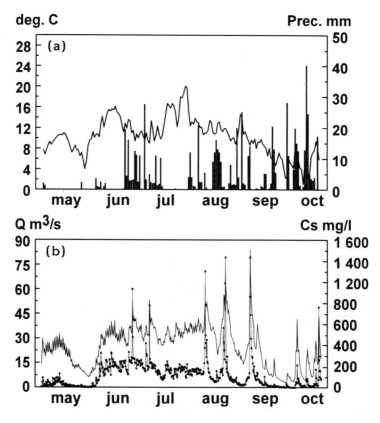

Figure 23.9 (a) Mean daily air temperature (full curve) and daily precipitation (bars) at the meteorological station Bjørkehaug in Jostedalen, 30 April to 18 October 1990. (b) Water discharge (upper full curve) and sediment concentration (full squares) at Haukåsgjelet in the River Jostedøla in the same period

glacier melting is reduced and the subglacial drainage system may be closed, so that the availability of sediment is low.

Whereas the suspended sediment moves through the river basin in less than a day, the bedload may take several decades to travel the whole length of the river basin. The glacial valley shape is of great importance to the movement of the bedload. A distinct feature within glacial valleys is the steps and overdeepened basins of the longitudinal profile (Figure 23.10). Some basins form lakes where the sediment load is deposited. Others have been infilled with sediments and form alluvial reaches with floodplains. In between the alluvial plains are steep reaches where channels are shaped in bedrock or

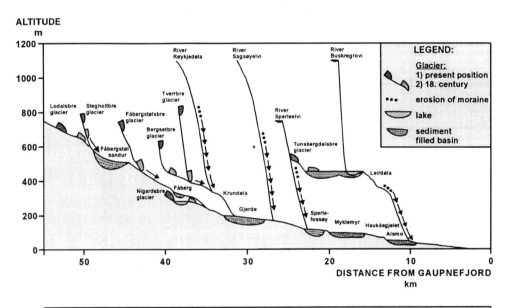

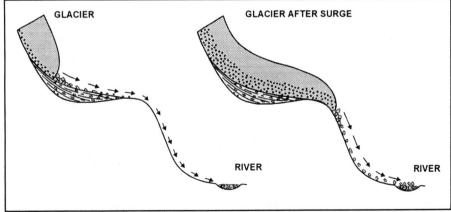

Figure 23.10 Longitudinal profile of the Jostedøla River and its tributaries. The location of major sediment sources is indicated. It is suggested that the fluctuation of the glaciers influenced the sediment delivery ratio

immobile moraine material. Some of the tributaries are rather steep, while others occur in wide glacial hanging valleys.

The position where the glaciers introduce their sediments into the main course of the river is of importance to the further movement of the sediment and the stability of river channels. At present, a large part of the coarse fraction is deposited in front of the glaciers and often far away from the main river stem. In the more advanced glacier stages, sediments were more easily transferred into the main stem of the proglacial river system. One example is the glacier Nigardsbreen, which reached its greatest advance around 1750. The subsequent retreat of the glacier uncovered lake Nigardsvatn during the years 1937–67. On average 1500 t or 75% of the total load of the glacier Nigardsbreen is at present deposited in the lake.

The glacier volumes in the Jostedal area have been subject to large variations in postglacial time (e.g. Bogen et al., 1988). Long-term changes in sediment transport dynamics of the river have certainly been the result.

REFERENCES

Bogen, J. (1976) Sedimentasjonsprosessens dynamikk i et delta—system (The dynamics of deltaic depositional processes) Unpublished thesis, University of Oslo, 103 pp. (in Norwegian).

Bogen, J. (1979) Sedimentasjon av partikulært materiale i innsjøer (The sedimentation of particulate matter in lakes). In: Hansen, E. (ed.), *Innsjøers Dynamikk (The Dynamics of Lakes)*, Norsk Hydrologisk Komite Internal Report no. 79/02, pp. 459–478 (in Norwegian).

Bogen, J. (1980) The hysteresis effect of sediment transport systems. *Norsk Geogr. Tidsskrq.* **34**, 45–54.

Bogen, J. (1983a) Morphology and sedimentology of deltas in fjord and fjord valley lakes. *Sediment. Geol.* **36**, 245–262.

Bogen, J. (1983b) *Atnas Delta i Atnsjøen. En Fluvialgeomorfologisk Undersøkelse (The Atna Delta)*. Kontaktutvalget for Vassdragsreguleringer, University of Oslo Report no. 83-70, 44 pp. (in Norwegian).

Bogen, J. (1986a) Transport of suspended sediments in streams. In: Hasholt, B. (ed.), *Partikulært Bundet Stofftransport i Vann og Jorderosjon*, NHP Report no. 14, KOHYNO, pp. 9–21.

Bogen, J. (1986b) *Erosjonsprosesser og Sedimenttransport i Norske Vassdrag. Utredning av Forvaltningsansvar, Faglig Status og Forskningsbehov (Erosion and Sediment Transport in Norwegian Rivers. River Management, Responsibilities, Status and Research Requirements)*, NHK Report no. 20, 109 pp. (in Norwegian with extended abstract in English).

Bogen, J. (1988) A monitoring programme of sediment transport in Norwegian rivers. In: Bordas, M.P. and Walling, D.E. (eds), *Sediment Budgets*, IAHS Publication no. 174, pp. 149–159.

Bogen, J. (1989a) Glacial sediment production and development of hydroelectric power in glacierized areas. *Ann. Glaciol.* **13**, 6–11.

Bogen, J. (1989b) *Transport av Suspendert Materiale og Substratforhold i Atnavassdraget. Forsknings og Referansevassdrag—Atna (Sediment Transport and Substrates in the Research and Reference River Basin Atna)*, Miljøvirkninger av Vassdragsutbygging, MVU Report B52, Oslo, 27 pp. (in Norwegian with abstract in English).

Bogen, J. (1992) Monitoring grain size of suspended sediments in rivers. In: Bogen, J., Walling, D.E. and Day, T. (eds), *Erosion and Sediment Transport Programmes in River Basins*, IAHS Publication no. 210, pp. 183–190.

Bogen, J. (1993) Fluviale prosesser og inngrep i vassdrag. Konsekvenser og tiltak—en kunnskapsoppsummering (Fluvial processes and man made effects—the state of knowledge).

In: Faugli, P.E., Erlandsen, A.H. and Eikenes, O. (eds), Inngrep i Vassdrag: Konsekvenser og tiltak – En Kunnskapsoppsummering. (Human interference with rivers – assessments of impacts and remedial measures. The state of knowledge) NVE Publication no. 13/93, (pp. 96–124. (in Norwegian).

Bogen, J., Wold, B. and Østrem, G. (1988) Historic glacier variations in Scandinavia. In: Oerleman, J. (ed.), *Glacier Fluctuation and Climatic Change*, Kluwer Academic, Dordrecht, pp. 109–128.

Church, M. (1972) *Baffin Island Sandurs. A Study of Arctic Fluvial Processes*, Geological Survey of Canada Bulletin no. 216, 208 pp.

Gilbert, R. (1973) Processes of underflow and sediment transport in a British Columbia mountain lake. *Natl. Res. Council of Canada, Assoc. Comm. Geodesy and Geophysics, Subcomm. Hydrology,, Proc. Hydr. Symp.* pp. 493–507.

Gjessing, Y. and Wold, B. (1980) Flommen i Jostedalen 14–15 August 1979 (The flood in river Jostedøla 14–15 August 1979). *Været* **1**, 29–34 (in Norwegian).

Harsten, S. (1979) *Fluvialgeomorfologiske Prosesser i Jostedalsvassdraget (Fluvial Processes in River Jostedøla)*. Kontaktutvalget for Vassdragsreguleringer, University of Oslo Report no. 09/79, 114 pp. (in Norwegian).

Hatling, J. (1967) Slamtransport og geomorfologi i Leirdølas nedbørfelt (Suspended sediment transport and geomorphology in the Leirdøla river basin). Unpublished thesis, University of Oslo, 106 pp. (in Norwegian).

Heede, B.H. (1980) Gully erosion—a soil failure: possibilities and limits of control. *Interpraevent 1980 (Proceedings of the International Symposium in Bad Ischl)* vol. 1, FGVH, Klagenfurt, pp. 317–330.

Krigstrøm, A. (1962) Geomorphological studies of sandur plains and their braided rivers in Iceland. *Geogr. Ann.* **44**, pp. 328–346.

Nilsson, B. (1972) *Sedimenttransport i Svenska Vattendrag. Et IHD Projekt. Del 1*, Metodik. Naturgeogr. Inst. University of Uppsala Report no. 2, 18 pp.

Nordseth, K. (1974) *Sedimenttransport i Norske Vassdrag (Sediment Transport in Norwegian Rivers)*, Department of Geography, University of Oslo, 175 pp. (in Norwegian).

Østrem, G. (1975) Sediment transport in glacier meltwater streams. In: Jopling, A.V. and McDonald, B.C. (eds), *Glaciofluvial and Glaciolacustrine Sedimentation*, Special Publ. Soc. Econ. Pal. Min. Tulsa, no. 23, pp. 101–122.

Østrem, G. and Olsen, H.C. (1987) Sedimentation in a glacier lake. *Geogr. Ann.* **68A**(1), 125–137.

Rapp, A. and Strømquist, L. (1976) Slope erosion due to extreme rainfall in the Scandinavian mountains. *Geogr. Ann*, **58A**(3), 193–200.

Raubakken, V. (1982) Suspensjonstransport og sedimentasjon i Erdalsvassdraget, Stryn. Suspended sediment transport and sedimentation in the Erdal river (Stryn). Unpublished thesis, University of Oslo, 106 pp. (in Norwegian).

24 Sediment Dynamics in the Polish Flysch Carpathians

WOJCIECH FROEHLICH
Institute of Geography, Polish Academy of Sciences, Cracow, Poland

INTRODUCTION

The Polish Flysch Carpathians are characterised by low mountain relief and extensive foothills elevated to between 300 and 1500 m above sea level (a.s.l.). They are built of sandstone–shale flysch, which is usually well weathered and of low erosional resistance. Slopes at high elevations range in angle from 15° to 35°, and include both convex and linear segments. In the foothills, more gentle slopes of between 5° and 15° occur. The high altitude areas are covered by regolith, whereas, on the lower slopes, thick solifluction, slopewash and aeolian sediments have been deposited. The headwater zones and steep slopes are characterised by intensively exploited forests, which are accessed by a dense network of unmetalled roads. Much of the original forest cover has been removed by deforestation.

On the lower slopes, much of the land is given over to arable cultivation. These cultivated slopes are characterised by a mosaic of field plots of various sizes, separated by terraces, and intersected by networks of unmetalled roads, which frequently extend to the stream channels. Slope foot and valley floor areas are occupied by meadows and pasture. The valley floors of third-order and higher streams are flat and are covered by alluvium. The river channels on these flat alluvial plains usually have no direct contact with the valley side slopes.

Unmetalled roads are a characteristic feature of the Carpathian landscape and date back to the original clearance and cultivation of the land. The road network is directly related to the field boundary pattern. In narrow valley bottoms unmetalled roads often run along stream channels which, in forested areas, often serve for the transport of felled timber. During storm events, the dense network of unmetalled roads and gullies promote rapid surface runoff, which transports large amounts of sediment to the channels (Froehlich and Slupik, 1984; Froehlich, 1982, 1986, 1991).

In situations when event rainfall totals exceed 20 mm, and rainfall intensities reach 1 mm min^{-1}, overland flow starts on the loamy soils. Soil flows and earthflows are generated by these events, which also cause intensive erosion along unmetalled roads. Local rill erosion can occur during heavy rainfall, and earth flows have been observed on some potato fields (Gil and Slupik, 1972; Gerlach, 1976). The rates of soil erosion under potato crops are as high as 22 t ha^{-1} yr^{-1}, whilst typical values for winter crops, meadows and forest are 2.4, 0.1 and 0.03 t ha^{-1} yr^{-1} respectively (Gil, 1986). Annual

---

Sediment and Water Quality in River Catchments. Edited by I.D.L. Foster, A.M. Gurnell and B.W. Webb.
© 1995 John Wiley & Sons Ltd.

sediment yields for the larger river basins lie in the range 90–1000 t km^2 (cf. Branski, 1968; Lajczak, 1989).

This chapter summarises results from the Homerka experimental drainage basin, where different monitoring techniques have been used to study sediment dynamics over the past 20 years. Investigations have been carried out in subcatchments of different geographical scale and on experimental slopes. Research into the slope and channel subsystems has focused on the spatial variability in both sediment sources and sediment delivery processes (Froehlich, 1982; Froehlich and Walling, 1992; Froehlich *et al.*, 1993; Higgitt *et al.*, 1992).

THE STUDY AREA

The drainage basin of the Homerka stream (19.6 km^2) is representative of the largely deforested Carpathian basins and lies at an altitude of between 375 and 1060 m a.s.l. It is underlain by flysch rocks of varying resistance to weathering. The catchment can be subdivided into two parts; the first represents the montane headwater and the second includes the lower foothill zones. The montane headwater is underlain by resistant sandstones and shales. That part of the basin with steep slopes (15–30°) and more permeable and shallow skeletal soils is forested. The forests, which occupy 52% of the drainage basin, are presently under intensive exploitation and are accessed by a dense network of unmetalled roads and lumber tracks.

The foothills rise up to 650 m a.s.l. and are underlain by the shale–sandstone flysch series. The silty–clayey soils that have developed on the flysch series have a permeability that decreases with increasing depth in the soil profile and are generally used for farming. The mosaic of arable fields is crossed by a network of unmetalled roads, which usually lie below the level of adjacent fields (sunken roads). The lower valley floors are occupied by meadows and permanent pasture.

In the lower part of the basin, the mean annual precipitation amounts to about 900 mm, whereas in the headwater region it exceeds 1000 mm. The equivalent values for mean annual air temperature are 7.5°C and 5.0°C respectively.

THE EXPERIMENTAL SLOPE

In order to conduct a detailed investigation of erosion and sediment delivery from a cultivated part of the basin, an area of 26.5 ha, located on the boundary of the forest and the agricultural region, was designated as an 'experimental slope' (Figure 24.1). It is 500–700 m long and convexo-concave in form. The slope ranges in altitude from 458 to 608 m a.s.l. and the silty–clayey soils increase in depth towards the foot of the slope.

The slope is subdivided into numerous field plots, which are cultivated across the slope. The field plots are of various sizes and are separated either by terraces and furrows or by unmetalled roads, which traverse the area from the watershed to the stream channel. In many places the roads are deeply incised into the slope and bedrock is exposed along them. The length of unmetalled roads traversing the experimental slope is 3.3 km, giving a road density of 11.9 km km^2. This is somewhat higher than the road density for the whole basin, which is 5.3 km km², and has important implications for sediment production, since drainage density has often been shown to be a critical factor in determining sediment yield (Campbell, 1983).

The experimental slope is composed of several sub-basins representing the main Carpathian sediment sources (Figure 24.1). These are: first, the drainage basin of a Holocene gully; secondly, drainage basins of unmetalled roads; and, thirdly, drainage basins of the interchannel areas. The two former areas supply sediment from the slope

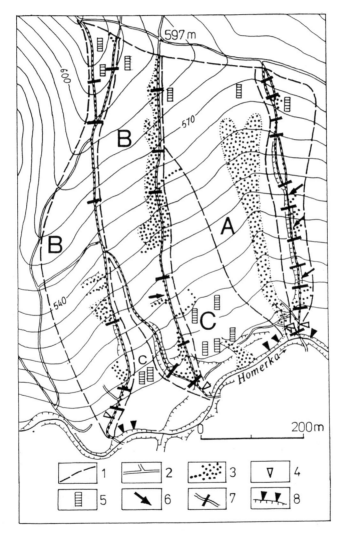

Figure 24.1 The experimental slope in the Homerka drainage basin: (A) drainage basin of the Holocene gully; (B) drainage basin of the unmetalled roads; (C) drainage basin of the interchannel areas; (1) drainage divides; (2) unmetalled roads; (3) contributing areas; (4) locations for the measurement of concentrated flow and water sampling; (5) location of containers for measuring sheet flow and water sample collection; (6) outflow of furrows; (7) sites for measuring unmetalled road surface changes; (8) sites for measuring channel erosion rates (based on Froehlich, 1991)

to the stream channel in the form of concentrated flow, while the third area delivers sediment as sheet flow and subsurface flow. Each of these drainage basins was instrumented in order to determine their relative contribution to the total sediment budget.

SEDIMENT PRODUCTION FROM UNMETALLED ROAD SURFACES

As suggested above, the unmetalled roads provide a substantial increase in drainage density and, since the permeability of the road surface is low, their response to rainfall is rapid. Even after insignificant precipitation of only a few millimetres, an increase in water turbidity is observed. Each high water stage is marked by a distinct hysteresis loop in the relationship between water discharge and suspended sediment concentration (Froehlich, 1982). During rising stages the suspended sediment concentrations are always higher on the unmetalled roads than in the Homerka stream channel, except in the case of local flash floods and snowmelt floods generated in the headwater region of the drainage basin (Figure 24.2).

The maximum recorded suspended sediment concentrations from the unmetalled roads were about 1.5×10^5 mg l^{-1}. Suspended sediment concentrations in runoff from the unmetalled roads are strongly related to the intensity of soil splash caused by raindrop impact. Research into soil splash on experimental plots on a ploughed field and an unmetalled road indicate that the intensity of the splash process can be 30 times greater on an unmetalled road than on a ploughed field. The rapid formation of a surface detention layer on an unmetalled road results in an increased intensity of soil

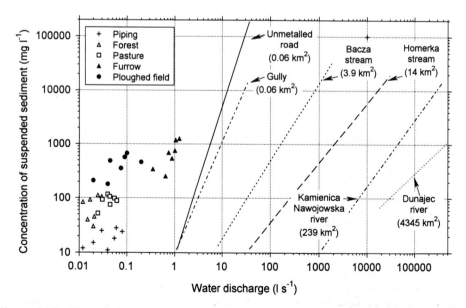

Figure 24.2 The relationship between water discharge and the concentration of suspended sediment in drainage basins of different size and with different sediment sources and land use

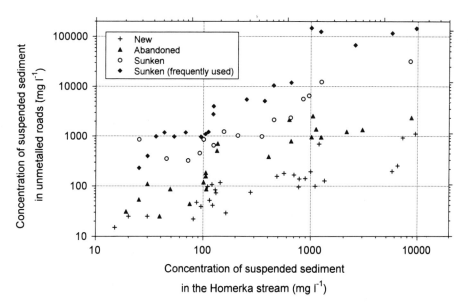

Figure 24.3 The relationship between the concentration of suspended sediment in the Homerka drainage basin and the concentration of suspended sediment in different types of unmetalled roads

splash. The results indicate that a considerably greater supply of sediment comes from unmetalled roads when compared to ploughed fields (Froehlich and Slupik, 1980).

Simultaneous measurement of the concentration of suspended sediment during different types of high water stage indicate that concentrations are always much lower in water draining the Holocene gully than in water flowing down the roads. The concentrations observed for particular roads are often variable, and the magnitude of suspended sediment concentrations is related to particular features of the roads, namely to their depth of incision, the grain-size composition of the regolith exposed in the road ravine, the moisture content of the ravine regolith and the frequency of traffic (Figure 24.3).

SEDIMENT DELIVERY TO THE HOMERKA CHANNEL

Sediment source areas vary from one rainfall event to another, and their precise location varies with the intensity and duration of precipitation, soil moisture and soil permeability. These contributing areas form narrow bands along the main stream channel, tributary channels and linear incisions and depressions of concentrated flow during periods of snowmelt and precipitation (Figure 24.2). The width of the zone varies along the length of the channel (Froehlich, 1982; Walling, 1971, 1983).

Most of the sediment is supplied from unmetalled roads to the Homerka channels, and comes from the erosion of road ravines, while only a small proportion is derived from the erosion of field furrows. Alonso *et al.* (1988), for example, have shown that soil loss from furrows is primarily dependent on the furrow gradient. Only during times of overland flow and wash was sediment supplied from field furrows, and this was always significantly less than that supplied from unmetalled roads (Figure 24.2). Moreover, the frequency of sheet wash on fields is also significantly lower than on unmetalled roads.

Similar parameters of suspended sediment transport were found to be important for the unmetalled roads in the forested part of the Homerka drainage basin. It is generally recognised that wash processes in forested areas of the lower subalpine region are quantitatively unimportant (Gerlach, 1976), which would suggest, therefore, that the largest proportion of suspended sediment comes immediately from unmetalled roads (cf. Reid and Dunne, 1984).

The concave section of the experimental slope is drained by a Holocene gully with a permanently wet zone at its outlet (Figure 24.1). Although this feature responds quickly to rainfall, suspended sediment concentrations are much lower than those associated with the unmetalled roads because of the permanent grass cover.

The concentration of suspended sediment in the overland flow derived from the pasture areas ranges from 25 to 117 $mg l^{-1}$, and was always many times lower than that observed in channels. The interchannel areas on the footslopes and on the valley floors support a high density of grass cover and are natural 'traps' for sediment derived from the fields by overland flow.

Concentrations of suspended sediment increase in direct proportion to discharge for individual source components, but Figure 24.2 also shows that, for the same discharge, sediment concentrations decrease with increasing drainage basin area. In this model the concentrations of suspended sediment in the smaller basins generally increase more rapidly in response to increasing discharge than in the drainage basins of a larger scale. This model demonstrates the complexity of the sediment delivery process in the Homerka catchment and is probably representative of the Carpathian flysch region as a whole.

Direct supply of sediment to the channel varies according to both rainfall duration and intensity. About 98% of the suspended sediment is supplied from unmetalled roads within the interchannel area during precipitation events that do not trigger overland flow. Sediment originating from the channel constitutes *ca*. 2% of the total sediment load during these low-intensity events. Sediment is deposited during the longer inter-flood periods (Froehlich, 1982, 1986).

Sediment supplied from unmetalled roads and ploughed fields during annual floods constitutes *ca*. 60–70% of the total load. The Holocene gullies supply *ca*. 10–15%, while bank and bed erosion of channels contributes 15–30%. Direct supply by overland flow from interchannel areas does not exceed 1%.

Erosion studies employing ^{137}Cs suggest that a considerable amount of soil is eroded within the field plots of the Homerka drainage basin and is deposited on terrace edges at the base of the plots. Samples collected from immediately below the terraces have similar total ^{137}Cs inventories to mid-plot samples, though they contain slightly higher amounts of ^{137}Cs. It is likely that during major runoff events some of the sediment eroded cascades over the terrace into the adjacent downslope plot, but the radiocaesium inventories suggest that this is not a major process. The magnitude of radiocaesium transfer from plot to terrace and its vertical distribution in terrace profiles suggests an accumulation rate of about 4 $mm yr^{-1}$. The agricultural terraces effectively restrict transport of sediment downslope, and the cultivated areas are not a significant

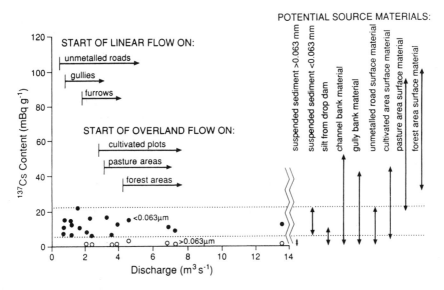

Figure 24.4 The relationship between the ^{137}Cs content of suspended sediment and discharge, and the discharge thresholds associated with the occurrence of storm runoff from various sources within the basin. The ranges of ^{137}Cs concentrations associated with suspended sediment and potential source materials are also illustrated (based on Froehlich and Walling 1992)

contributor to the overall suspended sediment output from the basin (Froehlich and Walling, 1992; Froehlich et al., 1993; Higgitt et al., 1992).

Measurements of the ^{137}Cs content of the <0.063 mm fraction of suspended sediment collected from the Homerka stream at the main gauging station in the pre-Chernobyl period indicated a range in activity of between 6.3 and 22.6 mBq g^{-1}, with a mean of 11.9 mBq g^{-1}. Comparison of these values with typical values for potential source materials (Figure 24.4) suggests that they closely match those associated with material collected from the surface of unmetalled roads. Sediment eroded from the surface of forest and pasture areas is very unlikely to represent an important sediment source since its ^{137}Cs content is substantially higher. Material eroded from cultivated areas and from channel and gully banks could represent a source of suspended sediment, but is thought unlikely to constitute a major source because the range of ^{137}Cs levels associated with these materials extends well above that which is representative of suspended sediment (Froehlich and Walling, 1992; Froehlich et al., 1993).

The evidence provided by the radiocaesium fingerprints suggests that the major source of the suspended sediment transported by the Homerka stream is the unmetalled roads that occur throughout both the forested and the agricultural zones of the basin.

CONCLUSION

At present, linear supplies of suspended sediment to the main channel via unmetalled roads are far more important than sediment supplied via dispersed overland flow from interchannel areas, which contribute less than 1% to the total sediment budget in these small Carpathian drainage basins. About 98% of the suspended sediment is supplied from unmetalled roads. The agricultural terraces effectively restrict transport of sediment downslope and the cultivated areas are not a significant contributor to the overall suspended sediment output from the basin. There is some connection between the field plots and the stream, through the network of furrows, gullies and unmetalled roads, but this linkage is limited and only operates over limited areas during extreme runoff events.

ACKNOWLEDGEMENTS

Many thanks are due to Ian Foster for comments on an earlier version of the manuscript and for improving the English.

REFERENCES

Alonso, C.V., Meyer, L.D. and Harmon, W.C. (1988) Sediment losses from cropland furrows. In: Bordas, M.P. and Walling, D,E, (eds), *Sediment Budgets*, IAHS Publ. no. 174, pp. 3–9.
Branski, J. (1968) Zmacenie wody i transport rumowiska unoszonego w rzekach polskich. *Prace PIHM*, **95**, 49–67.
Campbell, I.A. (1983) The partial area concept and its application to the problem of sediment source areas. In: El-Swaify, S.A., Moldenhauer, W.C. and Lo, A. (eds), *Soil Erosion and Conservation*, Soil Conservation Society of America, pp. 128–137.
Froehlich, W. (1982) Mechanizm transportu fluwialnego i dostawy zwietrzelin do koryta w górskiej zlewni fliszowej. *Prace Geogr. IG i PZ PAN* **143**, 1–144.
Froehlich, W. (1986) Sediment delivery model for the Homerka drainage basin. In: *Drainage Basin Sediment Delivery*, IAHS Publ. no. 159, pp. 403–412.
Froehlich, W. (1991) Sediment production from unmetalled road surfaces. In: *Sediment and Water Quality in a Changing Environment: Trends and Explanation*, IAHS Publ. no. 203, pp. 21–29.
Froehlich, W. and Slupik, J. (1980) Importance of splash in erosion process within small flysch catchment basin. *Stud. Geomorphol. Carpatho-Balcanica*, **14**, 77–112.
Froehlich, W. and Slupik, J. (1984) Water and sediment dynamics of the Homerka catchment. In: Burt, T.P and Walling, D.E. (eds), *Catchment Experiments in Fluvial Geomorphology*, Geo Books, Norwich, pp. 265–276.
Froehlich, W. and Walling, D.E. (1992) The use of fallout radionuclides in investigations of erosion and sediment delivery in the Polish Flysch Carpathians. In: *Erosion, Debris Flows and Environment in Mountain Regions*, IAHS Publ. no. 209, pp. 61–76.
Froehlich, W., Higgitt, D.L. and Walling, D.E. (1993) The use of caesium-137 to investigate soil erosion and sediment delivery from cultivated slopes in the Polish Carpathians. In: Wicherek, S. (ed.), *Farm Land Erosion in Temperate Plains Environments and Hills*, Elsevier Science, Amsterdam, pp. 271–283.
Gerlach, T. (1976) Wspolczesny rozwoj stokow w Polskich Karpatach Fliszowych. *Prace Geogr. IG i PZ PAN* **122**, 1–116.
Gil, E. (1986) Ruissellement et erosion sur le versants du flysh d'apres les resultats de parcelles experimentales. *Bull. Assoc. Geogr. Fr.* **63**, 357–361.
Gil, E. and Slupik, J (1972) Hydroclimatic conditions of slope wash during snow melt in the Flysch Carpathians. *Symposium International de Geomorphologie (University of Liege)*, vol. 67, pp. 75–90.
Higgitt, D.L., Froehlich, W. and Walling, D.E. (1992) Applications and limitations of Chernobyl radiocaesium measurements in a Carpathian erosion investigation, Poland. *Land Degrad. Rehabil.* **3**, 15–26.

Lajczak, A. (1989) Zroznicowanie transportu zawiesiny w karpackiej czesci dorzecza Wisly. *Dokum. Geogr. IG i PZ PAN* **5**, 1–85.

Reid, L. M. and Dunne, T. (1984) Sediment production from forest road surfaces. *Water Resour. Res.* **20**(11), 1753–1761.

Walling, D.E. (1971) Streamflow from instrumented catchments in southeast Devon. In: Gregory, K.J. and Ravenhill, W.L.D. (eds), *Exeter Essays in Geography*, Exeter University Press, Exeter, pp. 55–81.

Walling, D.E. (1983) The sediment delivery problem. *J. Hydrol.* **65**, 209–237.

Index

accelerated erosion 136
Agricultural Non Point Source Pollution Model 41–2
agriculture
  impact on water quality 41–2
  importance of tillage redistribution 318–19
  and nitrate pollution 49–51
  results of increased cereal cultivation 274
  see also cattle; cultivated soils; land use
Albania 378
  threat of water erosion 374, *375*, 376
Algeria, reservoir siltation problems 384–5
Alpine environment
  complexity of sediment behaviour 16
  see also alpine glacier basins; mountain rivers; Polish flysch Carpathians
alpine glacier basins, sediment yield from 407–32
ammonia volatilisation 53
ANSWERS 43
antecedent precipitation indices (API) 112–13
Arc/Info 40, 42
armour layer, mountain streams 441–2
Arolla glacier basin 414, *415*, 416
  drainage beneath glacier tongue 427
  patterns of water and sediment yield 419, *420*, *421*
  proglacial river, purging of sediment traps 416, *417*, 418
  sediment availability *421*, *422*, 422–3
  sediment pulses from 427
  sediment yield contrasted with Ferpècle basin 423, *425*, 425
  suspended sediment concentration 423
  suspended sediment productivity 430–1
asset management 44
Aswan High Dam, effects of 383
atmospheric fallout *see* fallout
attribute files 37

back-ponding, passive 254, 256, 260, 261
bacteria, catalysing 147–8
badland development, Mediterranean region 365–6, 379–82
Balkan Peninsula, rill and gully erosion on cultivated land *376*, 376

bank collapse 233–6, 238
  providing metal-contaminated sediments 167, 169
  removal of eroded material 236–7
bank erosion
  and mobilisation of radionuclides 216–17
  patterns of through time 237
  processes 232–7
    sediment generated 242–4
  Southcombe catchment 125–6
  and suspended sediment load, R Culm 229–44
bank material characteristics, R Culm 232
bank retreat *see* bank erosion
barytes, Teign valley 156–7
Bas Arolla basin *see* Arolla glacier basin
basin size
  and ^{137}Cs loss 346, *347*
  and sediment delivery 267
bed character
  influence on erosion and sediment transport 437, 441–3
  influence on sediment yield 423, 425, 428
  see also lithology
bedload deposits, mineral composition of 145–6
bedload transport
  Arolla and Tsidjiore basins 418
  from glacier basins 412, 414
  and glacial valley shape *449*, 449–50
  Nigardsbreen meltwater river 440, *441*, 441
bioavailability of metals 162
  Derwent River 174, *175*, 176
  Manifold catchment *173–4*, 174
bioturbation, redistributing ^{137}Cs 277
block stabilisation, after collapse 236
'blue line', digitising of 37–8
bottom sediments
  meaningful sediment yield data from 279
  use of in sediment fingerprinting 181, 183
braided river systems (sandurs) 443–6
braided rivers 437

^{137}Cs 2, 10, 11–13
  accumulation/depletion cycle 321, *322–3*

^{137}C (cont.)
  adsorbed by soil particles 208, 311, 314–15, 331
  adsorbed on clay and colloidal organic matter 288
  application in soil erosion studies 289–92, *293*
  budget for the Exe basin 342–5
  causes of release into the environment 311
  data revealing longer-term trends 252, 254
  environmental mobility of 289
  erosional transport of in catchment systems 331–48
  export rates 346–7
  in identification of catchment sediment sources 192–7
  intra-storm variability in 340–2
  in investigation of sediment delivery from the catchment 275–9
  in investigation of sediment sources, Hekouzhen–Longmen Basin (China) 353–61
  loss through harvesting and leaching 320–1
  routeing of to catchment outlet 336–7
  as a sediment tracer 278–92
  sediment-associated, inter-storm variability in 338–40
  showing recent floodplain accumulation 169, 172, 176
  as a source discriminator 208–24
  tracer of surficial soil movement 288, 289
^{137}Cs inventories
  cf. surface concentrations 258
  depleted (Columbjohn site) 254, 258
  excess (Columbjohn site) 254, 256, *257*, 258
  Exe catchment 333–4, *335*
  Homerka drainage basin 458–9
  impact of tillage redistribution 318–20
  impact of water erosion on 315–18
  quantification of erosion rates 300–1
  variation in 297–300
^{137}Cs loss, and basin size 346, *347*
^{137}Cs loss–soil loss relationship 321, *325*, 326
^{137}Cs profiles
  in floodplain sediments 12
  Old Mill Reservoir 278
^{137}Cs redistribution 326
  related to soil loss 289
  spatial variation of 288–9
  tagging the movement of fines 292
^{137}Cs technique
  implications of particle-size sorting 292–7
  logistic advantage of 301
  in soil erosion studies 289–92
  stages in application of 307–8

^{137}Cs transport
  annual and seasonal patterns 338
  at the catchment scale 345–8
^{137}Cs/^{7}Be ratio 198
calibration
  an adaptable approach to 321–6
  directly proportional relationship 308–9
  the problem 307–9
  use of a mass-balance model 309
calibration curves 321, *324*
carbon, organic, in sediment fingerprinting *182*, 182, 184
C/N ratio, in sediment fingerprinting *182*, 182, 183, 184
Carpathian watersheds
  analysis of rainfall–runoff–sediment yield data 101–5
  see also Polish flysch Carpathians
cartography, computer-assisted 35–6
catchment characteristics
  in hydrological modelling 38–40
  impact on hydrology 2
catchment denudation rates, calculation of 3
catchment sediment studies, use of ^{137}Cs in 11–13
catchment sediment yield, and maximum catchment elevation 374
catchment systems, erosional transport of ^{137}Cs in 331–48
catchment(s)
  defined 33
  lowland agricultural, important controls on sediment delivery 273–5
  mineralised, dispersal of metals in 161–76
  as a single linear store 39
cattle stocking density, results of increase in 274
cattle trampling/poaching 278
  hydrological effect of 274, 275
cereal cultivation, result of increase in 274
channel bank sediments
  ^{137}Cs content 210, 212
  metal concentrations in 166–7
  radionuclide content 216
  radionuclides from *219*, 219, *221*, 222
  see also bank collapse; bank erosion
channel banks, sediment eroded from 238–9
channel reach budget, Southcombe catchment 125
channels
  capacity changed through land-use changes 136–7
  proglacial, effects of changes in 427
chemical extractants, used to investigate speciation of metals 172–6
Chernobyl reactor accident 12, 288, 301

INDEX

chlorite 156
clay mineralogy
　sediments from R. Teign and tributaries *152*, 153, 155–6
　Southcombe catchment 129, 133
climate regime, sediment supply and delivery, Philippines 393–5
climatic conditions, and N–CATCH model 52–4
Columbjohn site, River Culm 252
　flood events 254–8
　isocaesium plots 254
complexing agents, in river water 148–9
concentrated flow 455–6
concentration–load relationships, use of 60–1
conduit drainage system, Arolla glacier 429–31
corrasion 238
　and undercutting 232–3
CORRINE 39–40
Corston Brook catchment/Upper Lake, Newton Park 108–9
　close correspondence, core and simulated flood index *115*, *116*, 117–18
　lake sediment cores 109–11
　rainfall record 112
　stream response simulation program 112–15
Culm River
　contribution of bank erosion to suspended sediment load 229–44
　overbank deposition on a lowland floodplain 251–61
　suspended sediment provenance case study 219–24
cultivated soils
　Polish flysch Carpathians 453, 457–8, 458–9
　predicting soil loss from ^{137}Cs measurements 197

Dalicot Farm, Shropshire 292, *294*, *296*
　implications of particle-size sorting for ^{137}Cs technique 292–7
　intra-field erosion (interpolated maps) *298*, *299*, 299–300
Dart River basin, ^{137}Cs content of potential source materials 211–13
database management systems (DBMS) 36
decision support systems (DSS) 43
deforestation 453
'delayed removal', ^{137}Cs dispersal 337
denitrification 53, 59
deposition 318
　and erosion, Rif Mountains 378–9
　of fallout 12, 311, *312*, 333
　from overbank flow 240
　and particle-size selectivity 318
　in reservoirs 366, 382–5, 397–8
　see also flood events; overbank deposition; sediment traps
Derwent River, metal levels in floodplain sediments 169, *170*, *171*
developing countries, potential value of ^{137}Cs measurements 290
digital elevation models (DEM) 37–8
digital terrain models (DTM) 37–8
dimensionless sediment concentration distribution (DSCD) 98, *99*
'direct runoff', ^{137}Cs dispersal 336–7
direct runoff hydrograph 97, 101–5
　see also instantaneous unit hydrograph
discharge regime, proglacial 407
dispersion, and water temperature recovery 85
distributed drainage system, Arolla glacier 429–31
downstream recovery 81–7
　recovery process 85–7
　reservoir releases 83–5
　tributary and mainstream changes 81–3
drainage basins
　^{137}Cs studies 11
　Homerka drainage basin 454–9
　sediment budgets 10
　soil erosion/sediment transport in 266–7
　sources of sediment within 189
　see also alpine glacier basins; catchments; Dart River basin; Exe basin; Hekouzhen–Longmen basin; river basins
drainage network 37, 38

earth flows 453
Ebro River, sediment deposition in reservoirs 383–4
EC Directive on Drinking Water 49
EC Nitrate Directive (1991) 42, 49, 70
　Nitrate Sensitive Areas 59–60
Ecclesbourne catchment, lead and zinc concentrations in *165*, 165–7, *166*
environmental controls, on ^{137}Cs fallout 333, 335–6
environmental damage, from mining 145
environmental problems, Mediterranean basin 366–7
EPIC 41
erosion
　assessment by aerial photographs 190–1
　incised erosion 317, 357
　Jostedøla river basin 437, 442

erosion (*cont.*)
  selectivity of in mountainous areas 381
  and suspended sediment yield,
    Mediterranean rivers 365–86
  water
    impact on ^{137}Cs inventories 315–18
    threat of, Albania 374, *375*, 376
    a threat to Mediterranean soils 369, *371*, 372
  see also bank erosion; gully erosion; rill erosion; sheet erosion; soil erosion
erosion pins 191, 231, 292
erosion plots 192
  Philippines study *398*
  quantification of net soil loss from ^{137}Cs measurements 194–7
erosion rate-relative relief relationship 377–8
erosion rates, estimation of from ^{137}Cs data, calibration question 307–26
'erosion season' 316
erosional removal factor (nuclides) 343
Exe basin
  ^{137}Cs budget for 342–5
  erosional transport of ^{137}Cs in 331–48
  fallout survey 333–7
  influence of Wimbleball scheme 82–3, 85
  intra-storm variability in ^{137}Cs 340–2
  routeing ^{137}Cs to catchment outlet 336–7
  suspended sediment sampling programme 337–42
  variability in suspended sediment particle-size characteristics 9, 9
Exe basin monitoring network 3, 5, 66–7
  used in Wimbleball study 67–78

fallout 288, *322–3*, 333
  distribution and environmental controls, Exe catchment 334–6
  variability of deposition 311, *312*
fallout–adsorption–tillage cycle 311, 314–15
farmers, and soil erosion awareness 139
Fe/Mn coatings, trapping metals 149
Ferpècle basin 414, *415*, 416
  sediment yield contrasted with Haut Arolla basin 423, *425*, 425
fertiliser use vs. set-aside 60
field boundary length 274
field furrows, soil loss from 457–8
fire/land management practices, impact of, Mediterranean environment 374
flash floods, meltwater rivers 440–1
flocculation 157
flooding/flood events
  affecting suspended sediment 109–10, *114*, 115, *117*, 118
  inundation and deposition 254–8
Jostedøla basin
  effects of 448–9
  non–glacier areas 442
floodplain aggradation 169
floodplains
  ^{137}Cs profiles 12
  ^{137}Cs concentrations in 344–5
  lowland, rates and patterns of overbank deposition 247–61
  as sink and store for pollutants 251
  storage of metalliferous sediments 167–72
flysch
  and badland development 379, 380–1
  giving high sediment yields 377, 378, *379*
  see also Polish flysch Carpathians
forests, Philippines 397
  forest clearance damaging 404
fuelwood gathering 381

Geographical Information Systems (GIS) 33–45
  in catchment-scale nitrate modelling 61
  coupled with hydrological models 41–3
  digital terrain and elevation models 37–8
  water resources management 43–4
geology, underlying
  importance of 374
  see also lithology
geothermal energy resources 397
glacial sediment production 439–41
glacier activity, and sediment yield 428
Glacier de Ferpècle basin *see* Ferpècle basin
Glacier de Tsidjiore Nouve basin *see* Tsidjiore Nouve basin
glacier drainage evolution, and sediment delivery 410–11, 428–32
glacier hydrology, and sediment transfer 407–9
glacier long profile, regularity influencing subglacial drainage network 425
gold, tracer in sediment studies 199–200
GRASS 40, 42
groundwater circulation, disrupted *see* springflow
groundwater-balance submodel, N-CATCH 55–6, *56*
gully areas, Hekouzhen–Longmen basin 356, 357, 359, 360
gully erosion 317, 318, *368*, *376*, 376, 381, 442

Haddeo, River (regulated) 67
  biological impact of thermal modification 87–90

causes of thermal modification 78–81
flow-related responses 74–8
inter-annual variability 73–4, *75*, 80–1
summer water temperature 71, 90
thermal regime altered 69, 71
Hamps catchment, contrasting metal concentrations 168–9
Harmonized Monitoring Scheme 13–14
Haut Arolla basin *see* Arolla glacier basin
heat exchange processes, and water temperature recovery 85, 87
heavy metals 146, 251
heavy minerals
    chemical concentration of 147–8
    mechanical concentration of 147
    in Teign Valley stream sediments *152*, 153–8
Hekouzhen–Longmen basin, Middle Yellow River (China) 356
    ^{137}Cs content of source materials and suspended sediments 356–9
    relative contributions of sediment, hill and gully areas 359–60
Highland Brook 123
    forest area and bankfull discharge 138
hill slope erosion 124, 127
Homerka drainage basin 454
    experimental slope 454–6
    sediment delivery to Homerka channel 457–9
    terraces affecting sediment transport 458–9
Hong Kong, fingerprinting sediment sources 179–85
human catchment disturbance 365, 367–8, 379
    and badland formation 381
    see also agriculture
hydro-electric power
    development in the Philippines 397
    Switzerland
        affecting seasonal pattern of sediment yield 423
        sediment traps in meltwater intakes *413*, 414, 416, *417*
hydrographs, direct runoff 101
hydrological component, and nitrate leaching 50–1
hydrological modelling
    hydrological submodel, IUSG 97, *98*
    use of GIS 38–43
hydrology, diverse themes within 41

IDRISI GIS 41
illite 133, 155–6
Imagine 40

infiltration capacities, permanent grazing land 274
instantaneous unit hydrograph 97, *98*, *99*, 100
instantaneous unit sedimentgraph 97–105
integrated GIS 44
inter-rill erosion, size selectivity of 318
intertropical convergence zone (ICZ) 394
invertebrate development, in a modified thermal regime 87–8
IUH *see* instantaneous unit hydrograph
IUSG *see* Instantaneous Unit Sedimentgraph

Jackmoor Brook catchment, ^{137}Cs content of potential source materials 211–13
Jostedøla river basin 437–9
    areas without glaciers 441–3
    a composite basin 447–50
    the lakes 446–7
    the sandurs 443–6

kaolinite 156
karst, denuded, former Yugoslavia 374

lag time, in rainfall–runoff modelling 100–1, *103*, 104–5
lake sediment cores, Upper Lake, Newton Park 109–11
    sedimentary sequence of 117
lakes
    as sediment sinks 10–11, 265
    as sedimentation basins 437
    see also reservoir and recent lake bottom sediments
Lam Tsuen River, Hong Kong 181–5
land degradation
    through human catchment disturbance 379
    through overgrazing 366–7, 381
    see also soil degradation
land use
    Culm catchment 230
    Philippines, and fragility of the environment 395, 397
    and sediment yield, significant relationships 272, 274
    up-stream change affecting areas downstream 136–8
lead 161
    see also Pb
lead pollution 162, 164
linear erosion/elevation measuring instrument (LEMI) 191
linear reservoirs, Haut Glacier d'Arolla 429–31, *432*

lithology
  affecting mean annual runoff–suspended sediment yield relationship 376–7, *377*
  and badland development *379*, 379–81, *380*
  of Exe basin 333
  a major control on sediment yield 377–8
Loess Plateau, China 356
logging 397
loss on ignition, in sediment fingerprinting *182*, 182, 183, 184

Magat reservoir, potential sedimentation rates 399
Magat study, Philippines 399–401
Maghreb 374, 376–7, *377*
magnetic measurements, for sediment tracing 200
Maluna Creek catchment (Australia), runoff-erosion (soil-loss) plots 194–7
Manifold catchment
  bioavailability of metals *173–4*, 174
  contrasting metal concentrations 168–9
mass movement, in high mountain areas 442–3
mass-balance approach, quantification of erosion rates 300–1
mass-balance model, in calibration 309
Mediterranean drainage basin 14, 369, 370
  natural vulnerability to erosion processes 365, 385
  reservoir siltation within 382–5
Mediterranean rivers, rates of erosion and suspended sediment yields 369, *371*, 372–9
meltwater 411, 439–40
  alpine glacier basins 407–9
  see also subglacial drainage system
Merevale catchment
  ^{137}Cs inventory *276*, 277
  reconstructed sediment yields *270*, 271
metal salts, effects of efflorescence 148
metals
  benefits and health risks 161
  bioavailability of 162, *173–4*, 174, *175*, 176
  chemical modification of in sedimentary deposits 172–6
  environmental impact of 161–2
  variation in concentrations, channel sediments 165–7
microtopography, affecting overbank deposition 256, *259*
Midland reservoir catchments, regional sediment yields *271*, 271

Midlands, atmospheric influx of ^{137}Cs 275, *276*, 277
Milford Haven, long-term fallout monitoring programme 343
MINDER 42
Minimata disease 161
mining
  environmental impact, Teign Valley 150–8
  impacts on mineralised catchments 162–7
models
  digital terrain and elevation models 37–8
  distributed 33, 39
  empirical and deterministic 33
  hydrological, coupled with GIS 41–3
  for nitrate leaching (plot-scale) 49–51
  see also N-CATCH model (case study)
Morocco, mean annual precipitation related to suspended sediment yield 377, *378*
mountain rivers, sediment transport and deposition in 437–50

N-CATCH model (case study) 51–62
  concentration calculation 54–5
  effect of annual climatic conditions 52–4
  effect of catchment hydrogeology 55–6
  model development 51–2
  model testing 56–60
N-CYCLE model 51, *52*
natural events, flow changes from 78
networking 40–1
neutron-activation technique 199
Nile River, effect of impoundment on suspended sediment loading 383
nitrate leaching 49–51
  effect of weather conditions 52–3
  effect of winter drainage 53–4, *54*
nitrate modelling, catchment-scale 50–1
  further developments 60–1
  use of GIS 61
  see also N-CATCH model (case study)
nitrate pollution 42, 49–50
Nitrate Sensitive Areas 59–60
nitrate vulnerability model 42–3
nitrogen cycling in catchments 5
nitrogen (N)
  mineralisation of 49, 52–3
  in the soil 50
  total, in sediment fingerprinting *182*, 182, 184
  see also N-CATCH model
NO_3-N concentrations 54–5, 60–1
  measured/predicted comparison *59*, 59
numerical mixing model 10
  for suspended sediment source ascription 213–19

INDEX 469

nutrients 8
  and water quality 41–2
  see also nitrate leaching

Old Mill Reservoir
  ^{137}Cs inventory of sediments *276*, 277
  ^{137}Cs profile 278
Old Mill Reservoir catchment
  ^{137}Cs fingerprinting 275
  grassland susceptible to erosion 278–9
  impact of land use on hydrological response *274*, 274
  reconstructed suspended sediment record *272*, 273
organic matter, in suspended sediments *127*, 127, 133
overbank deposition, rates and patterns of, on a lowland floodplain 247–61
overbank deposition rates
  contemporary and recent *248–9*
  not uniformly distributed 250, 256, *259*
  variability of 258–60
overbank flooding 109
overbank flow 240
  effect on sediment transport 240
overbank sediments
  ^{137}Cs concentrations in 344–5, *345*
  sandur systems 443, 445–6
overburden, mountainous areas, affecting erosion and sediment transport 437, 441–3
overgrazing 366–7, 381
overhang failure *see* bank collapse
overland flow 402, 404, 453, 458
  saturated 109, 113, 118
oxidation, of sulphides 147–8

particle aggregation 8
particle size
  in overbank deposition 260
  and radionuclide concentration 214–16
  and soil adsorption of ^{137}Cs 315
  and variation in ^{137}Cs content 358–9
particle size analysis
  lake core 110–11, *111*
  in sediment fingerprinting 129–33
particle-size selectivity, and ^{137}Cs content of deposits 318, *319*
particle-size sorting, implications for ^{137}Cs technique 292–7
pasture areas, overland flow from 458
^{210}Pb
  in situ and atmospheric derivation 210
  unsupported, in suspended sediment provenance determination 207–24

peak discharge
  an environmental control on erosion 135
  ratio of instantaneous to mean 395
pedogenic processes, influencing dispersed metalliferous sediments 172, 174
Philippines, the, soil erosion and sediment yield 391–405
  important physical, social and economic factors 391–8
  Magat study 399–401
photoelectric erosion pins (PEEPs) 191
photogrammetry, for identification of sediment sources 190–1
Polish flysch Carpathians, sediment dynamics in 453–60
pollution
  by nutrients 41–2
  from mining processes 145–58
  Wales, by toxic metals 162, 164
  in water quality analysis 41
population expansion, an environmental problem 366
Postlebury Wood catchment *122*, 124
  land–use change and surface runoff 137–8
precipitation
  and ^{137}Cs fallout 311, *313*
  Corston Brook catchment 112–15
  and increase in sediment yield, Seeswood Pool and Old Mill catchments 273
  intensity and duration, Carpathian watersheds 101–3
  mean annual
    controlling deposition of ^{137}Cs 333, 335–6
    Philippines 393
    related to suspended sediment yield, Morocco 377, *378*
profile simulation model (calibration) 321–4
  limitations of 324, 326
profilometers 191
proglacial rivers 416
  sediment loads 14, 16
  sediment transport and discharge in 409–10
  suspended sediment transport in 411–14
  see also Jostedøla river basin; subglacial drainage system
pseudovarves 446
Pulham, River (unregulated) 67
  flow-related responses 74–8
  inter-annual variability 73–4, *75*, 80–1
pyrite, oxidation of 147–8

^{226}Ra
  particle-size behaviour of 214, 216
  as a source discriminator 208–24

^{226}Ra/^{232}Th ratio 198, 210
radioactive decay 309, 311, *322–3*
radiocaesium *see* ^{137}Cs
radioisotopes, used to trace sediments 198
radionuclide enrichment/depletion ratio 218, 223
radionuclide studies, catchment-scale *346*
radionuclides 251, 343
   concentrations in source materials 216–17
   grain–size behaviour of in soils 214–16
   radionuclide studies 11–13
   as source discriminators 208–13, 218
   see also ^{137}Cs; ^{210}Pb; ^{226}Ra
rainfall *see* precipitation
rainfall–runoff events 97, 101
raster models 35
reforestation, reduces runoff 399
REGIS 42
relational DBMS (RDBMS) 36
remote sensing 36
reservoir and recent lake bottom sediments, proxy hydrological data from 267–75
reservoir releases
   and downstream recovery 83–5
   extent of downstream persistence 83–5
   flow-related responses 74–8
reservoir sedimentation/siltation
   Mediterranean drainage basin 366, 382–5
   Philippines 397–8
reservoirs
   ^{137}Cs activities/concentrations in sediments 277
   construction of, and regulation of downstream watercourse 65
   life-spans shortened by sediment accumulation 384
   thermal impact of 65–6
Rif Mountains (Morocco) 378
   badlands in 381–2
   soil loss through water erosion 372, 386
rill erosion 198, 200, 317, 318, 357, *368, 376,* 376, 453
rilling, measurement of soil loss 192
river basins
   Mediterranean, patterns of erosion and suspended sediment yield 365–86
   nested, Magat study, annual water discharge and sediment yield 399, *401,* 401
river stage, and shear failure 234–5
rivers
   flashy discharges, Philippines 395
   mountain, sediment transport and deposition in 437–50
runoff
   and change in land use 137–8
   Jostedøla river basin, variations in 447–8

SACFARM 51, 62
Saharan North Africa 368
sample size, influence of *298, 299,* 299–300
sampling, problems of 38–9
sandurs 443–6
scour path 258
scour remobilisation 256, 260
'sediment', defined 121
sediment availability
   beneath glacier to meltwater 422–5, *440*
   Jostedøla river basin 437, 439, 440, 441–2
sediment concentration, meltwater streams 439–40
sediment delivery problem 287
sediment erosion budget, Southcombe catchment 124–6, *127,* 127
   organic and inorganic contributions *127,* 127
sediment exhaustion effect 378
sediment fingerprinting 8, 129–33, 197–200, 207
   Hong Kong 179–85
   and sediment sources 9–10
   see also ^{137}Cs
sediment flushes, Tsidjiore Nouve basin 425, 427
sediment mineralogy, and mining impact 145–58
sediment production
   environmental controls on 134–6
   from channel bank erosion *239,* 239
sediment provenance, major control on measured activity 340, *341,* 342
sediment quality 7–9
sediment rating curve 7
sediment remobilisation 265
sediment routeing coefficient, empirical estimation of 100–5
sediment sinks 10–11
sediment sources
   ^{137}Cs in investigation of, Hekouzhen–Longmen basin 353–61
   Carpathian *455,* 455–6
   catchment, identification of 189–201
   identified through fingerprinting 179–85
   importance of intra-flood variation in 359
   Jostedøla river basin 437, 439
   proglacial 409
   properties used for fingerprinting *182,* 182
   sandur systems as 443, 445
   and sediment delivery 275–9
   and sinks 9–13
   subglacial 431, *440,* 440
   surveying methods for identification of 190–2
   tapped by draining meltwater 408–9
   and their environmental controls 121–39

INDEX 471

sediment stores 265
   floodplain/river bed, Philippines 397, 402
   important for sediment transfer rates 410
   loss of 275
   within the drainage basin 10, *11*
   see also sediment traps
sediment transfer rates
   fluvial, regional contrasts, Mediterranean environment *373*, 374, *375*
   role of sediment availability and storage, glacier basins 410–11
   and subglacial drainage systems 428
sediment transport
   fluvial, magnitude and frequency of 107–18
   increased due to changes in channel network *445*, 445–6
   and mass movement in mountain areas 442–3
   and meltwater discharge 440
   in mountain environments 437
   Nigardsbreen meltwater river 440–1
   proglacial rivers 409–10
   significance of intrinsic and extrinsic controls 266–7
sediment traps 251
   Astroturf mats 252, 254
   in HEP meltwater intakes *413*, 414, 422–3
   purging of 416, *417*, 418–19, 421, 422
   lakes as 446–7
   sandur systems as 443, 445
sediment yield(s)
   estimated, comparisons problematic *268*, 268, 271
   from alpine glacier basins 407–32
      Ferpècle cf. Haut Arolla basins 423, *425*, 425
   global *15*, 16
   not reflecting soil loss–sedimentation problem 378–9
   rivers, controlling factors 367–8
sedimentary couplets 268
sedimentation
   problems of, in Hong Kong 179–80
   rates in the Philippines 397–8
sedimentology submodel, IUSG 97, *98*
sediments
   bank-derived, loss through overbank flow 240–2
   deposition/storage of, Rif Mountains 378–9
   eroded, controls on ^{137}Cs content of *310*
   fine
      polluted, impoundment checking downstream transfer of 384
      preferential removal of 317–18

      laminated 268
      losses from the channel 240–2
      paraglacial 410
      supply linked to rainfall duration and intensity 458
seepage patterns, and bank collapse 236
Seeswood Pool
   ^{137}Cs inventory *276*, 277
   soil erosion maintaining ^{137}Cs supply 277–8
Seeswood Pool catchment 270
   impact of land use on hydrological response *274*, 274
   rising sediment yields *270*, 271
shear failure, and bank collapse 233, 235
sheet erosion 192, 198, 200, 357, *368*
sheet flow 456
shifting agriculture 397
siltation
   coastal and lake, Philippines 395, 397
   see also reservoir sedimentation/siltation
Slapton catchments 274
slumping, and undercutting 233
social and economic factors, affecting the Philippines 391
soil conservation, in the UK 138–9
Soil Conservation Service (SCS) curve number 36
soil contamination, by mining 146
soil degradation, Philippines 397
soil erosion 109, 453
   annual variation in 315–16
   by wind 318
   depth and spatial extent of 316–17
   environments for 366
   mapping using ^{137}Cs technique 192–3
   seasonal timing of 316
   and sediment transport 266–7
   size selectivity of 317–18
soil erosion studies, application of ^{137}Cs measurements 289–92
soil flows 453
soil moisture content, and bank shear strength 235–6
soil moisture deficit, and N leaching *53*, 53
soil profile
   ^{137}Cs in 275, 277
   ^{137}Cs and unsupported ^{210}Pb distribution in 208–10
soil profile reconstruction 192
soil shear strength, and bank collapse 234
soil splash intensity, and suspended sediment concentrations 456–7
soil structure
   affecting NO_3-N concentration 54–5, *55*
   and bank collapse 235

soil type(s)
and ^{137}Cs depth distribution 195–6
Philippines, rapidly degrading 393
soil water content, and bank collapse 234
soils
^{137}Cs content of topsoil *211*, 212
adsorption of ^{137}Cs 208, 311, 314–15, 331
grain size behaviour of radionuclides in 214–16
removal outstripping formation 378
uncultivated, radionuclides from *219, 221, 222*
solutes 3, *4*
solute yields 14
source area control, ^{137}Cs levels 338, 340
Southcombe catchment *122*
geology and land use 123–4
sediment yields and budgets 124–6
silt content related to sediment concentration *130*, 130, *132*
SPANS 42–3
spoil heaps *see* waste/spoil heaps
springflow, and water temperature in regulated streams 79–80, 81, 90
storm events 6
expansion of sediment-gathering area 130, 133
influencing sediment output 242–3
and removal of eroded bank material 236–7
suspended sediment sampling 337–42
storm runoff volume 135
stream response simulation program
data compared with lake core particle size analysis *115, 116*, 117–18
development of 112–15
stream sediments
historical perspective of studies 149–50
nature and constituents of 145–6
stream solute behaviour 3
in Exe basin monitoring network 3, 5
streamflow discharge
R Culm 231
and rate of particle entrainment 239–40
Structured Query Language (SQL) 36
subglacial drainage system 428–31, *432*
changes in generating sediment flushes 427
influence of glacier profile 425
seasonal development of 429–31, 439–40
substrate material, a sediment source 181, 184
'surface-eroded percentage' 316
suspended bed load 399, 401
and cyclone events 401–2
suspended sediment concentration
after flooding, streams with low glacier cover 442
alpine glacier basins 418–19

non-glacial streams, high mountain areas *442*, 442–3
proglacial rivers 409, 411–12
reduced in lakes 446
sandur systems 443, 445
on unmetalled roads 456
variation in, Jostedøla river 448–9
suspended sediment load
alpine glacier basins 418
contributions to (R Culm) 242–4
influential factors 6
input from bank erosion (R Culm) 239–40
Nigardsbreen meltwater river 440–1
a non-capacity load 237–8
proglacial rivers 409, 414
suspended sediment provenance, using radionuclides 207–24
suspended sediment sources
ascription using a numerical mixing model 213–19
Homerka drainage basin 456–7, 458–60
identification of 210–13
in river systems 8
suspended sediment transport 6–7, 231, 409
alpine glacier basins 416
decline in following impoundments 383
related to maximum discharge of flood events *114*, 115
suspended sediment transport rate 265
suspended sediment yield(s)
Alpine glacier basins 14, 16
comparative estimates, bottom sediment reconstruction and river monitoring *268*, 268, 271
Ferpècle and Haut Arolla basin 423, *425*, 425
increased, Seeswood Pool and Old Mill catchments, explanation 273
Mediterranean rivers 372, *373*, 373
natural and human components of 366, *367*
problems for estimation of 6–7
regional variations in 372–9
related to annual mean runoff 376–7, *377*
suspended sediments
in British rivers 14
cf. source materials 8
concentration–discharge relationships 5
enriched in fine soil particles 222–3
in sediment source tracing 181–2
sourced by bank erosion 237–42
storm-period dynamics 5–6
Upper Lake, Newton Park 109–11
variability in particle-size characteristics 8, *9*
sustainable simple agriculture 397
Système Hydrologique Européen (SHE) model 41

INDEX 473

tectonic activity 369, 381, 385, 393
Teign Valley, impact of mining (case study) 150–8
tensile failure, and bank collapse 233
tension cracks 235
terraces, agricultural, effect of 458–9, 460
thermal modification
   biological impact of 87–90
   causes of 78–81
thermal regimes 5
   modification of 65, 66, 71
Thiobacillus ferroxidans 147–8
3D geoscientific information systems (3D GGIS) 44
thunderstorms 402
   Philippines 394, *395*
tillage mixing 311, *322–3*
tillage redistribution 318–20, *322–3*, 326
tin mining, Dartmoor 150
TIN models 37, 40
TOPMODEL 42–3
tracers
   ^{137}Cs as a sediment tracer 287–302, 353
   used to identify catchment sediment sources 197–200
trap efficiencies 107, 384
   see also sediment traps
tributaries, as sediment sources 127–8
tropical cyclones, Philippines 393–4
   mass soil movement in cyclone events 399
   seen as main sediment-transporting events 402
   water discharge and sediment yields 401–4
trout development, in a modified thermal regime 88–90
Tsidjiore Nouve basin 414, *415*, 416, 428
   frequent sediment pulses 425, 427
   high sediment yield, early ablation season *421, 422*, 422–3
   patterns of water and sediment yield 419, *420*, 421
   sediment trap purging 416, *417*, 418
Tunisia, storage loss in Chiba Dam 385
turbidity peak, early, R Exe 340, 342
turbulent flow, suspended sediment concentration in 411–12

undercutting 233
   and bank collapse 234
unmetalled roads, Polish flysch Carpathians 453, 454–5
   sediment production from 456–7, 458, 459

varves 267
   in estimation of sediment load 446–7
   in proglacial lakes 446
vector models 35
vegetation cover
   effects of changes in catchment cover 381
   and erosion rates 374
   importance of on erodible lithologies 386
   and suspended sediment yields 183–4
   would mitigate thunderstorm and cyclone damage 404
visualisations 35–6

Wales, water pollution by toxic metals 162
wash-load 399, 401–2
waste/spoil heaps
   chemical changes within 148
   point source of metals 164
Water Information System (WIS) 42
water quality
   catchment-wide view 3, *4*
   effects of mine effluents 162, 164
   and nitrate leaching 50, 52
water resources management, use of GIS 43–4
water temperature change, delay in annual cycle of 71
watershed lag time 101
Waunafon mine spoil, Gwent 292, *294*
   implications of particle-size sorting for ^{137}Cs technique 292–7
weather conditions, effects on nitrate leaching 52–4
weathering, and bank erosion 232–3
Wimbleball Lake, characteristics of 67, *69*
Wimbleball study 67–78
   biological impacts 87–90
   causes of thermal modification 78–81
   downstream recovery 81–7
winter drainage, and nitrate leaching 53–4, *54*
woodland clearance, effect of 367, *368*
woodland/forest soils
   ^{137}Cs concentrations 212
   depth distribution of ^{137}Cs 208, *209*

Yellow River 353, *354*
   and Qinjian River
      ^{137}Cs contents of flood sediments *358*
      runoff and sediment characteristics *355*
Yugoslavia (former), denuded karst in 374

Index compiled by C. H. Tyler